CW00606870

DISPATCHES FROM THE
VACCINE WARS

DISPATCHES FROM THE
VACCINE WARS

FIGHTING FOR HUMAN FREEDOM DURING THE GREAT RESET

CHRISTOPHER A. SHAW

FOREWORD BY ROBERT F. KENNEDY JR.

Skyhorse Publishing

Skyhorse Publishing books may be purchased in bulk at special discounts for sales promotion, corporate gifts, fund-raising, or educational purposes. Special editions can also be created to specifications. For details, contact the Special Sales Department, Skyhorse Publishing, 307 West 36th Street, 11th Floor, New York, NY 10018 or info@skyhorsepublishing.com.

Skyhorse® and Skyhorse Publishing® are registered trademarks of Skyhorse Publishing, Inc.®, a Delaware corporation.

Visit our website at www.skyhorsepublishing.com.

10 9 8 7 6 5 4 3 2 1

Library of Congress Cataloging-in-Publication Data is available on file.

Print ISBN: 978-1-5107-5850-6
Ebook ISBN: 978-1-5107-5851-3

Printed in the United States of America

"The assumption that what currently exists must necessarily exist is the acid that corrodes all visionary thinking."

—Murray Bookchin[*]

Dedication

For my new son, Lucien:

ה׳ יִשְׁמָר־צֵאתְךָ וּבוֹאֶךָ מֵעַתָּה וְעַד עוֹלָם

*The LORD shall guard thy going out and thy coming in,
from this time forth and forever.*

Psalms 121:8

And in memory of my dearest friend, Lewis Dauber:

אוֹר זָרֻעַ לַצַּדִּיק וּלְיִשְׁרֵי לֵב שִׂמְחָה

Light is sown for the righteous and gladness for the upright of heart.

Psalms 97:11

---

[*] Murray Bookchin (1921–2006) was an influential American political philosopher. The quote is taken from "The Meaning of Confederalism," *Green Perspectives*, no. 20 (1990).

Contents

Acknowledgments

There are literally too many people to thank for their help with various aspects of this book in this short space. Some of those who helped are named here; others who equally helped, or helped in understanding some aspect of this very complex story, may not be named, since the very nature of the book might put their careers at risk. Of the first group, my deepest thanks to my laboratory colleagues Michael Kuo, Suresh Bairwa, and Janice Yoo, and Drs. Jess Morrice and Housam Eidi. Truly, the book would not have occurred without their help. Rabbis Dick Ettelson and Zev Epstein gave valuable critiques and religious information that were essential. Professor John Oller and Amy Newhook provided extremely valuable comments on a draft of the manuscript, and I owe them both my deepest gratitude for their careful and cogent suggestions. Drs. David Lewis and James Lyons-Weiler also provided much-needed critiques. Aaron Siri, Alan Cassels, Courtenay Stellar, Katrin Geist, and Bruce Cahan all provided feedback and encouragement. Thanks also to Leah Rosenberg, Dr. Mateja Cernic, Dr. Alvin Moss and Dr. Robert Sears, Ambra Fedrigo, Micheal Vonn, Jill McEachern, and Darcy Fysh. My thanks also to Tony Lyons for accepting the book's initial proposal and to Caroline Russomanno for her incredibly helpful copy editing. Annaka Cox designed an early version of the book cover. I thank Danika Surm for taking on the bulk of parenting of my smaller children while much of this book was being written. Also, thank-yous to those I cannot name: you've each contributed bits that have helped move this project to completion. Next, I need to offer a somewhat backward thanks to some former colleagues on the "left": watching some of you turn yourselves into pretzels to accommodate "progressive views" while kowtowing to the pharmaceutical cartel was a wonder to behold.

As this book goes to press, I want to acknowledge the passing of a friend and comrade-in-arms, Alex Moreau (Şervan): You fought against monsters, Alex. Your job is done; be at peace. Her biji!

Last, but definitely not least, my love to all of my children, Ariel, Emma, Caius, Tevah, and Lucien, for putting up with me being mentally absent much of the last year: I know this was tough; I can only hope that one day you will see that it might all have been worthwhile.

Preface

Nothing in life is to be feared, it is only to be understood. Now is the time to understand more, so that we may fear less.
—Melvin A. Benarde[1]

The present book arose from diverse circumstances that were nevertheless related by a common theme: vaccine safety.

Vaccine safety, like the pejorative term "anti-vaxxer," is a suitcase phrase in that within two simple words are a number of subthemes that span an enormous range. These include the concepts of what science is, and isn't; what those who are pro- or anti-vaccine—not to mention a vast middle-of-the-road group—actually believe; the fears that people in both pro- and anti-camps harbor; how the questions raised by the various groupings have impacted and, in turn, been impacted by politicians; and, not least, the elephants in the room, the interlocking roles of the pharmaceutical industry (the "pharma"), the Bill and Melinda Gates Foundation, the World Health Organization (WHO), and the World Economic Forum (WEF).

The "pharma" is often seen in some circles as somehow a benign player. People, especially those whose stances tend to be highly on the pro-"vax" side, may acknowledge the serious money the industry makes from vaccines, acknowledge the very clear evidence that the same industry is rife with corruption and preventable disasters like Vioxx, and yet fail to see the possibility that money and corruption play a role in how vaccines are developed and rolled out to a public that has been trained to trust vaccine doctrine completely. Governmental agencies in the United States, such as the Centers for Disease Control and Prevention (CDC), and internationally, such as WHO, are often seen as neutral and generally beneficial bodies, even by people normally distrustful of pretty much anything governments do. This odd phenomenon crosses the political divide, often in some very peculiar ways.

I came into research in the vaccine safety area quite by chance. First, I should point out that I am not an ophthalmologist regardless of the medical department I am in at my university. In actual fact, I am a neuroscientist by training and profession who happens to be in an ophthalmology department solely because I once did basic research into eye disorders. Indeed, for much of my career as a scientist, I had pretty much no views on vaccines at all, apart from what I had learned as an undergraduate and graduate student that vaccines were uniformly safe and effective. This was all it seemed that I needed to know for many years. That view did not change until about 2005.

That was the year when a graduate student in my laboratory and I decided to seek another cluster of Lou Gehrig's disease (amyotrophic lateral sclerosis, or ALS). In brief, my laboratory had been studying the cluster of ALS on Guam and decided that, to find more clues to this disorder, we had to broaden our search. To do so, we sought another cluster that might serve to diminish the number of potential causes of the disease.

In due course, we found one in Gulf War Syndrome, the mysterious multisystem disorder that emerged after the American Coalition's 1991 war against Iraq. In this syndrome, ALS incidence in Coalition soldiers appeared to occur at a much higher incidence, and at a much younger age, than in the general population.

In turn, our reading of the published literature led to the emerging epidemiology on the syndrome that, in some cases, pointed the finger at the anthrax vaccine that most soldiers had received. The correlation with this vaccine seemed to be independent of whether the soldiers actually deployed to the Gulf or not. This fact alone seemed to rule out environmental factors that arose during the war such as exposure to oil well fires or anti-nerve gas agents.

With this as a background, we attempted to purchase the anthrax vaccine, made at that time by a company called BioPort. BioPort refused to sell us the vaccines, so we decided to simply look at the listed ingredients and try the components individually that, based on the scientific literature, seemed most likely to be involved. Two such ingredients stood out, both adjuvants, or helpers, to the vaccine: aluminum salts, such as aluminum hydroxide; and squalene, a triptertene. The first was acknowledged to be in the vaccine; the second was not, but other investigators were able to show that it was there in at least some of the anthrax vaccine vials. Aluminum was recognized as a neurotoxin even then.

We conducted a typical in vivo animal model study in which we injected young male mice with a weight-adjusted amount of aluminum hydroxide or squalene, versus both, and all compared to control mice getting only saline. At this time, we felt that we would fairly rapidly discover that there were no negative effects and go back to look for other possible causal factors for Gulf War Syndrome.

To our surprise, we found that the aluminum, in particular, had a significantly negative impact on motor functions and reflexes. Further, histological examinations showed that the motor cortex and spinal cords of the aluminum-treated mice had significant increases in motor neuron degeneration.[2]

Now intrigued, from that point on we did what scientists are supposed to do and kept following the leads. The emerging data in adult and young mice supported the general notion that aluminum was harmful to the central nervous system (CNS). This alone was not particularly surprising, as we were to discover when we began a detailed survey of the existing experimental literature.

We went on from this early work to publish a number of reviews, experimental studies, and other commentaries on aluminum (see Chapter 5).

In science, provocative results are supposed to be met with attempts by others to replicate the findings in order to see if the data hold up to scrutiny. Indeed, various researchers were finding the same things we had seen. In contrast, agencies like WHO did not have experimental data, but rather simply dismissed our work out of hand using one of their sub-bodies, the Global Advisory Committee on Vaccine Safety (GACVS). Here is their comment in reference to several of our studies:

> The GACVS reviewed 2 published papers alleging that aluminium in vaccines is associated with autism spectrum disorders 3,4 and the evidence generated from quantitative risk assessment by a US FDA pharmacokinetic model of aluminium-containing vaccines. GACVS considers that these 2 studies 3, 4 are seriously flawed. The core argument made in these studies is based on ecological comparisons of aluminium content in vaccines and rates of autism spectrum disorders in several countries. In general, ecological studies cannot be used to assert a causal association because they do not link exposure to outcome in individuals, and only make correlations of exposure and outcomes on population averages. Therefore, their value is primarily for hypothesis generation. However, there are additional concerns with those studies that limit any potential value for hypothesis generation. These include: incorrect assumptions about known associations of aluminium with neurological disease, uncertainty of the accuracy of the autism spectrum disorder prevalence rates in different countries, and accuracy of vaccination schedules and resulting calculations of aluminium doses in different countries."[3]

In Chapter 5, we will see if the WHO/GAVCS comments are valid or not.

The second convergent event in my personal trajectory into vaccine research was actually a series of events that began to suggest to me that we were not wading into just any "typical" medical controversy such as those that populate ALS or other neurological disease research areas, but rather one that had frankly religious overtones. Actually, as we came to see, it was more cult-like than simply religious.

I had also begun to realize that "talking truth to power" was not sufficient.* Power in this case either knew what we knew, that is, that aluminum vaccine adjuvants are harmful, or simply didn't care. In either case, two possible

---

* The notion of "talking truth to power" is a phrase often used by those on the various ends of the political spectrum. In essence, it means that if you simply tell those in power what is true about a particular situation or problem, they might respond in a way designed to correct whatever that problem is. The concept presumes those in power don't already know about the problem and/or their role in it and, further, that they care.

reasons for the lack of response became clear: dogma and money. The first had served to convince most of the world's medical professionals that we had to be wrong because, after all, "the science was settled." And behind much of this was the naked fact of how much money vaccines brought into the pharmaceutical industry's profit margin.

The combination of these two, in turn, led to a series of actions that I believe have the fingerprints of the various companies smudged all over the question of vaccine safety. These included the demonization of both scientists and lay scholars who raised even the tamest questions about safety and the push for vaccine mandates around the world.

In the first case, we have seen this before when various industries find their products threatened, as will be discussed at length in later chapters. In other words, the attack on independent scientists studying vaccine safety was nothing new.

As so often in history, attempts to suppress people, either with mandates or anything else, tend to have predicable consequences, namely, pushback and outright resistance, actions that were emerging even before the COVID-19 pandemic rocked the world.

What had started as a fight for vaccine safety has since rapidly emerged as a fight for basic human rights, in this case that of security of the person.

Whatever the politicians and their pharma backers thought they might achieve by pushing for mandates has hit a wall of resistance, resistance that seemed to be growing then, and even more now, as this book goes to press: the more those in power push, the more they threaten and demean those critical of any aspect of vaccine safety, the greater the resistance grows. Anyone who has ever studied counterinsurgency warfare knows precisely how this process works and what the end result is likely to be. Fear can only be maintained for so long, even if it is increased by pandemics real or imagined.

What was emerging pre-COVID-19 and since can be described, fairly accurately I think, as a war. Not a war involving weapons, thankfully, at least not yet, but one of ideas and about rights. Hence, the title of this book. In the following pages, I will attempt to dissect the various issues that have emerged, and continue to emerge.

Certainly the most dramatic event to emerge since this book was started has been the COVID-19 pandemic, which is ongoing as I write and which will certainly be with us as the book goes to press and beyond. COVID-19 as a disease and the social and political responses to it, fanned by very accommodating media, are likely to be some of the long range aspects of the "new normal."

COVID-19, from its origins to the future, is the subject of a separate chapter that was not planned when this book was begun. However, I think readers will see in the pre-COVID-19 history what should have been clues to future events that have since transpired.

There are various books critical of vaccines, of course. And there are many books taking the opposite tack. Instead of trying to put myself into either camp, I have chosen to go back to basics and try to see what history and science actually tell us about vaccine safety. In so doing, I expect to find opponents from both camps. Some will think I am too "anti-vax" (many already do) for pointing out the flaws in vaccine theory, development, and administration. Others will find me not critical enough. My feeling is that getting vitriol from both sides is the right place to be. Individual readers will decide for themselves.

Needless to say, none of what follows should be considered to reflect the views of my current employer, the University of British Columbia, as it most certainly does not. I will discuss this point in some detail as I think it illustrates how heavily the pharmaceutical industry influences academia.

There are many people to thank, whose contributions I have already acknowledged. The discerning reader will note that some entities and individuals are deliberately not mentioned.

Last issues: In a book attempting to cover so much territory, there will be omissions and gaps, and not everything that might be cited has been. The subject of vaccines and vaccine safety with all of the associated scientific and social ramifications is just too broad. I apologize for any items that I missed and mean no slight to any of the authors of such articles or books.

Additionally, early on in the process of writing this book, I solicited opinions from a range of individuals on topics such as vaccine mandates. By no means was this a rigorous selection process: I simply sent a questionnaire to people I knew. I viewed this as a "temperature check" on various issues. The verbatim responses are shown after the last chapter.

Finally, needless to say, any errors of fact or interpretation are solely mine.

A selection of supplementary material, including appendices, a glossary of terms, and a questionnaire with people from various fields, on particular topics in the months before COVID-19, can be found online here: **dispatchesfromthevaccinewars.com**.

"Fight the power; do no harm," the slogan of the late Black Cross Medical Collective, expresses the basics of my social and medical beliefs, and most of what follows in the rest of the book is from this perspective.[*]

—Christopher A. Shaw
Victoria, British Columbia
January 31, 2021

---

[*] The Black Cross Health Collective was a volunteer radical medic community based in Portland, Oregon. They are not now operational, but back when I was developing my street medic skills, they did a lot of training for new medics.

Foreword

by Robert F. Kennedy Jr.

In *Dispatches from the Vaccine Wars: Fighting for Human Freedom During the Great Reset*, Dr. Christopher Shaw chronicles the long and troubled history of vaccination culminating in the raging global controversies over COVID-19 jabs. Shaw's book offers important new insights for the growing cohort of Americans who still love science and critical thinking and who feel growing discomfort with the mainstream media routine of force feeding Americans pharmaceutical industry pablum and state-sponsored propaganda which aggressively censors skepticism and dissent and abolishes debate altogether.

Shaw shows how official vaccine doctrine is almost entirely reliant upon appeals to authority—a feature of religion, not science—and crooked and fatally flawed studies ginned up by industry biostitutes. He introduces us to the high priests of *Vaccinology,* a coterie of richly compensated charlatans, flakes, trolls, and medical mercenaries. Shaw systematically obliterates the key canons of their orthodoxy.

The COVID pandemic has made the once exotic subject of vaccines required learning for the many Americans who still value our democracy and love freedom more than they fear disease. The media portrays *Vaccinology* as a benevolent medical discipline where "science rules" and where white-coated physicians and researchers commit their selfless lives to fighting disease and safeguarding public health. Their "miraculous" vaccines are always "safe and effective." Shaw exposes this narrative as the self-serving mythology of a venal and homicidal Pharma/Medical cartel ruthlessly focused on profiting from the generously stoked fears of infectious diseases. It's a polite fiction, Shaw demonstrates, to claim that our captive public health agencies do public health. Their real gig is pushing vaccines.

Shaw demonstrates how *Vaccinology* only survives by suppressing empiricism, stifling debate, enforcing dogma. Using its hundreds of billions of dollars in annual advertising expenditures, Pharma has transformed the once independent media into a quasi-religious inquisition that silences heresy, and burns heretics. Evidence-based research under the Pharma rubric has become a foreign language that exposes any scientist with fluency as a dangerous subversive subject to demotion, retraction, censorship, and bankruptcy. All orthodoxies are tyrannical, cruel, and often murderous and the vaccine orthodoxy has left a wide wake of human carnage.

In Shaw's words, "Simply talking about the possible dangers of vaccine adjuvants is speaking a language that the medical cartel does not comprehend and cannot tolerate."

Shaw shows how almost four hundred years after Galileo, the perils of challenging the "scientific" hegemony still has devastating costs. "The mind-set in the mainstream medical community is pretty clear: It is preferable to censor and self-censor data about vaccine safety than to do real science that invariably challenges official dogma precipitates career suicide."

My father once said, "Few men are willing to brave the disapproval of their peers, the censure of their colleagues, the wrath of their society. Moral courage is a rarer commodity than bravery in battle or great intelligence. Yet it is the one essential, vital quality for those who seek to change a world that yields most painfully to change."

In *Dispatches from the Vaccine Wars,* Dr. Christopher Shaw demonstrates true moral courage.

CHAPTER 1

Dispatches from the Vaccine Wars: An Introduction

One of the saddest lessons of history is this: If we've been bamboozled long enough, we tend to reject any evidence of the bamboozle. We're no longer interested in finding out the truth. The bamboozle has captured us. It is simply too painful to acknowledge—even to ourselves—that we've been fooled.

—Carl Sagan[1]

In the Beginning . . .

Early in 2019, I began to think about the sabbatical year that I was originally authorized to take in the fall of 2020. For those who don't know, a sabbatical for academics is a period, up to a year long, in which one can take leave of the university and most of the duties associated with a faculty position: the endless grant writing if one runs a laboratory, teaching of various types, daily supervision of graduate students and postdoctoral fellows, and the like. It's not that these are individually or cumulatively necessarily onerous tasks, but merely that they take their toll on one's time and freedom to explore new ideas, try different things, and perhaps launch thoughts or actual research in novel directions. Best of all, at least at my university, a sabbatical leave is mostly paid, making it economically feasible to take the time away.

I didn't know where I wanted to go then but did know that I wanted to write with two possible projects holding the most interest for me. One of these was the ongoing drama in northeast Syria in a region called Rojava by the mainly Kurdish population. It was here that the Kurds, Yezidis, and Syrian Christian communities had carved out an autonomous region in the midst of the chaos of the Syrian civil war. It was also here that these same populations had defeated the Islamic State and at the same time had begun the process of creating a very different political entity, one that actually rejected the notion of statehood, choosing instead a form of voluntary confederation of the different peoples of the region. The emerging revolutionary society embodied the equality of women and all ethnic groups, fostered bottom-up democratic decision making, and at the same time embraced what the American social philosopher Murray Bookchin had termed "social ecology."[2]

I had been fascinated by this social experiment since late in 2015, and by the end of 2018, I was fortunate to have been able to visit Rojava twice. Telling the story of Rojava to the Western world that largely did not understand what was happening there seemed an important and even possible project.

But there were serious problems with the actuality of carrying out this project. Not least, Rojava was, and remains, a war zone and is far from safe. So *not* safe, in fact, that I could not envision taking my family with young children into the area. Thus safety was the primary factor in my decision, but almost as crucial was the underlying assumption that my university attached to a sabbatical, namely, that the sabbatical year was to further one's knowledge or other capability *in one's own field*. As I am a neuroscientist, not a political scientist, it seemed a stretch to get the university to see that a sabbatical year in Rojava would further my ability to do better neuroscience research or medical school teaching. There were ways around this, of course, such as the notion of helping with the establishment of a medical school in Rojava, or teaching emergency first aid. While such solutions were plausible, the first problem still remained: safety.

The final concern was that a number of books have already been written about Rojava and its unique social experiment, many of them quite good.[3]

In the end, I was not sure that any contribution I might make to the revolutionary literature on Rojava as a very biased observer would overcome the other concerns already noted.

It was at about this time that the ongoing worries of some parents concerning vaccine safety erupted into a significant social movement in the face of vaccine mandates for schools being forced down people's throats in various places, notably in some US states.

The struggle in Rojava was all about people being able to make their own decisions on how they wanted to live their lives and to be free from the tyranny they had fought. In many ways, the struggle for health freedom was not all that dissimilar in a general sense, at least to me. Very few on the vaccine resistance side saw in the proposed mandates similarities to the utter brutality of the Islamic State or the Syrian regime. And yet, removal of freedom of choice about one's body, or the bodies of one's children, had all the hallmarks of totalitarianism even if the iron fist of the government was cushioned by the velvet glove of what has been described as "evidence-based medicine."[4]

Whether this view is right or wrong, and some would argue that it is wrong, it nevertheless led me to view the fight against vaccine mandates and all that they represent as part of a global freedom struggle: freedom from political and religious repression in Syria, freedom for Native people in Canada and elsewhere to control their lands and destinies, and freedom for parents to choose to exercise bodily autonomy for themselves and their children without fear of State reprisal.

This is the view that then led to the decision to write this book, not merely as a scientific treatise on the pros and cons of vaccination theory and practice, but also in light of a broader struggle for human freedom.

What followed this decision was that I had to consider a range of pluses and minuses that might be involved in this decision. The key item on the plus side included the notion that this was a book that needed writing and that maybe I could do justice to the maze of conflicting pieces of information that seemed to abound when the subject of vaccination comes up in scientific circles, or even in social settings. As a scientist, I hoped my training would help sift through what can only be described as a scientific mess, in spite of the prevailing notion in the media and official entities such as the Centers for Disease Control and Prevention (CDC) and many others, particularly in the mainstream media, that the "science is settled," a decidedly nonscientific thing to say.

I knew from my work on aluminum adjuvants that this area of inquiry at least was very far from being "settled." In fact, if it was settled at all, then it was trending toward being settled in the opposite way from what the official narrative maintained. If the prevailing medical view is that aluminum in vaccines is harmless, and if this is almost certainly wrong (see Chapter 5), how much else in the official story is also wrong? Maybe nothing, but is that likely?

A second issue had to be whether I could bring to bear any aspect of impartiality to the vaccine issue that I could never do for the Rojava story. For the latter, I have been firmly committed to one side and have failed to see the merit in any arguments against. On vaccines, however, I decided I could at least be fair enough to examine critically the evidence in favor of, or against, the notion that vaccines are universally safe and effective if I approached it solely from my science background. To do so would necessitate my going back to much of the primary literature to evaluate the various published papers and the assertions that flowed from them. In addition, I would have to look at various reviews and meta-analyses, as well as the statements from entities such as the CDC and the World Health Organization.

To be even fairer, I would have to commit to doing the review as impartially as possible, keeping to the notion that I really had, or should have, no dog in this fight, at least from a purely scientific perspective.

The problem with such an approach was really two-fold. First, I was not sure if I could do so given what I had learned in the years since I first became involved in vaccine safety research and if I could ignore the prejudices that I have likely picked up over the years. Second, I realized that for some on the pro side, it really wouldn't matter at all how much I committed to trying to be neutral. For some people, *any* critique to *any* vaccine under *any* circumstances was proof that I was an "anti-vaxxer" and that I would remain so regardless of what the science actually showed.

There really was, and is, no answer to this problem, apart maybe from noting that it exists.

Taking a neutral perspective, one has to ask if the literature really needs another "anti-vax" book? Certainly there are enough books classified in this manner,[5] some of which make little effort to be neutral. Of course, the opposite is true, as well: the pro-vaccine books mostly look to be variants on one another in their praise of all things vaccine-related. Many of each will be cited in the chapters that follow.

Neutrality itself can be problematic, in practice if not in theory, and I note that a recent book by the Danish physician/scientist Peter C. Gøtzsche[6] plainly showed the problems with trying to hew too close to the center of the road. In brief, in his short book, Gøtzsche seemed to be almost bipolar in trying to both praise and damn vaccines and their respective proponents in virtually every chapter.

Particularly telling in the Gøtzsche book was the opprobrium heaped on Dr. Andrew Wakefield (discussed in detail in later chapters of the current book), the former medical doctor often considered to be fully disgraced with his work "debunked." Oddly, in discussing Wakefield, Gøtzsche went on to praise the evidence about Wakefield from the nonscientist and highly conflicted journalist Brian Deer. From that, Gøtzsche did another 180-degree turn and critiqued the CDC (and WHO) as being almost completely untrustworthy while relying on their information on vaccine safety.

Clearly, there are dangers in any approach, pro-, anti-, or neutral, and any single one of these seemed likely to draw kudos or brickbats from the predictable voices on any side of the divide. Neutrality seemed to simply offer the chance to be clobbered by both sides.

It was in fact this last point that finally swayed me to do this book. Namely, that if I was going to draw flack no matter what I wrote, there really was nothing to lose if I wanted to do the book at all.

And I decided that I did: freedom for the people of Rojava was no different in kind from the freedom of people to make their own choices about health and what goes into their bodies. Human rights issues are human rights issues whatever the intensity of the violations and wherever they occur.

All of the above will come up again and again in what follows and will serve as the central theme of this book on the benefits versus harms of vaccination or any other medical treatment that may be on offer.

In the context of such human rights concerns, I will consider the basis of the widespread medical belief that vaccines are the best health defense yet invented against infectious diseases. To do so, in the following pages I will critically examine the scientific studies by the proponents of this view. I will also critically review the key evidence put forward by those who do not share this perspective, or those who hold a middle-of-the-road position.

But before I do so, I want to consider the basis of what is called the "scientific method" and what it can, and can't, do and more specifically in regard to vaccines. In this consideration, we will see the value of the evidence from all sides.

The Scientific Method and What Science Can and Can't Do

The scientific method is considered to have arisen from the writings of Francis Bacon (1561–1626) in the seventeenth century. Bacon is often considered to be the first to formalize the concept, but the concept was in reality extant before this time, and Bacon and others were highly influenced by the earlier work of Copernicus (1473–1543) and Galileo (1564–1642). The Oxford online dictionary describes the scientific method as "A method of procedure that has characterized natural science since the 17th century, consisting in systematic observation, measurement, and experiment, and the formulation, testing, and modification of hypotheses,"[7] a definition that most scholars would generally agree upon.

It is important to keep this definition in mind in all that follows here, as it is clear that there is considerable misunderstanding about what is involved in the scientific method, both by some in the medical establishment and by lay people on the various sides of the vaccine issue.

The first thing to stress is that the scientific method is just that: a method for attempting to understand the natural world.

As per the above definition, the scientific starts with observation, then proceeds to the development of a hypothesis, the latter merely a more formal statement about how the person doing the observation thinks nature behaves. In the context of this book, I will use the oft-told story of how Dr. Edward Jenner came to be considered the father of vaccinology. This status is linked to how Jenner came up with his notion in the last decade of the eighteenth century to use pus from cowpox virus sores on milkmaids to inoculate people against the related smallpox virus.

Jenner, so the story goes, observed that the milkmaids who had been infected with cowpox did not later become susceptible to smallpox. Although he did not state a formal hypothesis as such, he apparently believed (hypothesized) that he could duplicate the apparent immunity to smallpox by exposing others to cowpox.

Figure 1.1. Composite of historical and current figures involved in describing the methodology of science regarding vaccine issues (shown chronologically): William of Ockham, Francis Bacon, Edward Jenner, Karl Popper, Thomas Kuhn, Carl Sagan, Andrew Wakefield.

Jenner's work will be examined in more detail in Chapter 2, but for now I will continue with the "official" story to note that Jenner then did an experiment that tested the hypothesis by deliberately exposing people to cowpox and then later observing if the subjects developed smallpox if challenged by the actual disease. The results convinced Jenner that his working hypothesis was correct and that inoculation with a substance that mimicked smallpox without actually giving smallpox would prevent a later appearance of the disease. The evidence that this worked seemed to confirm the hypothesis and led to the first vaccines and the widespread use of vaccination to provide immunity from many infectious diseases.

It is important at this stage to clearly define what a hypothesis is versus what a theory is. The terms are often used interchangeably, even by those who should know better, but they are very different, albeit related, things.

The best definition of theory that I have found comes from Wikipedia. I quote from it in detail, as it is important to get this correct right from the start:

> In science, the term "theory'" refers to "a well-substantiated explanation of some aspect of the natural world, based on a body of facts that have been repeatedly confirmed through observation and experiment."

Theories must also meet further requirements, such as the ability to make falsifiable predictions with consistent accuracy across a broad area of scientific inquiry, and production of strong evidence in favor of the theory from multiple independent sources (consilience).

The strength of a scientific theory is related to the diversity of phenomena it can explain, which is measured by its ability to make falsifiable predictions with respect to those phenomena. Theories are improved (or replaced by better theories) as more evidence is gathered, so that accuracy in prediction improves over time; this increased accuracy corresponds to an increase in scientific knowledge. Scientists use theories as a foundation to gain further scientific knowledge, as well as to accomplish goals such as inventing technology or curing diseases.[8] [For emphasis, italics are mine.]

In contrast, a hypothesis is defined by the Oxford online dictionary as "A supposition or proposed explanation made on the basis of limited evidence as a starting point for further investigation."[9]

In other words, a hypothesis is a hunch or a guess based on observation that depends on experiment for validation or rejection. A theory is based on a collection of outcomes from related hypotheses that create a general body of knowledge about some topic. It is important to keep in mind that both hypotheses and theories are the products of the human mind, not self-generating entities.

Note that a key part of the definition of theory requires that it can be falsified by experiment. The same holds true at the hypothesis level in that a good hypothesis must also be able to be rejected based on the outcome of the experiment. So in the stages of the scientific method, the design of the hypothesis must lead to an experiment in which the hypothesis can be supported or rejected. A hypothesis that is validated by an experiment, however, is not "proof," nor for that matter is a theory built up of various observations proof. Both, in fact, are merely probability statements. For the first, this is where the statistical methods used in science come into play: they provide the probability that a given outcome to the testing of a hypothesis is likely to be correct. The cumulative probabilities of the various experiments that give rise to theories make the theory more likely, overall, to be also correct.

So how does one apply statistical inference to hypothesis testing? Basically, this is done by making two statistical hypotheses, the first being the *null* hypothesis (*Ho*); the second is the *alternative* hypothesis (*H1*) in the comparison of bodies of data in some experiment. For example, if testing whether a particular drug will deliver a benefit (or an adverse effect) to a treatment group of subjects compared to untreated true placebo control subjects, one gathers experimental measurements for both groups and then analyzes these using various statistical methods that account for the variation in the data from some mean value (standard deviation or error). Testing the null hypothesis is basically testing the notion that there is no difference between groups. The probability value that results (the *p value*) tells you how strong the null hypothesis is. If the probability is very low, it means that the hypothesis is probably not correct. Typically, one rejects the null hypothesis if the p value shows that that it could be true at or less or equal than five times in a hundred. Thus the lower the p value, the more likely that the null hypothesis is wrong. If the null hypothesis is rejected, the alternative hypothesis is accepted. P value measurements are typically used in most biological experiments in comparisons between controls and one or more separate groups. In many vaccine safety trials, however, as we will see in Chapter 3, real controls are not typically used nor is it atypical to compare one vaccine against another or the vaccine against the adjuvant. In many epidemiological studies, researchers use confidence internals (CI) rather than p values. However, like p values, CIs measure the degree of uncertainty or certainty in a sampled population. Basically, a CI of 95 percent is the same as a p value of less than 5 percent.

In both cases, it is important to realize that we are speaking of probabilities. With this in mind, what does the frequently heard comment about vaccine safety that "the science is settled" actually mean in regard to the scientific method? The answer is that it is a meaningless and even nonsensical statement because being settled would imply that something had been proven. Rather, proof lies in the domain of mathematics and formal logic where a theorem can indeed be proven to be correct.

In the philosophy of science, there are several main views about how science progresses. As noted by Dr. James Lyons-Weiler,[10] objective science follows from the work of Karl Popper (1902–1994), who criticized what is termed "positivism," that is, the collection of facts that tend to support our own inferences or hypotheses. Popper proposed instead a form of "hypothetico-deductive" science based on a clearly stated hypothesis and what he called a "critical" test of that hypothesis, one in which the critical test is only such if the hypothesis can be falsified. A positive outcome to a critical test supports the hypothesis going forward; a negative outcome moves the science away from the hypothesis. This last, in turn, reinforces a view that the scientific method cannot *prove* anything but actually advances more by disproof, that, is the failure of a hypothesis based on the evidence, and thus the probability that it is not correct.

A somewhat alternative view of science was provided by Thomas Kuhn in his *Structure of Scientific Revolutions*,[11] which suggested that scientific theories about nature depend to a great measure on a majority view of those in the field that only change when enough experimental anomalies have arisen to make the majority view untenable. It is at this point that a new view arises that better addresses the anomalies and this new view becomes the dominant narrative.

Kuhn provides various examples of how "revolutions" in our understanding of nature arise and shifts the established scientific "paradigm" to another one. Such paradigm shifts are, in fact, relatively common. These are then, in turn, subject to further revisions. There are numerous examples of prevailing theories in medicine and in neuroscience and other disciplines where paradigm shifts have happened and then been subject to additional changes.[12]

It is against this backdrop that we have to evaluate the claims for and against any aspect of vaccination "theory," practice, and policy. For example, does an objective view of the experiments done to date support the majority view on vaccination safety, or not?

In all the examples to be considered, it will be important to keep the lessons of the above clearly in mind: Is any experiment designed to be a critical experiment with a clear conclusion? Can the hypothesis to be tested be falsified by a negative outcome? Does the body of evidence suggest the probability that the existing theory is correct, or not? And finally, if the evidence is not clear and/or unambiguous, and if anomalies have arisen, what further information would we need to reject or modify the underlying theory?

In this regard, I want to quote a statement made about a controversy that involved two writers: Malcolm Gladwell, author *of Outliers: The Story of Success*,[13] and David Epstein, who in his book *The Sports Gene*[14] disputed Gladwell's notion that intense specialization is key to mastering any skill. As Epstein noted about the subsequent debate with Gladwell:

> He [Gladwell] could have viewed our ideas as in zero-sum competition. But he didn't. He viewed it as an opportunity to engage in more discussion—often politely antagonistic but very productive discussion—and consequently we learned from one another. [This] set in motion what became not only a really productive intellectual relationship for me, but also a model of how two people publicly associated with certain ideas can engage without forcing zero-sum competition.[15]

In brief, scientific controversies don't have to be zero-sum events.[16] And indeed, one question that will be implicit throughout the rest of this book is whether the contentions about vaccine safety and effectiveness are indeed a zero-sum game or instead amenable to civilized scientific and social dialogue. The evidence to date that the latter is possible remains to be demonstrated, but in my view, it is the only way out of what seem to be rapidly solidifying positions that are in some senses solitudes.

The notion that the competition of ideas can prove beneficial for both sides is well entrenched in Stoic philosophy.[17] With this in mind, in the following pages I will present the evidence by both mainstream vaccine proponents and the skeptical opposition.

To be as fair as possible, I will let the mainstream view on vaccine safety and effectiveness go first in Chapter 2 and follow it with the more skeptical vaccine narrative in Chapter 3. This is only reasonable, since (a) the pro-view of vaccine safety is the dominant view and (b) any gaps, real or imagined, in the claims about vaccines will serve to introduce the likely positions of the other side.

Some ground rules:

1. I will only consider in what follows the published and peer-reviewed articles cited as evidence for or against vaccine safety or effectiveness. In this regard, I do not consider statements or declarations from entities such as the CDC, Food and Drug Administration (FDA), WHO, or others to be informative to this discussion, unless, as above, primary references and sometimes meta- and systematic reviews in the peer-reviewed literature are cited.

2. The opinions of bloggers on any side will not be viewed as valid scientific arguments for the simple reason that such opinions are just that, opinions.

3. Articles that contain a lot of unsupported statements, such as "vaccines are the most effective medical treatment of all time" or "unvaccinated kids are healthier than vaccinated kids" will be discounted. Such statements may well be true but if so deserve to be fully referenced to the scientific literature. Anything else is opinion and/or speculation.

4. Logical fallacies such as the appeal to authority or ad hominem comments are automatic fails.

Occam's Razor and the Role of External Players

A long-established scientific principle is termed Occam's Razor (variant spelling: Ockham), also more formally termed the "**law of economy**" or the "**law of parsimony.**"

In brief, Occam's Razor was first clearly stated by the philosopher William of Ockham as "*pluralitas non est ponenda sine necessitate,*" or "plurality should not be posited without necessity." In other words, simplicity is best when comparing different hypotheses or explanations for any phenomenon such that the simpler one, if equally able to explain the phenomenon, is more likely to be correct.[18]

It is not always true that the simpler explanation is the best, but it is true often enough to give some weight to the notion that nature prefers simpler solutions.

Occam's Razor is also used in criminal investigations, even if it is not called by this name. For example, police and others know that most crimes of violence such as murder are not committed by perfect strangers, but rather by someone the victim knew, if only in passing. It is for this reason that if a murder is committed, investigators first seek to rule out the family and friends of the deceased.

If we apply Occam's Razor to the question of vaccine mandates and the obdurate refusal of much of the medical profession to accept the reality of vaccine damage, the principle serves to shortlist the likely culprits for this state of affairs. Could the vaccine-injured patients or their relatives be responsible for such legislation? Maybe, but such would require an active conspiracy of people who likely don't know one another to force on themselves something that most of them profess to believe is an outrage. Could it be the legislators themselves are seemingly independently convinced that this is something they must do for the common good? Again, maybe. But anyone who has known people in any legislative body knows full well that most legislators are decent, hard-working people, and thus the notion that one or more of them would independently seek to potentially estrange their electors seems highly unlikely. What's left?

This is where, if we had followed police procedure, we would have asked ourselves, cui bono, who benefits? Now the likely answer becomes clearer: the pharmaceutical industry that makes the very vaccines that mandates are designed to foist on the population. We now have motive: money (or greed, one of the seven deadly sins; maybe pride as well and a vast host of others[19]). Occam's Razor and good police procedure would now zero in on those who had motive (greed) and opportunity (control of legislators through donations), namely, the same industry.

Occam's Razor doesn't always work, of course, but in this case it likely does. The hypothesis is vastly simpler than the alternatives while explaining all the facts: the endless push for legislated mandates, the lobbying of Congress in the United States, and all the other items so well documented in *Trust Us, We're Experts*, a book discussed in more detail later.

The Benefits versus the Adverse Effects of Vaccines

Weighing the benefits versus harms of any medical procedure, just as in any human endeavor, is key to deciding whether or not that procedure should be implemented. This consideration applies whether one is contemplating a treatment for an individual or a population. The two are often intertwined, but not always. For example, the Framingham Heart Study (FHS) of cardiovascular disease monitored 5,209 of people aged twenty-eight to sixty-two years old over decades documenting a range of cardiovascular and biochemical measurements. In so doing, the FHS provided a clear list of risk factors for such disease as the population aged. This study is now working with the grandchildren of the original participants.[20]

Based on these data, individuals can now be evaluated by their health-care provider for their own individual risk of developing cardiovascular disease, heart attacks, or stroke and advised on some steps to take to diminish the risk factors.

When it comes to vaccines, however, the issue is less clear-cut. At a population level, the various phase trials of any vaccine can evaluate the overall effectiveness of a new vaccine in a subject population and detail some of the identified adverse effects, if any, in that same population. So, for example, with the current measles, mumps, and rubella (MMR) vaccine, the conventional view in the mainstream medical community is that two doses of MMR confer 97 percent protection in the population at large.[21]

A problem that can arise lies in the selection process for this: Is this 97 percent effectiveness for all ages both sexes, and for all racial groups, or is it not? Further, how would you know if you were one of the 3 percent where the vaccine did not work?

Obviously, one way to find out would be if you got measles. Another way would be to be tested for antibody levels for measles, but very few people do this unless they are in the health professions and need to know if they need to receive a booster shot or not. The general population is not likely to do this. In the face of this, many doctors are likely to simply recommend another booster from the perspective that "it can't do any harm given that adverse effects are extremely rare," an often-heard medical opinion. If adverse reactions are not as rare as stated, it will significantly alter the benefits-to-harms ratio.

In broad strokes, the benefit of receiving the MMR vaccine is that for those not in the 3 percent with primary vaccine failure, the vaccinated are now thought to be protected from the three diseases. The protection afforded can now be compared to the negative outcomes of the disease itself, which can include all of the symptoms of the disease, but also the potential for encephalitis, hospitalization, and death, mostly in the youngest and oldest segments of the population.

How many of those who get measles have one of these outcomes? The numbers are actually very unclear and largely depend on whom one asks. The CDC makes a claim that one (or more) people in one thousand people who get measles will die.[22] In North America, this number is likely to be highly inflated,[23] thus leading medical organizations to promote vaccination as an alternative. However, in order to do the full benefit-risk calculation, one needs the numbers per thousand (or whatever number of the population) who have significant adverse vaccine outcomes. The actual number is harder to calculate, as it depends on the various surveillance systems designed to monitor negative vaccine outcomes, as described in Chapter 3. One such is the sometimes-maligned Vaccine Adverse Events Reporting System, VAERS, maintained jointly by the CDC and FDA.

A major complaint about the system is that anyone can file a report. However, those same complainants tend to forget that all reports are vetted by physicians working for the CDC who can reject reports that appear (to them) to be spurious. However, since reporting adverse events is not mandatory, in many cases adverse reports are not filed at all. In part, the lack of complete reporting stems from two main factors: one, doctors who make up the bulk of those submitting reports are often too busy to do so routinely; and two, medical doctors trained to believe that adverse vaccine events are extremely rare may simply not believe a patient's complaint. In Chapter 3, I will show that while it is convenient for those defending vaccines to critique the VAERS system, it remains for many the primary repository of vaccine adverse effects and is often used by those same researchers in attempts to provide evidence for the safety of various vaccines.

To determine just how effective VAERS is in general, a private company, Harvard Pilgrim, was contracted by the CDC. Harvard Pilgrim concluded that less than 1 percent of adverse reactions were reported.[24] The extent to which this is true means that getting the correct ratio of benefits to risks is going to be inaccurate, maybe wildly so by overestimating the benefits and underestimating the risks. The same problem with evaluating the risks attends many vaccination recommendations, notably for the human papilloma virus (HPV) vaccines and many others.

All of the above simply illustrates that, in the absence of accurate information, risk-benefit ratios are actually hard to calculate and, with an apparently increasing skeptical population of most things medical, harder to get people to comply with vaccine recommendations.

The risk-benefit analysis for vaccines actually forms the core of the controversy about vaccine safety and effectiveness, and it is here in this arena that the two very unequal adversaries battle for the scientific, social, and moral high ground

Although aspects of this risk-benefit analysis will be presented throughout the following pages, let me start by presenting, as I understand it, the concerns of differing sides in the "debate" that one side, the mainstream medical side, typically refuses to have.

Let me emphasize at this point that to many from the mainstream perspective, my analysis is likely to be biased by being considered to be an "anti-vaxxer." I don't consider myself to be anti-vaccine, but then those who make this claim would go on to say that anti-vaxxers never do.

Fair enough. Let's see what a summary of the data shows when we get to Chapter 3.

What Evidence Should Any Side in the Vaccine Wars Present To Best Support Their Position? (Part 1)

With these rules in mind, what sort of evidence would be needed to make the case for or against vaccine safety or effectiveness?

In regard to safety, the pro side would have to demonstrate with actual verifiable numbers vaccine-preventable damage versus injuries arising from the vaccines developed as protection against the actual disease. For example, with measles outbreaks being heavily covered in the news in the United States and Canada in 2019, a quantitative evaluation of these numbers of deaths or serious injuries to the disease compared with the same for the vaccine would be essential. Ideally, it would have to provide this evaluation for each vaccine in the CDC's schedule. Further, the evaluations would have to be done in the same general populations such that potentially vaccine-preventable deaths or serious injuries to measles in Kenya, for example, would not be juxtaposed with serious measles vaccine adverse events in the United States. Vaccine-induced herd immunity, or the more politically correct form of what is called "community immunity," if evoked, would need to be validated by studies in actual animal herds or in humans. The concept of herd/community immunity that occurs in nature and is purported to be possible with vaccination will be addressed in Chapter 4. Additionally, any reliance on computer modeling without any adjustment for possible secondary vaccine failure would not be acceptable. Further, the time period of the study should be of sufficient length to allow for slowly emerging negative outcomes, such as those associated with autoimmunity, to be recognized and documented.

In terms of vaccine effectiveness, the ideal data would consist of challenge studies in humans in which a randomized control test (RCT), the "gold" standard of practice, was in place. In an RCT, two clear groups of subjects, randomly assigned, are given either the actual vaccine or a real placebo control. By placebo, I mean that the placebo treatment should be something that is deemed to have no possible effect, such as saline. The use of another vaccine as a control, or an aluminum adjuvant as the control, is simply unacceptable, in spite of a great deal of the literature used to justify such treatments as valid. The study should state a hypothesis that in the study design was falsifiable. The study would have to be adequately powered to detect what it intended to detect, and appropriate statistical methods would have to be used.

In the challenge part of the study, the people or animals in both groups would have to be exposed to the pathogen in the same way and for the same duration, and both groups would have to be representative of the actual populations being considered for vaccination. For example, excluding people with certain conditions such as autoimmune disease would not be acceptable generally, at least without some commentary in the final report that the study did not apply to that part of the overall population nor could vaccine safety for that subpopulation be assumed in vaccination recommendations. Since vaccines are in fact mostly given to populations at large without any screening for conditions such as autoimmunity, without such stipulation there would be no evidence that this fraction of the population might not be at risk and should be exempt from this vaccine.

It is a widespread method in pharmaceutical company trials to compare an experimental drug that is intended to address some medical condition with an older drug for the same condition. In this case, the comparison is designed to compare drugs with one another for efficacy, not against a "no" drug real placebo control. In vaccine safety or efficacy trials, however, as cited above, the use of another vaccine as a control for the novel vaccine is problematic, since it presumes that the older vaccine has a demonstrated safety profile that may not actually be in evidence, particularly if that older vaccine was tested in the same manner. In addition, different vaccines are designed for specific diseases, not diseases in general, and are developed on different platforms and with different antigens and hence may vary significantly in their immunogenicity and the types and levels of adjuvants used.

Further, the use of the adjuvant alone, usually one of the traditional aluminum salts or newer proprietary formulations with aluminum, does not constitute a real control, merely a test of whether the whole vaccine with antigen and aluminum and other excipients is different from the adjuvant alone. If, for example, the aluminum adjuvant was shown to be a prime factor in adverse reactions in some vaccines, then no difference would be expected between health outcomes in the two conditions. Indeed, this tends to be what is found in many studies and points to the obvious fact that such a study design is probably the best way to *not* find adverse outcomes. Such study designs may be standard in the field, but that fact does not make the methods here any less careless or even deceptive.

Let's take another example, not vaccine-related, to see how such a study might play out in a study of the impact of toxins on the nervous system. For example, perhaps I want to know if a toxin (Toxin "A") that has been described in the scientific literature will generate motor neuron loss in our laboratory mice if presented in diet or by injection. If I use as the comparator group another toxin (Toxin "B") that also might trigger motor neuron loss, all I know at the end of the study is that toxin "A" is more, less, or the same in toxicity compared to toxin "B." If it turns out that the outcomes are the same, I am not justified in concluding that toxin A is not toxic.

In academic studies, any master's or doctoral student who tried this without a control arm would rapidly find their thesis rejected if they tried to justify this study design by claiming that past studies of the older toxin had once upon a time been compared to controls. Given how much might have changed since such an older study was completed, an assumption that the previous control could carry over to the newer study would certainly be discounted.

The lack of controls thus looms as a significant methodological flaw no matter who performs the experiment. As an example, a relatively recent article by the Exley group that claimed that brain samples from people with autism had high levels of aluminum deposits was roundly, and I think somewhat justifiably, critiqued for the lack of control samples by many of the same bloggers and mainstream sources who routinely accept the lack of real controls in vaccine trials. Such a dichotomy is more likely to be driven by some agenda than by a desire to protect the scientific method.

A second caveat, to be explored in more detail in the next chapter, concerns the use of "surrogate" markers of the stages of a disease or of the strength of an immune response. This is understandable in vaccine trials, since actual challenge experiments in human trials are quite rare. The reason this is so is that they might require exposing test subjects to the actual disease-causing agent.

The Combatants and Bystanders in the Vaccine Wars

In this section, I will consider who the players are in the "vaccine wars." The actual history of vaccination and the stages in its evolution as the dominant paradigm for disease control in modern medicine will be presented in Chapter 2.

The "Pro-Vaccine" Camp

The first grouping can be described as the "pro-vaccine" group. It comprises most medical doctors trained in any medical school that might be described as "allopathic." The Merriam-Webster dictionary definition of allopathy:

> relating to or being a system of medicine that aims to combat disease by using remedies (such as drugs or surgery) which produce effects that are different from or incompatible with those of the disease being treated."

In other words, allopathic medicine is what is conventionally considered modern medicine, ideally evidence-based by being rooted in peer-reviewed scientific studies.

It would be safe to say that most graduates of allopathic medical schools believe, or profess to believe, that the practice of vaccination against various infectious diseases is one of the fundamental and most important advances in medical history. Such doctors (and nurses) credit vaccination with saving millions of lives and much suffering since its initiation in the late eighteenth century.

These medical professionals also usually hold that any adverse effects of vaccination are mostly minor (i.e., a sore arm) and thus inconsequential to general health. Similarly, it appears to be a widespread belief that serious adverse events, for example, seizures, neurological disorders (including autism spectrum disorder, or ASD), and those impacting other organ systems are *extremely* rare. Thus, for those trained in this perspective, the risk-to-benefit equation vastly favors vaccinating over not vaccinating. Also, since in this view the risk is microscopic for any serious adverse event for any one vaccine, increasing the number of vaccines does not measurably change the equation.

Medical doctors have spent considerable time (and money) getting their credentials, first in whatever premedicine program they followed, then in medical school, later in their medical residencies. They consider themselves to be experts on things medical, as they often are in their own particular specialties, and usually very much resent lay people disagreeing with such hard-won expertise: eight or more years in medical training is likely to be vastly superior to studying at "Google University," and thus medical doctors have little time for those whose "training" is from the Internet. In this regard, doctors are not dissimilar to those in other professions: plumbers don't like nonplumbers telling them how to solder pipes; lawyers don't usually like lay persons explaining torts to them, etc.

Some medical doctors have even seen the ghastly impact of "vaccine-preventable" deaths, especially in young children, and such events are highly likely to influence how they view people who choose not to welcome what they may consider the life-saving impacts of the various vaccines on offer.

In other words, summing up, the mainstream allopathic view toward those who don't, or won't, vaccinate would be something like the following:

> I am the expert in this, not you; vaccines save lives and I've seen it the preventable deaths that a vaccine would have saved, you haven't. In fact, vaccines work so well that they are victims of their own success. You have never seen the real consequences of not vaccinating. Your sources concerning vaccine safety are pseudoscience since vaccines are highly tested and screened for adverse effects, most of which are trivial. Those who disagree are cranks whether they hold advanced degrees or not, and/or grifters out for a fast buck. You are endangering your child by refusing to accept my expertise and the wealth of science that proves what I am saying. Thus, those who choose to reject the science, which is "settled," are the reason why we need vaccine mandates.

This may not be the views of all allopathic doctors and nurses, but I suspect the vast majority of the same would agree with most of this.

I should mention that I actually understand the annoyance that many medical doctors feel when challenged by patients or laypersons about vaccines or anything else. Certainly, most neuroscientists I know are not particularly welcoming to lay people who want to try out their Internet-derived hypotheses on us. I get more than a few of these hypotheses every year, often cloaked in a potpourri of neuroscience terms that make the hypothesis (usually described as a theory) sound superficially more credible than it actually is. The worst part is not the hypotheses/theories per se, rather the inability to accept that some/most of the premises and "evidence" leading to the theory are simply wrong. I spent four years getting my bachelor's degree, another three doing my master's, five years in a PhD program (although some of this time was eaten up by army service), and eight years as a postdoctoral fellow/research associate. It's a lot of time and lots of dues paid to even get to an assistant professorship. For this reason, I get how doctors feel about being challenged, I really do.[25]

What's the difference between basic science people and some MDs? One thing I think is true is that scientists, maybe especially neuroscientists, as arrogant and Type A as many can be, do mostly realize that what we understand about the brain in health and disease is always changing. Maybe this fact doesn't provide better behavior, but it at least encourages more humility in regard to the reality of changing evidence and the interpretations of this evidence. In contrast, at least when it comes to the entire subject of vaccines, many MDs approach the topic as though it were part of Holy Scripture. Thus, challenging any part of it is not merely annoying and an attack on their expertise, but rather an attack on their "faith." (See Chapter 8.)

It is worth remembering in this regard that the training basic researchers in any field receive versus that of MDs is quite different. Admission to graduate school and admission to medical school are in general fairly challenging, but once one is admitted, the expectations and paths vary considerably. Graduate school candidates are not usually looked on as particularly special or elite people just because they got into grad school. Medical students, however, are, and this sense of specialness permeates much of their self-image that emerges over the course of their training from the first years on through residency and beyond. In medical school, challenging your instructors on medical issues is not usually a path to career success.

In contrast, in graduate school challenging accepted paradigms may well be a good choice, since it is accepted that your role is to provide some original insight into your chosen scientific field. Indeed, throughout their careers, basic scientists are used to having their hypotheses challenged and to finding out that these hypotheses are often, in fact usually, incorrect. Scientists are used to their field evolving such that many of the things they learned in graduate school or as postdoctoral fellows will eventually be shown to be incomplete, or just frankly wrong. Make no mistake, no one likes abandoning a pet hypothesis or finding out that some core concept they have carried for years is incorrect. However, in my experience, with minor exceptions, few neuroscientists are so enamored of their hypotheses that losing a pet hypothesis creates much professional pain.

Medical doctors, in contrast, *are* their profession. They are used to the social prestige that being a doctor conveys and to upholding the teachings they have received. Challenging their training tends to be seen as challenging them as people.

Of interest, a number of MDs also hold PhDs, and this makes for an odd mix: unwilling to be challenged professionally as doctors, but quite used to being challenged as scientists.

This is not to say that medicine doesn't evolve in its knowledge of human health and disease and in treatment options. It does. However, the process by which it does is different. Some examples will be cited later in the book.

Of course, all of this is a generality. Not all MDs are the same, nor are all basic scientists the opposite. However, having known more than a few of both professions, I think it fairly safe to say that, in general, these observations apply.

One other key aspect distinguishes medical doctors from basic scientists: the public perception of what each does and how this is often portrayed in popular culture. For example, there are numerous television shows about doctors, few to none about bench scientists. Doctors are seen to inhabit a rarefied world of life-and-death decisions, the stuff of drama, where their training and experience make the difference about who lives and who dies. In contrast, scientists tend to be personified as sort of nerdy, often chasing weird, improbable hypotheses. The television show *House, MD* versus the movie *Back to the Future* comes to mind, and these are, I think, pretty typical depictions of what is portrayed to the general public about what each profession does. Given this, whom are lay people expected to trust with their health? It's obvious, and in news interviews or courtrooms, scientists rarely carry the respect and gravitas of medical doctors on any given medical topic.

I will come back to this in the next sections, but I really think that this is the root of much of the disconnect between parents and the medical profession that the vaccine wars have generated, namely, prestige and type of training.

True "Anti-Vaccine" Proponents

A distinctly contrasting view to the ideologically pro-vaccine mainstream medical profession is held by those who could legitimately be described as "anti-vaccine" in that they are simply against the practise of vaccination. Those who fit this description are much like those opposed to genetically modified foods in that they don't believe the claims made by the proponents in regard to the science behind the practice, the safety profiles of the products, or the reality of the benefits.

In regard to anti-vaccine positions, true anti-vaccine proponents tend to flatly disbelieve that vaccines provide any level of immunity to the infectious diseases for which they were developed; if they believe in germ theory at all, find the published or reported scientific literature wanting; and frankly believe that vaccines can only cause harm, rather than any benefit. A corollary view is that vaccination and vaccines are somehow unnatural and hence violate the sacred nature of the body. Some of the anti-vaccine stance in some opponents is religious in nature, and some aspect of this has existed since the earliest attempts to vaccinate human populations. Indeed, the pro-vaccine side tends to dredge up this now-ancient vaccine meme from the years after Jenner's seminal reports and endlessly attempts to tar the current vaccine-hesitant with the same brush.[26]

What fraction of those typically described as "anti-vaxxers" do these individuals comprise? It is hard to say precisely, but a best guess is that they are a distinct minority, although a minority that may be growing. As alluded to elsewhere in this book, the growth of true anti-vaccine sentiment may be driven in part by a "pushback" to the oftentimes-condescending attitudes and behaviors of the pro-vaccine camp. Some in this grouping even have degrees in the related subjects of immunology and base their objections to vaccination based on their understanding of immunology and how the immune system operates in health and disease.

The Vaccine-Hesitant or Resistant

By far, most of those who have any objections to vaccines in any aspect of efficacy or safety fall into this camp. Many of these, at least in my experience, are quite knowledgeable about vaccinology theory and practice and tend to be relatively up-to-date on emerging studies on vaccine safety in particular. They may also be quite adept at seizing on and critiquing the statements of pro-vaccine proponents, particularly medical professionals who may not be as firmly grounded in vaccine theory as they should be and hence not equipped to defuse, let alone debate, the knowledgeable vaccine-hesitant. The status that medical doctors once seemingly enjoyed by right of their medical degrees, that is, the "trust us, we're experts" mantra, has fallen flat with this group.

The numbers of this grouping have been swelled by several key factors. One is that much of this group originates in those who believe that people they are close to, often their own children, have been injured by vaccines. Or, they are simply tired of feeling bullied, pressured, and dismissed by the by pro-vaccine side and the latter's unwillingness to engage in debate/discussion except to denigrate them as tin foil (or aluminum) hat conspiracy theorists, a position taken by far too many on that side of the issue.

An argument could be made, as various commentators have done, that this grouping represents the bulk of opposition to vaccine mandates and reflects the very predictable response to pressure, particularly pressure coming from those who may have the *official* credentials but actually may know less about vaccines than those they are belittling.

One interesting aspect of the categorization of this group as anti-vax is that it tends to sweep in medical doctors and research scientists who have apparently deviated from the true faith (see Chapter 11). The assumption, often stated by bloggers such as Orac or Skeptical Raptor (see Chapter 9), is that doctors who ask questions have forgotten, or abandoned, the vaccine wisdom they learned in medical school. Indeed, such comments suggest that medical doctors in this category should lose their licenses. Particular targets of such ire are Drs. Bob Sears and Yehuda Shoenfeld, the first a pediatrician in Southern California from a well-known family of other pediatric specialists, the latter a lead researcher on autoimmunity. There are many others, in particular the now legendary Andrew Wakefield, briefly mentioned later in this chapter and in detail in Chapter 4. Wakefield is considered in some pro-vaccine circles, and much of the mainstream media, to be the virtual godfather of the anti-vaccine movement, whatever it is that movement really represents.

In much the same way, bench scientists who publish any data raising vaccine safety concerns have now joined the legion of "scientists who used to do good work until they went over to the dark side." Certainly, a number have fallen into this bracket, including this author, as well as Drs. Chris Exley, an aluminum expert at Keele University in the United Kingdom, and Romain Gherardi, a Paris-based researcher who has looked intensively at aluminum adjuvant transport into the nervous system. There is a host of others who will be mentioned in the following sections of this book.

By so attacking and demonizing those who dissent, the mainstream medical community and the bloggers simply demonstrate time and time again the essentially fundamentalist religious nature of vaccinology (Chapter 8).

The Remainder

The rest of the population, much as in most political issues, is made up of those who simply have no concrete opinion on the subject of vaccination, or simply don't care. They either go along with vaccine schedules because they have always done so or were told to do so by their physicians. Alternatively, others innately distrust any governmental organization and refuse to go along, regardless of the subject. The former tend, for the purposes of accounting, to fall in the pro-vaccine side; the others usually, but not always, into the vaccine-hesitant/skeptical camp. (In later chapters, I will explore in more detail the sometimes bipolar nature of what is termed the "left" when it comes to vaccine belief and compliance.)

All of the above should make clear that vaccine acceptance or rejection comes in many flavors and reflects a variety of attitudes, not all hostile to vaccination in general. Attempting to reduce this spectrum of attitudes to vaccination as a "you are either with us, or with the terrorists" polarity is simply incorrect in the same way that former President George W. Bush's statement after September 11 misrepresented the range of responses around the world to the events of that day and afterward. The end result of such simplistic thinking led inevitably to the US invasion of Iraq and the endless wars in other parts of the world. Trying to force people into one or the other extremes in the vaccine wars is just as likely to backfire.

Once an "Anti-Vaxxer," Always an "Anti-Vaxxer"

My colleagues and I sort of suspected that we had been branded as anti-vaxx early in 2011. In that year, a new postdoctoral fellow in my laboratory, Dr. Lucija Tomljenovic, and I had submitted a grant application to the Canadian Institutes for Health Research (CIHR), the Canadian equivalent of the US National Institutes of Health (NIH). The grant was designed to look at the toxicity of aluminum from various routes of administration: inhalation, diet, and injection. The word *vaccine* was mentioned only as an example of injected aluminum, without any comment by us about vaccines in general or in relation to autism.

The grant was reviewed and rejected. Not a great outcome, but not really an unexpected one given that CIHR review panels only fund about 12 percent of applications, often less. Perhaps it was just a poorly constructed application? Sure, maybe. However, of greater interest were the comments of two of the three reviewers, both of whom said that the proposed experiments added nothing novel, since, to paraphrase, Andrew Wakefield's original article had been retracted and thus all of his work "debunked."

It was in this rejection of our grant that one sees the power of the frame: a grant that was all about aluminum exposure by *any* route was rejected with comments about a study and vaccine that had nothing whatsoever to do with aluminum. In brief, in the view of the reviewers, Wakefield had been debunked, which then meant that any critique of vaccines of any form was also debunked.

The second major revelation to me about the frame came much more recently as I was just starting to write this book. One of my main interests outside of academia, as previously noted, is the struggle of the Kurds and other minorities in Rojava for autonomy.

In October 2019, the Turkish army invaded parts of Rojava in order to crush the nascent autonomous region. A medical group I helped found (Rojava Emergency Medical Service) had a member inside Rojava working to coordinate medical personnel and supplies in aid of those injured in the fighting.

Our member, who goes by the nom de guerre of Argeş, was the man on the ground coordinating the supplies and people that I was able to send into the country. One odd *Signal* message suddenly came from Argeş, something along the lines that some of the potential donors from a Chicago- based anarchist collective for our medical group were concerned about me being an "anti-vaxxer."

What Argeş wanted to know was if I really was an anti-vaxxer, although he stated that he didn't have an opinion about it one way or the other, apart from how it might impact donations. Needless to say, I gave Argeş a serious dressing-down that started with the words "Are you serious?"

The bottom line is that Argeş, an activist fighting for a political cause and a radical by most definitions, is not someone educated in, or even particularly interested in, medical issues. And yet, he had found that somehow anarchist donors had thought I was anti-vaccine and hence were wavering on providing medical assistance to Kurdish refugees fleeing an invasion. Seriously?

The point of these two vignettes is this: the demonization of Andrew Wakefield, and all those deemed to be his successors, spreads to all aspects of medical research and practice, no matter how remote or far away from any aspect of vaccination. Further, it spreads to people far removed from the very theory and practice of vaccination.

Science Literacy versus Illiteracy: It's Not Just Confined to Lay People

It's typical that those in the pro-vaccine camp tend to present science in a very nonscientific light. Included in those who do so are most mainstream medical health providers; the official medical organizations; entities in the United States such as the CDC and FDA and those internationally, including WHO, along with the pharmaceutical companies that make vaccines; and the freelance and for-hire Internet bloggers (Chapters 9 and 10 explore these links in more detail). By and large, most mainstream media tend to go along (see Chapter 10).

For each of the above entities, there seems to be a tendency to label the status of vaccine safety as "settled," an odd thing to state given that it is not the fundamental nature of science to be settled about much of anything. Such a statement is often uttered or written with a corollary statement that some study, for example, "proves" that vaccines don't cause autism.

I suspect that most scientists working on vaccines know all of this fully well and further know that stating that science is settled is nonsense. The statement about vaccine science being settled, however, is likely not meant for them, but for the lay public, whose overall grasp of the scientific method and what it can and can't do is often meager.

It is in this latter statement that we can see the level of confusion about what science really is, and isn't, as described earlier in this chapter. Proof, as cited previously, is the domain of formal logic and mathematics. In contrast, science does not prove anything, but merely builds by experimentation a body of data that makes certain assertions about nature more or less likely. In some cases, this body of knowledge can become pretty overwhelming, but this is still not proof, but rather probability.

Of all the sciences, perhaps the discipline where it can be said the least is in vaccinology. Hence, rather than being settled or proven, the very best that might possibly be said is that, based on a number of studies, a high probability exists that vaccines are safe, that they do not cause autism or any other developmental disorders, and that they are effective at preventing the diseases for which they were developed.

However, to anyone with both science training and an even partially open mind, such statements have to contend with a number of rather fundamental challenges. For example, what is one to make of the experimental data that contradict these notions? Further, how are the studies that support the "safe and effective" perspective chosen, and why are those studies that dispute it rejected? Which criteria are used for either? Even if true in one population or age group, what about other populations or age groups? What about temporal effects in which a vaccine may experimentally be considered safe on the day of administration, but not after a year? Are such vaccines equally safe for both sexes? What about combinations of vaccines given sequentially, or simultaneously, the latter incidentally a question never answered by any studies until that of Eidi et al., 2020?[27] All it would take to lower the probability that the blanket safe and effective designation for vaccines is true would be a contrary outcome to any of these questions.

In actual fact, it is not that such negative outcomes don't exist in the experimental literature, but rather that the pro-vaccine lobby is remarkably effective at cherry-picking its way past such studies to focus on only those that give the desired outcome.

Not only is this not science in any traditional sense, but rather is a form of pseudoscience in the pursuit of an agenda. Professor John Oller has termed such approaches akin to "calculated bias and loaded dice."[28] What that agenda is will be explored in more detail in Chapter 6. I suspect that most working scientists would tend to agree with this view, at least those who are able to leave behind their own self-interest and examine the issue rationally based on their training.

It should be mentioned that medical doctors are not necessarily scientists in the real sense of the word. Some are, of course, but most are not, and the process of scientific enquiry is not the fundamental part of their training compared to research scientists. True, most people who go on to medical school have basic undergraduate science backgrounds, and it is also true that many of the greatest medical scientists, past and present, have been primarily trained in medicine.

However, medicine, while increasingly backed by scientific inquiry, is not in many cases actually scientific. Rather, it is an assumed body of knowledge, some based in experiment and observation, some based on intuition and indeed the "art" of medicine. While most MDs have, in their training, been exposed to most of the broad strokes of medical theory and practice, few are likely to be masters of any but a small part of this.

How Much Vaccine Education Does One Get in Medical School?

Medical doctors oftentimes denigrate laypersons who disagree with them, particularly when the subject is vaccines and vaccination, by pointing out that they have the medical training to understand medical issues and the laypersons have only the "misinformation" provided by "Dr. Google" or some blog.

Let's imagine for a moment that this is true. If so, can we quantify how much formal education most MDs have on the subject coming out of either medical school (usually four years) or after a residency? At my university, the latter is typically five years in length for most specialties, with two years being the required time for "family practice."

Not all medical schools are equal in this regard, but those that have adopted a "problem-based learning" structure are likely relatively similar.[29] It is important to remember that the four years of primary medical training are jam-packed with information, and a lot of in-depth material simply cannot be presented in the time available.

However, given this, I want to explore some examples, starting with my own school, the University of British Columbia's Faculty of Medicine.

Based on a current list of the "blocks" of subjects in the first two years, there are none devoted to vaccines or vaccinology. However, within the various blocks are a number of lectures that touch on immunology, infectious disease, and related subjects. Here are the numbers:

Year One:

The number of lectures that seem to focus on vaccines (each lecture is fifty minutes long): one; The number of lectures that *might* touch on some aspect of vaccination theory or practice: seven.

In the former category is the following lecture in the fifth week of instruction: "Vaccines and Immune System Review." This follows three lectures two days earlier on: "Introduction to Immunology," "T cells," and "Bridging the innate and adaptive immune systems." The next day, the students get: "B cells" and "Immune dysregulation and deficiency." One assumes that this crash course on immunology, a rather complex subject, is designed to provide the basis for understanding how vaccines work, or are intended to work.

Year Two:

Zero and two, respectively. I note that in the first teaching block of Year 2, students have a case involving developmental neurological delay that turns out not to be due to vaccination, but due to a gene deletion. Nonetheless, the program inserts a short article from the Canadian Pediatric Association that assures the students that vaccines are safe and effective and that Andrew Wakefield's work was debunked. This document can be found online in Appendix 1.

Year Three (the clinical rotation or "clerkship" year):
In this year, the students rotate through four main clerkship blocks of ten weeks each that include:

Women's and children's health: This includes two weeks' worth of topics in pediatrics and obstetrics/gynecology. Included in this block are subjects such as infant developmental milestones, a consideration of autism spectrum disorder (ASD), attention deficit hyperactivity disorder (ADHD), and other causes of speech and language delays. In a session titled "The Newborn," students receive instruction on the "importance of vitamin K administration in [the] newborn period." [Brackets for addition are mine.] More specifically for vaccination, there is a block called "Pediatric Health Supervision," in which students are taught to "identify [the] routine immunization schedule in BC; [and] counsel parents on contraindications and side effects of common immunizations." [Brackets are mine.]

Surgical and perioperative care, which includes general surgical practices and exposure to some of the sub-specialties: nil.

Brain and body, including a week of psychiatry: nil.

Ambulatory care: nil.

In a sub-block week titled "Family Practice," one session is on "Immunization" and includes the following objectives:

1. Appraise evidence for various vaccinations, including appropriate timing and administration, immunization schedules, indications and contraindications;
2. Demonstrate the ability to counsel patients and parents on the risks and benefits of vaccinations, including *dispelling myths and misconceptions*;
3. Counsel parents and families on the risks and benefits of travel vaccination and medical planning;
4. Identify the patient vaccination status. [Italics are mine.]

Finally, by the end of Year Three, students are expected to be able to demonstrate a range of clinical skills. Relevant to vaccination in the Family Practice section of Ambulatory care is item 22, "injection administration."

Within this year, the students also get a week each of dermatology and ophthalmology.

Year Four is the year in which students in twenty-four weeks take "electives" from a list of 590 possibilities and some other courses to help them get ready for their residencies as they start to transition to the actual practice of medicine. Of the 590 electives, there are some nineteen general categories with some having more than one possible subsection. Examples of some of these electives are anesthesiology, dermatology, internal medicine, surgery, public health, pediatrics, and many more. The large number of listings reflects not only similar electives running within hospitals in Vancouver, but also others from other areas of the province and even out of the province or the country.

Clearly, in some of these areas, such as in pediatrics, there will be lectures and other instruction on vaccine-related issues. Whether such instruction serves to actually present the pros and cons of vaccination, rather than just check the box that "vaccines are safe and effective," is not clear.

These numbers seem to be broadly similar to the numbers from other medical schools examined in conversations with MDs elsewhere, but specific comparisons are difficult.

However, in regard to UBC's medical school, several points should be considered: First, the amount of material the students are expected to absorb and master is enormous, and clearly there is little time for going into any of the broad range of subjects in particular depth. Second, depending on how it's calculated, students at UBC may get more instruction on vaccine theory and practice than medical students at other universities. This, of course, is very good.

What is uncertain, however, as above, is just how critical such instruction is. The note above on the training for "dispelling myths and misconceptions" may give us some clue and may not be surprising given the general consensus in medical schools about vaccination and the uncritical view of its benefits in general.

At the end of the day, the goal of programs such as at UBC is to encourage the use of what is called "evidence-based medicine" (EBM), that is, the practice of medicine based on valid experimentation and evaluation. One definition of EBM is:

> Evidence based medicine (EBM) is the conscientious, explicit, judicious and reasonable use of modern, best evidence in making decisions about the care of individual patients. EBM integrates clinical experience and patient values with the best available research information.[30]

The "best available research information" should be, in turn, derived from actual research, rather than assumptions and statements of faith.

Medical schools and their students are officially committed to this concept. The challenge is to ensure that EBM actually occurs, and to do that requires that physicians know, and act on, the best research in any field. In order to effectively do this, they have to know all sides of the issue, whatever that issue might be, not merely parrot information that they have been told but not researched themselves.

Residencies

What about residencies at the University of British Columbia (UBC)? How much vaccine content is there?

This seems to depend on the program itself. For example, Family Practice and Pediatrics, in which the bulk of vaccine delivery actually occurs, will likely feature some of it. The question again is how much of this is critically evaluated with a concern for all of the peer-reviewed literature, rather than those parts cherry-picked to drive home a particular point.

Again, various medical programs may be quite different, and from talking to colleagues at other schools, it seems that UBC's level of instruction about vaccines and vaccination is far more extensive than most.

Overall, this training consists of the basic history of vaccines, what is in them, some of the purported success stories such as smallpox eradication and the control of polio, and the fact that vaccines in general are safe and an effective means of infectious disease control. As noted above, medical students in Years Three and Four get more information about immunology and vaccinology as they rotate through various medical departments, but again this knowledge is not extensive unless they later choose to specialize and do residencies that delve in more detail into such topics. For this reason, trying to convince the average MD that, for example, aluminum is neurotoxic is largely a fool's errand.

Back to What Science Is (and Isn't)

The same lack of a broad knowledge of subfields applies to scientists in any discipline. For example, I am trained as a neuroscientist particularly in neurophysiology, but neuroscience is a vast field, and, while I have likely heard in passing details about most of the subfields, I can't claim any expertise in the majority of them, let alone in various other branches of science.

As noted earlier in this chapter, science is primarily a method for exploring nature and attempting to establish probabilities that any of the hypotheses proposed and the experimental outcomes found are, in fact, likely to be correct. A large enough body of evidence, the collated body of experiment from tested hypotheses, can become a theory, but even so it remains a probability discussion. Discrepancies, anomalies, etc. in the testing of aspects of a theory can lead to its downfall if enough of them arise. As philosopher of science Thomas Kuhn, cited above, observed, science does not advance by proof, since absolute proof is unattainable; rather, it advances by disproof, and disproof only requires one negative result. This is key, as cited previously: a truly acceptable scientific hypothesis has to be falsifiable, that is, it has to be open to negation.

Much of the stated belief that the science of vaccine effectiveness and safety is settled can only arise from the deliberate denial of the actual scientific method in which only those who appear to agree with the conventional wisdom are accepted, while those who disagree are ignored or cast out. As such, the process by which scientific finds are accepted or censored has more to do with religion, a topic that will occur frequently in this book, than the scientific method.

As noted above, most working scientists speak in terms of probabilities and don't make definitive statements about much of anything being settled. This can be annoying for nonscientists who often expect that science will do what mathematics does: prove things. Thus, it is typical for a scientific paper to have statements like "The evidence thus indicates that 'molecule x' plays an important role in the regulation of 'protein y,'" or something like that. Non-scientists raised on Disney movies want more definitive statements, statements that working scientists are usually loathe to provide. And for good reason: one of the best ways to get your figurative butt handed to you on a platter is to "go beyond your data" in a grant application or a manuscript. This is particularly true given that the next scientist who does the same experiment might find something very different. Or, worse—but it happens—you might not even be able to duplicate your own original findings.

This last is termed the "decline effect," in which previously robust and statistically significant outcomes in an experiment simply become less and less significant with replication. This doesn't always happen, of course, but it happens enough to make it a real concern. The reasons for this are uncertain but may have to do with subtle changes in study design or execution, or simply the fact that even with what should be adequate statistical power, greater variations in measurements for whatever reason can collapse p values to the point where statistical significance vanishes.

For the lay public, pronouncements in the mainstream media are often taken as fact. And since the media tend to favor pro-vaccine statements over those more skeptical, the lay public, not necessarily particularly scientifically literate, goes along. The forcing of some narrative along the path of a particular belief or agenda is termed a "frame," a concept that will be revisited throughout the chapters that follow.

A Brief Introduction to the Wakefield Controversy

A common example of the public going along is the oft-repeated claim that the *Lancet* study of autistic children and gastrointestinal abnormalities of Dr. Andrew Wakefield and colleagues was "debunked." The Merriman-Webster dictionary defines "debunk" as "to show that something (such as a belief or theory) is not true: to show the falseness of a story, idea, statement, etc."[31]

I will deal with this case in greater detail in Chapter 4, but it is worthwhile briefly considering it here in order to understand how the use of this term has been deliberately manipulated to make Wakefield the whipping boy of the pro-vaccine camp and the mainstream media.

First, it would be safe to say that the majority of journalists and most of the public have never read the original Wakefield et al. article, because, if they had, they might all appreciate that the data in the article were not in reality "debunked" at all.

The original observations on which this case series was based were these: twelve children[32] were referred to the Royal Free Hospital in London at which Wakefield worked. The children, eight with and four without regressive autism, showed signs of intestinal pathology. The article's authors recorded the observation by the parents of the children that gastrointestinal (GI) abnormalities and regressive autism arose *after* the MMR vaccine. In regard to these points, various other investigators have confirmed that children with autism often have such GI abnormalities, so this observation and the data in the Wakefield paper have not been debunked at all. Brian Deer, a British journalist, claimed that these data in the Wakefield study were fake, but his evidence that this is so has been subject to claims that his evaluation of the study has itself been "debunked," as discussed in detail in Chapter 4.

Did the children in the study suffer from autism? Yes, they did, so this point has not been debunked, either. Did the MMR vaccine cause these negative GI and neural outcomes? Maybe not, but this cannot have been debunked, since Wakefield et al. made no such claim.

Hence, what has been constantly described as a rogue study being completely fake is simply wrong. The observations made in the study have not been contradicted by other researchers or by the authors themselves. The interpretation others put on these outcomes were never written by the authors. Yes, the paper was retracted by *Lancet*, but that is not the same as being shown to be fraudulent, and indeed the notion that the data were fake hinges completely on the interpretation of one man, Deer, a nonscientist.

What is possible here is that the GI analyses in this article were incorrect, but being wrong happens a lot of the time in biomedical research[33] and certainly does not involve scientific malfeasance in most cases.

So, in this most famous of "anti-vax" studies, the mainstream media reportage that has served to influence much of what the public thinks it knows about vaccines and vaccine injuries comprises a knowledge base that is distinctly false and blatantly misleading.

It is difficult to avoid the interpretation that such attempts to mislead the public at large are not random, and it is to this point that I will return in Chapter 8.

CHAPTER 2

Vaccination History, Theory, and Practice: A Brief Overview

Ere I proceed, let me be permitted to observe, that truth, in this and every other physiological inquiry that has occupied my attention, has ever been the object of my pursuit; and should it appear in the present instance, that I have been led into error, fond as I may appear to the offspring of my labours, I had rather see it perish at once, than exist and do a public injury.

—Edward Jenner[1]

Edward Jenner and the Formal Beginning of Vaccination

Dr. Edward Jenner is widely considered to be the founder of the theory and practice of vaccination based on his work and publications concerning smallpox in the late 1790s and later. In turn, Jenner's work, along with that of Dr. Louis Pasteur, fifty or so years later, is also believed to have laid the foundations for the science of immunology. Most of the published literature I have seen broadly agrees with this assessment.[2]

However, as noted by Gross and Sepkowitz, scientific breakthroughs in any field are rarely due to one person alone. As these authors write:

> However, actual examples of the lone genius phenomenon, in which an investigator single-handedly resolves a large problem, are few and far between. Much more commonly, developments represent the culmination of decades, if not centuries of work, conducted by hundreds of persons, complete with false starts, wild claims, and bitter rivalries. The breakthrough is really the latest in a series of small incremental advances, perhaps the one that has finally reached clinical relevance. Yet once a breakthrough is proclaimed, and the attendant hero identified, the work of the many others falls into distant shadow, far away from the adoring view of the public.[3]

The above is certainly true in Jenner's case, as the initial attempts at actual vaccination with cowpox to prevent smallpox may go back to another English physician, Dr. Benjamin Jesty, in 1774. To put this into context, let's consider Jenner's history.

Edward Jenner (1749–1823) was born in Berkeley, Gloucestershire, England, the son of Rev. Stephen Jenner. The young Jenner apprenticed to various physicians and seems to have shown talent in a variety of areas. Overall, he had a keen interest in natural sciences, in many ways resembling a near contemporary, Dr. James Parkinson, who first described the neurological disease that bears his name and also wrote three large volumes on paleontology.

Jenner returned from his apprenticeship and took up his practice as a physician in Berkeley and faced, as did his fellow physicians, the repeated scourge of smallpox epidemics, epidemics that can be traced back as far as ancient Egypt. Riedel, cited earlier, notes that the disease had very high levels of mortality with some four hundred thousand deaths per year in Europe in the eighteenth century. Of those infected, between 20 to 60 percent would die. The survivors often suffered from disfigurement or even blindness. It is worth noting that smallpox is one of the infectious diseases, along with measles, that was responsible for the massive number of deaths of the native populations of the Americas' post-European contact.[4]

Two methods to try to prevent the disease were known in Jenner's time and even before. One, from Chinese medicine, involved inhaling ground-up smallpox pustules from someone with the disease. The second method was termed "variolation," in which material from the smallpox sores of an infected person were applied to skin incisions made in another person. The term *variolation* comes from *variola*, or Latin for smallpox. This procedure in turn was sometimes termed "inoculation," also from the Latin term for "to graft" (*inoculare*).

Variolation was quite effective as a smallpox prophylactic, although some 1–3 percent of patients came down with the disease with the attendant mortality or disfigurements.

Jenner apparently knew about variolation and even used the procedure frequently. One reference mentions that he may have had the procedure performed on himself as a young boy.[5]

The breakthrough for which Jenner is credited, however, came in 1796, when he heard a young milkmaid, Sarah Nelmes (sometimes written as Nelms), claim that she would never get smallpox because she had had cowpox, another of the orthopox family of viruses to which both cowpox and smallpox belong.

Combining this observation with his previous knowledge of variolation, Jenner took some pus from Nelmes's cowpox lesions on her arms and placed this material into small superficial incisions on the arm of an eight-year-old boy, James Phipps.

Phipps had only minor signs of illness following this procedure and was soon well. Two months later, Jenner exposed Phipps in the same way, but this time using pus from an actual smallpox victim. Phipps did not contract the disease. Jenner followed this experiment up with twelve more people, eventually testing the procedure on twenty-five people in total. One of Jenner's observations was that immunity seemed to work both ways: those who had survived smallpox did not get cowpox, and, as with Sarah Nelmes, cowpox prevented her from contracting smallpox.

The obvious flaws in such studies from a modern study design perspective are that there were no controls to speak of, a trend that continues to the present in most vaccine trials. In addition, from an ethics perspective, performing what were essentially "challenge" experiments without true informed consent would be a problem. There is no indication that Jenner, or anyone else of the period, even thought of this problem.

The outcome of these studies led Jenner to write his first paper, *An inquiry into the Causes and Effects of the Variolae Vaccinae, a disease discovered in some of the western counties of England, particularly Gloucestershire and Known by the name of Cowpox,* at first rejected for publication and then self-published in 1798.[6]

Note that in his title, Jenner mixed the terms *Variolae* (smallpox) with *Vaccinae*, with the Latin *vacca* being the word for cow, from which his infectious material came.

In pushing for vaccination over variolation, Jenner ran into the established medical community of his time, some members of which were operating variolation clinics and fought his discoveries. The irony of this situation in regard to the virtual worship of vaccination in today's modern medicine and the disdain for any alternative views will be apparent to most readers.

Eventually, fame and wealth from grants to him by Parliament rewarded Jenner's work, but he always seems to have remained modest in the face of such accolades. Rather than selling his "vaccines," Jenner gave them to anyone who asked and even operated a free clinic where he would administer them.

Seventeen years after his death, in 1840, the British Parliament banned variolation, thus removing the main competitor to vaccination. It is tempting to wonder which events behind the scenes caused this law to come into force.

I will confess that before starting this book, my knowledge of Jenner and his work at the dawn of vaccination was fairly meager. In fact, the iconography that has grown up around Jenner's work from mainstream medical sources had pretty much convinced me that what I would find would be sloppy and unethical science and a man flawed by greed and ego.

What I found, to my surprise, was to a great extent the opposite. In regard to the first part, the sorts of scientific procedures and ethics of today really can't

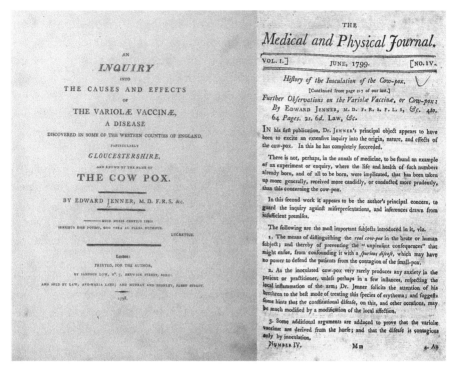

Figure 2.1. Cover pages for Edward Jenner's first two publications on smallpox and his use of cow pox as a prophylactic.

be used against Jenner over two hundred years earlier. Rather, Jenner, to his credit, saw a horrible disease and through observation, hypothesis, and experimentation, followed by replication, sought to control it. Certainly in terms of methods, the number of patients initially studied was tiny, and Jenner really had no idea which substances he might be transferring to the recipients, since this was long before either germ theory from Louis Pasteur[7] or Robert Koch and his famous postulates,[8] or even the discovery of viruses[9] that were still years in the future. In addition, Jenner had absolutely no way to control the dose of the virus administered.

In regard to the second part, there is nothing to suggest that Jenner was driven by either ego or greed. As noted above, he freely gave his vaccine away and treated patients for free.

After Jenner, the process of "vaccination" remained basically variolation with cowpox, as hypodermic needles were not invented until 1844 and the work of Irish physician Francis Rynd. The first syringes did not make an appearance until 1853.

To this day, smallpox vaccines are still made from cowpox-infected cow lymph or skin. Although smallpox was considered by WHO to have been eradicated in 1980, supplies of the vaccine ACAM2000 have been kept in storage in case of a future smallpox resurgence.

On the next page, I included a timeline of vaccine development beginning with Jenner's 1798–1799 publications and ending with today's rush for a COVID-19 vaccine (the major choke points are marked with*). Space limitations don't allow me to go into detail about most of the time points, but a few especially important ones should be mentioned. The first of these is Jenner's own work, which shifted the concept of immunization from variolation to vaccination, a rather key step in what would come later. Another important year was 1840 (not starred in the timeline figure), when the British Parliament banned variolation, essentially enshrining vaccination in its place. Other crucial years came with the experiments of Pasteur and the development of germ theory. In addition, Pasteur was the first to experiment on the use of attenuation of the suspected pathogen, although viruses had still not been identified. Pasteur used various methods to attenuate the pathogen (virus), including the first use of "passaging" by growing the viruses repeatedly in cell culture. In the process, Pasteur gave the world an effective vaccine against rabies.[10]

A critical step came with the research and publication in 1926 by Glenny et al. This was the first article to describe the enhanced antibody production of a vaccine by the use of aluminum adjuvants (see Chapter 5). The work arose when Glenny and his coresearchers at Burroughs-Welcome Laboratories in the United Kingdom realized that the antigen alone in the diphtheria vaccine tested in guinea pigs was not able to produce a sustained antibody response.

Some investigators would have stopped at this point, taken a deep breath, and then gone back to vaccinology and immunology basics to see why not. This was likely the most crucial T-junction in the road to many more vaccines, and the field would never be the same again. Instead, Glenny and colleagues searched high and low for a "Band-Aid" to the problem whose apparent solution was the addition of aluminum salts.[11]

A further crucial commercial milestone occurred when the National Vaccine Compensation Act of 1986 became law, essentially making vaccine companies in the United States liability proof. As a result, the Act opened the floodgates to a plethora of vaccines to follow along with the attendant increases in pharmaceutical company profits juxtaposed with the rapid rise of autism spectrum disorder (ASD).

In the above, I have not included the eradication of smallpox by the smallpox vaccine, nor the polio vaccines of the 1950s, since it is not clear that these alone led to the diminution of these diseases.

The latest jump has arisen as this book goes to press with the still-experimental mRNA vaccines being fast-tracked against COVID-19. This more recent development will be explored again in more detail in Chapter 13.

Key Events in the History of Vaccines

*1798 Edward Jenner publishes his first work on the smallpox vaccine.

1853 Compulsory Vaccination Act in England made smallpox vaccination mandatory for children.

1879 Louis Pasteur discovered that attenuated chicken cholera protects against challenge with virulent organisms.

*1885 Pasteur developed first live attenuated viral vaccine (rabies).

1902 The Biologics Control Act was the first law that required licensing and inspection of vaccine and antitoxin manufacturers in the United States.

1905 The US Supreme Court confirmed the right of states to enforce mandatory vaccination law (*Jacobson v. Massachusetts*).

1918 "Spanish" influenza pandemic killed twenty-fifty million worldwide.

*1920s Vaccines for diphtheria (toxoid), tetanus (toxoid), and pertussis were developed.

1932 Aluminum adjuvant was first used in human vaccines (diphtheria and tetanus).

1936 Max Theiler developed the yellow fever vaccine.

1944 The Public Health Service Act established the federal government's quarantine authority for the first time.

1945 An inactivated influenza vaccine was licensed in the United States.

1955 Albert Sabin developed orally-administered live polio vaccine.

1957 Asian influenza (H2N2) pandemic killed more than one million worldwide.

1961 Jonas Salk developed first polio vaccine (injected, inactivated). "Cutter incident"—defective polio vaccine manufactured by Cutter Laboratories—caused forty thousand cases of polio and 260 cases of poliomyelitis.

1962 President Kennedy signed the Vaccination Assistance Act into law, which allowed the CDC to support mass immunization campaigns and to initiate maintenance programs.

1963 The Federal Immunization Grant Program was established to provide states to purchase vaccines and to support basic functions of an immunization program. Maurice Hilleman developed the measles vaccine.

1964 The Immunization Practices Advisory Committee (IPAC) was formed to review and revise immunization schedule.

1971 The United States licenses Merck's combined trivalent MMR vaccine (developed by Hilleman).

1974 The first monovalent (group C) meningococcal polysaccharide vaccine (Merck) was licensed.

1977 The first pneumococcal vaccine developed by Robert Austrian was licensed.

1980 The World Health Assembly certified the eradication of smallpox. By 1980, all fifty US states had laws requiring vaccination for school entrance.

1981 The first hepatitis B (Recombivax HB by Merck) vaccine was licensed.

1985 The first Haemophilus influenza type b (Hib) polysaccharide vaccine was developed.

*1986 The Vaccine Adverse Event Reporting System and National Vaccine Injury Compensation Program were introduced.

1992 Japanese encephalitis vaccine (JEVax by BIKEN) was licensed.

1994 WHO certified the Western Hemisphere as "polio-free."

1995 The first varicella vaccine (Varivax by Merck) and hepatitis A vaccines (Havrix by GSK and Vaqta by Merck) were licensed.

1998 Rotavirus vaccine (live, oral, tetravalent; RotaShield by Wyeth) was licensed and taken off US market in 1999 because of an association with intussusception. The Children's Vaccine Program was established at WHO to provide vaccines to children in the developing world.

2000 Measles is declared no longer endemic in the United States.

2001 Thimerosal is taken out of childhood vaccines in the United States.

2005 CDC announces that rubella is no longer endemic in the United States.

2006 New rotavirus vaccines (RotaTeq by Merck in 2006 and Ratarix by GSK in 2008) were licensed. The first human papillomavirus (quadrivalent; Gardasil by Merck) was licensed.

2009 H1N1 swine flu pandemic infected as many as 1.4 billion and killed approximately a half a million worldwide. The Office of Special Masters of the US Court of Federal Claims (known as "vaccine court") ruled that the MMR vaccine administered with thimerosal containing vaccines does not cause autism.

2012 Institute for Safe Medication Practices (ISMP) launched new Vaccine Error Reporting Program.

2015 California eliminated all nonmedical vaccine exemptions in response to measles outbreak at Disneyland.

2016 National Vaccine Program Office released National Adult Immunization Plan.

2020 COVID-19 pandemic resulted in 7.5 million confirmed cases and 423,000 deaths worldwide (as of June 14, 2020) and by June 2020, 163 vaccine candidates were in development. World Health Assembly prepares to ratify the WHO document, Immunization Agenda 2030.

The development of the field over the last 221 years leads me to wonder if Jenner could see today's vaccine world, would he be surprised by the vast expansion of the branch of science that he founded? Would he have been pleased or horrified by the resulting industry that grew from his work? One suspects, and in all such cases it is merely speculation, that he would be gratified to find his work still extant all this time and pleased to know that what he started has likely saved many lives. But would a man who gave his vaccines and service away for free respect the rampant commercialization of the vaccination companies today? Based on what we can infer about Jenner's character, as partially suggested by the quote that opens this chapter, this is unlikely.

Sadly, in my opinion, the field that Jenner spawned is now the driving force behind the push for mandatory vaccination for school-aged children. Further, the still experimental COVID-19 vaccines being hailed as the way back to "normal" come with calls for mandatory vaccination for everyone.

Later chapters (10 and 13) will explore the nature of vaccine mandates and take a closer look at the politicians and media who call for them.

But before that, I want to consider what might be called the *ethics of vaccination*, a subject that very much arises from the work of Edward Jenner.

The Ethics of Vaccination

I think it will be generally conceded by virtually everyone, pro or not on vaccine issues, that preventing a disease is far superior to treating it after it has been acquired. An example where a prophylactic approach has worked well can be found in the Framingham Heart Study on cardiovascular disease[12] described in Chapter 1.

As another example, modern dentistry is exceptionally good at preventing the development of dental caries and gingivitis and in so doing not only preserves a person's teeth throughout life, but with that general health.

As a neuroscientist who mainly studies neurological diseases, I would be delighted if we could do the same thing for Lou Gehrig's disease (amyotrophic lateral sclerosis, ALS), Parkinson's disease, or Alzheimer's disease.[13] Alas, we cannot because the incidences of these diseases, particularly for Alzheimer's disease, are nowhere near as high as those for cardiovascular disease. For this reason, it remains difficult, maybe impossible, to do the equivalent of a neurological Framingham study,[14] without which susceptibility genes and their likely interactions with environmental risk factors can't be easily defined. And without a key understanding of potential risk and causal factors, preventative medical approaches don't know exactly what to prevent. Even preventing a collection of diseases, such as various cancers that may have much higher incidence than those of the progressive neurological disorders, is still difficult. And yet, if we could prevent any of these diseases, the lives of many people would be prolonged along with a greater quality of life.

With the above in mind, it is important to recognize that vaccination as a prophylactic approach to infectious diseases can be viewed in this same light, namely, as a good idea that was developed with the best of intentions by doctors and scientists. In some measure, this is precisely what vaccination may have done at both individual and population levels. In turn, such prophylaxis has resulted in numerous lives saved with much suffering prevented. Certainly, such was the goal of Edward Jenner. But then somewhere along the line, delivering vaccines became highly profitable and thus subject more to market, rather than medical/humanitarian, forces.

Let's go back to a consideration of any drug or procedure that might in the future be found to prevent any neurological disease and ask the following question: Would we give it to everyone regardless of age, sex, genetic background, race, etc.? No, we wouldn't simply because we know fully well that while some people might respond well to the drug, thus providing protection, others might experience negative outcomes in some measure.

Without knowing which persons would have adverse effects and in which proportions, there is no way that modern medicine should prescribe a drug for *everyone* regardless of medical history and demographic background. Nor would we prescribe most drugs at a "one-size-fits-all" dosage.

We wouldn't do this unless the profit motive became the dominant consideration. Alas, for too many drugs, from Alzheimer's treatments to vaccines, the corporate bottom line for many pharmaceutical companies (the pharma) holds sway, and the massive amount the pharma spends on slick television ads is a testament to the increasing role of the industry in our lives. The pharma seeks to extract wealth from our wallets by endless television advertisements that tell us to ask our family physicians, "If drug 'x' is right for you," while at the same time showing us all of the possible side effects these drugs may evoke. It's a confusing message for many, but clearly one that works.

The role of the pharmaceutical industry in literally pushing numerous drugs, and now vaccines, onto people has been the topic of various books, and these and other considerations will be addressed in more detail in Chapter 3 and later in Chapters 10 and 13, the latter in relation to COVID-19.

Before then, however, it is important to try to gain some understanding of what can best be called the "official" story about vaccines.

But even before that, let's consider how vaccines are supposed to work, both in theory and in practice, and the types of vaccines currently used. In Chapter 13, I'll consider the newer types of experimental genetic vaccines, notably those termed viral vector or messenger RNA (mRNA) vaccines now being touted as the solution to COVID-19.

Vaccine Theory and Practice after Jenner

For the purposes of understanding how vaccinology developed post-Jenner and to understand the overall trajectory of vaccine proliferation, it is important to consider how the physicians/scientists working on vaccines went from the fairly simple technique of variolation with smallpox material to variolation with cowpox.

The official rationale seems to have been that the former was too hazardous with the infection rate, whereas the use of cowpox-derived vaccines was considerably less so. It's not clear that this is entirely true, but it should be acknowledged that in the absence of any idea how much smallpox virus was being put into people by variolation, it must have seemed less hazardous to give the immune system what was essentially a surrogate pathogen that triggered an effective immune system memory, but one that could not actually lead to smallpox. From there, the path led to seeking ways to trigger an effective immune response without the disease. In other words, the attempt to fool the immune system into mounting a full response to a pathogen that was not actually there with the intention that the same system would later recognize and deal with the real pathogen.

Of the various vaccines developed, there are four main types, as illustrated in Table 2.1.

These include the following, starting with vaccines designed against viruses. Note: in the following I will consider viruses[15] and virions[16] as synonymous, although in virology they are not.

Live attenuated virus vaccines: These are basically viruses that have been weakened either by chemicals, heat, or by "passaging," which is being grown multiple times in cell culture such that each time the virus grows less able to infect other cells. In some circumstances, however, the inactivation is not sufficient, and viruses are still able to cause the disease.

Killed virus vaccines (called inactivated): As the name implies, these viruses are not able to infect another cell.

Recombinant protein vaccines: Recombinant protein vaccines fall into two categories: subunit and virus-like particle vaccines.

For subunit protein vaccines, the gene making the desired protein is inserted into a host cell to trigger the production of the protein in quantity using the host cell's own machinery. Hosts can include a bacterium, plant, mammal, or insect cell. The resulting protein is then harvested and used in the vaccine to trigger an immune response. In the case of recombinant vaccines being designed for COVID-19, the gene inserted is for the spike protein of the

	Type of Vaccines	Description	Examples
1	Live, attenuated	These contain whole bacteria or viruses that have been weakened after repeated passages in the laboratory in culture; they are not able to produce clinical disease, but active enough to induce immunity.	• Measles, mumps, rubella • Rotavirus • Smallpox • Chickenpox • Yellow fever • Shingles • Nasal flu • Oral typhoid
2	Whole, inactivated	These contain whole parts of bacteria or viruses (proteins or sugars) that have been killed by heat, chemicals, or radiation and can no longer cause disease.	• Hepatitis A • Influenza • Inactivated polio • Rabies • Japanese encephalitis

3	Subunit, recombinant, polysaccharide, and conjugate vaccines	These are specific parts of the bacteria such as protein, sugar, or capsid.
		• Recombinant vaccine:
		  o Made using bacterial or yeast cells
		  o A small piece of DNA is isolated from the disease virus or bacterium
		  o Inserted into other cells to produce active ingredients (a single protein or sugar) in large quantities
		For example: to produce hepatitis B vaccine, part of the DNA from the hepatitis B virus is isolated and inserted into the DNA of yeast cells. These yeast cells produce one of the surface proteins from the hepatitis B virus, which is then purified and used as the active ingredient in the vaccine.
			• Hib (Haemophilus influenzae type b) disease
			• Hepatitis B
			• HPV (Human papillomavirus)
			• Whooping cough (part of the DTaP combined vaccine)
			• Pneumococcal disease
			• Meningococcal disease
			• Shingles
			• Injected typhoid

4	Toxoids	Produced by treating bacterial toxins with chemicals until they are not able to produce clinical disease, but active enough to induce immunity.	• Diphtheria • Tetanus • Pertussis (whooping cough)

Table 2.1. The main types of vaccines currently used.

virus, specifically the "receptor binding domain" that allows the protein to attach to the ACE-2 receptor.

The next type, virus-like particle vaccines, are composed of a set of viral proteins that mimic the shape of the virus. This particle "pseudo-virus" is an empty shell, devoid of genetic material and noninfectious, but this does not prevent the immune system from recognizing it and ideally responding appropriately.

Recombinant virus vaccines: These use viral subunits, or even a "capsid," the virus protein shell. Examples include the HPV vaccines conjugated to compounds that are themselves capable of generating an immune response, such as aluminum adjuvants.

Of note, various vaccines for viral infections can use different strains of the virus, for example, those for measles or HPV.

Viral vector vaccines: This approach is based on using a virus that is nonpathogenic or of little danger to humans. In many cases for vaccines under development, particularly for COVID-19, viral vectors are mostly adenoviruses, the latter family responsible for some cases of the common cold. A viral vector vaccine is, in essence, the molecular equivalent of a piggyback ride: The adenovirus is modified to carry the COVID-19 DNA for the spike protein as a means of carrying it into the body, where it is intended to stimulate an immune response. Viral vector vaccines are a recent strategy and have been used in the development of the Ebola virus vaccine.

mRNA vaccines: These experimental vaccines will be discussed in detail in Chapter 13. See Table 2.1 for a list of viral vaccines for different diseases.

Bacterial/toxoid vaccines: These are vaccines using a killed or subunit protein fragment of bacterial origin as the antigen.

Finally, there are the toxoid vaccines intended to combat some endotoxin released by bacteria. Bacterial/toxoid vaccines include those for the three diseases in the DPT and variant vaccine formulations for diphtheria, tetanus, and pertussis (whooping cough). These use either parts of the bacterium's coat protein or target parts of the endotoxin. The tetanus part of the DPT vaccine targets tetanus toxoid, for example. The typhoid fever vaccine is also part of this group.

These types of vaccines are typically administered subcutaneously (*s.c.*) or intramuscularly (*i.m.*) in the majority of cases. There is also an intradermal form of injection (*i.d.*) that goes to shallower skin layer than the *s.c.* vaccines, which are injected below the dermis into the fatty tissues.

The reasons why there are these variations seem to be the products of trial and error rather than any robust parametric science. For example, adjuvanted vaccines are supposed to be injected *i.m.*, the notion being that there is greater blood supply and thus absorption. However, the anthrax vaccine, adjuvanted with aluminum hydroxide, is delivered *s.c.*; the MMR vaccine, unadjuvanted, is given *s.c.* The tuberculosis vaccine that targets the tuberculosis bacilli is given *i.d.* Meningitis vaccines are usually given *i.m.* And, finally, there are oral vaccines, such as those for polio, and intranasal vaccines, such as some for influenza.

Table 2.2 summarizes a number of these vaccines and their routes of administration. For those directed against viruses, Table 2.2 also lists the vaccines and the viruses and their taxonomical groupings.

Vaccinology: The Methods and Practice of Making Vaccines

Vaccinology is, simply, the science of making vaccines. It is a detailed branch of science from a materials science perspective, but hardly one that most, not even its practitioners, would likely describe as particularly theoretical. Perhaps after a drink or two at vaccinology conventions, the attendees might wonder about whether Jenner's vaccine with cowpox in place of variolation with smallpox was a useful turning point in the control of infectious disease, but I am inclined to doubt it.

It is pretty clear that the materials part dominates, largely consisting of ways to isolate and grow what will be the antigen, whether to use a bit of it, or the whole pathogen; if they need to attenuate it, or kill it; how to boost the antigenic signal to get a longer-term immune response; which excipients (other molecules, described soon) to add to enhance the vaccine's impact; and if they need to use an adjuvant. In regard to the last: Which one? If aluminum, should they use one of the conventional aluminum salts or a novel construct? How does the antigen stick (adsorb) to the aluminum?, etc.

All of this is crucial to making a vaccine that works to generate at least a surrogate signal, that is, antibodies or other immune cells (see next section).

It is no doubt a very demanding field, one in which creative solutions may often be needed to get a product that seems to work. Considering it more or less like a scientific form of Lego building would probably be the best way to describe it.

But does vaccinology worry about adverse effects of any of the materials used? Maybe, but mostly no, as we will see when we consider some comments in the Simpsonwood document of Chapter 5. Rather, it seems that vaccinologists simply kick such problems down the road to the clinicians who will do the phase trials. In some ways, this relationship by vaccinologists to clinicians resembles the lyrics of the satirical song about the German rocket scientist,

Virus	Diseases or conditions	Species	Genus, Family	Vaccine(s)	Brand Name(s)
Dengue virus	Dengue fever	Dengue virus	Flavivirus, Flaviviridae	Dengue vaccine	Dengvaxia
Ebolavirus	Ebola	Zaire ebolavirus	Ebolavirus, Filoviridae	Ebola vaccine	Ervebo, Mvabea, Zabdeno
H1N1 virus	Swine flu	Influenza A virus	Alphainfluenzavirus, Orthomyxoviridae	H1N1 vaccine	Panvax
Hantavirus	Hantavirus hemorrhagic fever with renal syndrome, Hantavirus pulmonary syndrome	Hantaan virus, Sin Nombre virus, Andes virus, Puumala virus	Hantavirus, Bunyaviridae	Hantavirus vaccine	Hantavax
Hepatitis A virus	Hepatitis A	Hepatovirus A	Hepatovirus, Picornaviridae	Hepatitis A vaccine	Avaxim, Biovac-A, Epaxal, Havrix, Twinrix, VAQTA
Hepatitis B virus	Hepatitis B	Hepatitis B virus	Orthohepadnavirus, Hepadnaviridae	Hepatitis B vaccine	Comvax, ComBE Five, Easyfive TT, Elovac B, Engerix-B, Genevac B, Pediarix, Pentabio, Pentavac PFS, Quinvaxem, Recombivax HB, Sci-B-Vac, Shan-5, Shanvac B, Twinrix
Hepatitis E virus	Hepatitis E	Orthohepevirus A	Hepevirus, Hepeviridae	Hepatitis E vaccine	Hecolin
Human papillomavirus	Cervical cancer, Genital warts, anogenital cancers	*Too many	*Too many, Papillomaviridae	HPV vaccine	Cervarix, Gardasil
Influenza virus	Influenza	Influenza A virus and Influenza B virus	Alphainfluenzavirus and Betainfluenzavirus, Orthomyxoviridae	Influenza vaccine	Agriflu, Fluarix, Flubio, FluLaval, FluMist, Fluvirin, Fluzone, Influvac, Vaxigrip
Japanese encephalitis virus	Japanese encephalitis	Japanese encephalitis virus	Flavivirus, Flaviviridae	Japanese encephalitis vaccine	Encevac, Imojev, Ixiaro, Jeev, Jenvac, Jespect, JEvax
Measles virus	Measles	Measles morbillivirus	Morbillivirus, Paramyxoviridae	Measles vaccine, MMR vaccine, MMRV vaccine	MMR II, Priorix, Priorix Tetra, ProQuad, Tresivac, Trimovax
Mumps virus	Mumps	Mumps orthorubulavirus	Rubulavirus, Paramyxoviridae	Mumps vaccine, MMR vaccine, MMRV vaccine	MMR II, Priorix, Priorix Tetra, ProQuad, Tresivac, Trimovax
Polio virus	Poliomyelitis	Enterovirus C	Enterovirus, Picornaviridae	Polio vaccine	Ipol, Kinrix, Pediacel, Pediarix, Pentacel, Quadracel
Rabies virus	Rabies	Rabies lyssavirus	Lyssavirus, Rhabdoviridae	Rabies vaccine	Abhayrab, Imovax, RabAvert, Rabipur, Rabivax, Speeda, Verovab
Rotavirus	Rotaviral gastroenteritis	Rotavirus A (most common)	Rotavirus, Reoviridae	Rotavirus vaccine	Rotarix, Rotateq
Rubella virus	Rubella	Rubella virus	Rubivirus, Togaviridae	Rubella vaccine, MMR vaccine, MMRV vaccine	MMR II, Priorix, ProQuad, Tresivac, Trimovax
Tick-borne encephalitis virus	Tick-borne encephalitis	Tick-borne encephalitis virus	Flavivirus, Flaviviridae	Tick-borne encephalitis vaccine	Encepur, EnceVi, FSME-Immun, TBE-Moscow
Varicella zoster virus	Chickenpox, Shingles	Human alphaherpesvirus 3	Varicellovirus, Herpesviridae	Varicella vaccine, Shingles vaccine, MMRV vaccine	Priorix Tetra, ProQuad, Varilrix, Varivax, Zostavax
Variola virus	Smallpox	Variola virus	Orthopoxvirus, Poxviridae	Smallpox vaccine	ACAM2000, Dryvax, Imvanex
Yellow fever virus	Yellow fever	Yellow fever virus	Flavivirus, Flaviviridae	Yellow fever vaccine	Stamaril, YF-VAX

Table 2.2. Existing vaccines characterized by identified pathogen, the disease that pathogen is supposed to induce, the vaccine developed with type, and whether it is adjuvanted or not.

Verner von Braun, by Tom Lehrer, which claims that it's not the scientist's department to care where the rockets come down once they've been sent up.[17]

Vaccinology is sort of like that. And is likely one reason why developing a vaccine, even with the typically minimal postlicensure surveillance, still can take a number of years. It's sort of a back-and-forth process between the vaccinologists and the clinicians in which a vaccine that shows safety concerns in phase trials goes back for a makeover. Sometimes a vaccine candidate, such as an mRNA vaccine, never makes it out of the process due to safety concerns that can't be fixed.

Until now, that is, with the push for a COVID-19 vaccine during "Operation Warp Speed," the Trump administration's ambitious effort to get an effective vaccine out to the American people en masse in 2020. I'll return to COVID-19 vaccines and the speed of their development in Chapter 13.

So, with that as a preamble, let's look the basic steps in vaccine development and manufacturing. The following is not a detailed survey of the methods but is rather intended to give a broad overview:

1. Antigen generation, by first isolating the infectious agent, viral, bacterial, or other. This can be a frustrating and time-consuming process.
2. Generating enough of the pathogen:
    a. Generation of a pathogen in quantity. This consists of:
        i. Growing virus in cells such as primary cells, i.e., chicken fibroblast (example: yellow fever vaccine), or continuous cell lines, i.e., MRC-5 (hepatitis A vaccine);
        ii. Growing the bacterial pathogen in bioreactors, which are devices that allow the bacteria to multiply.
    b. Generation of recombinant proteins from the pathogen:
        i. Can be produced in bacteria, yeast, or cell cultures, as discussed above.
    The first step here is to establish a master cell bank, and from this, working cell banks are prepared that are used as routine starting cultures for mass production.
3. Release the antigen from the substrate material and isolate it.
    a. Isolation of free virus, or;
    b. Isolation of secreted protein from the cells, or;
    c. Isolation of cells containing antigen from the medium.
4. Purification of the antigen
    a. For vaccines containing recombinant proteins:
        i. Column chromatography and ultrapurification;
    b. For inactivated vaccines:
        i. Inactivation of isolated virus.

5. Formulation
   a. This step may include an adjuvant, such as one of the aluminum salts or proprietary formulation based on aluminum or other molecules. This step may also include preservatives or other chemicals (excipients). Past preservatives in multi-user vials to control bacterial contamination due to repeat needle insertion have included Thimerosal, an ethyl mercury compound, discussed in detail in Chapter 4. Common excipients can be found in various publications.[18]

Different vaccine preparations are the hoped-for result of the above steps with vaccines being classified as one of the following:

1. Live, attenuated vaccines: contain whole bacteria or viruses that have been weakened after repeated passages in the laboratory in cell culture. The intention is that these attenuated pathogens are not able to produce clinical disease, but active enough to induce immunity. Examples include: measles, mumps, rubella; Rotavirus; smallpox (still derived from cowpox); chicken pox; yellow fever; shingles; nasal influenza vaccines; oral typhoid.
2. Whole, inactivated vaccines: contain whole part of bacteria or viruses (proteins or sugars) that have been killed by heat, chemicals, or radiation and can no longer cause disease. Examples include: Hepatitis A; Influenza; Inactivated polio; Rabies; Japanese encephalitis.
3. Subunit, recombinant, polysaccharide, and conjugate vaccines: they are specific parts of the bacteria, such as a particular protein, sugar, or capsid (protein shell of a virus).
   a. Recombinant vaccines:
      i. Made using bacterial or yeast cells;
      ii. A small piece of DNA is isolated from the disease virus or bacterium;
      iii. Inserted into other cells to produce active ingredients (e.g., a single protein or sugar) in large quantities.
   For example, to produce a hepatitis B vaccine, part of the DNA from the hepatitis B virus is isolated and inserted into the DNA of yeast cells. These yeast cells produce one of the surface proteins from the hepatitis B virus, which is then purified and used as the active ingredient in the vaccine. Examples: Hib (Haemophilus influenzae type B); Hepatitis B; HPV (Human papillomavirus); Whooping cough or pertussis (part of the DTaP combined vaccine); Pneumococcal disease; Meningococcal disease; Shingles; Injected typhoid.

   b. Toxoid vaccines: produced by treating bacterial toxins with chemicals until they are not able to produce clinical disease, but active enough to induce immunity.
   Examples: Diphtheria; Tetanus; Pertussis (whooping cough).[19]

As stated at the beginning of this section, all of the above involves a lot of detailed materials science applications, "state of the art" expertise with similar past vaccines, and even some innovative methods such as with mRNA vaccines.

Note concerning Excipients in Vaccines

Excipients are basically ingredients added to a vaccine apart from the antigenic material that is derived from the above processes. These can include things like adjuvants, whose sometimes adverse effects are discussed in detail in Chapter 5, preservatives such as Thimerosal (also in Chapter 5), and a range of other molecules including amino acids, antibiotics, and others. The Appendix, found online, shows a current list of excipients from the FDA website.[20]

The "Official" Story of Vaccines

The mainstream medical establishment today controls the dominant narrative that can be summarized by several key and endlessly repeated phrases. One is that "vaccines are safe and effective," typically bolstered by stating that "the science is settled" on this issue. Period. The narrative continues by stressing that only "anti-vaxxers" and those trained by "Dr. Google," or medical/scientific quacks, dispute this.

Additionally, the narrative holds that vaccines are thoroughly tested for both safety and effectiveness, that the surveillance systems in place to detect safety issues are well established and effective, and that most vaccines can safely be taken by most people, including infants, young children, pregnant women, and the elderly. The only allowed exceptions to who can have the various vaccines are those who may be immune-compromised or who have a serious allergy to some vaccine development by-product, for example, egg proteins, or some component of the vaccine itself.

In this narrative, it is true that clear failures to achieve some vaccine-dependent end state are acknowledged, usually grudgingly. Such admissions, or rather attempts to appear open and transparent, are typical. For example, various influenza vaccines over the years have failed to match the type of circulating pathogen, the influenza strain itself or subtype, and hence have had limited effectiveness.[21] Or, in regard to safety, the swine flu vaccine in the 1970s that was recognized to increase the risk of Guillain-Barré syndrome, a neurological disease in which antibodies are generated against the myelin covering of long axons in the periphery of the body. In this last case, even medical schools such as the University of British Columbia, where I teach, recognize that the 1970 "flu" vaccine was likely responsible for increased cases of Guillain-Barré and cited speculation about "molecular mimicry," where the immune system of the victims misidentified parts of the myelin sheath as being similar to some amino acid sequences on the coat proteins of the virus and attacked both.[22]

This sort of medical establishment assertiveness about vaccines tends to spill over into politics, often in some very odd ways. Or at least odd from the perspective of why accepting vaccines or not accepting vaccines is a political issue at all.

As a fairly recent example, back in 2015 as various politicians began the long and costly process leading to nominations for president, Hillary Clinton, later the Democratic Party nominee, tweeted, "The science is clear: The earth is round, the sky is blue, and #vaccineswork. Let's protect all our kids. #GrandmothersKnowBest."[23]

This tweet followed comments made in a Republican primary debate in which several candidates, e.g., New Jersey Governor Chris Christie and US Senator Rand Paul, had noted concerns about parental choice and the possibility that vaccines might contribute to autism.[24] Similar concerns were later raised by other Republican candidates, including Dr. Ben Carson, later Trump's secretary of Health and Human Services, and Donald Trump, later president. All of the comments were thoroughly castigated by various professional medical organizations and the media as being without any merit because, after all, "the science was settled."

Most of the wannabe presidents toned down their comments, or even did backflips to reverse what they had said. Donald Trump was the exception to such backpedaling, a stance that would later help him consolidate support with many of the vaccine-hesitant by doubling down on concerns about a possible vaccine-autism connection.

Early in 2017, then-President-Elect Trump had a meeting with Robert F. Kennedy Jr. that Kennedy reported had discussed forming a panel to investigate the possible vaccine-autism connection.[25] (I will discuss Kennedy in more detail later in Chapter 7). Many of the anti-vaccine/vaccine-hesitant who had supported Trump's journey to the White House were ecstatic. That ecstasy should have been short-lived, but hope never dies apparently, as it rapidly became clear that no such panel would come to pass.

Added to this, President Trump appointed pharmaceutical industry insiders into the various parts of the government who might have been able to affect such a study.[26]

None of what should have been seen as an outright betrayal of this constituency moved a number of these vaccine opponents away from their utter conviction that Trump would eventually do the right thing by them and their vaccine-injured children. One commentator even opined that the reason why Trump had not followed through on what had been seen as election promises was that he was biding his time and the hoped-for actions would come in his second term.

It is worth noting in regard to the above that Clinton had been ambiguous about the role of vaccines in autism in her earlier run for president in 2008, but the opportunity to score points against potential Republican adversaries must have been too great to pass. Hence the tweet.

However, the tweet itself is revealing, not only for the false equivalences that Clinton or her media person stated, but for what it demonstrated about her grasp of science and the "frame" that vaccines have been put into in order to package them as selling points.

Let's take the statement apart to see how it cuts corners to make a point that was casual in its wording, not to mention lacking in any serious understanding of the science about vaccines.

First, take the comment that "the earth is round." Well, broadly speaking, this is true, if not fully accurate. The Earth is actually an oblate spheroid as proposed by Isaac Newton[27] with deformations at the poles and equator, not to mention the bulges created by mountain masses and troughs in the oceans.

Second, "the sky is blue." Again, not fully accurate. On a clear summer day in Kansas, the sky certainly appears to be blue. But in desert areas it can appear to be white. In the same Kansas sky in winter, the sky will appear gray. At sunrise or sunset, the sky appears to be red or orange. In reality, the sky is black, as it appears at night and what we see with the various colors is due to refracted light in the atmosphere, which on a sunny day appears blue because the blue wavelength of white light is shorter and thus is more scattered.[28]

Now the last: "and vaccines work." As with the Earth being round and the sky blue, this last is a nonscience oversimplification, in this case vastly more than the others. "Vaccines work," really? What do they work to do? Prevent infectious diseases? Yes, sometimes. Increase autoimmune disorders? Yes, sometimes. Generate vast profits for the manufacturers? Yes, always.

This sort of applying a sloppy political talking point to a complex issue highlights why we probably don't want politicians voting on vaccine mandates if their knowledge of vaccines is at the level of Clinton's.

Sadly, as we will see in the following chapters, this sophomoric level of analysis continues to permeate both politics and the mainstream media, both topics addressed in more detail in later chapters.

What Evidence Should Any Side in the Vaccine Wars Present to Best Support Their Position? (Part 2)

Before wading into the claims by the mainstream proponents of vaccination versus those on the skeptical side, let's start with the following: The science on vaccines and their impact on human health, beneficial and harmful, are not settled, and it is highly likely that they never will be. As discussed earlier in Chapter 1, science is not about proof, but rather probability.

The key question, therefore, is how close is the pro or the skeptical side to a greater level of probability? Essentially, it's a question about scientific methods and the published scientific literature, and the rigor of both applied to the questions about vaccine safety and effectiveness. It is not, as sometimes framed, a contest between political beliefs, as in red versus blue states, Democrats versus Republicans, or leftists versus the alt-right. All attempts by some to phrase vaccine controversies into political "us versus them" boxes have nothing really to do with the debate, particularly about vaccine safety. In actuality, such stratification says more about the people making derogatory statements about the "other" than it does about the quality of the science on either side.

I should mention at this point that one recurring theme, only from the mainstream side, is that there really is no debate to have because, after all, as already cited numerous times, the "science is settled." Given that this is not the case here or in virtually any other branch of science, the refusal to engage others with different views is not a sign that those who advocate such withdrawal can comfortably relax since the other side has nothing worth discussing. Rather, it seems, at least to me, a sign that those making this argument are not confident enough in their position to survive an intellectual encounter with those who feel differently.

As this book was being written, I ran into precisely this situation on a Facebook encounter with a "friend." This person, K. P., had denigrated a woman with an autistic child, the latter now an adult. The woman, C. S., had simply made a comment in a thread about the COVID-19 pandemic that she was not willing to get a future COVID-19 vaccine, as she felt her son's autism was the result of childhood vaccines. Given this, she wasn't willing to take the risk of future adverse reactions. K. P.'s response had been to call C. S. ignorant and state the hoary platitude that "Vaccines don't cause autism, period."

Truly, I had not had any intention to get into this on Facebook, but such rudeness seemed to call for a reply, especially since the thread had originated with me. My reply to K. P. was that ad hominem comments were really not acceptable. Further, how did he know that vaccines never caused autism? K. P. replied that he had "devoted hundreds of hours" studying the subject and that it was "settled."

I should have quit at this point but decided on a fairly benign rejoinder: Did he know the Taylor et al. (2014) meta-analysis or the Madsen et al. 2002 studies on the subject of some vaccines and autism (both covered in Chapter 3), and, if so, how rigorous did he think these studies were? The reply surprised me but perhaps shouldn't have: "I don't have time to go down the anti-vax rabbit hole," K. P. wrote back. And that was that. He eventually unfriended me on Facebook after a later exchange where I asked for references to back up some of his other assertions.

The above turned out to be a typical response, one that is often heard from some of the online bloggers, but also, as I will soon detail, occurs routinely when someone on the skeptical vaccine side tries to engage with someone promoting the mainstream view. In my experience, almost every time, the mainstream side withdraws from the field, whether that field be scheduled discussions or on social media. One can mark this up to really not wanting to engage in what they may honestly think is a nonsensical, time-wasting debate. Or it could be a lack of confidence in their own positions, as noted above. Or, even more likely, it may arise from the almost religious belief in vaccines. This latter frame of mind will be discussed in greater detail in Chapter 8.

Regardless of the actual reason(s), from the perspective that science is rarely settled for all the reasons stated previously, the likelihood is that there is indeed some middle ground, hard as that ground may be to find.

In other words, it is important to remember that any conversation about vaccines is not a zero-sum game in which one side is totally right and the other totally wrong. Thus, for vaccines to be useful medical adjuncts, it is not required that they never have adverse effects. And vice versa.

At the same time, which side is right and which wrong is not a function of how many trees died to produce the most papers or how many electronic words there are in cyberspace. The issue is not, nor ever has been, about quantity, but rather quality.

In this regard, almost any working scientist will tell you that a publication in *Nature* does a lot more for your job prospects than ten papers in a predatory journal. It's the same here: it is the quality and rigor of the science that is up for debate, not the volume of the material presented.

It will be important to remember this when we get into the competing arguments about whether vaccines are safe or unsafe, or a bit of both, and for whom; and, independently of this, how effective vaccines may be at either an individual or population level. These are topics that are the subject of the next two chapters.

Before we do this, however, there are some other topics that need to be fleshed out in the remainder of this chapter. The first of these is how to sift correlation from causality, a problem that bedevils both sides of the vaccine discussion.

Coincidence versus Causality and the Hill Criteria

One area where people on both sides of the vaccine controversy get hung up is on the issues of coincidence versus causality when deciding if a vaccine does, or does not, do what it intends to do; or if it does do the required action, does it cause any harm?

Briefly, something is causal if it causes something else, i.e., ". . . a relationship, link, etc. between two things in which one causes the other."[29]

Coincidence refers to:

> . . . a concurrence of events that occur without an apparent causal connection, that is, an occasion when two or more similar things happen at the same time, especially in a way that is unlikely and surprising.[30]

While it might seem that it should be easy to distinguish coincidence from causality, it can actually be quite hard to do, and in the highly charged world of vaccine safety it can be more difficult still.

For example, much is made by both sides about declining rates of infectious diseases during the twentieth century. Looking at the historical time line, it is quite clear that most of the major infectious diseases have dramatically decreased during this time period. Depending on where one starts the time line, it can appear that a particular vaccine is responsible for the decline in incidence or death. Or, one can take a longer time frame and demonstrate that the decline preceded the vaccine, in which case the vaccine may not have had all that much to do with the decline at all.

Similarly, one can plot the rate of increase in autism spectrum disorder (ASD) over time and plot a similar time frame curve with the appearance and increased use of glyphosate, an herbicide.[31] As per the definitions above, just because two events seem to occur in lockstep does not imply causality, although often people, even scientists, may sometimes assert that it does.

To me, one of the more glaring examples of confusing coincidence with causality actually was published by a botanist, Dr. Paul Cox, and a very prominent neurologist, Dr. Oliver Sacks, the latter now sadly deceased.[32] These scientists were attempting to link the rise and fall of the neurological disease spectrum on Guam, ALS-parkinsonism dementia complex (ALS-PDC) to the Chamorro consumption of fruit bats, the latter allegedly containing a particular neurotoxin that was supposedly biomagnified up the food chain to humans. The plotted data seemed to show that both curves went up, and then down, more or less in general synchrony. Much was made of this in some neurological disease circles,[33] although it was viewed as spurious in others.[34] A huge part of the problem in this paper, and some of the following ones from this group, was that a large percentage of the fruit bat population levels in various years were guesses based on assumptions that were clearly not correct. In this example, to make their case, Cox and Sacks had taken calculations of the fruit bat population per acre of forest in the years just before their study and extrapolated the numbers in previous years for which they did not have data. They based their calculations of bat numbers on the amount of forest at those earlier times. Wildlife biologists on Guam had told them that this was spurious, but Cox and Sacks used these numbers anyway. The article's outcome based on these nonexistent numbers seemed to support their attempt to claim a causal relationship, but the real numbers would likely have given a very different picture.

In regard to a potential vaccine-autism connection, various bloggers have posted some deliberately spurious correlations such as an increase in ASD in the general population with the increased consumption of organic produce. They are right, of course, about such links not being causal but often fail to realize that the misuse of correlation is highly problematic generally and may be particularly so in some attempts to claim vaccine benefits that may not actually exist.

So how does one distinguish between the two? Is there a way to do this? In fact, there are several ways, one of which is by doing the actual experiments where only one variable is changed at a time. In this way, if the relationship between events really exists, the impact of the one variable on the outcome will occur if the variable is present, and not if it is not. However, such demonstrations in humans or in animal models are time-consuming and often extremely costly to do.

The simpler method, which is not a replacement for actually doing the experiments, involves the use of a series of questions first proposed in 1954 by Prof. A. B. Hill, a British statistician.[35]

Called the Hill Criteria, or Standards, these criteria are used to determine the possibility of causality and require at least about half of the following to be present in order to be able to suggest a causal relationship:

1. The *strength* of the association between events;
2. The *consistency* of the association of these events on repeated examination;
3. The *specificity* of the association, e.g., whatever it is produces one outcome, not many;
4. The *temporality* of the association, in brief, the purported causal event precedes the outcome;
5. Does a *biological gradient* occur, that is, a dose-response function?
6. The *plausibility* of the association, meaning does the association make sense (in a way that increased organic food consumption with ASD would not)?;
7. *Coherence,* in that the hypothesized cause and effect do not conflict with what is known in that field;
8. That there be *experimental evidence* in favor of the association;

And last:

9. *Analogy*: Does the effect and the outcome look and act like other cause-effect relationships?

In general, satisfying at least four of these is considered to lend support to the notion of causality. Failing most of these criteria, particularly on temporality, is thought to discount an association.

It is worth noting that satisfying some or even most of these criteria does *not* prove that a causal relationship exists, but rather merely strengthens the rationale for actually doing a falsifiable experiment. In other words, the Hill Criteria are not themselves the end of the process, but merely a step in the direction of establishing a causal relationship. Even Hill recognized the limits of these criteria/standards.

With all of this in mind, the Hill Criteria are still a useful tool and one that we can employ in the service of determining the relative strengths of the data that claim that vaccines are safe versus those that suggest the opposite. Further, the Hill Criteria can be used to determine the effectiveness, or lack thereof, of the studies that have tackled this issue. And while they cannot provide concrete proof of *any* assertion about vaccines pro or con, they can provide, like the scientific method itself, a sense of the probability of various claims being correct.

Model Systems

In both basic research studies, as well as a lead-in to human trials, experiments on "model systems" are employed.[36] These include so-called in silico studies, which are essentially computer models; in vitro studies, for example, employing various cell types in a tissue culture preparation; and in vivo studies on whole animals.

All model systems approaches have some advantages to their use; at the same time, they all have problems. In silico studies are less messy and cheaper than doing actual "wet" experiments but, depending on the algorithms used, and particularly on the data inputted, can yield results that are not in line with biological reality.

Overall, in terms of validity and reproducibility, the work of Dr. John Ioannidis[37] suggests that the majority, some two-thirds, of biomedical studies are incorrect.[38] They are incorrect, in this view, for a variety of reasons. Sometimes the results are due to actual fraud such as data manipulation and the like. Most often, however, the problem lies with poor study design, incorrect experimental methods, statistical errors, or interpretations of the results that are unwarranted. The net result of such issues is that models based on incorrect data, or unwarranted assumptions, cannot help but give incorrect model outcomes.

Computer Modeling

The phrase "Garbage in, garbage out" comes to mind and applies to in silico models in particular, but the others to follow, as well.

In Vitro Modeling

Cell culture models, the in vitro approach, can work as long as any differences in cell types and their potential interactions are accounted for. In my view, this concern would be most problematic in central nervous system (CNS) studies. The reason for this is that any region of the nervous system is composed of various distinct types of cells. For example, there are neurons, glial support cells, other structural cells, etc., all of which interact in complex ways. For this reason, trying to model whole organism behaviors by looking at only one cell type in your culture preparation is likely to provide a very misleading outcome. If one wants to see how motor neurons in culture respond to a particular toxin, a failure to have supporting glial cells such as astrocytes in the same culture will not give results that are relevant to the actual condition being modeled. In addition to being mostly a single type of cell, in vitro models of brain function are also usually comprised of cells taken near birth. This is a problem as well, since neurons do not usually behave the same way in intact adult animals.

In Vivo Animal Studies

In a hierarchy of methods, the highest on the list are the in vivo (live animals) studies. These avoid the problems of the other types of studies but have their own issues. Not least of these, animal systems and responses may not really resemble those of humans at all at any level, starting from the responses at a genetic level all the way through to behavior.[39] This last is actually a huge problem, at least in principle.

A good example can be taken from some studies now ongoing in my laboratory: in order to model how toxins might kill motor neurons as a way to understand Lou Gehrig's disease (amyotrophic lateral sclerosis, or ALS),[40] we use a zebrafish model in which the embryonic or very young fish are exposed to particular toxins. We then evaluate motor function and motor neuron survival.[41] The pro side of using this model is that as embryonic zebrafish are transparent, we can look at the outcomes by microscope pretty much in real time. The negative side, however, is that zebrafish don't really get ALS and their brains lack a major area where ALS occurs in humans, the motor cortex. In addition, the use of young fish makes problematic the interpretation of what the same toxins might do in older fish. It is therefore difficult to distinguish whether the toxins are actually damaging motor neurons or simply interfering with normal development. Finally, the toxin we mainly use, Bisphenol-A, an endocrine disrupter, is unlikely to be causal to motor neuron death in ALS.

There are some escapes from these concerns, yet we always have to be aware that they exist and keep this in mind when interpreting results.

In addition to all of the above, there are other factors to be aware of, namely, the age and sex of the animals being used, as these can have major impacts on the results. And finally, to be able to be sure that any studies achieve statistical significance, one has to do what is called power analysis to determine the minimum number of animals you need to study. This number, called "N," can in turn reflect the number of animals in different groups, or the number of litters if doing a study of young animals. The calculations by power analysis tell you how many animals you need in each group, for example, to achieve a statistically significant outcome, if one exists.[42] Generally speaking, one wants to have enough of an N so as to be able to detect a p value of less than 5 percent ($p < 0.05$) with 80 percent reliability. What this means is that if you get a <0.05 value, you can relatively sure that it means what you think it means, but you still have a 5 percent chance of being wrong. A better and safer value is 0.01, or a 1 percent chance of being wrong, but most researchers use the 0.05 level.[43]

A lot of in vivo studies don't accomplish this, of course. As an example, we recently published a study of a newer therapy for ALS with only five or so animals per group. This was essentially the outcome of cost and logistics issues; there was little else we could do apart from acknowledging the small numbers and list the paper, as we did, as a pilot study.[44]

Human trials of any drug or vaccine treatment are supposed to start with animal trials before going to studies with humans, but they don't always do so, as with the current COVID-19 experimental mRNA vaccines being developed.[45] Prior to this, pharmaceutical companies often did animal studies, but these could be underpowered in the extreme and seemed often to be designed to fail and routinely did not find any potential problems.[46]

Types of Clinical Trials in Medicine

Clinical trials in medicine (sometimes referred to as clinical studies) are studies involving research with human subjects. Based on a number of established ethical principles that supposedly govern human experimentation (see the next sections) these studies are supposed to use volunteers who have been thoroughly briefed on all aspects of the possible benefits and harms involved in being part of these trials. In other words, the volunteers have been able to provide what is called "informed consent."

A number of types of clinical trials can be performed. In brief, the first distinction is between *observational* and *interventional* trials. The first is exactly what it sounds like: patients are simply observed over time for changes in health status. In contrast, interventional studies are designed to evaluate the safety and effectiveness of any proposed treatment/drug/vaccines, etc.[47]

As noted by Singal et al, intervention studies can be placed on a continuum that moves from efficacy trials defined as the performance of some procedure or drug under "ideal and controlled circumstances," to effectiveness trials, the latter referring to "performance under 'real world' conditions."[48] These authors note that "the distinction between the two types of trial is a continuum rather than a dichotomy, as it is likely impossible to perform a pure efficacy study or pure effectiveness study."[49]

In the literature, these terms are sometimes used interchangeably, which they should not be.

Subgroups of clinical trials based on the above can be one of the following:

Treatment trials and *prevention* trials, which will, respectively, evaluate possible treatments or look at a range of options designed to lower the risk of someone developing a disease.

The other subgroups include *screening* trials to find ways to detect the disease in question at early stages and *quality of life* trials designed to explore just that.

Finally, in this group may be *genetics* trials that may take place to seek out genetic factors the patients may have that can predispose to the disease.

In regard to treatment trials as considered by the Michael J. Fox Foundation for potential treatments of Parkinson's disease, these trials are designed to assess the safety of any treatment or drug proposed to alleviate Parkinson's disease, but also at a more advanced stage to determine if the treatment/drug actually accomplishes the goal of preventing or treating the disease. The same applies to clinical treatment trials for any disease, and such trials are a vital part of the medical research process that is essential for developing better treatments for disease patients.[50]

Typically, such clinical trials have multiple *phases*, Phases I though IV in human subjects and often, but not always, an animal *preclinical* phase. The latter will be described in more detail below and elsewhere in the book, but it is notable that these have been skipped in some of the COVID-19 vaccines that have been developed and are still being developed by various companies and research groups. The implications of not doing such preclinical studies will be addressed in Chapter 13 in relation to COVID-19 vaccines.

A Phase I involves a relatively small number of volunteers from ten to about a hundred or so and is designed to detect any safety signals or side effects for the drug or vaccine, a preliminary screen of different dosages should be done, as well as an evaluation of the best means of administering the treatment.

Phase II trials add more subjects, up to several hundred, and continue the safety surveillance of Phase I.

Phase III trials can involve thousands of subjects, are often evaluated at multiple treatment centers, and are mainly designed to assess the effectiveness of the drug or treatment while continuing to assess overall safety. This phase is when the treatment/drug/vaccine is supposed to compare the benefits versus the risks. Such trials can take years to complete and evaluate.

As such, Phase III trials have, or are supposed to have, a control arm, but what actually constitutes a "control" is often different from what the same term means in animal studies. In the latter, any experimental group receiving a drug or treatment is supposed to be compared to another equivalent group of animals receiving no treatment or a "placebo" treatment that should not have any biological function. For example, if evaluating the impact of an injected aluminum adjuvant, an animal study would necessarily be required to have an adjuvant group (or more than one with different doses or administration schedules) versus a group that received an injection of saline. In the case of human trials, the "control" is often not a control in the same way, particularly in drug or vaccine trials. In the latter case, it is typical for such trials to use another vaccine as the control, or the adjuvant, if there is one. The implications of such controls for evaluating safety will be discussed in Chapter 5.

Phase IV trials occur, if they occur at all, after the licencing of the treatment/drug/vaccine by the relevant health authority, for example, the FDA in the United States, Health Canada in Canada, and equivalent agencies in different countries. Primarily, Phase IV trials are intended to provide a longer time frame safety assessment.

Clinical trials of the treatment variety are often part of a design described as being a Randomized Control Trial (RCT), usually considered the "gold standard" for the evaluation of any type of treatments.

In an RCT, patients are first evaluated for eligibility to participate in the trial by criteria that are designed to prevent "bias" in the trials and their outcomes (see following section). For example, in a Parkinson's disease drug therapy RCT, eligibility criteria might be designed to reject participants of particular age groups or those who have comorbid conditions unrelated to Parkinson's disease. As Parkinson's disease is primarily a disease of middle and old age, the inclusion of teenagers might skew the results; similarly, if patients had other conditions or were taking other medications for such conditions, those aspects alone might confound any treatment.

Following eligibility screening, potential participants are randomly assigned to the various treatment arms, the random nature of this designed to remove *selection bias*. Various methods are used to achieve randomization. In the *parallel group* RCT study design, the most common, one or more groups actually receive whichever treatment is being evaluated; the other is a control group that should get a placebo treatment, but this last is often not a true placebo, but another form of pseudo control, as noted earlier. In vaccine trials, for example, it is not unusual for the control group to be another vaccine not related to the one being tested, or to some component of the vaccine, e.g., the aluminum adjuvants. Some guidelines consider these last controls to be acceptable (see CIOMS, Guideline 11, as cited in the next sections); others disagree, including this author. There are other RCT designs in use, as well.

To eliminate other forms of potential bias, RCTs are typically *blinded* in that the participants in either treatment group do not know which group they are in, nor do the researchers actually doing the studies know who is who. Both forms of blinding, in this case a so-called *double blind* method, are designed to reduce subject and experimenter bias.

As with all clinical trials of any kind, conflicts of interest, such as disclosed or undisclosed ties to the pharmaceutical industry, can bias outcomes and thus make the results of any study unreliable. I will deal with this last point in detail in Chapter 3, in which I consider the various clinical trials or meta-analyses designed to evaluate potential vaccine adverse reactions.

As noted, RCTs are considered the top of the heap in terms of quality of clinical trials and studies, but these are usually very expensive to run and take a considerable amount of time. For this and other quite valid reasons, other types of clinical studies are often performed that do not actually require any current treatment procedures.

More or less in order of impact, these are:

Ecological studies, essentially observational studies in which population data, compared to individual data, are collected. Ecological studies are often used to measure prevalence, the overall numbers, and incidence, the number per a set period of time, of disease. One of our studies looking at a possible correlation between the number of aluminum adjuvanted vaccines in the CDC's vaccine schedule versus the incidence of autism spectrum disorder (ASD) over time was an ecological study.[51]

There are also what are called *case-control* studies (also called *case-referent* or *retrospective* studies). By whichever title, such studies simply compare patients with some disease against controls who do not have the disease or medical condition and assign a probability that some potential risk factor is responsible for any differences observed between the groups. In case-control studies, there is no treatment arm involved and no attempt to alter disease course. One such study could be one that hypothesizes that a vaccine, or vaccines generally, contributes to neurodevelopmental delays or disorders such as with ASD. The widely criticized study by Mawson is one such example.[52]

As cited by Levin, there are both advantages and disadvantages to using case-control studies.[53] Extracted from Levin are the following concerns.

Under advantages: We find that such studies are good for studying rare conditions or diseases; less time is needed to conduct the study because the condition or disease has already occurred; this design lets one simultaneously look at multiple risk factors; the studies can be useful as preliminary to establish an association; and such studies can answer questions that could not be answered through other study designs.

Under disadvantages: The author cites problems with data quality because they rely on memory and may be subject to recall bias (which will be discussed in more detail); such studies are not good at evaluating diagnostic tests, and it may be hard to find a real control group, the latter a problem often cited as a barrier to performing a vaccinated versus unvaccinated study.

Levin also notes that *confounding* bias should be watched out for, such as the presence of a third variable, as well as the validity of the inclusion and exclusion criteria for picking the participants of both the risk factor and control groups.

Next down the chain are *cohort* studies in which patients are followed over time and evaluated for health status at different time points. The goal here is to identify possible initial risk factors that may contribute to the emerging disease. The advantages of cohort studies: matched cohorts, including a control group, to limit confounding factors; criteria and outcomes can be standardized; and they are much easier and cheaper to perform than an RCT. Disadvantages listed are that cohorts can be difficult to identify; randomization is impossible; blinding of the study is difficult; and the outcomes can take a lot of time to emerge.

One clear aspect to control for is whether the cohorts reflect similar, but separate, populations of patients.

Next on the list are *case series* in which:

> A group or series of case reports involving patients who were given similar treatment. Reports of case series usually contain detailed information about the individual patients. This includes demographic information (for example, age, sex, ethnic origin) and information on diagnosis, treatment, response to treatment, and follow-up after treatment.[54]

The MMR study of Wakefield et al. in 1998 was a case series, not a cohort study, as often erroneously stated.[55] A key distinction is that case series and case reports do not have controls, nor do they need any.

And, finally, a case series is a compilation of individual *case reports*, that is, the medical history of one particular condition, often a rare or unusual one, or some variant on a known condition, both of which may reveal novel information such as a new disease entity or the response to a drug or treatment. As per the hierarchy of evidence, case reports are considered to be the lowest but still the *first line* of evidence in human clinical trials. Naturally, with an N of one, case reports might be anomalies and/or utterly spurious.

Figure 2.2 shows what has been called the "evidence pyramid,"[56] with the top considered to have the highest scientific value, and the bottom the least. Note that at the top of the pyramid are systematic reviews and meta-analyses. Taking the latter first, a *meta-analysis* typically is the product of examining data from a number of other studies.

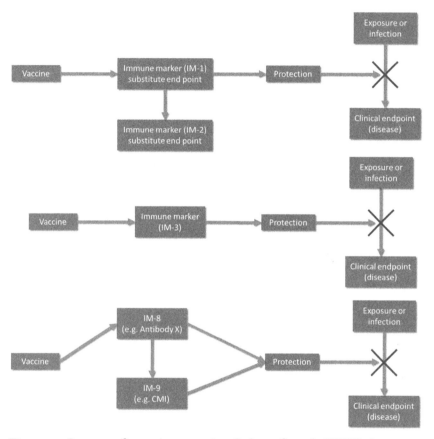

Figure 2.2. Surrogates for vaccine protection. Redrawn from the WHO's document.

The goal of these latter studies is to see if independent studies show similar outcomes. Meta-analyses usually involve a statistical approach for combining data from the various sources. A *systematic review* is the result of a process designed to select, evaluate, and synthesize all evidence, or at least the evidence deemed relevant to a particular topic. Needless to say, for any of the above levels of evidence from the bottom to the top of the pyramid, errors and biases of various types can negate any conclusions reached.

We will see examples of all of these types of studies and the problems that can arise when we evaluate some of the literature from the CDC and other entities in Chapter 3.

Some Ethical Concerns about Clinical Studies

As noted above for case-control studies, many in the vaccine-hesitant group-ings believe the issues of vaccine harms versus benefits would be resolved by one basic experiment, albeit one that is not necessarily simple from a demographics perspective, namely, a back-to-back comparison of health outcomes of fully vaccinated (to the CDC schedule), partially vaccinated, and totally unvaccinated children. Various people on the pro-vaccine side make the claim that to conduct such as study would be "unethical," pre-sumably based on the assumptions that vaccines, individual ones and the whole schedule, are life-saving procedures. In this view, any lack of ethics arises from depriving a number of children of the health benefits that vac-cines would confer. This view seems to be supported by Guideline 11 and the discussion thereof in the CIOMS guidelines cited in Chapter 8. Note that there is no consideration concerning any adverse effects that the vac-cination schedule might also have for some children in this guideline and from those who rely on this guideline. The media tend to simply state that such a study would be unethical, use a talking head to reaffirm it, and leave it at that.

On the face of it, this objection seems to be at the very least conflicted by apparently starting with the assumption of what the experiment would show. In other words, if the hypothesis to be tested is that vaccinated kids are health-ier than unvaccinated kids (or the opposite) and you already are sure you know which one is true, then why do the experiment at all? This behavior would simply violate clinical equipoise.[57]

Not doing the "unethical" experiment because the science is already "set-tled" is very much like the Trump impeachment trial when the Republican senators voted not to hear witnesses because they already knew how they would vote. Hence, this whole rationale really is not about ethics at all, rather the cre-ation of a nonfalsifiable hypothesis that one is afraid to test with a real hypoth-esis lest the outcome is one that is somehow not acceptable. It may make for good talking points for media interviews, but whatever it is, it isn't science, or at least real science.

Let's consider briefly how such an experiment might differ from others of the same sort and see if the notion that such a study is unethical stands up to examination. First, what do the people who take this view imagine that most vaccine trials are? Typically, vaccine trails for safety or effectiveness pit a new vaccine against either an older established vaccine, or against some component of the vaccine, for example, the aluminum adjuvant. Whether this is actually a valid control has been considered earlier and will be addressed in more detail in Chapter 5. Further, it is worth noting here that only very rarely are true placebo controls used.

In such a study of whichever design, RCT or other, one group gets the presumably life-saving vaccine and the control group gets something that is supposed to be neutral to the disease state. For example, in trials of Gardasil, one of the antihuman papilloma virus (HPV) vaccines,[58] outcomes measured are adverse effects and the number of precancerous lesions that the Gardasil vaccine is designed to prevent. One group therefore has no protection against the viral strains; the other, the study authors hope, will be one in which the vaccine will prevent these lesions. Leaving one group unprotected in this case is no different from those individuals unprotected in the vaccine versus nonvaccine case. Nor is this example any different from a cancer RCT or one involving Alzheimer's disease, comparing patients receiving some experimental drug from the controls who do not.

The short answer is that it is not: these are all the same sort of study in which the experimental question is whether a certain drug or procedure is more effective than nothing. One is not more unethical than another, especially where the study design allows the untreated to move into the treatment group if the results show a clear benefit for those in the latter.

Of course, all of this assumes that all involved in this study have given informed consent and are fully aware of any risks, that is, those in the control nonvaccinated group have been told of the risk of being exposed to various pathogens from which they are not vaccine-protected, and those in the vaccinated group are aware that vaccines may have adverse effects of greater or lesser severity. This is not a *challenge* experiment in which the participants are deliberately exposed to the prospective pathogens; rather, in this case, it is left to chance.

The answer to whether it is unethical to do this study thus now depends on an estimation of the level of risk that the participants may experience. It is noteworthy at this stage of COVID-19 vaccine development that various governmental agencies and the vaccine companies themselves are contemplating challenge experiments rather than simply waiting for COVID-19 exposure to occur, in order to more rapidly evaluate the efficacy of their vaccines.

In actuality, however, the argument that such an experiment is unethical is the ultimate straw-man argument because one does not have to randomly assign kids to an unvaccinated group, although one could, because such children already exist and in sufficient numbers to satisfy even the most rigorous statistical requirements. To be clear, such a study would not be an RCT, as the populations are self-selecting, at least the nonvaccinated population; and this can be seen as lessening the value of the study for the reasons described above in the section on the hierarchy of values for clinical trials.

Of note, Dr. Anthony Mawson has already done a smaller version of this using children in his home state of Mississippi who are unvaccinated compared to children who are.[59] This study has flaws for sure, one of which being that the study is basically a survey that asks participants to evaluate health status retro-spectively. However, the notion that one either can't find the unvaccinated or that it is somehow unethical to allow them to be in an experiment that merely attempts to quantify an ongoing life choice is simply nonsense. Or, more likely, it is playing to an agenda and presupposing unknown outcomes.

What is doubly odd about this all is that this experimental comparison of vaccinated and unvaccinated children might well be the way to resolve the basic question of whether vaccines are, on average, beneficial, harmful, or neutral to a child's health. The powers that be should welcome such a study, fund it adequately, and get it done as expeditiously as possible so that they can demonstrate conclusively that vaccines make children, and thus maybe all of us, healthier. The fact that they don't, while hiding behind bogus claims that such studies are unethical or impossible to do, simply illustrates what many interpret as a fear of what the studies might show.

Bias in Clinical Trials and Particularly in relation to Vaccine Studies

Bias in any experimental paradigm is a significant problem for evaluating the veracity of any findings. Nowhere is this more problematic than when trying to evaluate safety studies of vaccines, as will be detailed in Chapter 3. In prepara-tion for a detailed look at such studies, the following will provide an introduc-tion to the problems that can arise from different kinds of bias.

These are the following:

Information bias (also called *observation* or *measurement* bias) reflects a lack of accurate measurement or classification of key variables in a study. An exam-ple would be if information was collected differently from members of the various groups that make up a study, or if only partial information that might be relevant to the study is collected.

Selection bias arises when the selection of study populations, or their reten-tion in the study, gives a result different from what might have occurred had the whole population affected been considered. Since whole populations cannot usually be involved in a given study, representative samples of the population should instead be used. If one is examining the potential impact of a given vac-cine on ASD, for example, and excluding those with any family history of neu-rodevelopmental delays or autoimmune conditions, then selection bias might be at work, as it would diminish the possible impact of genetic background that might contribute to an adverse vaccine outcome. Selection bias, as we will see in the next chapter, tends to be a significant problem in vaccine safety studies.

Confounding bias can occur when a group exposed to a drug or treatment has a risk of some background disease different from that of a group not exposed. For example, again using the ASD example, one group has gastrointestinal abnormalities, and the other does not. Such background conditions are considered to be *confounders*. Such a confounder may bias the potential association, moving it closer or farther away from the true association. Another one would be comparing the relative health of vaccinated versus unvaccinated children if the latter had a more nutritious diet and more time in the outdoors, or came from wealthier families.

Ascertainment bias arises when data for a study or an analysis are collected (or surveyed, screened, or recorded) such that some members of the target population are less likely to be included in the final results than others.

Finally, there is what is termed *healthy user* bias, which is a form of sampling bias. Healthy user bias occurs when subjects enrolled in a study are more likely to follow the study protocols, for example, being more likely to take a particular drug or treatment, since they are more likely to follow medical advice compared to the more general population.

In general, while many clinical studies attempt by their design to minimize and/or control the various forms of bias, these are not always successful, and the extent to which they are not can very much impact the experimental outcome of the study and thus its interpretation.

Surrogate Markers in Vaccines

When most people think about vaccine effectiveness against some infectious disease, they usually consider only what happens to vaccinated and unvaccinated people when exposed to the pathogen. This is the percent reduction in risk that a vaccine may provide. The outcome may indeed suggest that some vaccine is effective. But how do the companies and other interested parties know that a vaccine in development, or after, is effective? The usual answer is that vaccine efficacy/effectiveness is determined by the use of antibody production against that disease. Antibody levels are thus accepted substitutes, *surrogates*, to the actual vaccine-induced immune response that should, in principle, prevent the disease.

In vaccine development, particularly vaccine effectiveness studies, a surrogate is a particular type of molecular marker, such as a specific antibody or level of antibodies, that is presumably on the direct causal pathway to the desired immune response, i.e., protection from the pathogen. Thus, the antibody serves to take the place of the actual demonstration of a vaccine's efficacy.

The key idea here is to use *seroconversion*, that is, the time period during which a specific antibody develops and becomes detectable in the blood, as the measure of future disease prevention.[60]

A 2013 WHO document lays all of this out in detail: *Correlates of vaccine-induced protection: methods and implications*, a report commissioned by the Initiative for Vaccine Research inside of WHO.[61]

The document makes for interesting reading overall, when even right away from page ii the authors give: ". . . thanks [to] the donors whose unspecified financial support has made the production of this document possible."

Who these donors are is not specified.

The authors begin by saying that "[t]he ability to assess the protective efficacy of a vaccine by measuring the proportion of vaccines who [*sic*] generate a particular immune response, without having to measure clinical outcomes, has significant advantages."[62] And "[v]arious statistical tools have been developed to evaluate these endpoints, but few epidemiologists are familiar with the details of these methods."[63]

This last part, that "few epidemiologists are familiar" with this, seems odd, but I'm going to guess WHO knows what they are talking about in this case.

Figure 2.2 (as seen on page 60) illustrates the types of surrogate markers from a simple case where IM-(immune marker) 1 is a surrogate marker, while IM-2 is what they term a "correlate," as it is not on the path to the protected response. The document defines this "correlate of protection" as "an attribute that is statistically associated with an endpoint."[64]

The authors write:

> In the context of vaccines, *protection* implies an immunological mechanism to prevent or reduce severity of infection or disease. The mechanism can involve both humoral and cellular arms of the immune system. Many aspects of these mechanisms are not yet understood.[65] [Italics are theirs.]

Whether the mechanism in question really involves both cellular and humoral protection is debatable, but the authors are correct in noting that the mechanisms are not understood.

The document provides for different mechanisms of immune response and the relation of the surrogate marker to this response. Some of the proposed mechanisms:

A single pathway with one immune marker, as in Fig. 2.2; two or more markers, either dependent or independent of each other; and immune markers that have no role in protection. Some of these are given in Fig. 2.2.

What is clear from the examples shown in the article is that antibody status is not always tied to protection postvaccination for an individual. This point is clearly made when the authors write:

> In ecological studies, associations are drawn between vaccination status, the substitute endpoint (e.g., antibody titre) and the clinical endpoint only at the population level. It is therefore not possible to draw conclusions about the associations of interest at the individual level.[66]

This is very true and perhaps one reason why some health officials occasionally let this point slip out: antibodies do not necessarily equal immunity for individuals, or even always for larger groups depending on the statistical significance. This point has been problematic during the COVID-19 events when an official at WHO and Canada's federal health officer said the same thing,[67] then just as rapidly walked it back, since skeptics took the statements to mean that antibodies as surrogate markers were not proof of immunity. This observation had been in the vaccine-skeptical literature for quite some time, and the health officials surely realized that their loose, albeit truthful, lips were going to endanger the carefully cultivated mainstream narrative.

The WHO document is actually quite intriguing in that it provides the comprehensive basis for surrogate markers in vaccination. Within the document are considerations of various modeling methods, including the Prentice criteria for a successful surrogate, which requires that the protection against a clinical endpoint (having or not having the disease) is significantly related to having received the vaccine for that disease; that the surrogate endpoint is significantly related to the vaccination status; that the substitute endpoint is significantly related to protection against the clinical endpoint; and finally that the full effect of the vaccine on the frequency of the clinical endpoint is explained by the substitution endpoint if it lies on the "sole" causal pathway.

There is also a Qin model that is sometimes used for validation.

Both models are based on a host of assumptions. Mentioning only the assumptions on antigens, for example, the WHO authors write:

> Most of the statistical methods used to predict VE [vaccine effectiveness] based on substitute endpoints *assume that the relationship between the immunological marker and clinical protection is similar among the vaccinated and unvaccinated groups.* This assumption ignores the fact that any immunity observed in unvaccinated individuals may be derived from natural exposure to antigens that are not in the vaccines (i.e., from exposure to the pathogen itself, to related organisms or to other antigens in the environment that share epitopes with the infectious agent in question), and can thus be qualitatively and/or quantitatively different from that observed among those vaccinated.[68]
> [Brackets for addition are mine.]

This is, at least, honest but misses the point: the same immunity thought to be conferred solely by the vaccine in vaccinated persons may also reflect the same "natural exposure to antigens."

All of the above is going to be very relevant when we consider the testing done for COVID-19 in Chapter 13. Some of these considerations taken from this document, mostly ignored by rather hysterical media, include: a definition of the clinical endpoint, e.g., sniffles versus severe respiratory distress; exposure intensity, e.g., how much virus exposure over which time frame; host factors such as age, with the elderly being more vulnerable; socioeconomic status with people lacking decent nutrition and access to medical services; antigen factors; type of antibody; kinetics of the immune response; and errors of measurement.

Modestly, the authors of the WHO document note that:

> The use of immunological markers as substitute endpoints for vaccine evaluation is important, but complicated. It is not straightforward to identify such markers or to ensure that the estimated VE derived from their use predicts accurately the vaccine effectiveness that would have been observed if the clinical endpoints were recorded. Differences between infectious agents, vaccines, immune responses and population contexts provide many opportunities for heterogeneities and complex relationships.[69]

All very true. In other words, evaluating any COVID-19 vaccine is going to be complicated in the extreme, particularly if it is rushed due to some social/political agenda, whether that agenda is driven by Bill Gates or others.

In many ways, WHO and regulatory agencies are stuck. It makes sense to use a surrogate marker in vaccine efficacy studies, since the alternative would be to do a true challenge study. Since the latter carries risks (and in the case of the proposed mRNA vaccines against COVID-19, perhaps major risks) to the health of the study subject, an antibody titer is considered to represent the likelihood that the vaccine will, or will not, provide disease protection. Whether this is always true is still open to question for COVID-19, as it has been for diseases to which surrogate markers are the only marker available at present.

In other cases, surrogate markers are even more problematic, for example, in human papilloma virus vaccines designed to prevent cervical cancer, where the surrogate marker of efficacy is the presence of precancerous lesions that are thought to be a requisite stage in the development of the cancer.[70] Typically, HPV vaccine makers claim that the various vaccines in use, or planned, will indeed prevent the development of cervical cancer in women. The problem the vaccine companies have in this regard is that the vaccine was first tested on young women, typically aged nine to twenty-four, while cervical cancer does not usually show up until women are much older and the girls and women in the vaccine trials have not (yet) reached the ages at which the disease might emerge. In terms of the pathogenesis of the disease, infection with the HPV virus strains can do one of several things. One is that the viruses do nothing. Another is that some of them can lead to cervical intraepithelial neoplasia

(CIN) lesions (stages CIN 2 and 3) on the cervix that *may* progress to cervical cancer.

In some 90 percent of women, this does not happen, as the immune system resolves the infection on its own. In spite of this, the two companies making HPV vaccines, Merck (Gardasil) and GSK (Cevarix), have decided that they don't want to wait the decades for the disease to emerge on its own, but rather adopt as surrogate markers the presence of the CIN2/3 lesions as precursors to cervical cancer.

The problem is similar to that often considered in other branches of medicine, for example, looking for "biomarkers" for neurological disease before the disease has emerged at a clinical level. In both cases, the use of surrogate markers may be the best one can do, but using such surrogates requires some restraint in reporting, and advertising, outcomes. For example, it might be perfectly fair to say that the vaccines diminished the appearance of CIN2/3 lesions, if true, but not that the vaccines prevented cervical cancer.

In terms of safety, as noted above, one can't even say that the vaccine is safe without a placebo control arm against which to check adverse effects. Neither Gardasil nor Cevarix has a significant control population, although one study of the former had a small true control population, but this was later blended into the aluminum group.[71]

The ultimate problem with surrogate markers is that they are essentially correlational in nature rather than demonstrating causality for the outcome. For this precise reason, stating that Gardasil or Cevarix are "safe and effective" is not an accurate statement, at least not at this time.

In general, due to these issues, the use of surrogate markers has come under greater observation. Some of the issues arising with the use of surrogate markers (or endpoints) in clinical trials generally include: the risk of false negatives or false positives; the surrogate does not provide knowledge about the magnitude and duration of the effect on the pathway of the disease process; or the surrogate does not capture side effects. This last point is particularly important, since it can lead to a greater chance that safety signals, described in Chapter 3, will not be discovered until postmarketing studies are completed, if the latter ever are.[72]

In regard to vaccine trials, other issues with surrogate markers arise. These include, apart from those noted above:

- The need to identify a standardized and reproducible "correlate of protection." While some vaccines accomplish this goal to some extent, others do not.[73]
- Specificity of the measured antibody response in that a vaccine may generate a range of antibodies measured by an ELISA antibody assay.

Further, there is an important difference between the total amount of antibodies versus binding and neutralizing antibodies, the latter being the ones that actually attack the pathogen. In a case where the latter are not determined, an overestimation of protection may result.[74]

- The accuracy of the correlate of protection is crucial, yet the correlate is absolute, since there are multiple innate and adaptive immune system parts that influence protection and because of the inherent variability of protective immune responses across individuals.[75]

With these caveats in mind, below is part of the FDA's table of surrogate endpoints for various pediatric vaccines that were the basis for vaccine approval or licensure. From section 507c(9) of the FD&C Act:

> . . . the term 'surrogate endpoint' means a marker, such as a laboratory measurement, radiographic image, physical sign, or other measure, that is not itself a direct measurement of clinical benefit, and—
>
> Is known to predict clinical benefit and could be used to support traditional approval of a drug or biological product; or
>
> Is reasonably likely to predict clinical benefit and could be used to support the accelerated approval of a drug or biological product in accordance with section 506(c).[76]
>
> The full list of FDA's surrogate endpoints is found on their website.[77]

It will be important to keep the information of the last three sections in mind as we take an in-depth look at aspects of the official narrative on vaccine safety and effectiveness as presented in the next chapter.

But first, a brief overview of the very complex field of immunology is in order.

CHAPTER 3

Health Consequences of Vaccination and the "Official" Story

Throw out your conceited opinions, for it is impossible for a person to begin to learn what he thinks he already knows.

—Epictetus, *Discourses*, 2.17.1[1]

Immunology and the Nervous System

Up until now, I have only skimmed the surface concerning the actual science of vaccines, vaccinology, and the role vaccines play in the control of infectious disease and how they do so. In order to be more comprehensive, it will help for me to present a basic understanding of the broader field of immunology, the science of the immune system. And, tying into this, it will be important to consider the interactions of the immune system, on which vaccines are designed to work, with the nervous system, particularly the central nervous system (CNS). Some of the following section is excerpted from a previous book of mine that examined some of the precursors to neurological diseases and, in order to do this, attempted to put these diseases into the context of immune function and dysfunction.[2]

This last aspect will be particularly important when considering any role that vaccines have as possible etiological factors in ASD, particularly in context to whether any immune function/dysfunction plays a significant part in any neurological disorder, ASD being a key one to consider in the following pages.

There are several things to mention at the outset. The first is that until at least the late 1980s, immune and CNS functions were considered separate systems with few to no interactions. Certainly, that's what I had learned in graduate school. Since then a lot has changed, and the bilateral roles of immune-CNS interactions are now widely accepted. Some of these will be detailed but only cover some of what is rapidly becoming a vast literature. For this reason, what follows can only provide a brief incursion into the rapidly developing subject, particularly as it concerns the developing CNS.

The second thing to note is that the immune system is very, very complex; and it would be safe to say that the field, for all of its progress, is still far from knowing some very basic things, one being how vaccine adjuvants actually work, a topic that I will address in more detail in Chapter 4.

The CNS is also very complex, and I can fairly confidently state as a neuroscientist that there is vastly more to figure out than we know now. As in science in general, some of what is believed now will have to be cleared away, much like earlier programming of software, to give rise to newer and better standards of understanding that are already emerging. It should be clear that if you combine the interactions of two very complex systems, the immune and the nervous systems, you get still more complexity. Adding on the changes to both during early development, it will become apparent that anything that perturbs either cannot but help to have an effect on both. Indeed, as documented below, the mutual interactions continue throughout life and may have considerable roles in neurological diseases across the lifespan.

As cited by Dunn (1995), these are some of the observations that reflect this connection: (1) Alterations in immune function can be conditioned (and conditioning in the classical sense is a CNS function); (2) Electrical stimulation/lesions to various regions in the CNS can alter immune responses; (3) Stress alters immune responses, an observation that your grandmother could have told you; and (4) The activation of the immune system can be correlated with changing properties in neural cells.

Dunn also discusses a key point in the interactions between the two systems, namely, that cytokines, common to both, can be considered as "immune—transmitters" whose actions are bidirectional. Similarly, some traditional neurotransmitters and various hormones can signal within both systems and bidirectionally, as well.

Postponing my explicit discussion of the CNS and its interactions with the genome, before getting into those details let's first consider the different presumed and studied components of the immune system.[3]

First, the *innate* ("natural") immune system is believed to be a nonspecific first line of defense against infectious diseases, composed of various cells and molecules that can recognize invading pathogens. These cells comprise a shopping list of cell types, including eosinophils, monocytes, macrophages, natural killer cells, dendritic cells, toll-like receptors, and complement system mediators.[4] In this system, the first response to a given pathogen is relatively slow, but it becomes more rapid with a secondary exposure to the same entity.

The *adaptive* immune system is also directed against invading pathogenic entities such as bacteria, viruses, and even toxins. These are somewhat loosely termed "antigens," from the French *antigène*, and indicate that they act against

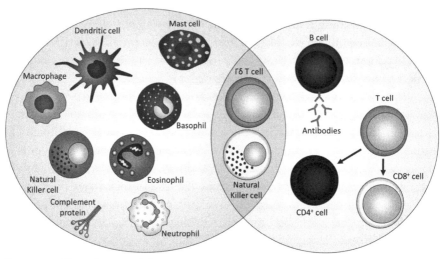

Figure 3.1. Key cell types of the immune system.

something but may be the attacker or target of the action, in this case something inside the body that is foreign, such as a virus.

The adaptive immune system contains highly specialized cells such as T (thymus-derived) and B lymphocyte (bone marrow-derived) cells, generating, respectively, cellular and humoral types of immune response. T cells, also termed T-helper, T4, or CD4 cells, are white blood cells that are essential for the adaptive immune response. "CD4" refers to a glycoprotein (cluster of differentiation four) found on the surface of T cells and other cell types (e.g., monocytes, macrophages, and dendritic cells). T-helper cells do not themselves destroy invading pathogens, as they have no phagocytic or cytotoxic capabilities, but they enable other cells such as CD8 killer cells, also called T-killer cells, to do so.

Two types of T-helper cell are recognized, Th1 and Th2, each of them known to be involved in eliminating, attacking, or neutralizing various types of pathogens. Th1 cells produce interferon γ (gamma) and act to activate the bactericidal actions of macrophages and induce B cells to make complement-fixing antibodies. These responses are the basis of cell-mediated immunity. The Th2 response involves the release of interleukin 5 (IL-5), acting to induce eosinophils to clear parasites. Th2 also produces IL-4, which facilitates B-cell isotype switching. In general, Th1 responses are usually directed against intracellular pathogens (viruses and bacteria), while Th2 responses act against extracellular bacteria, other pathogenic parasites, and toxins. In regard to Th1 and 2 responses, an intriguing recent finding from my laboratory is that Il-5 responses are elevated by almost 300 percent in young mice following the administration of the recommended vaccines in the CDC's schedule from zero to five years of age,[5] supporting a shift, at least at this stage of development, from a typical vaccine-induced Th1 response to a Th2 response. These results are discussed in more detail in Chapter 14.

A second crucial aspect of the adaptive immune system, particularly in response to future pathogen presentation, is the production of antibodies against that pathogen.

Antibodies are proteins called immunoglobulins (Ig's), large Y-shaped proteins produced by plasma B cells, a type of white blood cell. Ig's come in a variety of isotypes: in placental mammals, there are five, termed IgA, IgD, IgE, IgG, and IgM. Ig isotypes are classified based on the various types of heavy chain making up the protein. For example, heavy-chain classes named, with Greek letters, α, γ, δ, ϵ, and μ, give rise to IgA, IgG, IgD, IgE, and IgM, respectively. These isotypes differ in their biological properties, functional locations, and power to deal with different antigens. The main Ig is IgG, with four variations that provide the majority of Ig-based immunity against invading pathogens. It is the only antibody capable of crossing the placenta. IgM is associated with B cell-mediated cellular immunity. IgD acts mainly as an antigen receptor on B cells that have not been exposed to antigens and activates other immune cells, basophils and mast cells to produce antimicrobial factors. IgE acts primarily as an antigen receptor on B cells that have not been exposed to antigens. Finally, IgA is found in mucosal membranes of the gut and the respiratory and urogenital tracts and acts to prevent colonization of these structures by pathogens.

Antibodies contain a variable region on the tip of the Y structure called a paratope. Each paratope can recognize a unique molecule of the pathogen, the antigen. In this way, an antibody can mark a pathogen or an infected cell to be attacked by other cells in the immune system, such as T-helper and T-killer cells, or can serve to neutralize the pathogen. The so-called Fc region at the base of the Y protein structure allows it to communicate with other immune cells. The production of antibodies describes the primary function of the humoral immune system.

Antibodies are released by B cells, specifically differentiated B cells (B plasma cells). They can occur in two physical forms: a soluble form that is secreted from the cell and a membrane-bound form that is attached to the surface of the B cell (a B-cell receptor, or BCR). Soluble they are released into the blood and tissue fluids. The BCR facilitates the activation of these cells and their differentiation into plasma cells or memory B cells. Both types of cell will remain in the body and "remember" that same antigen at a future presentation.

In regard to antibodies, it is important to recognize that they come in two major categories in relation to how and to what they bind, and what happens when they do. One of these is "neutralizing" antibodies that bind to a site on the pathogen and prevent it from infecting a targeted cell. This is the type of antibody that the various companies are hoping to elicit in the efforts against COVID-19. In contrast, there are "binding" antibodies that bind to some site on the pathogen but do not interfere with its action on the targeted cell. What binding antibodies may do, however, is mark the pathogen for destruction by other immune system cells.

The response of the adaptive system is slower than that of the innate system but is more potent and specific. Further, the memory of past pathogen exposure resides in the adaptive immune system. Unlike the innate immune system, adaptive immunity is thought to be found only in vertebrates.

When describing features of the innate immune system, it is necessary to consider the role of the "inflammasome." The inflammasome is an intracellular, multiprotein complex that controls the activation of proinflammatory caspases (those molecules inducing programmed cell death or apoptosis), primarily caspase-1. The complex generally has three main components: a cytosolic pattern-recognition receptor known as the nucleotide-binding oligomerization domain (NOD)-like receptor (NLR), the enzyme caspase-1 (part of the apoptosis pathway), and an adaptor protein known as apoptosis-associated speck-like protein (ASC), which facilitates the interaction between the NLR and caspase-1.[6] The NLR subfamilies include NLRP3, the best-studied part of this group.

The NLRP3 inflammasome is activated by various stimuli, including pathogenic signals (e.g., bacterial, fungal, viral),[7] endogenous danger signals (adenosine triphosphate, or ATP), amyloid β (Aβ) (a hallmark molecule of Alzheimer's disease), uric acid crystals,[8] and environmental microparticles (e.g., silica crystals, aluminum salts).[9]

The latter are of obvious importance in consideration of the impact of aluminum salts used as adjuvants that may also gain ingress into the CNS. The role of adjuvants in vaccines will be considered in more detail in the next chapter.

NLRP3 activation is a two-step process. A first signal, such as the presence of microbial toll-like receptor ligands, primes cells by producing pro-IL-1β expression. A second signal, such as the molecule adenosine triphosphate, activates caspase-1 and leads it to process pro-IL-1β and pro-IL-18.[10]

The activation of NLRP3 is not completely understood, but three upstream mechanisms of activation have been proposed: ion fluxes (potassium and other ions), mitochondrial-derived reactive oxygen species, and phagosome destabilization and the release of lysosomal enzymes (cathepsins), both of which digest proteins after cell death.[11] The effects of NLRP3 inflammasome activation within the CNS remain unknown in many cases, but recent evidence suggests it has a role in neurological diseases and, as already mentioned, in the context of adjuvant aluminum salts.

Some neurological disease-specific molecules, such as Aβ, prion protein, and α-synuclein, have been reported to activate the NLRP3 inflammasome in CNS microglia and macrophages generally.[12]

A crucial component part of the innate immune system as it specifically applies to the CNS concerns the action of microglia, the resident macrophages of the CNS. Microglia are activated in response to insult and disease and are thus key components of what is termed the non-cell-autonomous aspects of neurological diseases. The term *non-cell-autonomous* refers to the role of such glial cells in the destruction of neurons in diseases such as ALS and microglial activation, widely considered to be part of this pathological process.[13]

In order to understand the role of microglia in neurological disease, including in ASD, it is important to understand the types of microglial activation that can occur.

Activated microglia are broadly categorized into two different states: the "standard" activated M1 state and the alternatively activated M2 state. Generally, the M1 state promotes a neurotoxic T-cell response and is cytotoxic due to the secretion of reactive oxygen species and proinflammatory cytokines such as IL-1, IL-6, and tumor ad factor alpha (TNF-α) and because of a reduction in protective trophic factors. These actions of M1 microglia serve to destroy foreign pathogens and galvanize T cells to mount an adaptive immune response. The M2 microglial state produces high levels of anti-inflammatory cytokines and neurotrophic factors (IL-4, IL-10, IGF-1, CD200, and fractalkine).[14] The M2 phenotype is generally associated with suppressing inflammation, conducting repairs, and restoring cellular and system homeostasis.

It is increasingly clear that the activation states for microglia are more a spectrum of responses than two static conditions.[15] In fact, these conditions are quite mutable. In this regard, various studies have shown that individual microglia can express both pro- and anti-inflammatory signatures at the same time,[16] in part due to the role of the local microenvironment.[17] Further, there is now an increasing recognition that the M2 phenotype is not a single entity, but rather comprises various subtypes, termed M2a, M2b, M2c, and Mox, each with unique features and functions that remain poorly characterized.[18]

* * *

So what might it mean if as part of the immune system microglia were activated inside the CNS? Quite a lot, as it turns out. Not surprisingly, microglial activation has been observed in virtually all neurological diseases and can be activated by aluminum in its various forms, not least as a vaccine adjuvant. That topic will be addressed in full in Chapter 5, which considers aluminum in all of its manifestations as a driver of immune disorders as well as disorders of the nervous system, the two of these linked in a variety of ways. With the above as a background in basic immunology, let's consider the data on vaccine safety from the perspective of the CDC, FDA, and other mainstream organizations.

The Safety and Effectiveness of Vaccines: The "Official" Narrative

The well-worn mantra that is continually repeated is that "vaccines are safe and effective," as already noted in earlier chapters. One hears this statement in any official comment on vaccines by the CDC, FDA, and National Institutes of Health (NIH), as well as by the health agencies of most countries. It is repeated as a standard health catechism by much of the media whenever questions about vaccine safety and effectiveness are raised. As with any other form of religious training, most people are taught to believe this and not question just how solid the foundations of such claims may be.

Given this, the goal of this chapter is to look at such claims in some detail to see just how firmly they may, or may not, actually be based on the very sources that lead to such claims being made in the first place. Needless to say, this evaluation is crucial before we begin to evaluate the countervailing literature later in this chapter and in the next.

In the following, I will start with the question of safety. In part, this is because there is a lot more literature on vaccine safety with much of the effectiveness of vaccines more or less taken for granted. This latter point will be addressed in the sections that follow.

Are Vaccines Safe?

In some sense, this is the key question in the vaccine wars. While some anti-vaccine proponents maintain that vaccines simply don't work, most people hesitant about vaccines don't appear to hold this view. Nor, for that matter, do these people oppose other people getting vaccines. Rather, the opposition is against vaccines being forced *on them or their children.* If vaccines were shown to be completely safe (not possible, since no drug is), the issue of forced vaccination or mandates would still not be addressed, because it is really not a health discussion. Rather, the issue of mandates raises that of bodily autonomy and security of the person, essentially the most fundamental of natural rights that humans should expect to enjoy. But before we delve into the rights issue as it plays out against a chorus of "state of exception" proposals from various legislative bodies (considered in detail in Chapter 11), let's first consider the actual, rather than the assumed, safety record for various vaccines.

The first question to ask is how do we know that vaccines are safe? One way is through the various vaccine surveillance systems that will be described and critiqued at the end of this chapter. The other way is through the various phase trials, as described in Chapter 2, that the manufacturers of any given vaccine perform, either on their own or with various academic or regulatory body collaborators. Unlike in experimental science in other fields, the results of a given experiment by one group, particularly if it is viewed as important to the field, are supposed to be replicated by other researchers elsewhere. The ability to replicate the findings of others is considered to be a hallmark of the scientific method, although it does not always, or even often, happen in the present.[19]

Even though many commentators state that vaccines are highly evaluated for safety, indeed, paraphrase "more highly checked than most drugs," the reality is somewhat different. In this regard, it is important to keep in mind that none of the regulatory bodies actually do basic research in this area, although an agency like the NIH could do so, unlike the CDC and FDA. I note here also that our key regulatory agency in Canada, Health Canada, is a ministry of the federal government that also does no direct testing of vaccine safety claims.

What these agencies do, however, is evaluate the safety data based on the various phase trials conducted by the company making the vaccine (or drug). These companies, however, have a vested interest in positive outcomes for effectiveness and negative outcomes concerning harms. In essence, the regulatory bodies have to trust in the honesty of the companies.

It is, of course, one thing to trust a PhD candidate's work when going into a thesis defense, and this is where the scientific honor system works best. However, in the years before the defense, the candidate has had their research scrutinized frequently by their thesis supervisor, has had numerous thesis committee meetings to check progress and evaluate the ongoing experiments and data, and finally, all is subject to the final evaluation before and at the actual thesis exam.

In contrast, this does not happen when a company seeks to license a drug or vaccine. In the case of a company's data, evaluators can only examine what is put in front of them and to a very great degree rely on the scientific integrity of the company scientists. This trust may be warranted, but it is also a potential conflict of interest for the company given that their future profits clearly depend on licensing their product, which, in turn, depends on what they give the evaluators. We have seen this play out recently with evaluations of the efficacy and safety data for the mRNA vaccines of Moderna and Pfizer. These evaluations will be subject to further scrutiny in Chapter 13.

In the peer review process for grant applications and submitted articles, the reviewers also have to trust the integrity of the scientists submitting the grant or article. In the case of grants, there is money potentially involved for the applicants; for authors of journal articles, the money is indirect in that getting a grant approved usually depends on publications. For these reasons, conflicts of interest occur in such circumstances as they do for licensing reviews, but the amounts are vastly different.

One other point to note is that in the following evaluation of the literature, the reader should keep in mind that if indeed vaccines are more highly tested for safety than most drugs, it might be worthwhile worrying about how safe the other drugs actually are, mirroring some points made by Dr. Marcia Angell cited in Chapter 12.

The evaluation of the evidence for the safety of vaccines alone or in the broader context of various recommended vaccine "schedules," such as that of the CDC vaccine schedule, is based on two key sources: the CDC itself and the American Academy of Pediatrics (AAP), both of which have highlighted on their respective websites the peer-reviewed published and non-peer-reviewed articles that they think make the best case for vaccine safety, particularly in relation to any concerns about autism and other neurodevelopmental disorders. There are other websites featuring the same or other articles from the scientific literature.

Rather than simply accepting the cited articles as supporting vaccine safety without question, I want to look at each in turn, with the AAP going first, as it is by far the shorter list.

In all of my comments on the articles cited for both entities, the reader should keep in mind the following: it is vastly easier to critique someone else's work than to provide your own, sort of along the lines of the adage attributed to various persons that "Those who can, do. Those who can't, criticize." While this is true, the reader can decide if I have misused my critical role.

However, unlike bloggers who are often inclined to see their trade as one that has no need for accountability, I've tried in what follows to offer a fair critique that avoids ad hominem commentary, because, as much fun as it can be to trash others without apparent consequence (and I used to be a political blogger), one goal of the current book is to rise above the standard set by various "science" bloggers. Readers will note that I have applied the same standards to the articles critical of vaccine safety discussed in Chapter 4.

Overall, wading through all of the key studies was labor-intensive, but necessary. Keep in mind that while the articles examined are not all of the literature on vaccine safety from the mainstream perspective, the material cited in this chapter is from the primary lists provided by the AAP and CDC.

To be as comprehensive as possible for both data sets, I have evaluated the articles based on: methodology, the presence of controls, which diseases were evaluated, which vaccine(s)/ compounds were evaluated, the experimental outcomes, and the conflicts of interests, if any. In virtually all of the articles to be discussed, the various authors had tested hypotheses about possible causality between either the mercury compound Thimerosal (discussed in Chapter 5 in more detail) or the measles, mumps, and rubella vaccine (MMR). The exception to this is the IOC 2012 report, which looked at vaccine adverse events for a range of vaccines and patient outcomes.

I will adopt two primary strategies for the evaluations. The first is sort of a layered pyramid in which I start with the cited articles on the AAP website, the top of the pyramid, and then pick the one I think most comprehensive for further analysis. From the second-tier articles, I will examine the studies that the authors suggest were the most useful to their analysis, pick what seems to be the best of these, evaluate them, then go on to the next tier, etc. In this way, I hope to get a better sense of the claims made by the articles at the top by examining the underpinnings of each at the other levels. Whether this process strikes gold or coal will be determined in the evaluation at the end of this section.

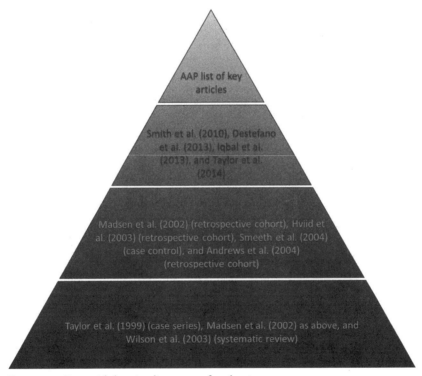

Figure 3.2. Pyramid showing literature of evaluation.

The second method will be more of a brute-force approach: take the 584 articles on vaccine safety from 2000 to 2020 on the CDC website; pick the 101 that seem most likely to be directly related to actual safety issues (rather than methodology), again primarily on Thimerosal and the MMR vaccine; and evaluate them year by year. Once again, a summary evaluation will be provided after this section.

Before I do this evaluation, however, it is important to first consider the nature of experimental "controls," as a lot of what follows hinges on just how important controls are, or aren't, in evaluating vaccine safety studies.

The Nature of Experimental Controls

It may serve us to better understand the nature of what constitutes an experimental control and whether or not this is actually essential.

To begin, just what can control populations do, both in terms of science generally and, more particularly, in the present cases where vaccine safety is being evaluated? For one thing, a control population is not always needed, as in the much vilified article by Wakefield et al. in 1998. This study allegedly linked the MMR vaccine to gastrointestinal pathology, and crucially to regressive autism.[20]

In reality, as discussed later in the book, no such claim was made by Wakefield and his colleagues. However, the work was attacked for various reasons, among which was the claim that the study actually had no control patients, a point that is irrelevant in a case series. Just to be clear, as described in Chapter 2, a case study is one in which a single example of a disease or a disease presentation is reported. Strictly speaking, such a study does not require a control, since the purpose is to describe a new finding with the general background knowledge of medical practice serving as a guideline to the new observations. A *case series* is much the same in that it reports on a number of similar disease presentations, again relying on the general field to distinguish these cases from the general population, and no controls are needed.

Many pharmaceutical studies also dispense with real controls when trying to compare the outcomes of a particular treatment, such as a drug or vaccine. In such cases, it is common to compare one drug with another, usually an older one of similar general function, that is, a "standard of care drug." For example, it may be considered acceptable to compare a new drug to treat Alzheimer's disease with another used for the same purpose, the assumption being that the older one already had a real placebo treatment group at some point, although such is not always the case.[21] Other companies clearly seem to understand the need for true placebo controls and provide them in their phase trials.[22]

Much the same occurs with vaccine safety (or effectiveness) trials run by either the pharma or by scientists from outside the pharma, in which the identified control group is often another vaccine or the adjuvant that may be part of the vaccine formulation. In this regard, it seems as if the "standard of care" provision were taken well beyond the original intent when comparing a new vaccine with an existing one, the latter often used for a different disease. Similarly, the use of the aluminum adjuvant as a control in safety trials for a vaccine containing the same adjuvant is at best bad science based on the view that adjuvants themselves pose no health risk, a point easily refuted in Chapter 5. It's either that, or a deliberate attempt to diminish a possible adverse effect signal. The reader can decide which of these possibilities are most likely.

Let me take an example from our own work on ALS/Lou Gehrig's disease in which we looked at various toxins to see how they might affect motor neuron health and survival. Over the last decade and more, we have used a particular compound called BSSG (short for β-sitosteryl β-D-glucoside), a sterile gluco-side, mixed into a food pellet and fed to mice to induce motor neuron death. This toxin is one that we originally isolated from the seeds of the cycad tree, the latter linked to the Guamanian form of ALS, termed ALS-PDC.[23]

Typically, when we do these experiments, we have a control group that gets nothing besides a regular mouse chow pellet, a group that gets BSSG mixed into the mouse chow, a group that may get some potential therapeutic agent, and a group that gets the BSSG and the therapeutic. In this set of groups, the controls establish the baseline for mice of this strain, sex, and age. The BSSG group establishes the impact of the toxin against the control group. The treatment group establishes if the treatment has any impact alone compared to the control group and the mixed group, then demonstrates if the therapeutic prevents the expected impact of the BSSG.

What would happen if we added a second toxin to the study but didn't evaluate it independently and simply combined it with BSSG? Could we then determine the relative contribution of both toxins? Yes, maybe, but only if each was tested alone and if each contributed less than 50 percent of the total change from control values. If, however, BSSG at the concentration we normally use gave 95 percent of the change in motor neuron survival from control and the second toxin added the rest to a total of 100 percent, we would still not know how much impact compound two had really contributed since the impact of the BSSG had largely saturated the total impact. The same sorts of concerns arise if we were to do a similar experiment using two, or more, compounds as therapeutics. There are ways to deal with all of this, of course, but they involve a range of additional experiments that, while doable, add to the overall complexity, time, and cost of the experiments.

This now brings us back to the question of the adequacy of the safety studies that attempt to evaluate MMR, or any vaccine or other drug, for safety or effectiveness. Thus, a study in which the MMR vaccine receipt is compared to those who didn't receive the MMR but did receive the other vaccines on the various schedules, such as that of Madsen et al. as cited soon, really has no way to demonstrate that the MMR vaccine is absolved of any causal relationship to developmental neurological disorders unless and until all of the other vaccines in the schedules that the participants may have received are independently evaluated. To paraphrase Del Bigtree, the vaccine activist discussed in Chapter 7, it would be like saying we were going to compare drinkers in a bar having five shots of Patrón versus another group of age- and sex-matched (ideally) drinkers having the five tequila shots plus one shot of Jameson.[24] If you do this experiment in your pub, you will find that a certain percentage of people in each group fall to the floor before or after the last shot. If your statistics show that the numbers are approximately the same, are you justified in concluding that Jameson does nothing? In this example, and in the real life studies cited by the AAP and CDC (and others), comparing MMR to no MMR vaccine receipt against a large background of noise created by the other vaccines received more or less guarantees of *not* finding a safety signal. If one doesn't want to find such a signal, then this would be the way to go. However, if this was intended to be an unbiased study, then it would not.[25]

The clear answer to the above question is no, and that doesn't really change if you add more participants to each group. In other words, the experiment in the pub has been designed to minimize the chance of finding a difference, and increasing the number of pub patrons won't solve the core problem. It's the same if you try to make another vaccine your control, or your adjuvant your control, in that in either case you are simply minimizing the prospects of seeing a difference. This is done all the time, as cited over and over in both the AAP's and the CDC's lists (both presented next) of vaccine safety articles; and thus, while this may be standard practice, that alone does not make it scientifically valid. This overall lack of controls, as we will see, will plague many of the studies cited by both organizations (and elsewhere) and renders invalid any claims to provide definitive evidence proving vaccine safety.

The AAP List of Vaccine Safety Studies

The AAP website features a number of the main studies they believe best make the case for vaccine safety. Some of the key and most recent ones include:

Taylor et al. (2014), DeStefano et al. 2013), Iqbal et al. (2013), and Smith and Woods (2010).[26]

The DeStefano and Iqbal articles are from CDC scientists. Other articles that make up this first tier include: Kaye et al. (2001), Mäkelä et al. (2002), Fambonne and Chakrabarti (2001), Fambonne et al. (2006), Budzyn et al. (2010), Klein et al. (2011), and Jain et al. (2015). Note the website lists one article as Claytoni et al. (2011), but this is actually Stratton et al.'s 2012 Institute of Medicine (IOM) full-scale review.[27] This last document is highly important and will be addressed independently after the others.

I want to start with four of the articles on the AAP's first tier list as cited, given that the overall analysis of the AAP on their website is that these articles provide a firm basis to reject vaccine links to ASD or other neurodevelopmental disorders.

Smith and Woods (2010): Let's just start with the declaration of conflicts of interest that these authors, more honest than some, have made:

Drs Smith and Woods are or have been unfunded sub-investigators for cross-coverage purposes on vaccine clinical trials for which their colleagues receive funding from Wyeth, Sanofi Pasteur, GSK, MedImmune, and Novartis; and Dr Woods has received honoraria for speaking engagements from Merck, Sanofi Pasteur, Pfizer, and MedImmune and has received research funding from Wyeth and Sanofi Pasteur.[28]

It's hard to have more conflicts than this, but let's give these authors the benefit of the doubt and see what they have found in their study. The authors looked at 1047 children born between 1993 and 1997 and, of this total, compared the children over the first year who received the entire vaccine schedule within one month of the recommended time to children who did not receive it in the same time frame, versus a third group that did not receive all the vaccines within the first year. Neurobehavioral outcomes were assessed at seven to ten years of age. Concerning differences between groups, the authors report better scores for twelve of forty-two measures in the timely vaccine group. If one wanted to make the case that vaccine delivery to the recommended schedule had no negative impact on cognitive development, this might be reasonable. Trying to make the case that children who are in this group actually are higher performing makes little biological sense, unless one also wants to postulate that vaccination on time according to essentially an arbitrary schedule increases brain function by some unknown mechanism. One assumes that the corollary would be that because some of the vaccines in the schedule contain aluminum adjuvants, then aluminum must be good for brain development. That such an assertion is blatantly incorrect will be discussed in detail in Chapter 5.

DeStefano et al. (2013): This is a curious article that appeared in the *Journal of Pediatrics* that seems to be a secondary analysis of data from an original case control study that was designed to investigate the impact of Thimerosal, although all the children in the study had some Thimerosal exposure. In the current study, the authors examined the number of "antibody stimulating proteins and polysaccharide antigens" in six-to-thirteen-year-old children who had ASD/ASD with regression or were deemed to be controls without such disorders. The children's vaccine history was divided up into blocks: zero to three months, zero to seven months, and zero to two years. No significant differences were found between groups in antigen levels, leading the authors to conclude that the exposure to more antigens didn't change the neurological outcomes. If one wanted to hypothesize that antigen levels as a function of increased numbers of vaccines were causal to ASD and the other related disorders, this might be a tenable study. However, since the study really had no controls from the perspective of having unvaccinated children, it is difficult to know if this interpretation is correct. For reference to the study, here is what the authors concluded in the first paragraph of the discussion:

> We found no evidence of an adverse association between the number of anti-gens received from vaccines in the first 2 years of life and any of the categories of neuropsychological outcomes evaluated.[29]

Iqbal et al. (2013): The study here by members of the same group was pub-lished in the same year in the journal *Pharmacoepidemiology and Drug Safety.* It appears to be a close variant of the DeStefano article. Here is what the authors concluded:

> In conclusion, we did not find any statistical evidence of an association between adverse neuropsychological outcomes later in childhood with level of antigen exposure from vaccines in the first 2 years of life.[30]

In other words, apart from some additional data rendering, it's the same arti-cle based on the same data set. One might be tempted to call this a case of self-plagiarism, but I doubt the CDC cared then or would care now as they had about DeStefano's back then, and then again when the William Thompson data fudging story came out years later. This last item is discussed in Chapter 5. The AAP, however, might want to be a bit more rigorous in their self-plagiarism surveillance.

Taylor et al. (2014): I will return to other articles in this list at the end of this sec-tion, but before I do I want to first consider that of Taylor et al. (2014), through which we can see the levels of analysis that go into making the judgments about vaccines and autism, especially as these judgments apply to Thimerosal and the MMR vaccine. Further, of these first four articles, only Taylor et al. actually directly address the issues of autism in context to the measles, mumps, and rubella (MMR) vaccine or Thimerosal.

In this paper, Taylor et al. provided a meta-analysis of one of the most recent of the AAP-cited studies about a possible role of the MMR vaccine or Thimerosal as triggers for autism.[31]

In their study, these authors started with a potential list of 929 articles to evaluate based on a web search of various medical and scientific data bases. After scrutinizing the abstracts of each of these, Taylor and colleagues finally winnowed the number of articles that met their core standards down to ten, five each of case-control and cohort studies. From these ten, the authors noted that seven of the ten showed moderate-to-high selection bias in choosing participants for the studies. The use of these seven studies might not be fatal, and the authors seemed to think they could deal with such prevalent bias in completing their analysis.

However, as a second-tier evaluation, let's consider in some detail four of the articles that Taylor et al. cited as *relatively* free from bias. In the case-control studies, the low risk of bias study cited is only that of Smeeth et al. (2004),[32] which I will examine in greater detail soon.

First, I want to take a look at the cohort studies of Andrews et al. (low risk), Hviid et al. (low risk), and Madsen et al., the latter considered "moderate" risk of bias but included because this article is frequently quoted in attempts to disparage any link between the MMR vaccine and autism.

Andrews et al. (2004) examined 109,863 children from six to ninety-six months old in the United Kingdom who received the Thimerosal-containing DPT (whole cell diphtheria, pertussis, and tetanus) vaccine or the DT vaccine from zero to three doses before three to four months of age or cumulatively by six months against a range of neurodevelopmental disorders. These disorders included general developmental disorders, behavioural problems, tics, attention deficit disorder (ADD), language/speech delay, and autism. Of these, only tics were significantly increased at four months of age when considering all the dosages, but not at three months, even though the p value approaches significance. Tics as a negative outcome are interesting, since they are defined as "sudden, repetitive, non-rhythmic motor movement or vocalization involving discrete muscle groups."[33]

They are often included as a feature of ASD, as well. (This observation will come up again when we consider Thimerosal in more detail in Chapter 5.)

Some other points about this article: It makes no mention of other vaccines administered, so we should probably conclude that the children in the study had received other vaccines on the United Kingdom schedule at the time.

Several quotes from the article are worth considering. The first concerns the exclusion criteria on page 585 in the Methods. The authors write:

> Children with Read and OXMIS codes relating to a variety of prenatal, peri-
> natal, and postnatal conditions that occurred before 6 months of age were
> excluded as were children who were recorded as having an outcome event in
> the first 6 months of life. These children were excluded from the main analysis
> because the presence of such a condition is likely to affect both vaccination
> and future neurodevelopmental outcomes. Examples of exclusions were birth
> asphyxia, Down syndrome, cerebral palsy, meningitis, encephalitis, and head
> injury. Children were also excluded when they received either hepatitis B or
> influenza vaccination in the first 6 months of life because such children are
> likely to be an atypical subgroup.[34]

As we will see repeatedly in the evaluations of much of the literature on vaccine injuries, exclusion criteria remove some of the groups who will certainly go on to receive the vaccines in the future and thus are innately biased toward a certain "healthier" cohort.

In regard to the tic issue, the authors also seem to have removed some data that didn't seem to fit:

> The main analysis included all children whether recorded as receiving 0, 1, 2, or 3 doses of DTP/DT at any age. However, it seemed possible that, as a result of socioeconomic or other confounding factors, children who did not complete vaccination in the first year of life would form a biased group. The data therefore also were analyzed after excluding all children who did not receive 3 doses of vaccination by age 366 days.[35]

Indeed, it seems likely that including this group would have biased the results, but not as Andrews et al. hoped.

The study was funded by WHO.

The article by Hviid et al. is next on the list, also testing the notion of Thimerosal in relation to autism or related ASD disorders. The authors examined children in a much larger retrospective cohort study than that of Andrews et al., looking at 467,450 children who had received Thimerosal-containing DPT vaccines in zero to three doses. They found no association between any dose of Thimerosal and the appearance of either autism or ASD. Further, increasing the doses did not show any dose-response increase in these disorders with dose. No details about exclusion criteria were reported in this article, nor is there any evaluation of what other vaccines those who were in the zero group may have received. These issues will come up in the evaluation of Madsen et al. next, as both Hviid and Madsen follow the same protocols for their studies.

The article by Madsen et al. will be discussed in detail, as it is part of the next tier analysis arising from the following article on Taylor's list, namely, that of Smeeth et al.

In this third-tier list is Smeeth et al., on the MMR vaccine and developmental disorders.[36] The Smeeth paper begins with the premise that the 1998 Wakefield et al.[37] article is one that urgently needed a rebuttal from formal epidemiological studies to demonstrate the safety of the MMR vaccine.

Smeeth et al. examined people born in 1973 or later with a diagnosis of pervasive developmental disorders (PDD), including autism, registered between the years 1987 and 2001, with 4,469 persons in total derived from the UK General Practice Research Database (GPRD). They start with 1,410 people with a diagnosis of PDD, later lowered to 1,294 cases after the removal of cases that did not match their criteria. They identified 6,465 "potential" controls, from which 4,469 remained after adjusting for the criteria. The groups were further categorized by sex and age, the latter from one to two years of age to greater than ten years of age.

The fundamental problem here, although disguised in the article, is that there are no real controls in the sense that all participants received the MMR vaccine in either one or two doses. In addition, since measles, mumps, and rubella vaccines could still be separate vaccines during the early time period covered, if individuals had received the three separate vaccines in a twenty-one-day period, these cases were classified as having received MMR. Comparing those who received MMR versus the separate single vaccines is also a confounding variable, as initially noted by Wakefield et al. in regard to the retrograde autism his group described in their 1998 paper.

The lack of real controls is, in general, highly problematic from a scientific methods perspective and is therefore my main complaint, because, in essence, the study is looking at PDD versus non-PDD when all have received the MMR vaccine or "equivalent" in separate vaccinations. Let's make an analogy similar to the one made previously with tequila: All patrons in a pub have five pints of Guinness. Some are hungover the next day; some are not. The numbers of those hungover and those not are not significantly different. Conclusion: Guinness does not induce hangovers.

Smeeth et al. go on to cite three earlier studies in their introduction, also performed to refute any link purportedly made by Wakefield et al. on the MMR vaccine and autism. Alas, as Smeeth notes, these studies did not "reassure the public," hence the need for the newer study.

Most telling is Smeeth's Figure 2 under the caption title of "Meta-analysis of studies that compared risk of autism or other PDDs among vaccinated and unvaccinated individuals." This is highly misleading, since Smeeth et al. have no unvaccinated patients in the study. Nonetheless, Smeeth et al. conclude that all of these studies give the same general outcome demonstrating no MMR relationship to PDDs.

What do the other three studies say in detail in our fourth-tier analysis? Let's see:

Taylor et al., 1999[38] (different from Taylor et al., 2014, from earlier): This is a small case-series study that looked at changes in the rate of autism of various kinds in the Northeast Thames region of the United Kingdom, in a time period spanning 1979 to 1988, the latter when the MMR vaccine was introduced into the United Kingdom. The authors observed a general increase in autism diagnoses in the period under study, but no clustering of autism cases and no "spike" in numbers in 1988. This seems reasonable, until one looks at their plotted curves, which appear to show a change in rate beginning in 1988/89. Nevertheless, the authors conclude that their results do "not support the hypothesis that MMR vaccination is causally related to autism, either in initiation or to the onset of regression-the main symptom mentioned in the paper by Wakefield and others."[39]

This is a reasonable conclusion, if limited by what the curves appear to show, since Wakefield et al. made no such claims, either.

The paper by Danish epidemiologists Madsen et al. (2002)[40] is often used as the presumptive counterweight to Wakefield and colleagues. In this article, Madsen et al. used a retrospective cohort study of all children born in Denmark in the period of 1991 to 1998 for a total of 537,303 children, of whom 440,655 had received a single MMR vaccine by age fifteen.[41] Madsen cited a number of previous studies, including the earlier Taylor et al., as providing only weak evidence against an MMR-autism link. Overall, the Madsen study did not support a causal relationship between the vaccine and the various forms of autism, at least in this population.

The problems here are twofold: First, the latter Taylor et al. paper lists Madsen et al. as a retrospective cohort in the category of cohort studies, noting, however, that it is of "moderate risk" of selection bias. Madsen et al. also had no real control population, not a trivial problem.

In the third study cited by Smeeth et al., Wilson et al. (2003)[42] conducted a systematic review to see if there was any association between the MMR vaccine and ASD. The authors claimed that there was no such association, but in the studies they deemed suitable for their analysis, only one, Madsen et Al. (2002), made their list.

Wilson et al. advance four hypotheses to be tested. The first is that ASD "rates are higher in individuals who received the MMR vaccine compared with those [who] have not been vaccinated." They cite only Madsen et al. 2002, with the problems in this study as already discussed. Second, "increasing rates of ASD are occurring as a consequence of the MMR vaccine." Of the six studies examined for this hypothesis, these will be meaningless if those considered to be the controls (not receiving the MMR) were vaccinated with other vaccines, the latter a likely possibility. Hypothesis 3: "Development of ASD is temporally associated with receiving the MMR vaccine." However, the authors interpret this hypothesis to basically suggest that the development of ASD in MMR-vaccinated children should be different from that of those not vaccinated with MMR. However, since ASD is typically diagnosed at certain ages, the age of diagnosis is not going to change. And, as above, if some other vaccine or combination of vaccines is included in the non-MMR group, it essentially renders this hypothesis absurd. Hypothesis 4: "A new variant form of ASD may be associated with the MMR vaccine." Here, variant form ASD was identified by the presence of developmental regression or GI tract symptoms. There were four studies evaluated here to make this claim. All these data say is that there is no association between the two symptoms, a conclusion with which the original Wakefield would, and did, partially agree, based on the observation that the gastrointestinal (GI) pathology linked to eight of twelve of the children.

Apart from these methodological concerns, two of the seven authors on the Smeeth paper declared a potential conflict of interest with ties to the pharmaceutical industry; one author had received funding from Merck, the manufacturer of the MMR vaccine.

The Other Articles Cited by AAP

In the following, I will briefly review several other papers in chronological order that were cited by the AAP as evidence for overall vaccine safety. Following this, I will go to the one comprehensive study performed by reviewers at the then-IOM (now the National Academy of Medicine, after July 2015). This latter document, Stratton et al. (2012), is well worth reviewing in some detail, as it may hold a key to considerations of vaccine adverse effects and suggest the basis for future studies. This latter aspect will be considered further in the last chapter of this book.

Kaye et al. (2001): These authors examined the possible relation between the MMR vaccine and autism looking at children under twelve years of age over the period of 1988 to 1999. The key finding is that while MMR uptake was constant at about 95 percent, the rate of autism increased by fourfold in two-to-five-year-olds in this period. Seemingly, this is a very straightforward result, but curiously this article did not meet the criteria of the Stratton (IOM) report, discussed soon.

The authors reported funding from various pharmaceutical companies.

Chakrabarti and Fombonne (2001): This is a very small sample of ninety-six children immunized with MMR and compared to other small samples both pre-MMR (ninety-eight children) and post-MMR (sixty-eight children) looking for evidence of pervasive developmental disorders (PDD). No significant differences were found; however, this study was also not included as part of the effective evidence in the IOM report.

Mäkelä et al. (2002): This Finnish study looked at three disorders in relation to the MMR vaccine—encephalitis, aseptic meningitis, and, of course, autism—in 515,544 children from one to seven years old. The children were assessed based on hospital admission for these various ailments. While the first two would clearly warrant hospital admission, it is far from clear why autism would. The authors also note on their page 926 that ". . . incidences of autism could not be defined with our approach . . ."

They also report that records of MMR vaccination were not accurate. The IOM report also did not use this article in their evaluations. Given this, what the AAP thinks this study adds to their argument is uncertain.

Fombonne et al. (2006): This study looked at PDD (including autism) in 27,749 children from Montreal in the period of 1987 to 1998. PDD was assessed in relation to exposure to Thimerosal or the MMR vaccine. The authors noted that PDD continued to increase while MMR coverage slightly declined. The authors view the PDD increase as due to better detection and awareness, not to an actual increase, per se, in the disorder. In addition, no relation was found to the cumulative amount of Thimerosal. The IOC report did not include this study in their evidence.

Mrozek-Budzyn et al. (2010): This Polish case-control study looked at autism versus the MMR vaccine or a measles-alone vaccine. The study had ninety-six cases of "atypical" autism in children two to fifteen years old compared to 192 age- and sex-matched "controls." Vaccine history was not available for any of the children. In other words, we have no idea if the controls were true controls.

Klein et al. (2012): These authors examined seizures in relation to receipt of either the MMR vaccine or the MMRV, the latter also against chicken pox (varicella). They found no association. Their listed conflicts of interest read like a virtual shopping list of the pharmaceutical industry, including Merck, GlaxoSmithKline, and six others.

Jain et al. (2015): This was a retrospective cohort study that examined 95,727 US children who had received the MMR vaccine and looked at outcomes stratified by whether or not the children had an older sibling diagnosed with ASD. In other words, this study is not about MMR in regard to ASD, but about a linkage to an undetermined cause of ASD in siblings. Although this study came out after the IOC report, it is difficult to believe that the report would have considered it to be informative.

Overall Evaluation of the AAP's Vaccine Safety List of Publications

Based on all of the above, a reasonable conclusion is that most of the cited studies, at each level, have rather serious errors of bias or study design and as shown next are underpowered to actually detect a real signal. For these reasons, from the bottom tier to the top, such errors propagate upward, making the entire foundation unstable, at least from a purpose of demonstrating in general that vaccines or their ingredients don't cause autism.

The exception to the above, and a savior from an AAP perspective, may lie in the 2012 IOC report of Stratton et al. that addressed eight vaccines and a range of adverse reactions, neurological and other, that had been linked to them.

The IOM's Stratton et al. Report on Adverse Effects of Vaccines: Evidence and Causality

This document, provided by the Committee to Review Adverse Effects of Vaccines, was tasked by the Board on Population Health and Public Health Practice of the IOM to provide a comprehensive review of all available data on vaccine-related adverse events.[43] This was the eleventh of similar documents provided by the IOM in twenty-five years after being mandated to do so by the 1986 National Childhood Vaccine Injury Act. It was also the first comprehensive review undertaken by the IOC since 1994. (The others are cited on pages 24–25 of the report.)

The resulting evaluations making up the final document were compiled into an 865-page book published by the National Academies Press in 2012. The editors were Kathleen Stratton, Andrew Ford, Erin Rusch, and Ellen Wright Clayton, the latter a professor of both pediatrics and law.

The report was built upon the work of eighteen scientific committee members, including Clayton, many from the fields of epidemiology, pediatrics, and public health.

Fourteen reviewers also read and commented on a draft of the document but apparently had no say in the final version. Of these fourteen, one was the well-known Stanley Plotkin, discussed in Chapter 6; the other was a member of the Bill and Melinda Gates Foundation. What role the latter might have served is not clear, but the presence of this person simply reinforces the growing evidence that anything at all vaccine-related has the fingerprints of the Gates Foundation on it somewhere. (Bill Gates and his foundation are topics for Chapter 12.)

In the Preface, the committee stated that:

> Following in this tradition [of IOC reports], the task of this committee was to assess dispassionately the scientific evidence about whether eight different vaccines cause adverse events [AE], a total of 158 vaccine-AE pairs, the largest study undertaken to date, and the first comprehensive review since 1994. The committee had a herculean task, requiring long and thoughtful discussions of our approach to analyzing the studies culled from more than 12,000 peer-reviewed articles in order to reach our conclusions, which are spelled out in the chapters that follow. In the process, we learned some lessons that may be of value for future efforts to evaluate vaccine safety. *One is that some issues*

simply cannot be resolved with currently available epidemiologic data, excellent as some of the collections and studies are.[44] [Italics for emphasis and brackets for addition are mine.]

An initial caveat to the overall mandate being pursued is also provided later in the Preface:

> . . . the limitations of the currently available peer-reviewed data meant that, more often not, we did not have sufficient scientific information to conclude whether a particular vaccine caused a specific rare adverse event. Where the data were inadequate to reach a scientifically defensible conclusion about causation, the committee specifically chose not to say which way the evidence 'leaned,' reasoning that such indications would violate our analytic framework. Some readers doubtless will be disappointed by this level of rigor. The committee particularly counsels readers not to interpret a conclusion of inadequate data to accept or reject causation as evidence either that causation is either present or absent. *Inadequate data to accept or reject causation means just that- inadequate.* It is also important to recognize what our task was not. *We were not charged with assessing the benefits of vaccines, with weighing benefits and costs, or with deciding how, when, and to whom vaccines should be administered.* The committee was not charged with making vaccine policy.[45] [Italics are mine.]

In regard to these italicized sections, the committee is correct that data judged inadequate to reach a conclusion about causality should not be misinterpreted by either side of the vaccine wars: those on the pro-vaccines side should not interpret absent data or insufficiency to make a decision as supportive of safety; those on the other side should not interpret insufficiency to indicate vaccine harms.

The second point is equally pertinent: this was not a committee whose mandate included an overall evaluation of all the foundations of vaccine theory and practice; it was vastly more limited in scope. An analogy can be made to a group of high-ranking members of the Catholic Church gathered together to discuss some fundamental tenant of Christian faith: they would be most unlikely to include Jewish or Muslim scholars in their deliberations. In vaccine studies, the faith remains the purview of faithful, as will be detailed in Chapter 8.

The committee essentially demonstrated this point quite well in the document by starting with a restatement of the catechism:

> Vaccines are widely recognized as one of the greatest public health successes of the last century, significantly reducing morbidity and mortality from a

variety of bacteria and viruses. Diseases that were once the cause of many outbreaks, common causes of loss of health and life, are now rarely seen, because they have been prevented by vaccines. However, vaccines can in rare cases themselves cause illness.[46]

From that initial doctrinal affirmation, the committee then actually undertook their analysis of the peer-reviewed scientific literature with some rigor. They first defined their inclusion and exclusion criteria and then launched into an analysis of eight vaccines. For each, the committee evaluated the existing data based on three streams:

> The first assessment applies to the weight of evidence from the epidemiologic literature; the second applies to the weight of evidence from the biological and clinical (mechanistic) literature. The third assessment is the committee's conclusion about causality.[47]

In weighing the evidence, the committee assessed each of the thousands of articles viewed for strengths and weaknesses. These outcomes were evaluated in context to the above "streams" to compute a "weight of evidence" for all the papers in order to summarize both the quality and quantity of the data for both the epidemiology and mechanistic streams. In turn, these evaluations led them to assign levels of probability as *high, medium, low,* or *insufficient to judge* for either accepting or rejecting any link of the vaccines to any of the conditions considered. Some of the terms used were "convincingly supports, favors acceptance, favors rejection, or inadequate to accept or reject."

In deciding how to assign the last category of "inadequate to accept or reject," the committee noted that this could arise in one of two ways. First, when the epidemiological evidence was limited or insufficient and the mechanistic category was weak or lacking in evidence. The other was when the epidemiology was of moderate certainty, but the mechanistic category was intermediate.

These vaccines and the outcome of evaluations over a range of disorders were:

MMR (measles, mumps, and rubella) targeted against the following viruses as noted by the document: Measles (RNA virus of the genus *Morbillivirus* and the family Paramyxoviridae); mumps (RNA virus of the genus *Rubulavirus*); rubella (positive-sense RNA togavirus of the genus *Rubivirus*).

Conclusions: after looking at a potential list of twenty-nine VEs (vaccine-related events), including autism, febrile seizures, Guillain-Barré, etc., the committee concluded that the evidence was high to reject any relationship between the MMR and autism or Type I diabetes, specifically the former: "Conclusion 4.8: The evidence favors rejection of a causal relationship between MMR vaccine and autism."[48]

In contrast, the committee assigned a high level of certainty to accept a link to febrile seizures, and moderate for things such as anaphylaxis. Others of the categories were listed as being "insufficient" or "limited."

Varicella (chicken pox) against the human alpha herpesvirus varicella zoster virus (VZV).

Conclusions: Of sixteen conditions examined, only that of anaphylaxis was rated as convincing. The others showed limited or insufficient data.

Influenza: As described arising from the enveloped viruses of the family Orthomyxoviridae of types A, B, and C.

Conclusions: Of twenty-seven conditions evaluated, the data convincingly supported an association only between some of the influenza vaccines with anaphylaxis and oculorespiratory syndrome, the latter defined generally as having some or all of the following symptoms: red eyes, acute respiratory symptoms of varying severity, and facial swelling. The committee rejected any link to asthma.

Hepatitis A: A spherical RNA virus that primarily infects the liver.

Conclusions: Of eight conditions considered, the committee decided the evidence was insufficient to accept or reject any linkage to the vaccine.

Hepatitis B: Also a spherical RNA virus replicating in the liver.

Conclusions: Of twenty-six conditions surveyed, all but one were deemed as of insufficient evidence to support a linkage. The sole exception was to anaphylaxis with strong support.

Human Papilloma Virus (HPV): The report describes HPV viruses as ". . . a family of more than 100 non-enveloped, double-stranded DNA viruses uniquely targeted to the human epithelial cells."

Conclusions: Of thirteen conditions evaluated, only the link to transient arthralgia (joint pain) was supported. It is important to keep in mind that this evaluation was conducted only a few years after the HPV vaccines began to be widely distributed, and many articles critical of HPV vaccine safety have emerged since.[49]

Diphtheria Toxoid-, Tetanus Toxoid-, and Acellular Pertussis-Containing Vaccines: From the summary by the reviewers:

> Diphtheria is an acute upper respiratory illness caused by *Corynebacterium diphtheria* . . . tetanus is a disease caused by the gram-positive spore form-ing bacillus *Clostridium tetani* . . . pertussis is an upper respiratory infection caused by *Bordetella pertussis*, a gram-negative, pleomorphic bacillus.[50]

Of the diseases and vaccine directed against them thus far in the evaluation, these are the first that involve bacterial pathogenic agents.

Conclusions: Of twenty-four conditions, the report found evidence for a role of the tetanus toxoid in anaphylaxis, but not for the other formulations. It rejected any link to Type I diabetes. The rest of the conditions had insufficient evidence to accept or reject the suggested links.

Meningococcal disease: Caused by the gram-negative bacteria *Neisseria meningitides*.

Conclusions: Of nine conditions, eight were listed as insufficient, and the only condition supported as linked to the vaccine was anaphylaxis.

Overall Evaluation of Stratton et al.

What can be concluded here? First and foremost, it is important to stress that there can be no empirical basis for affirming a null hypothesis, no matter how many individual articles are cited. In this regard, it is important to remember that one article showing evidence of harm due to any vaccine is sufficient to destroy the null hypothesis. The fact that there were so many instances of data insufficiency supports this last point. This is a logical error that arises in the entire report (and others). The level of detail provided in Stratton et al. is to be applauded and it seems clear that the authors approached their study in a professional manner, but the latter cannot overcome the lapse in logic.

In addition, the semireligious nature of the starting assumptions is a problem and one that we will see permeates much of the reports on vaccine safety.

It was, however, notable that the committee was fully aware of the range of factors and conditions that might make someone susceptible to a vaccine adverse event. A Venn diagram that they provided in the document has been redrawn here to show these various relationships,[51] all of which are highly relative to a full assessment of vaccine safety, particularly the likely cases where vaccines are mandated for the entire population.

A second point to note, one that mirrors my own assessment of most of the references cited by AAP, is that a lot of the literature that addresses some aspect of vaccine safety, perhaps particularly in relation to autism/ASD, is simply not very well done and thus not sufficient to enable firm conclusions to be drawn.

The Stratton report evaluated over twelve thousand peer-reviewed articles, but the ones from which they drew their various conclusions were far fewer. Thus, those who say that there are literally thousands of studies supporting vaccine safety are technically correct. But, as Stratton et al. have clearly shown, such views are only partially correct in that the actual number of good studies that demonstrate the safety of vaccines is far more limited. Even with the acceptable studies that Stratton et al. allowed, there were a number of clearly identified vaccine-related adverse events that render the idea that vaccines are always completely safe false.

In addition, the number of instances where the reviewers concluded that there was insufficient evidence to judge either way was quite striking. This

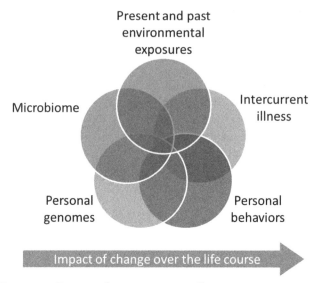

Figure 3.3. Factors influencing vaccine effectiveness.

last point means that far from resolving the issue of vaccines safety generally, Stratton et al. did not do so, nor, as they acknowledge, were they able to do so.

Now that we have examined the AAP's evidence that vaccines are safe, I want to look to see if the CDC citations are any better at reaching the same conclusion.

The CDC's List of Vaccine Safety Studies

The following contains the CDC's shopping list of articles proclaiming vaccine safety across the years from 2000 to 2020.[52] The website, as of summer 2020, listed 584 articles. I presume that prior to 2000, there are many more studies that the CDC could also cite.[53]

From the total list of articles on the CDC website, I selected a smaller subset using the titles and abstracts of all 584 articles totalling 101 (some 17 percent) for further analysis. This smaller group of articles spanned the full range of years in question. The reason for selecting these articles was that they seemed primarily to be concerned with vaccine safety studies versus mainly methodology. To some extent, this was a judgment call on my part in selecting which articles I thought would best represent the very best arguments for vaccine safety. Some may call it cherry-picking, which it may well be, as it always is no matter who is doing the selections.

The 101 articles chosen were all epidemiological studies that employed various epidemiological methods and relied primarily on the VAERS database[54]; a number used CDC's Vaccine Safety Datalink (VSD). These 101 articles are listed in Appendix 4, which can be found here: dispatchesfromthevaccinewars.com.

Most of the articles were peer-reviewed, but I also included several of the CDC's own non-peer-reviewed in-house journal, *Morbidity and Mortality Weekly Report* (*MMWR*) from the early years in the evaluations.[55]

Were there any obvious trends to these 101 articles? Yes, there were. First, there were no experimental studies at all, no in vivo animal experiments, nothing of the sort. Next, the topics by year tend to come in waves and fit a pattern: attempts to "reassure" readers that vaccines are safe, starting with studies of the MMR vaccine, Thimerosal in various vaccines, and later zeroing in on particular vaccines such as those for HPV. As one goes back in time, one senses that in 2000 and for a few years afterward, authors were genuinely concerned about vaccine safety, such as in the two *MMWR* studies (see Appendix 4), and trying to get to the root of potential adverse effects. After this, "reassurance" becomes the dominant theme. The increasing weight of the pharmaceutical industry becomes very clear from 2000 to 2020, with the majority of authors reporting funding from one or more company. The implications of this latter concern will be discussed in Chapter 12, which highlights the book by Dr. Marcia Angell, who has demonstrated the extent of the capture of the media and regulatory bodies by the pharmaceutical industry.

One of the most interesting articles, which is termed a Round Table Discussion,[56] is actually a collection of letters from some commentators responding to an article about vaccine safety surveillance systems in general, and VAERS in particular, the latter by Ward.[57]

In this latter paper, Ward concluded with a statement that:

> As the pace of vaccine development accelerates, it is crucial that we find the political will and financial resources to ensure that surveillance systems and support for basic and epidemiological studies of vaccine-associated adverse effects keep pace.[58]

The comments that follow are illustrative of the clear conflict within the ranks of WHO advisors about how to approach the problem of vaccine adverse reactions.

Dawish, who authored the lead to the commentary in the Round Table Discussion, wrote that:

> I believe the reticence of government about inaugurating surveillance systems has several causes. First, although WHO encourages its member states to develop and sustain a monitoring system, they still do not see the need for such a system or the value of it very clearly. Second, health departments are concerned that systematic searching for adverse events related to vaccination might have a negative impact on their immunization programmes, by giving

the impression to immunization teams that the vaccines are not safe, contrary to what they had learnt during training.[59]

Those concerns mirror the same concerns that John Clemens, then the director of the International Vaccine Institute, in Seoul, South Korea, was to make that same year at the Simpsonwood conference on Thimerosal (see Chapter 5).[60]
Clemens had this to say:

> This means that Phase 3 trials will usually not be large enough to detect such side effects. Moreover, the results of a Phase 3 trial may not pertain to all other populations.[61]

This last point is highly relevant to the COVID-19 vaccines that were rushed to market before this book went to press.
Another commentator, Phyllida Brown, a journalist, wrote that

> An increasingly informed and sophisticated public is the future, whether in high-income or low-income countries. It will no longer be enough for scientists merely to say 'trust me' or for newspapers to present sensational testimonies as the whole story.[62]

And yet, twenty years later, CDC and others in the medical establishment are saying precisely the same "trust me" things. What has changed, however, has been the capture of the media by the industry, as documented in Chapter 10.
The last commentator in the Roundtable was A. J. Ivinson, who wrote,

> Under the circumstances, surely the question is not whether vaccine surveillance is important but whether it is more important than the deployment of existing vaccines and the development of new vaccines.[63]

This last comment perfectly encapsulates the essentially religious nature of vaccinology and its adherents; this topic will be the focus of Chapter 8.
As I will also show in Chapter 10, WHO's shift to a closer relationship with the pharmaceutical industry began to accelerate in 2000 with the election of a new director-general, Dr. Gro Harlem Bundtland.
As for the rest of these 101 articles, here are some that stand out because they seem to really be concerned about safety versus vaccine reassurance. The former, of course, is the very reason the various surveillance systems were set up; the latter seems clearly to be an attempt by the pharma to put the best possible spin on vaccine adverse events.

Going by year, there is Takahashi in 2000 called for better surveillance,[64] mirroring Ward; Chen et al., 2000, providing a surprisingly balanced review of the limited data on adverse events[65]; Lindsey et al. (2008) correctly noted in relation to yellow fever that the any potential harms by the vaccine has to be balanced by the actual threat of the disease for which it was developed.[66]

The article by Loughlin et al. (2012) provided an interesting analysis of the VAERS database and WHO criteria for adverse vaccine reactions in which they put the number of cases that were definitely vaccine related at 3 percent with 40 percent more probable or possible.[67] This result, in concert with the Harvard Pilgrim's study of the percentage of adverse reactions that actually get reported to VAERS (less than 1 percent; see next section), may now allow us to actually calculate the real number of cases of vaccine adverse reactions that likely can be attributed to vaccines.

In 2011, there was Barile et al. essentially confirming the Verstraeten presentation at the Simpsonwood conference (see Chapter 5) about the increased incidence of tics in boys following receipt of vaccines with Thimerosal.[68]

Finally, in 2018, we have Donahue et al., who showed a significant jump in the number of spontaneous abortions following receipt of the influenza vaccine during pregnancy.[69]

Each of the above studies is cited in Appendix 4, which can be found online.

Overall Evaluation of the CDC's Vaccine Safety List of Publications

The reader may note some key points in the above evaluation. First, like the AAP's list, the studies listed do not clearly demonstrate that vaccines with Thimerosal are not involved in ASD. This point will be revisited in Chapter 5 with some of the more critical studies on this compound. In regard to the MMR vaccine, the Stratton et al. MMR conclusion, while not on CDC's list, may be the best comprehensive evidence against a link to autism/ASD although the caveats to "proving" a null hypothesis, as noted before, still apply.

In general, the problems demonstrated earlier for the AAP-cited studies are simply repeated in larger volume in the cited CDC's list. Neither entity's list, apart from Stratton et al., includes any studies that are truly critical of vaccine safety in relation to autism.

Curiously, the lists of the top articles cited by CDC don't match entirely those of the AAP, in that while one may cite a particular article, another may ignore it. For example, AAP cites Glanz et al. (2018) and Taylor et al. (2014) near the top of their list, but CDC doesn't mention either. At a guess, there will be more such examples.

This discrepancy in citations is odd, since one would have assumed that the CDC had a clear interest in providing all such references as the strongest arguments for vaccine safety. This situation is much like getting reviews back from a journal editor in which one reviewer says something about your paper and the second reviewer says the opposite. In such cases, the authors of the manuscript, and the editor, usually consider that opposite reviews cancel each other out. In the present case of trying to evaluate the validity of the various studies cited by both organizations, if one ignores a study due to insufficiency of data or analysis that the other cites, does this cancel out?

Overall Evaluation of AAP and CDC Vaccine Safety Studies

The studies cited by both entities appear to be the basis for the mainstream medical (and media) view that there is no relationship whatsoever between the two conventional bogeymen of vaccine safety concerns, Thimerosal and the MMR vaccine, to autism or other neurodevelopmental delays. While it is certainly true, as claimed, that there are hundreds of studies that support this view, the devil, as in so many things, is indeed in the details.

If the studies have not been done well, or honestly, can one really reach such a conclusion? Maybe we can reach such a conclusion about MMR, but this is one vaccine alone out of a number that young children receive and, further, are mostly mandated to receive.

I am inclined to think that the jury is still out on many of the considerations about vaccines and neurological disorders, including ASD, although obviously many, maybe most, will disagree with my assessment. To disagree effectively, however, would require someone with the opposing perspective to undertake the same sort of analysis as provided above. And if this is not done, then what any reasonable lay person or even scientist would have to conclude is that the evidence for vaccine safety is more about flag waving with two particular flags, Thimerosal and MMR, than a serious evaluation of the scientific literature.

It may well be that neither Thimerosal nor MMR has a role in ASD. Indeed, my opinion is that on their own they rarely do, which is a view that will be explored in greater detail when I consider the literature on Thimerosal in Chapter 5.

One last note: my short list of studies on the CDC list that actually seem to be trying to get to the root of any vaccine adverse events is small. However, these articles do exist and demonstrate two things: First, in all the verbiage from the CDC, some interesting and informative information comes out. Second, you may have to search for it, but it is there.

Taylor et al.'s Curious *Epilogue*

Let me conclude this discussion by quoting from what I found to be an odd addendum to the article by Taylor et al. (2014). In a final paragraph titled *Epilogue*, Taylor himself had this remarkably frank comment, which I quote verbatim, as I find it very revealing of the internal conflicts that even those on the pro-vaccine side may have:

> As an epidemiologist I believe the data that is presented in this meta-analysis. However, as the parent of three children I have some understanding of the fears associated with reactions and effects of vaccines. My first two children have had febrile seizures after routine vaccinations, *one of them a serious event.* These events did not stop me from vaccinating my third child, however, I did take some proactive measures to reduce the risk of similar adverse effects. I vaccinated my child in the morning so that we were aware of any early adverse reaction during the day and I also gave my child a dose of paracetamol 30 min before the vaccination was given to reduce any fever that might develop after the injection. *As a parent, I know my children better than anyone* and I equate their seizures to the effects of the vaccination by increasing their body temperature. For parents who do notice a significant change in their child's cognitive function and behaviour after a vaccination I encourage you to report these events immediately to your family physician and to the "Vaccine Adverse Event Reporting System."[70] [Italics are mine.]

What Taylor expressed is indeed the same dilemma most parents face when trying to make the best decision possible for their child(ren). For this reason, I applaud Taylor for stating the nature of his conflict so clearly at the end of his article (although had I been present when he did vaccinate his children, I'd have been inclined to call the equivalent of Child Protection Services). There is little doubt that most parents want to trust their medical care providers, yet some have reason to fear adverse effects of any given vaccine, in many cases based on prior outcomes in another of their children. What are they to do? Taylor believes that he as a parent knows his children better than anyone and exercises the right to decide to vaccinate or not, and to take the steps he feels have to be made to ensure that the vaccine cannot harm his child. My only commentary, apart from the critique of the article, is that the path this parent has chosen in the exercise of his right to choose is precisely the nature of the demand of parents who choose not to vaccinate at all or to curtail some vaccines. Choice remains choice, for everyone, or no one.

One study that is like Taylor et al. above in noting concerns yet still supporting the vaccination agenda is that of Yamamoto-Hanada et al. (2020).[71] These authors raised concerns about the amount of vaccine adjuvant exposure in infancy with the appearance of allergies later in life. This study was questionnaire-based with all the caveats that go with a study using such methods. The authors here used a multicenter analysis based on the Japanese Environment and Children's Study birth cohort of 103,099 pregnant women and their children.

The resulting analysis was based on 56,277 children. Physician-diagnosed asthma was significantly associated with receiving three to five adjuvanted vaccines compared to only one such vaccine by age twelve months. Similar statistically significant results were seen for wheezing and eczema. These were extremely clear results, yet the authors felt they had to make the following statement of faith:

> Despite this association, we strongly support vaccination strategy and *do not recommend that immunisation* be halted. The results of this research support the recommendation of better vaccination development in the future. Nonetheless, *we highly encourage immunisation and support its current global strategy*. The present results should be cautiously interpreted, but they suggest that the development of better vaccines and/or adjuvants could be beneficial.[72] [Italics are mine.]

Like Taylor et al., the clear recognition of a major potential problem with vaccines is overridden by the almost religious-cult like belief in vaccine theory and practice. This latter topic will be addressed in detail in Chapter 8.

Another Critique from an Independent Scientist on Vaccine Safety Studies

The previous section highlights my analysis of the vaccine safety data, or at least the portion of them that I have examined in detail, namely, some of the articles cited by the AAP, as well as the trove of articles listed by the CDC over the last twenty years.

To see just how reasonable my efforts have been, let's consider a similar, albeit smaller, evaluation provided by Dr. James Lyons-Weiler, a PhD scientist specializing in biomedical research design whose curriculum vitae, as of this writing, listed fifty-seven articles published in the peer-reviewed literature. Lyons-Weiler heads the Institute for Pure and Applied Knowledge, IPAK, an entity he created as an independent research think tank.

In much of his recent work, Lyons-Weiler has expressed concerns about the validity of many of the articles, such as those discussed above, in their ability to actually determine the level of vaccine safety as claimed by the CDC and other agencies. He has also been critical of the CDC and others that have cited studies that seem at first glance to dismiss any link between vaccines and ASD. Additionally, his reevaluation of aluminum toxicity in vaccines has suggested that much of what CDC and others rely on to demonstrate that aluminum is safe is simply incorrect. Needless to say, his studies have led him, like so many others, to be branded as an "anti-vaxxer" without any consideration of whether what he has written or said is correct.[73]

Lyons-Weiler's analysis of the first two items is presented below. His evaluation of aluminum toxicity in relation to the papers by the CDC's standard go-to articles, those by Keith et al. and Mitkus et al., extends my own concern discussed elsewhere in the book and is presented in that section on aluminum toxicity in Chapter 5.

In a paper titled "Systematic review of historical epidemiologic studies influencing public health policies on vaccination,"[74] Lyons-Weiler takes aim at the key studies cited by the various public health agencies and professional organizations, AAP and CDC, respectively, to determine their level of scientific rigor. In order to do so, Lyons-Weiler has created what he terms an "objective evaluation score" (OES) of eleven subscores to evaluate forty-eight such studies, most concerning the question of vaccines and autism or the safety of Thimerosal in vaccines. A positive score suggests that a study was adequately designed and evaluated. A negative score suggests the opposite.

Of the forty-eight studies evaluated, Lyons-Weiler notes that only two studies had an OES score that was positive; the rest were negative. Compared to studies not on the cited forty-eight list, Lyons-Weiler states that these latter showed an OES of 4.01, with many showing evidence for various forms of bias, including selection and healthy user bias, as discussed in Chapter 2.

In brief, the conclusion of the Lyons-Weiler article is that thirty-two of the studies only looked at one vaccine in making their statement about vaccine safety, twenty compared Thimerosal vs. non-Thimerosal vaccines, twenty-one lacked adequate statistical power, thirty-three had a flawed study design (of these, twenty-eight had flaws in analysis), and thirty-four contained "unwarranted" conclusions for negative results. Further, forty-seven of the studies were retrospective studies rather than blinded, prospective RCTs. All the studies evaluated were correlational. And, as cited above, my own evaluation shows that none of the studies cited actually had real placebo controls.

Some of the studies analyzing the potential impact of Thimerosal suggested a beneficial effect of the compound, a result that is biologically implausible given the large literature on mercury toxicity, even for ethyl mercury, which vaccine proponents consider to be safe compared to the well-established toxicity of methyl mercury.[75]

Taken together, none of these forty-eight studies, or some others that I have considered in addition, were actually able to answer questions about overall vaccine or vaccine ingredient safety.

Lyons-Weiler also noted, as I have, that the AAP and CDC have studiously avoided citing articles that find evidence for vaccine-induced ASD or other neurodevelopmental disorders in what is clearly a case of extreme "cherry-picking," presumably done to support an agenda favored by both entities.

In a personal comment to me describing this collection of studies that are designed to reassure the public about vaccine safety, Lyons-Weiler used an army slang phrase to describe what he considered to be an utter mess.[76]

It would probably be fair to categorize this collection of studies, in less harsh terms, as a papering over of the very real cracks in vaccine safety studies by masses of journal articles that cost thousands of trees their lives. Clearly, there are going to be many in the medical community who find this weight of studies acceptable, as if weight alone carried credibility, but it is indeed questionable how many of the list—the forty-eight cited by Lyons-Weiler, or those from the AAP website, or the 584 papers on the 2020 CDC website—have been read by those who adopt this view.

Lyons-Weiler has now provided a recent review of some of the studies used by the IOM in concluding that neither the MMR vaccine nor those containing Thimerosal have a role in autism.[77] In his analysis, Lyons-Weiler focuses on the five studies that the IOM accepted, out of twenty-two, as being able to answer the question of any association.[78]

Looking at these five studies, which I have considered previously, Lyons-Weiler shows that they are all underpowered to be able to detect a positive association, thus rendering the IOM's 2004 report as useless. Here is what Lyons-Weiler concluded, much of which mirrors my own previous comments:

> The association study paradigm does not provide a sufficiently well-defined test of causality; therefore, *negative results cannot demonstrate lack of causality.* Resources such as VAERS include warnings that their contents are not systematically collected, and therefore they cannot be used to infer causality. Because post-marketing surveillance systems do not represent a critical test of causality, negative results cannot be relied upon to conclude 'no effect.' *The association study paradigm has further failed on the question of a role of vaccines in contributing to autism risk in part due to the unseen effects of insufficient statistical power.* The study authors, the reviewing journal editorial boards, and the IOM should have insisted on a priori power analysis for each of these studies. Alternatively, they should have conducted the power analyses themselves and helped society understand the limits of knowledge from small epidemiological studies. *Exposure bias, previous adverse event bias, healthy user bias and cohort effects and fraud may all have played a role in the near-ubiquitous negative OR results in retrospective vaccine studies on autism.* Critically, three of the four studies (Mrozek-Budzyn et al., 2010; Smeeth et al., 2004; Madsen et al., 2002) not rejected by the IOM Committee in 2012 are now shown to have been underpowered, and could not have detected a realistic positive association. Had power analyses been required of these studies prior to publication, it is highly unlikely that the IOM Committee would have found them useful

for their assessment. Taylor et al. (1999), the remaining SCCS study (which as noted may have used erroneous MMR prevalence data), would seem to have overdrawn its conclusions, and it further demonstrated, given its conclusions, that at the time the study was done, it was extremely difficult to identify the age of first onset of symptoms. This is consistent with Brian Deer's reports on parent's uncertainty of timing of first diagnoses, which fueled the fear of suspicion of Dr. Wakefield. Rather than confirm Deer's conclusions, however, if Dr. Farrington and colleagues had a difficult time discerning age at first diagnosis, their conclusions would seem to further exonerate Dr. Wakefield from wrong-doing.[79] [Italics are mine.]

For all of these reasons, mine and Lyons-Weiler's, accepting the "official" version of the "science is settled" mantra is hardly adequate.

Other Critiques of Mainstream Studies

Lyons-Weiler is hardly alone in his critiques of some of what passes for mainstream science on vaccines. Hooker et al. (2014) critiqued the conclusions of five CDC-sponsored studies on Thimerosal and one other that was not associated with the CDC. The associated ones, some discussed earlier or in Chapter 5, were Madsen et al. (2003), Stehr-Green et al. (2003), Andrews et al. (2004), Verstraeten et al. (2003), and Price et al. (2010). The relatively independent study was by Hviid et al. (2003), all cited in detail by Hooker et al.[80] In the article, Hooker et al. discuss the gaps in methodology in each of these studies to critique the CDC's own internal data on a possible relationship between Thimerosal and autism. Similar critiques will be fully amplified in Chapter 5.

In regard to a possible MMR-autism connection, Goldman and Yazbak (2004) mirror many of my own concerns about the Madsen et al. (2002) study.[81]

All of the above should serve to make clear that it is not correct that there are no countervailing voices about vaccine safety, but rather that such voices are routinely dismissed by the CDC and various bloggers alike.

Studies of Vaccine Effectiveness

To reiterate the definitions of vaccine efficacy versus effectiveness from the previous chapter: the first refers to how effective any vaccine is under ideal circumstances with 100 percent uptake; the latter refers to real world conditions and less than complete uptake. For this reason, in the next discussion we will be considering vaccine effectiveness.

The "vaccines are safe and effective" mantra is repeated at every opportunity when the media or those in the medical profession seek to balance what they consider to be anti-vaccine statements about a particular vaccine, or even all vaccines.

Like a lot of other stock phrases, for example, "the science is settled," the ubiquity of these has to raise the question about who comes up with such phrases and promotes them. It may be that using such phrases is the simplest way to convey the outcomes of various studies in a way that is immediately understandable by the general public. It may also be that such phrases reflect less a grammatical shorthand than industry-derived "talking points," a topic I will address in later chapters.

To begin, let me pose the question: what would an effective vaccine look like? At an individual level, it would suggest that the inoculated individual had generated neutralizing antibodies to the pathogen of a particular disease. These antibodies could then be quantified as a measure of seroconversion with the notion that the more antibodies produced, the less likely the individual would be to develop the disease. Such measurements would be at best surrogate markers of potential disease protection, as discussed in Chapter 2, since an actual immune response is more complex, as also cited previously.

Another way to measure the effectiveness of a given vaccine to provide protection from a disease would be to "challenge" a population of vaccinated and unvaccinated individuals with the actual disease and determine what percentage of each became ill. This latter was done passively in the Moderna and Pfizer trials of their COVID-19 mRNA efficacy studies, which will be described in Chapter 13.

Ideally, the seroconversion levels of the vaccinated group would match, more or less, the absence of the disease in the same group. Similarly, the unvaccinated should have fewer if any antibodies and more disease. Such studies are routinely done in animal studies of vaccine effectiveness, for example, using ferrets to evaluate the effectiveness of any year's influenza vaccine.[82] However, deliberately exposing a human population to a disease pathogen would today be considered highly unethical, although such scruples were not always observed in the past,[83] nor are they always in the current push for a vaccine against COVID-19.[84]

Perhaps the best way to evaluate effectiveness, from both a statistical and moral perspective, would be to simply passively monitor the number of individuals in a population that naturally contract the disease evaluated over a long enough time period and see if that relative number changes after a vaccine against that disease is introduced. Such retrospective studies in human populations, by definition ecological, were discussed in the previous chapter. While not considered to be the strongest form of clinical evaluation, ecological studies are perfectly adequate to assess effectiveness for a population at large. It is interesting to note in this regard that the CDC and other groups who decry ecological studies that find fault with vaccines seem perfectly able to use the same methods when it suits their own purposes.

More selective ecological studies can be designed to look at vaccine effectiveness in different demographic groups within the population, by sex, age or health status, etc.

Given that the "effective" portion of the "safe and effective" slogan hinges on doing precisely this, it is perhaps surprising how few times it has been done. Some published papers have looked at disease levels at some pre-vaccine point versus disease levels in the present day. The CDC is wont to do this, as are various other entities, but trying to see any trend in a two-point graph can be highly misleading for the simple reason that there may be other factors at play when a disease declines in a population.

The CDC acknowledges this indirectly, citing improved sanitation, living conditions and potable water, along with vaccines, as contributing factors.[85]

Other researchers have examined a number of sequential years, both pre- and post-vaccine, in an attempt to demonstrate effectiveness, but problems have arisen. One of these was that some authors had chosen a restricted number of years before the introduction of the vaccine and then showed the years following the vaccine. The difficulty with this approach is that the pre-vaccine period may be too short to show an overall temporal disease trend. For example, if the few years before the vaccine are relatively similar and if the vaccine is effective, it may appear that the vaccine had more dramatic impacts than if evaluated over a longer pre-vaccine period in which there may be a clear downward trend in disease cases regardless of whether a vaccine existed or not.

Those in the camp questioning vaccines tend to show extensive temporal scales in which there is a distinctive downward trend in disease expression and, from this alone, conclude that vaccines didn't do anything beyond what was happening already.

The problem with each approach is that both can be misleading and/or inaccurate, since the full temporal nature of disease incidence cannot be quantified by just "eyeballing" the resulting graphs, particularly if there are multiple effects in play that may cause the curves to trend downward.

None of the above is a trivial problem, indeed, much of what constitutes the vaccine wars hinges on a better understanding, along with safety aspects, of just how effective any vaccine actually is.

In regard to the actual literature, unlike the safety studies cited earlier, there are few studies that have actually looked at vaccine effectiveness in detail, particularly over an extended time period. The paucity of effectiveness studies is particularly striking in light of a widely held mainstream view that since vaccines for various infectious diseases have come on line, millions of lives have been saved. And while it would seem intuitive that this should be correct, saying it and actually demonstrating it are quite different things.

An older article by van Panhuis et al.[86] attempted to deal with the assumptions about vaccine effectiveness that seem to pervade so much of the official discourse. These authors examined smallpox, measles, polio, rubella, mumps, hepatitis A, diphtheria, and pertussis across the twentieth century by creating a database called Project Tycho.[87] However, in spite of the nice color graphics in this article, the authors do not attempt to correct for any other factors, expecting the reader to pretty much take their word for the general decline being due to vaccines for which they have data both pre- and post-vaccine. Nor do these authors apply any statistical analysis to the decline rates before and after any of the vaccines were introduced. In spite of this, the nice color graphics by van Panhuis et al. show quite clearly that infectious disease levels in the United States were generally falling during the twentieth century, observations that mirror what vaccine skeptics have been saying for years.

In order to address what my colleagues and I considered a serious shortfall, we revisited the issue of vaccine effectiveness using two data sets. The first was from Project Tycho. The second data set was the Historical Records of the United States, the latter also showing disease incidence by year for the years 1933 to 1998.[88]

Instead of just eyeballing the resulting curves, we linearized the data to make them straight lines from which we could derive slopes and then compared the slope of the line pre- and post-vaccine introduction for six diseases to which our methods could be applied for both data sets:

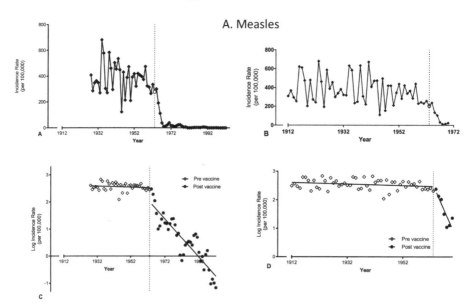

A. Measles

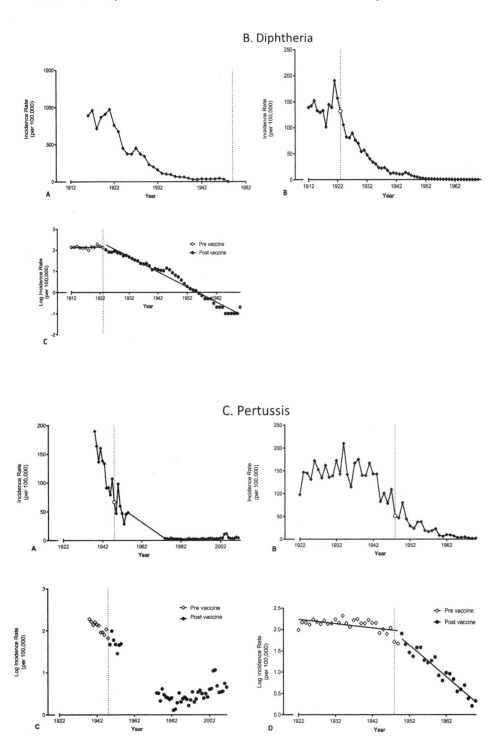

B. Diphtheria

C. Pertussis

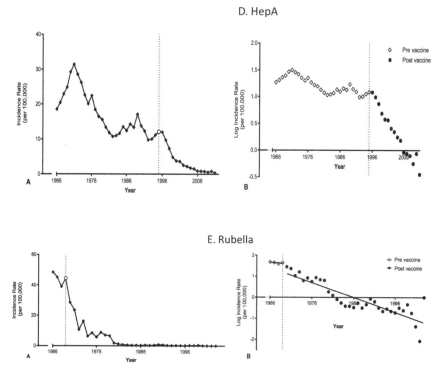

Figure 3.4. Effectiveness plots for various infectious diseases as labeled over the course of the twentieth century. Note data have been linearized by semilog plots in order to determine the slope of the change in disease rates following the delivery of the various vaccines. Two sources of these data are Project Tycho or the Historical Record of the United States.[89] These data have not been published previously (Morrice, J., and Shaw, C.A.).

Unlike van Panhuis et al., we decided not to evaluate smallpox since there were no pre-vaccine data, rather just estimates. Similarly, we did not attempt to evaluate polio, since the temporal incidence curves were not easy to fit into a simple model.

What we found was pretty straightforward: for all six infectious diseases we were able to study, all showed a temporal decline in incidence over the measured time period. As shown in Figure 3.4, five of six showed a significant additional decline once the vaccine for that disease was introduced. Amusingly, at least for me, the only vaccine that did not do so was the vaccine against rubella, a vaccine invented by the modern godfather of vaccinology, Dr. Stanley Plotkin (see Chapter 6).

We were reasonably pleased with our analysis, although at the time we failed to convince several journals that we had added a valuable interpretation to the question of effectiveness. Editors for both opined that we had added nothing new, since it was "obvious" that vaccines were effective. We tried to counter that it was *not* obvious until someone, such as us, had done the detailed statistical analysis. Such arguments were to no avail, and we dropped trying to get this study published. Apart from using some of the figures in some talks, the figures shown here are the first time these data have been presented in print form.[90]

In summary, we had answered our own question about vaccine effectiveness: vaccines can be effective and do contribute at least to part of the diminution of some infectious diseases over time. But this relative effectiveness is not without possible consequences, as discussed in the section on safety earlier.

Claims, Controversies, and the Danger of "Cherry-Picking"
From Rabbi Zev Epstein:

> Rabbi Yankel Cohen was in his study adjudicating a dispute between Sam and Morris, two of his congregants.
>
> Sam claimed that Morris owed him one hundred dollars. Morris hotly denied the charge.
>
> "Rabbi, it's like this . . ." Sam launches into a detailed account of a monetary dealing with Morris, concluding, ". . . so, you see, Rabbi, Morris owes me $100!"
>
> The Rabbi strokes his beard and looks gravely at Sam. "Sam," says the Rabbi, "You're right!"
>
> "But Rabbi," protests Morris. "You haven't heard my side of the story!" Morris gives his version of the events, concluding, "So you see, Rabbi, I don't owe Sam anything!"
>
> The Rabbi once again strokes his beard, and sits, deep in thought. "Morris," says the Rabbi, "You're right!"
>
> At this, the Rabbi's wife—who is in the room serving tea to the men—clangs down her tray and turns on her husband.
>
> "Yankel!" she blurts out. "How can Sam be right, and Morris be right at the same time!?"
>
> The Rabbi grins at his wife. "My dear," says he, "you're also right!"[91]

I use this version of a joke Rabbi Epstein told to legislators in New Jersey in the hearings leading to a vote on the removal of religious exemptions for vaccination in that state. I use it because I think it captures the essential nature of the vaccine "wars," figuring out which side is right and which side is wrong. One side, the other side, both, or neither? In other words, is it a zero-sum game, a term cited before? Or, as in the joke, are there elements of right and wrong on both sides?

I will try in these following pages to list what I think the evidence, the actual evidence, not the assumed evidence, actually shows. And in this regard, the fact that I know some of the participants and don't know others personally, or that some seem "nicer" people than others, will not figure into my analysis.

In regard to this, the argument can and surely will be made that I have "cherry-picked" the pieces of the science that I like and disregarded those that I don't. It's a fair critique and one with some merit. However, I would add to that charge that we, scientists and lay persons alike, all do this to a greater or lesser extent in regard to vaccines, or indeed to any subject whether it be in science or not. It plays out in politics, particularly in the United States at the moment as perhaps the most dominant example, but it also plays out in any scientific discipline.

As the curator of the Vancouver Museum of Anthropology once explained to me, it's pretty easy to tell any story you want to tell. As she explained, there were at least a million items in storage in the basement of the museum, but only a fraction of that million could be displayed at any one time. Hence, she, like other curators, had to decide which items to put on display. In turn, the choice was dictated to a great measure by first deciding which story they want to tell to visitors, that is, what the dominant narrative was supposed to be.

In this example, the curator said that if she wanted to frame the narrative that European powers, their settlers, and the descendants of these settlers had abused the Aboriginal people in British Columbia, it was easy to do so with the choice of artifacts available. If, instead, she wanted to frame the narrative as cooperation between the groups, she could do this, as well.

Similarly, when writing a book, this one for example, or my previous book on neurological diseases in the context of immune function and dysfunction,[92] when faced with competing views on the data or interpretation of various studies, one has to decide which ones are correct, or at least more correct. Often this is quite hard, made harder still by the realization that a lot of what is published in the peer-reviewed literature is simply incorrect.[93] Or worse, you may never really know which articles are or aren't correct. Also, many of the things that are thought to be correct haven't been validated further by replication for a host of reasons. Additionally, there is the problem that human emotions can come into play. For example, you think that some researcher is an arrogant snot and hence anything that researcher says or writes is not to be trusted compared to the work of the really nice person whom you met on a grant review panel.

In regard to one's own data, virtually all research scientists have confronted the problem of which figures to use to make a point in an article. Let's say you have ten photographs of some cellular process: one shows the outcome that sup-ports your hypothesis in crisp, lovely colors, and the other nine show the same thing, sort of, but not exactly. Maybe there are ugly spots in some photographs, or maybe the colors aren't as nice. Which photograph are you going to use, the nicer one that might warrant the figure being on the cover of a journal, or one of the dingier pictures that might actually resemble nature more accurately?

I think we all know the answer to this.

There are ways around these problems of selection, of course. In the first case of which sources to trust, stating that "the science is settled" is not one of them. In fact, one couldn't hope for a clearer example of selection bias and a pre-determined narrative at work. As for the second case, the guideline is honesty.

Both of these concerns spill over into the question of whether one should trust the cumulative weight of publications for vaccine safety on the official medical side versus the fewer articles that are critical. We will consider the latter in the next chapter, but if the evaluations of the mainstream output done are any guide, then the curation of vaccine safety studies, if it is going to be honest, has to look at the quality of the individual studies, not just their respective numbers.

The Calls for a "Vaxxed" versus "Unvaxxed" Study

In the vaccine wars, it is common to hear the vaccine-hesitant side say that a study that would resolve the issue of potential vaccine harms would be one in which the groups evaluated would be: 1. Fully vaccinated in some age range according to the CDC's recommended vaccination schedule (see Appendix 4); 2. Those partially vaccinated in the same age range; 3. Those in the same age range with no vaccines in their history.

Of course, in addition to age, a number of other factors would have to be considered as potential confounders: sex, family income, parental age, diet, other factors in the environment such as pollution or toxins, and a range of others.

To be sure, it might be difficult to find large enough numbers of fully unvaccinated persons to compare individual or multiple vaccines against given that otherwise the statistical power might be too low to draw firm conclusions. And it would be a challenge to adjust for all of the other possible factors that might go with decisions to vaccinate, or not, as cited above.

Could such a study be done?

The opinion sometimes expressed by the mainstream medical community that such a study would be "unethical" is dealt with in more detail elsewhere in this book. However, in brief, this view is based on a series of assumptions that are simply invalid. The first assumption, as in some drug trials, is that those in the true placebo arm of a trial are being denied the life-saving benefits of the drug/vaccine. This might be true if we were to conduct a RCT with volunteers in a cancer or neurological disease study and failed to allow those in the placebo group to cross over to the treatment group if the latter showed emerging benefits from the treatment. Such, however, is not the case with calls for a vaccinated versus unvaccinated study that would use people who have already made a choice in one or other direction before the study commenced. Yes, such a study would almost certainly have to be a retrospective analysis and might be considered to be inferior to a true RCT, but so is the Madsen et al. study that the CDC likes to highlight vis-à-vis the lack of evidence for a role of the MMR vaccine in ASD.

Such a study might well answer the questions that have arisen about vaccine safety, although whether it would satisfy either the pro or anti solitude is questionable. An outcome that implicated vaccines, or at least some of them, in either acute or chronic diseases in children would face enormous pushback from the mainstream medical community; from the pharmaceutical industry, which makes vaccines; from the mainstream media, which depends on pharma advertising dollars; and from a range of others who would see such an outcome as an attack on their various interests.

At the same time, those who are skeptical of vaccines or even truly "anti-vax" would not likely accept a result that said that vaccines don't produce harm to most of those who receive them. For these reasons, the choice of who would do such a study and how would be crucial, and both sides would have to agree, in advance, that if they had agreed to and trusted the process, they would accept the outcome, regardless of any prior position. Is this likely?

Is it even likely that, as in the joke by Rabbi Epstein, both pro and skeptical of vaccines sides might come out of such an exercise with a more balanced view on the whole subject? Something along the lines of: "vaccines, in some amounts, can be beneficial in some circumstances for some people, but not for others depending on a range biological factors"?

My guess is that the groups would not accept anything but a zero-sum outcome in that the vaccine wars that are really just beginning are not ready to see either side concede defeat or even compromise. I will go into this in more detail later in the book.

In actuality, the questions posed above may now be moot, as the proposed studies have been performed with three articles in the peer-reviewed literature by independent researchers, and one more pending acceptance. The first, as cited in Chapter 2, was by Mawson[94] and used a retrospective questionnaire to compare a range of health outcomes in vaccinated and unvaccinated children. Such a study can easily be subject to recall errors. Further, the numbers of children were small, and this alone is problematic with an N of 405 vaccinated children compared to 261 unvaccinated. However, in its favor, the Mawson study did attempt to adjust for things like family income, since the latter can determine general health outcomes. In spite of the above concerns, the Mawson study outcome was pretty clear: while as expected, unvaccinated children were more likely to have had chicken pox and pertussis, they were far less likely to have middle ear infections (otitis media), allergies, or neurodevelopmental delays. While this study has been critiqued on various methodological issues, as noted previously, the fact that it could be done at all was an important first step in refuting the notion that a vaccinated/unvaccinated study could not logistically or morally be done.

Since then, Hooker and Miller (2020) and Lyons-Weiler and Thomas (2020) have also published related studies on health issues in vaccinated versus unvaccinated children.

The first of these, Hooker and Miller,[95] examined data from three medical practices between 2005 and 2015, comparing unvaccinated children (345 male, 258 female) to vaccinated children (718 and 696, respectively) under one year of age and evaluated at age three or older. In classifying health status, Hooker and Miller used the International Classification of Diseases-9 and -10 codes by way of a medical chart review. The outcomes showed significant differences between the groups with increased odds of developmental delays, asthma, and ear infections in the vaccinated group.

A second study by these same authors examining the health status of vaccinated versus unvaccinated children in context to type of birth and breastfeeding is under review.

The study by Lyons-Weiler and Thomas[96] was a retrospective analysis designed to test the hypothesis that vaccinated children compared to unvaccinated children would have a different overall health status. The study encompassed ten years of the pediatric practice by Thomas at his clinic. The ages of the children in the study ranged from two months of age to 10.4 years, in a roughly equal male/female ratio. Of these, 2,763 were fully or partially vaccinated; 561 were unvaccinated. The authors compared a novel variable described as a Relative Incidence of Office Visits (RIOV) with a billable outcome and then stratified the condition identified in each visit for the groups of children to compute odds ratios and p values for the various conditions.

Out of seventeen conditions identified in the RIOV analysis—fever, ear pain, otitis media, conjunctivis and other eye disorders, asthma, allergic sinusitis, breathing issues, anemia, eczema, uticaria (hives), dermatitis, behavioural issues, gastroenteritis, weight and eating disorders, and seizures—all but seizures occurred more frequently in the vaccinated children.

In their discussion of these results, Lyons-Weiler and Thomas go to great lengths to discuss the experimental literature and the possible confounding factors and sources of bias in their analysis.

As predicted, all of the above studies were dismissed by the usual suspects and acclaimed by those on the vaccine-skeptical side. It must be said, however, that the limitations of all three of these studies equally apply to many of the studies cited by the CDC as evidence for vaccine safety.

The cumulative effect of these studies was summarized by Lyons-Weiler compared to some of those cited by the CDC:

> Past retrospective studies on the question of pediatric vaccine safety tended to focus on a single vaccine and did not compare never-vaccinated to vaccinated individuals. They also tended to focus on the singular question of the link between the MMR vaccine specifically, and autism. The methods used—espoused by a white paper not formally associated with but nevertheless published by the US Centers for Disease control— represent excellent examples of how to *not conduct objective epidemiological studies* . . . The three vaccinated vs. unvaccinated studies conducted by truly independent researchers in the US—Mawson et al. (2017), Hooker and Miller (2020), and Lyons-Weiler and Thomas (2020) all point to signals missed by past studies, mostly conducted by those who stand to gain from the public's perception of vaccine safety. Together the three studies point to atopy—a bodily condition of being prone to autoimmunity—in the airways, gastrointestinal tracts, and central nervous systems . . . The Hooker and Miller and Lyons-Weiler and Thomas studies dispel to a large degree the concerns over recall bias and other potential limitations of the Mawson et al. survey study.[97] [Italics are mine.]

Chapter 4 will look in more detail at how such studies may be used in relation to ASD. Before that, however, I want to briefly delve into two items that are directly related to how vaccine safety is actually evaluated after vaccines have been approved. The first will discuss the various vaccine safety surveillance systems set up in the United States and in Canada. The second will address the recourse that those injured by vaccines have, at least in the United States, in the so-called "vaccine court."

Vaccine Licensing and Surveillance Programs in the United States and Canada

Licensing in the United States

In the United States, vaccine licensing is under the control of the FDA, which, according to their website, is responsible for testing and evaluation.[98]

The FDA is the oldest comprehensive consumer protection agency of the American government with its origins in 1848 as an agency to test the safety of agricultural products. The FDA later formally assumed this role from the federal Department of Agriculture in 1862. The FDA's formal regulatory functions began in 1906 with the passage of the Pure Food and Drugs Act to prohibit interstate transport of adulterated and mislabeled food and drugs. The FDA assumed its present designation in 1930.

In regard to vaccine safety, the FDA notes:

> Throughout the process, FDA works closely with the company producing the vaccine to evaluate the vaccine's safety and effectiveness. All safety concerns must be addressed before FDA licenses a vaccine.[99]

What this means in reality is that the FDA, which does no actual bench science apart from testing vaccine batches for purity, relies on the company making the vaccine to do its own safety testing and file a report, which the FDA will review. The FDA is also supposed to inspect the facility where the vaccine is made, but this inspection is more designed to ensure that there are no obvious sources of contamination of the vaccines stocks. A recent report has questioned if this relatively low level of inspection is even plausible for the FDA in relation to the fast rollout of COVID-19 vaccines. However, the FDA's recent halt on vaccine production by Emergent, the company making the AstraZeneca vaccine for COVID-19 in the United States, shows that they can be effective when they choose to be so.[100]

In regard to just how effective such inspections might otherwise be, consider the range of contaminants recently reported by Italian researchers for vaccines licensed in Italy.[101]

These reports should obviously be confirmed by other researchers and even if found to be valid may not apply to other vaccines licensed in the United States. One question that does arise is whether vaccines made in foreign countries and then imported into the United States also have FDA officials visiting the manufacturing plants in these other countries. It would be nice to presume that the FDA indeed does so; if they do, how thorough are their inspections? Their website suggests that they do indeed conduct inspections of foreign facilities.

Nevertheless, it is important to remember that such inspections are not designed to determine the safety of vaccines after administration into humans. In other words, this crucial aspect is left to the companies themselves in their Phase 3-pre and Phase 4-post-marketing surveillance studies, the latter often not completed.

A problem with the FDA in relation to vaccine safety evaluations, and in particular in relation to the use of aluminum adjuvants discussed in the next chapter, was cited in a recent paper by Lyons-Weiler and Ricketson.[102]

The authors note that

> FDA regulations require safety testing of constituent ingredients in drugs (21 CFR 610.15). With the exception of extraneous proteins, no component safety testing is required for vaccines or vaccine schedules.

In regard to this, the authors go on to write that

> FDA regulations require that proteins in vaccines be tested for safety.
> Aluminum is a known neurotoxin and it is unfortunate that additives in vac-
> cines are not required to be subjected to animal safety studies prior to use on
> human subjects.

This is of course all very true as will be explored in greater detail in Chapter 5.

The recent rush to get the COVID-19 vaccines to market has also raised questions about the capability of the FDA to effectively monitor manufacturing facilities,[103] although the Emergent example is encouraging.

One final note about the FDA concerns the Center for Biologics Evaluation and Research (CBER), one of the Agency's subdivisions. CBER regulates biological products for human use under various federal laws, including the Public Health Service Act and the Federal Food, Drug, and Cosmetic Act. CBER:

> . . . protects and advances the public health by ensuring that biological prod-
> ucts are safe and effective and available to those who need them. CBER also
> provides the public with information to promote the safe and appropriate use
> of biological products.[104]

The problem that arises is that CBER considers vaccines by nature to be "biologics," not drugs, thus holding the evaluation of vaccines to a standard different from other drugs.

Adverse Effects of Vaccines

The CDC and FDA do acknowledge that adverse events can occur following vaccine administration. These are often called an Adverse Event Following Immunization (AEFI). Categorized by the WHO,[105] this includes: *serious events* which include death or is life threatening, requiring "in-patient hospitalization or prolongation of existing hospitalization," has a "congenital anomaly/birth defect," "requires intervention to prevent permanent impairment or damage." Next up are *severe events* categorized by intensity as mild, moderate, or severe. They also use the term *mild* to indicate things like a sore arm, redness at the injection site, minor fever, headache, loss of appetite, and all things resolving in a "short" time.

WHO goes into more detail, but the above provides the basics. Of note, Moderna, in their phase vaccine trials (see Chapter 13) used a variant scheme of mild, moderate, and severe; the latter, based on the outcomes in their trials, would seem to be more along the lines of a serious event.

How the regulatory agencies, CDC and FDA, determine if an AEFI has occurred is the subject of the next section.

Vaccine Safety Surveillance: FDA and CDC

As noted above, the FDA is responsible for the licensing of vaccines, but post-licensing monitoring is also a part of vaccine safety surveillance. The best known of these is the Vaccine Adverse Events Reporting System, or VAERS. Even though this and other surveillance entities exist, as described next, various US government websites make sure to convince us over and over again that "vaccines are safe and effective."[106]

VAERS came into existence in 1990, following on as a consequence of the 1986 National Childhood Vaccine Injury Act, discussed in detail in the following section. The 1986 legislation required monitoring of adverse vaccine events and the Department of Health and Human Services, in response, created VAERS. Prior to this, reports of adverse vaccine reactions were reported in two different systems, with one under the control of the FDA; a second system, the Monitoring System for Adverse Events Following Immunization, established in 1978, was administered by the CDC. As a result of the 1990 legislation, the two reporting systems were integrated into VAERS.

Once the FDA has done its job of licensing, the CDC[107] then takes over and decides whether or not to approve the vaccines for use in the United States. As with the FDA, the CDC does not conduct its own safety testing, but also relies on the filings of vaccine manufacturers. After a vaccine is in use, the CDC will monitor for so-called "safety signals," but always with a "risk-benefit" ratio in mind. Here is what CDC says it does if concerns arise:

> If a link is found between a possible side effect and a vaccine, public health officials take appropriate action by first weighing the benefits of the vaccine against its risks to determine if recommendations for using the vaccine should change.[108]

The surveillance systems that the CDC can use to determine if side effects are occurring are varied. The first one, as noted above, is VAERS.

While one might imagine that all adverse vaccine reactions are actually reported to VAERS, this is not, in fact, correct, as use of the system is voluntary.

In principle, anyone can report a vaccine adverse event (even vaccine companies, although one might be skeptical about how often this occurs). Any report received is first screened by VAERS staff to cull out those that appear to be manifestly unlikely as true adverse effects. The reports that are not culled are reviewed and scaled for the severity of the reaction by the "medical professionals" associated with the CDC and FDA "on a daily basis." Some of our own recent work has called into question just how the individual cases are evaluated.

Our results suggested that the medical professionals mentioned are more likely than independent physicians to rate cases as "nonserious."[109]

Because the system is voluntary, it does not capture all vaccine adverse events, and a recent report by a private consulting firm hired by the CDC, Harvard Pilgrim, contained numbers that made 1 percent look normal.[110] The CDC did not use this report.

Apparently, however, the CDC loves VAERS, as their website notes:

> VAERS data provide medical professionals at CDC and FDA with a signal of a potential adverse event. *Experience has shown that VAERS is an excellent tool for detecting potential adverse events.* Reports of adverse events that are unexpected, appear to happen more often than expected, or have unusual patterns are followed up with specific studies.[111] [Italics are mine.]

It will help in much of the following discussion to remember that CDC considers VAERS an "excellent tool," except, as we will see, when it actually reports a serious signal. In those cases, the CDC and others characterize VAERS as ". . . an imprecise system whose outcomes cannot be trusted."[112]

CDC goes on to state that "VAERS data alone usually cannot be used to answer the question, 'Does a certain vaccine cause a certain side effect?'"[113]

This is where the Advisory Committee on Immunization Practices (ACIP) comes in. According to the CDC, ACIP is:

> . . . a group of medical and public health experts, [who] carefully reviews safety and effectiveness data on vaccines as a part of its work to make recommendations for the use of vaccines. The ACIP modifies recommendations, if needed, based on safety monitoring.[114] [Brackets are mine.]

What the CDC does not say is that ACIP is rank with conflicts of interest including extensive ties to vaccine manufacturers and rarely rejects a vaccine for any reason.[115]

There are also other surveillance systems the CDC and FDA use. These include the VSD to:

> . . . do studies that help determine if possible side effects identified using VAERS are actually related to vaccination. VSD is a network of 8 managed care organizations across the United States. The combined population of these organizations is more than 24 million people.[116]

Another system is the Post-Licensure Rapid Immunization Safety Monitoring (PRISM) system used by the FDA. It is described as the "largest vaccine safety surveillance system" used in the United States. The scientists involved use PRISM to "actively monitor and analyze data from a representative subset of the general population. PRISM links data from health plans with data from state and city immunization registries. Because PRISM has access to information for over 190 million people, FDA is able to identify and analyze rare health outcomes that would otherwise be difficult to assess."[117]

Finally, there is the Clinical Immunization Safety Assessment (CISA). The CISA project is "a collaboration between CDC and 7 medical research centers. Vaccine safety experts conduct individual case reviews and clinical research studies about vaccine safety."[118]

All of these are listed by the CDC and FDA as "tools by which the CDC and FDA measure vaccine safety to fulfill their duty as regulatory agencies charged with protecting the public."

What, according to these agencies, are strengths of VAERS in particular? In summary, the strengths are listed as:

- VAERS is national in scope and able to "quickly" provide an early warning of a vaccine safety problem (note that underreporting is listed as a weakness);
- It is designed to "rapidly detect unusual or unexpected patterns of adverse events," i.e., safety signals;
- If such a signal is found, further studies can be directed to the other systems, which the CDC notes do not have the limitations of VAERS.
- What, according to this same source, are the limitations of VAERS?
- It cannot be used to "make definitive vaccine safety determinations" because it is voluntary and thus "subject to underreporting and bias."
- The way VAERS is designed, it "does not allow the definition of a study population" and therefore cannot be used to calculate accurate adverse events rates";
- Best of all:
  VAERS cannot be used for statistically valid analyses of direct links between vaccinations and health outcomes. The US system of medical record keeping also does not allow direct links between vaccination and health outcomes, because medical record data are not maintained electronically in a standard format.

According to the same source document, VSD and the others are better able to overcome the shortfalls of VAERS.

Let me attempt a rough translation from government bureaucratease into English: VAERS is an "excellent tool" to detect vaccine adverse events, except when it is not, which seems to be a lot of the time. The other systems for evaluating vaccine adverse reactions are great, mostly because they are not VAERS. Do they also have flaws? Given what we have seen with VAERS, it would be awfully naive to assume they do not.

Of note, a large number of the papers cited earlier in the 584 studies that show that vaccines are not associated with autism rely on VAERS. Of course, the opposite would likely be true, as well: It would be hard for independent researchers to use VAERS to demonstrate the opposite.

Before I fully comment on the utility of these surveillance systems, let's consider the oversight systems that neighboring Canada has in place.

Canada

Vaccine safety monitoring is under the overall control of the Biologics and Genetic Therapies Directorate of Health Canada (BGTD), one of the ministries in the federal government.

Health Canada is the "nickname" for the Ministry of Health. It was originally created in 1919 as the Department of Health and merged with the Department of Soldiers' Civil Re-establishment to form the Department of Pensions and National Health in 1928. It became the Department of National Health and Welfare in 1944. It became the Department of Health, a.k.a. Health Canada, in 1993 and is now termed the Ministry of Health.

Some authors have described Canada's vaccine safety system as having eight compartments or components.[119]

These are listed as:

1. an evidence-based pre-license review and approval process;
2. strong regulations for manufacturers;
3. independent evidence-based vaccine use recommendations;
4. immunization competency training and standards for health care providers;
5. pharmacovigilance programs to detect and;
6. determine causality of adverse events following immunization (AEFIs);
7. a program for vaccine safety and efficacy signal detection;
8. the Canadian Immunization Research Network's special immunization clinics for children who have experienced serious AEFIs.

Part of Health Canada's vaccine safety surveillance is done by a program called IMPACT (Immunization Monitoring Program ACTive) designed to "detect adverse events related to immunization and to monitor vaccine-preventable diseases."[120] The website for Immunize BC notes that

> Nurse monitors at 12 children's hospitals across Canada . . . review all admissions to the hospital for certain serious illnesses. If a child with any of these illnesses had recently received a vaccine, a report is sent to the local health unit and to the Public Health Agency of Canada to be investigated further. These events are reported even if another cause may ultimately be found for the illness.[121]
>
> The Public Health Agency of Canada is discussed soon and is a branch of Health Canada.
>
> What are the steps in the vaccine approval process in Canada?

As noted above, Health Canada's BGTD regulates biologic drugs (termed "Schedule D") for human use in Canada. These include vaccines. Before manufacturers or sponsors are eligible to market a product in Canada, they must file a "New Drug" submission. This submission contains ostensibly extensive information and data about the vaccine's safety, efficacy, and quality, including the results of the nonclinical and clinical studies, details regarding the production of the vaccine, packaging and labeling details, and information regarding therapeutic claims and side effects. This is *all* based on a company's own submission, not an independent determination by Health Canada. The next level of evaluation of the submission includes an onsite visit to the production facilities of the company, as well as laboratory testing of samples from three to five consecutive lots (or batches of vaccine production) to verify manufacturing consistency. The Health Canada website suggests that their visits to production facilities occur both in Canada and foreign countries and that they themselves do the vaccine batch analyses. These procedures seemed to have failed with the contamination problems discovered by the FDA for Emergent, as cited previously.

After BGTD determines that a vaccine is compliant with Canada's Food and Drugs Act and Regulations, Health Canada will issue a Notice of Compliance and a Drug Identification Number (DIN) for market authorization.

Health Canada requires that vaccine licensees adhere to the Food and Drugs Act and Regulations and further requires that licensees "analyze adverse drug reaction data for safety concerns and prepare an annual summary report which represents a comprehensive assessment of the worldwide safety data of the vaccine" as well as notify the ministry if they become aware of "significant changes in the vaccine's benefit-risk profile."[122]

Health Canada will assess such reports and "may" request more safety information.

In addition, Health Canada reviews what is called a Risk Management Plan (RMP) for new vaccines to be introduced into Canada.

For an approved vaccine to be recommended for routine use in Canada, a formal independent review separate from the licensing review is done by the National Advisory Committee on Immunization (NACI), described as a national advisory "committee of experts in the fields of pediatrics, infectious diseases, immunology, medical microbiology, internal medicine, and public health."

These members are joined by eleven "liaison" members representing entities that have an interest in immunization and by six ex officio members drawn from parts of the federal government. While the latter two groups participate in discussions and can serve on subcommittees, they have no voting privileges. All of the members of these groups are appointed for set terms by the chief public health officer of the country.[123]

The NACI committee reports to the assistant deputy minister of Infectious Disease Prevention and Control and works with PHAC to "provide ongoing and timely medical, scientific, and public health advice in the form of recommendations for the use of vaccines currently or newly approved in Canada."[124]

Overall, with the levels of bureaucracy and oversight, the Canadian system should be bulletproof, and one might believe that the safety of approved vaccines was assured. The fact that such safety cannot always be assured, as witnessed by the Emergent issue, speaks less to the structure of the surveillance system than the assumptions that may underlie it. The nature of such assumptions is detailed in various places in this book, many of which seem to operate from a core belief that vaccines are inherently safe when a detailed examination of the actual published literature would not support such an assumption.

The RMP itself is revealing in several respects. Basically, this is a document summarizing

> known important safety information about a health product; identifies gaps in knowledge; *outlines how known and potential safety concerns will be monitored by the market authorization holder* and provides a proposal to minimize any identified or potential risk.[125] [Italics are mine.]

> So once again, the necessary due diligence is left to the vaccine manufacturer.

> Health Canada's overall obligations include the following:
> • Monitor the safety of their vaccines;
> • Comply with the Food and Drugs Act and Regulations, including Good Laboratory Practice, Good Clinical Practice, and Good Manufacturing Practice;
> • Prepare Annual Summary reports.

> Other than the above, there is the Canadian Adverse Events Following Immunization Surveillance System (CAEFISS) which is described as a

- collaborative post-marketing federal/ provincial/territorial (F/P/T) passive surveillance system with the following objectives:
- to continuously monitor the safety of marketed vaccines in Canada;
- to identify increases in the frequency or severity of previously identified vaccine-related reactions;
- to identify previously unknown AEFI that could possibly be related to a vaccine (unexpected AEFI);
- to identify areas that require further investigation and/or research; and to provide timely information on AEFI reporting profiles for vaccines marketed in Canada that can help inform immunization-related decisions.

As with the US safety monitoring, it is important to cut through the bureaucratic jargon and focus on the fundamentals of how the regulation of vaccines actually occurs: The bottom line in the regulatory process at virtually every step is dependent on the material supplied by the vaccine makers to Health Canada or one of their subagencies.

In respect to the above, Health Canada is part of the Health Portfolio, the latter made up of a number of entities including the PHAC, the Canadian Institutes of Health Research (the equivalent to the US National Institutes of Health), the Patented Medicine Prices Review Board, and the Canadian Food Inspection Agency. Health Canada is overall the responsible government ministry.

It is worthwhile considering the role of PHAC in how vaccines are evaluated for safety.[126]

PHAC publishes an "immunization guide" that defines adverse reactions in terms of frequency, ranging from "very common" at 10 percent of cases to "very rare" at 0.01 percent.

In terms of assigning risk from any vaccines, PHAC uses two measures: The Vaccine Attributable Risk, basically a difference in outcome between those vaccinated versus not for that vaccine, and the Vaccine Safety Signal, which is

> any information that arises from one or multiple sources and suggests a new potentially causal association, or a new aspect of a known adverse reaction (increased severity and/or increased frequency), between immunization and an event or set of related events, and is judged to be of sufficient concern to justify verification and remedial action *if appropriate*. [Italics are mine.]

PHAC's website notes that it partners with a number of vaccine "safety" entities, including WHO, a branch of which is the Council for International Organizations of Medical Sciences (CIOMS)[127]; the Brighton Collaboration[128];

and the US Institute of Medicine, now called the Health and Medicine Division.[129]

In summary, at a superficial glance, both countries look like they have set up rigorous systems with overlapping monitoring and surveillance. A facile explanation is that these systems should work and work well. The question is if they do. The main flaw that I can see is that clinical safety testing is done by the vaccine companies themselves, most of which are, as described by Robert Kennedy Jr., "serial felons."[130]

Some gaps clearly remain. For example, if the FDA in the United States or Health Canada visits vaccine facilities, does that mean either entity conducts purity checks on vaccine batches? Maybe, but if so, then such checks are not universal, judging by the level of contaminants found in Italian vaccines by several groups of investigators.[131] Remember, too, that a number of vaccines licensed in the United States and Canada are not made in these countries, so the actual level of surveillance at the facility level may not be as rigorous as projected.

And there is one other thing: at least in the United States, both the CDC and the FDA hold patents on vaccines, and both are reliant on vaccine profits. Does this alone not constitute a major conflict of interest, as noted in a previous section?

Finally, it is difficult, at least for me, to come away from the evaluation of the surveillance programs, particularly in Canada, with the view that they have not been designed to make vaccines as safe as humanly possible. Indeed, I think that only the most obdurate opponent of vaccination could conclude that the systems are not trying their very best to protect people.

And yet, the systems fail. Not because they are not generally well designed—they are. Not because those who administer and evaluate them wish anyone harm—they do not. But because of a scientific-social-quasi-religious belief system that simply fails to see what it is not looking for, nor one that accepts what it is not trained to see.

I will return to these latter themes in Chapter 8, which will explore the religious nature of vaccination practice and policy, and in Chapter 14, which will look forward to what the future may hold.

The National Vaccine Injury Compensation Program

In 1986, the 99th US Congress passed H.R. 5184—the National Childhood Vaccine Injury Act. The lead sponsor of the bill was Representative Henry Waxman, a California Democrat, and it first passed reviews in the House Energy and Commerce and Ways and Means committees before moving on to the House of Representatives as H.R. 5184. The Act came into effect on October 1, 1988 (Public Law 99–660), whose provisions are shown in Appendix 6.

As described by the US government on their National Vaccine Injury Compensation Program website:

> Vaccines save lives by preventing disease in the people who receive them. Most people who get vaccines have no serious problems. However, vaccines, like any medicines, can rarely cause serious problems, such as severe allergic reactions. In those rare cases, the VICP, which is a Federal program, provides compensation to people found to be injured by certain vaccines. The US Court of Federal Claims decides who will be paid. Three Federal government offices have a role in the VICP:
> - The US Department of Health and Human Services (HHS);
> - The US Department of Justice (DOJ); and
> - The US Court of Federal Claims (the Court).
> - The VICP is located in the HHS, Health Resources and Services Administration, Healthcare Systems Bureau, Division of Injury Compensation Programs.[132]

This legislation establishing the NVIC Program (also colloquially called the Vaccine Court) was in response to concerns voiced by vaccine manufacturers in the United States that civil litigation arising from vaccine harms, mostly in relation to the DPT vaccine for diphtheria, pertussis, and tetanus, was going to pose a serious risk to their business. Given this concern, the vaccine makers requested legislation providing them immunity from civil actions, without which they threatened that they would cease making vaccines in the United States.

Appendix 7 shows details of the NCVI Act Part 1, whose key aspects include the various mandatory requirements for safer childhood vaccines. Of particular interest is Section 2127, which requires the secretary of the Department of Health and Human Services (HHS) to report to the Committee of Energy and Commerce of the House of Representatives and the Committee on Labor and Human Resources of the Senate describing the actions taken during the preceding two-year interval. As far as can be determined by Aaron Siri, a New York City lawyer, this was never done. One subsection of the Act deals with the DPT vaccine (section 312) with the requirement that a report by the secretary of HHS be made within three years. This was also not apparently done. A similar requirement was for a comparable report on the MMR vaccine (section 313). The secretary was also to study all other vaccine risks. Neither of these requirements was met.

The vaccines specifically covered by the NCVI include all that children in the United States are currently exposed to, and most were later evaluated by the IOC, as described above. The list includes:

Diphtheria, tetanus, pertussis (DTP, DTaP, Tdap, DT, Td, or TT), Haemophilus influenzae type b (Hib), Hepatitis A (HAV), Hepatitis B (HBV), Human papillomavirus (HPV), Influenza (TIV, LAIV) [given each year], Measles, mumps, rubella (MMR, MR, M, R), Meningococcal (MCV4, MPSV4, MenB-FHbp, MenB-4C), Polio (OPV or IPV),Pneumococcal conjugate (PCV), Rotavirus (RV), Varicella (VZV) and any combination of the vaccines above.[133] [Brackets, mine]

Claims by those alleging vaccine injury need to contain the following information:

- who was injured by the vaccine;
- which vaccine caused the injury;
- when the vaccine was given;
- the city and state or country where the vaccine was given;
- the type of injury;
- when the first symptom of the injury appeared; and
- how long the effects of the injury lasted.

In addition, the legislation notes that "your claim should also include your medical records and/or other appropriate documents, the court's cover sheet, and the $400.00 filing fee."

The guidelines go on to note that

The general filing deadlines are: For an *injury,* your claim must be filed within 3 years after the first symptom of the vaccine injury. For a *death,* your claim must be filed within 2 years of the death and 4 years after the start of first symptom of the vaccine-related injury from which the death occurred.

You must prove that:

- the injured person received a vaccine listed on the Table; and
- the first symptom of the injury/condition on the Table as defined in the Qualifications and Aids to Interpretation (QAIs) occurred within the time period listed on the Table; or
- the vaccine caused the injury; or
- the vaccine caused an existing illness to get worse (significantly aggravated).
- In addition, the Court must determine that the injury or death did not result from *any other possible causes.* [Italics are mine.]

Further, the alleged injury must be one of those listed and within timelines that are in the range of hours to weeks.

Part 2 of the Act details a number of additional points. Specifically, 300aa-12 deals with the appointment of the Special Masters, the judges who hear and adjudicate the cases presented.

Of particular interest, 300aa-14 provides a "Vaccine Injury Table" that details in which time frame after a vaccine administration an adverse event has to occur to be eligible for consideration. Many of these have time frames of twenty-four hours; the various DPT and DPT/P are up to three days; MMR encephalopathy is fifteen days. The Act defines encephalopathy as:

> (3)(A) The term 'encephalopathy' means any significant acquired abnormality of, or injury to, or impairment of function of the brain. Among the frequent manifestations of encephalopathy are focal and diffuse neurologic signs, increased intracranial pressure, or changes lasting at least 6 hours in level of consciousness, with or without convulsions. The neurological signs and symptoms of encephalopathy may be temporary with complete recovery, or may result in various degrees of permanent impairment. Signs and symptoms such as high pitched and unusual screaming, persistent unconsolable [sic] crying, and bulging fontanel are compatible with an encephalopathy, but in and of themselves are not conclusive evidence of encephalopathy. Encephalopathy usually can be documented by slow wave activity on an electroencephalogram.

Other sections of Part 2 define compensation (aa-15); "Limitations of actions" (aa-16), the period during which an injury must be claimed post-1988 (thirty-six months from first symptoms of adverse effects); the "Recording and reporting of information" (aa-25); "Vaccine information" (aa-26); and, crucially, the "Mandate for safer childhood vaccines" (aa-2 on 27).

This last is key because it requires the HHS secretary to establish a "task force" to examine safety issues composed of the director of the CDC, the director of the NIH, and the commissioner of the FDA. It further stipulates that the secretary will report every two years, as cited earlier. This reporting requirement was supposed to begin in 1987 on the progress being made.

Section 300aa-28 has a last requirement: "Manufacturer record keeping and reporting." It is far from clear if this was ever actually enforced.

There are a number of other considerations/caveats to accessing the NCVI for compensation from vaccine injury, as listed by Dr. Judy Mikovits in her recent book, *Plague of Corruption*.[134]

The first of these is that plaintiffs are not suing any vaccine company, as the NCVI has made the defendant the US government, whose legal team is supplied by the Department of Justice. Next, lawyers for parents in any case cannot rely on case law in the sense that if any vaccines have been found to be causal to damage in one case, this ruling may not be used in another case. A third point is that lawyers for plaintiffs cannot use the process of *discovery for examination* to acquire relevant documents from any vaccine company.

As Mikovits notes in summary:

> The plague of corruption is enormous and encompasses many areas of our
> scientific, medical, and daily political existence.
>
> The pharmaceutical companies have corrupted the laws regarding vacci-
> nations, and a corrupt media has poisoned the mind of the public. The public
> does not ask the simple question: if vaccines are as safe as sugar water, why do
> the pharmaceutical companies need to have complete financial immunity and
> be protected by a battalion of lawyers from the US Department of Justice.[135]

Overall, while a process does exist in the United States at the federal level, it
would be an overstatement to claim that the process is simple or evenly applied.
Nevertheless, that it exists at all is far better than in Canada, where only the
province of Quebec has a vaccine injury compensation program. This situation
has just changed at the Canadian federal level due to concerns about the immi-
nent release of the mRNA COVID-19 vaccine from Pfizer.[136]

As cited above, much of the job of the Vaccine Court in the United States
deals with cases of ASD, alleged by the complainants to arise from various parts
of the CDC's recommended vaccine schedule. For this reason, the next chapter
will first discuss the features and potential etiologies of ASD as a precursor to
a clearer understanding of how and why the vaccine skeptic movement arose
in modern times.

Concluding Remarks

The material in this chapter—from the studies cited cumulatively by the AAP
and CDC, combined with the comprehensive study of the IOC in 2012 in
the Stratton report and further bolstered by the various vaccine adverse effects
surveillance systems—should have finally nailed down the coffin lid on the
smelly corpse named "vaccines are causal to ASD." Clearly this is the conclusion of most of the medical profession and the mainstream media, whether
either knows this literature or not.

But is the debate really over? Is the science actually "settled"? Can I just
stop writing this book now and go back to trying to find a therapeutic for ALS?

To find out, I turn next to the evidence that maybe things in this most
contentious field are not as settled as many would like them to be. And after
that, we get to aluminum in vaccines—the point where the mantra of vaccines
always being safe and effective really begins to unravel.

CHAPTER 4

Vaccine Safety: The View from the Skeptical Side of the House

Let's be clear: the work of science has nothing whatever to do with consensus. Consensus is the business of politics. Science, on the contrary, requires only one investigator who happens to be right, which means that he or she has results that are verifiable by reference to the real world. In science consensus is irrelevant. What is relevant is reproducible results. The greatest scientists in history are great precisely because they broke with the consensus.

—Michael Crichton[1]

The Origins of Vaccine Skepticism

Resistance to, or at least skepticism about, vaccination goes back to at least the early years after Edward Jenner's initial studies on the use of cowpox to prevent smallpox. A range of concerns were raised then, as now, with key ones back in the early years being that the practice was likely to inflict harms of some manner on vaccine recipients.[2]

In subsequent years, vaccines against smallpox and other deadly diseases were largely credited with removing, or at least diminishing, such diseases from much of the world. Smallpox holds the title of a disease that was officially declared eradicated by the 33rd World Health Assembly in May, 1980. However, as we've seen in Chapter 3, the assertion that vaccines deserve the sole credit for the decline in many infectious diseases has to be weighed against other factors that surely played significant roles, as well: better sanitation and cleaner water, for example. One reason not to give vaccination the entire credit for disease eradication rests with the dubious quality of much of the official literature, as also detailed in Chapter 3.

An example from my early childhood will suffice to demonstrate the power that vaccination as a medical miracle had on most people my age and older: I vividly recall my parents telling me about the various polio vaccines, those of

Jonas Salk[3] and Albert Sabin,[4] and how these remarkable scientists had removed the scourge of polio from the world. Indeed, it may seem ironic to some readers that my desire to pursue health research originated with just such stories.

The bottom line is that while there were always those true "anti-vaxxers" opposed to vaccines and vaccination for religious or other reasons, the notion that vaccines were an unalloyed good was never questioned in the circles in which I grew up, nor in much of the world.

That widespread reverence to all things vaccine-related began to change slowly in the early-to-mid-1980s with what appeared to be a rise in the numbers of those afflicted with a spectrum of developmental neurological disorders. These disorders go by the Autism Spectrum Disorder (ASD) and include the classical form of autism first described by Dr. Leo Kanner in 1943.

It was into this emerging, or apparently emerging, concern that a small and seemingly inconsequential case series by a group of British physicians served to rekindle alarms about vaccines, in particular the MMR vaccine against measles, mumps, and rubella. The now legendary paper by Wakefield et al. (1998) will be discussed in detail in the next section. For now, it will serve to note that the authors of this study, led by Dr. Andrew Wakefield, who like the others was a physician specializing in gastroenterology,[5] likely never imagined what this fairly mundane study on GI abnormalities associated with a regressive autism would trigger.

In the years since, that little case series with its ever-so-vague suggestion of a link to the MMR vaccine would trigger protests that would eventually evolve into an actual movement. Against what would eventually be called the vaccine freedom movement were arrayed most of the medical community, most of the politicians in the various countries, and, as we will see, the dead hand of the pharmaceutical industry. Even before COVID-19 arrived on the scene, the vaccine war battle lines were drawn, and it became ever clearer that the schism between the two sides was not only about vaccines, but a range of social and political social beliefs through which vaccine concerns filtered.

If one had to summarize in a sentence the essence of the vaccine wars, it would be this: do I have agency over my body and that of my children, or does some government? Those who argue for the former have a range of freedom movements, past and present, from around the world to bolster their view. Those who argue for the opposite find that they cannot escape the embrace of the pharmaceutical industry, an embrace that increasingly makes clear that profit trumps everything else when it comes to human health. I will delve in greater detail into the pharmaceutical industry connection in Chapter 12.

One vague suggestion by Wakefield et al. may have been the spark. But it took governments, backed by pharma lobbying, to push mandates and thus fan that spark into something far bigger. It's a lesson well known to military planners: Counterinsurgency wars are hard to win because each action one takes to suppress an insurgency creates more insurgents. The lesson still does not seem to have been learned as we face COVID-19, a subject I will return to in the last chapter of the book.

I want to backtrack and look in more detail at the spark that may have triggered all that followed: the suggestion—and it was no more than this—that both gastrointestinal (GI) and neurological symptoms began after the children had been vaccinated with the MMR vaccine. It was a simple observation: the children had some form of autism, they had GI abnormalities, and the parents *reported* that it began after the MMR vaccine. Reading this paper, it seems obvious that Wakefield and his medical colleagues were not setting out to upset the apple cart of vaccination theory and practice. All they did, like their illustrious British forebear Edward Jenner, was to observe a medical condition and seek to understand and treat it.

The "zeitgeist," or spirit of the times, of the late 1990s should hardly have been conducive to the "perfect storm"[6] that developed, perhaps the opposite. After all, concerns about vaccines had not been notably accelerated by the various influenza pandemics of the 1960s and '70s with the attendant adverse vaccine reactions that had followed in a number of cases.

Yet something made the vaccine storm emerge. Something else must been percolating in the background. Perhaps it was an understated but pervasive belief that the medical establishment was increasingly under the thrall of the pharmaceutical industry, a point discussed in Chapter 12. Perhaps it was the growing notice of the rise of autism cases. Whatever the reason, the Wakefield article and the events that soon followed served as the unexpected catalyst.

As will be detailed in the next section, events kept cascading over the next two decades, many of them driven by the mainstream media. Indeed, it may well be the media that then, as now with COVID-19, own serious responsibility for taking what might have been a minor scientific squabble and turning it into a major confrontation in which many felt that the lines had been starkly drawn with no middle ground between opposing camps.

There are those in various domains who see Andrew Wakefield as the "father" of the "anti-vaccine" movement, hardly a title that I think Wakefield would endorse. I think it rather more likely that it was the media who went looking for a bogeyman, found a perfect candidate in Andrew Wakefield, then demonized and belittled those parents who turned to Wakefield and others hoping to understand how their children had descended into autism.

So if one really wants to know who fanned the flames of the vaccine wars, one needs look no farther than the British and American media with their extensive financial ties to the pharmaceutical industry.

Both the media and the pharma will be examined in the following chapters.

But for now, let's backtrack a bit further and consider in the next sections the proximal triggers of the vaccine wars: the rise of ASD and the curious, and widely misunderstood, history of Andrew Wakefield and his famous, or infamous, study.[7]

Autism and Autism Spectrum Disorder (ASD)

Classical Autism

The word "autism" comes from the Greek word *autos*, or "self," describing a condition in which a person appears removed from social interactions. The term was first used by Swiss psychiatrist Eugen Bleuler.[8]

The first recognized article in English on autism was published by Dr. Leo Kanner in 1943.[9] However, it should be noted that in 1925, Sukhareva, working with children in Moscow, published an earlier case series that clearly showed the features of autism.[10]

In his article, Kanner describes how between 1938 and the publication of his article in 1943, he saw eleven cases of what he termed autism, using Bleuler's term. In the article, two additional cases were noted in a footnote. Of these first eleven cases, eight were boys and three were girls Most of these children were very capable in cognitive skills, at least one having a high IQ of 140. The children's families, usually both parents, were often highly educated, many being from professions such as medicine, law, engineering, and the like.

What characterized these children as distinct from other children consisted of fairly uniform, stereotypical behaviors featuring repetitive actions. Speech abnormalities were common, such as the misuse of personal pronouns, and in general the development of language skills was often delayed. In particular, Kanner noted that the children seemed to avoid social interactions with others. There was a hint that the disorder in Kanner's patients had a familial basis, as at least one of the patients, a boy, seemed to have an older sister with the same behavioral symptoms. In most cases, the children seemed to improve in many of these defining features, particularly that of social interactions, as they aged.[11] GI issues, similar to those described later by Wakefield et al. in 1998, were also common.

Kanner's characterization of the children is instructive:

> All of the children's activities and utterances are governed rightly and consistently by the powerful desire for aloneness and sameness. Their world must seem to them to be made up of elements that, once they have been experienced in a certain setting or sequence, cannot be tolerated in any other setting or sequence; nor can the setting or sequence be tolerated without all of the original ingredients in the identical spatial or chronological order. Hence the

obsessive repetitiousness. Hence the reproduction of sentences without alter-
ing the pronouns to suit the occasion. Hence, perhaps, also the development
of a truly phenomenal memory that enables the child to recall and reproduce
complex 'nonsense' patterns, no matter how unorganized they are, in exactly
the same form as originally construed.[12]

In Kanner's view, these children were victims of a genetic disorder, and he
described the children as having "inborn autistic disturbances of affective
contact."

After Kanner's initial article, other researchers described the basic epidemiology of the disorder. This included the observation that it appears in all
industrialized countries with some variations, that it impacted those in various
ethnic groups across all income levels, and that it could be found in all regions
of such countries. Curiously, as is still reported, it affects males more than
females and by about the same ratio.

Many of the details cited previously concerning autism and the work of
Kanner were highlighted in a book by Oller and Oller.[13]

Autism Spectrum Disorder (ASD)

The form of autism first described by Kanner has broadened over the years to
recognize that autism per se includes some subsets of behavioral abnormalities.

In brief, ASD describes a range of developmental neurological disorders that
typically start in early childhood. The range of disorders runs from classical autism,
as above, of varying degrees of intensity and dysfunction that can include mental
retardation through to more socially dysfunctional states such as in Asperger's syndrome.[14] Included in the spectrum are also: a form of "regressive autism,"[15] more
formally termed "childhood disintegrative disorder," which may emerge after a
period of normal neurodevelopment; "pervasive developmental disorder not otherwise specified"; and Rett syndrome.[16] ASD symptoms normally appear before
thirty-six months of age. Regression or loss of skills occurs in 30 percent of affected
children, usually between eighteen and twenty-four months of age.[17]

Characteristics of all these disorders include a dysfunctional immune function and stereotypical behaviors such as arm flapping, toe walking, and head
banging. In addition, various degrees of impairment in social skills and verbal
communication (i.e., delayed language acquisition and usage) often occur.

Various neurological and other medical conditions frequently cooccur
with ASD, including mental retardation (30 percent of cases considered mild
to moderate; 40 percent are listed as serious to profound)[18] and epilepsy (40
percent of cases).[19] Behavioral and psychiatric diagnoses associated with the
core ASD symptoms include aggression, disruptive behaviors, hyperactivity,
self-injury, sensory abnormalities, anxiety, depression, and sleep disturbances.

The most frequent nonneurological conditions associated with ASD are GI abnormalities and underlying inflammation, as initially reported by Wakefield et al. (1998)[20] (and in more detail in the following section), food sensitivities, and feeding difficulties.[21]

At an anatomical level, the brains of people with autism show differences from neurotypical children in the types of connectivity between different brain regions, both locally and at a distance, the latter including the density of the corpus callosum, the fiber tract that links both halves of the cortex. In addition, changes in synaptic structure and type have been suggested.[22]

Temporal Increases in ASD Prevalence

One of the most striking features of ASD is that it seems to have changed dramatically in the years since Kanner's article, both in prevalence (total amount of the population with the disorder) and incidence (the amount in a unit of time for a portion of the population). Kanner, in a span of five years, saw eleven cases of the disorder. The current level in the US is one in fifty-four American children—the CDC's Autism and Developmental Disabilities Monitoring (ADDM) Network, as cited by Zablotsky et al., 2017[23]—but the numbers may be higher in 2021.[24] As also noted by the CDC, ASD occurs in "all racial, ethnic, and socioeconomic groups" with a male-to-female ratio of four to one.

These sex ratios are generally in keeping with Kanner's original report of almost a 3:1 ratio of males to females. As for the incidence, if it were the same as in Kanner's day, he, and others, would certainly have noted the disorder far earlier.

As Rogers notes in regard to the increase since the 1970s alone:

> This is a 27,000% increase from the first autism prevalence estimate in the US that established an autism prevalence rate of less than one per 10,000 people in the population (Treffert, 1970). ASD is almost 5 times more common among boys than among girls (Baio, 2014). Buescher, Cidav, Knapp, and Mandell (2014) estimate that more than 3.5 million Americans live with an autism spectrum disorder.[25] [Note: I have left these references in for readers to pursue should they wish to do so.]

Rogers also cites Edwards (2016)[26] with an estimate of seventy million people around the world with ASD.

The changing rates of ASD in the United States are shown as Fig. 4.2 later in the chapter in context to the controversy about the role of the mercury compound Thimerosal.

So, a key question to ask is if this increase is real or not and, if it is real, what is driving the increase?

In partial answer, the CDC seems to acknowledge the temporal increase as cited above.

The following are the most obvious possibilities to consider. If the increase is real, then similar to the search for causal factors in other age-related neurological diseases, the likely factors often come down to genes, including noncoding regions of DNA, environmental toxicants, or a combination of these. The contentious nature of ASD, however, adds in addition to the above the possibility that any change is due to shifting criteria for the disorder, or if it is a basic artifact due to increased general awareness of the disorder.

So let's consider these in the following order: genes, toxins/environmental generally, better diagnosis with changing DSM criteria, and finally greater social and medical awareness.

ASD and the Evidence for Genetic Causality

A trend over the last few decades has been to look for genetic causes to a range of neurological disorders, including ASD. These might include mutant genes, or parts of the so-called "dark DNA," that is, the noncoding regions of DNA such as single nucleotide polymorphisms (SNPs).[27]

What is the evidence to date? First, Mendelian genetics as a key driver for any increase in ASD can largely be ruled out because to date no causal mutation in any gene giving rise to a temporal increase in cases has been described in the literature in spite of the bulk of funding for ASD research being directed at finding precisely such a mutation. Nor has there been anything like a significant "gene drift"[28] in the US population, or anywhere else for that matter. Full genome screens in ASD have identified various chromosomal regions with SNPs linked to the condition, but a problem is that these identified regions do not usually match between studies.[29] One investigator described ASD patients as "snowflakes," with no two alike from a genetic perspective.[30] These data have now been further supported by a gene study of the well-known scientist with autism, Dr. Temple Grandin.[31]

Animal models of ASD are in accord with this latter interpretation, suggesting that genetic variation, rather than being strictly causal to the disorder, instead confers an altered vulnerability to exposure to environmental stressors.[32]

The situation for ASD is thus quite unlike the situation for the age-related neurological diseases—ALS, Parkinson's disease, and Alzheimer's disease—which all have something on the order of 10 percent of cases arising from various demonstrated gene mutations.

However, where ASD and the older age-related pathological neurodegenerative disorders may be similar is the very real potential for gene-toxin interactions. An emerging view of the latter is that with most of those with ALS, Parkinson's disease, and Alzheimer's disease versus ASD, it is less about the gene mutations than it is variants of the noncoding regions of the DNA combined with some toxic factor. In this view, ALS and the other adult onset disorders have quiescent nucleotide variants that, when combined with a toxicant, can trigger the onset of the disease state. From this perspective, various toxins, including the aluminum adjuvants to be discussed in Chapter 5, may be the final trigger to a pathological process.

In ASD, one gene, MTHFR (*methylenetetrahydrofolate reductase*), has been linked to ASD, but also to a host of other ailments, and the latter likely suggests that variants of this gene alone are not sufficient to cause ASD. It is in this sense that some of the SNPs described in the ASD literature, cited earlier, may be relevant for understanding the etiology of the disorder when combined with a toxic insult.

Finally, it seems likely that ASD involves, at some level, a mitochondrial disorder. In a paper published in 2014, Giulivi et al. concluded that ten children with autism compared to ten children without autism showed signs of mitochondrial dysfunction.[33] However, saying mitochondrial dysfunction is involved is a "motherhood" statement, since mitochondrial dysfunctions appear in many neurological disorders and is thus not specific to any of them.[34]

From the above, it seems that both groups are trying to play the gene card: those looking to exclude vaccines as the culprit aim at genetic errors that have nothing to do with environmental triggers such as vaccines; those aiming to damn vaccines look to genes for the opposite. If one were to bet, at least based on the gene-toxin studies for other neurological disorders, the latter are more likely to be correct.

Environmental Factors

Environmental factors are known to be damaging to the nervous system across the life span and figure prominently into hypotheses for the other age-related disorders cited earlier.[35] That such may occur in early neurological development leading to ASD is certainly not an outrageous scientific reach. For example, we live in an increasingly polluted world containing a huge number of potential toxicants which can, alone or in combination, impact the CNS. These include toxins/toxicants such as industrial chemicals of various types including pesticides, endocrine inhibitors, and those such as the heavy metals contained in biosolids,[36] those known to be present in food,[37] and, of course, various medicinal products, the latter including vaccines. Any of these might be reasonable based on the simple observation of increased rates of ASD correlated temporally with increased toxic loads from the various sources.

In regard to the last component of the above explanation, there is clear evidence for an immune system misregulation involved,[38] although whether such dysfunction is cause or consequence to the development of ASD is not clear, much the same way that the age-related neurological disorders have an associated immune system component that is not well understood.

Of course, where this becomes contentious is when some scientist proposes that some of these toxicants arise from vaccines. Much of Chapter 6 will explore this aspect in great detail, so I will leave the subject of environmental toxicants until then.

Correlation does not equal causation, as noted previously, but neither does correlation warrant being dismissed out of hand simply because it might be harmful to the bottom line of some corporate entity.

Changing DSM Criteria

One such argument is that ASD's *apparent* increase reflects a broadening of diagnostic criteria in the Diagnostic and Statistical Manual (DSM)[39] volumes 3 through 5. In these editions, the criteria defining those disorders included in ASD changed from version 3 to 4 in 1994, becoming broader, then changed from 4 to 5 in 2013, becoming more selective again. These changes from DSM 3 to 5 concerning ASD are shown in the schematic of Fig. 4.1.

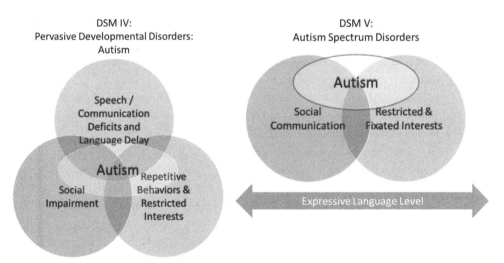

Figure 4.1. Schematic of changes in diagnostic criteria from DSM 4 to DSM 5.[40]

DSM-III	DSM-III-R	DSM-IV	DSM-5
Pervasive Developmental Disorders • Infantile Autism, full syndrome present • Infantile Autism, residual state • Childhood onset PDD, full syndrome • Childhood Onset PDD, residual state • Atypical PDD A	Pervasive Developmental Disorders • Autistic Disorder • Pervasive Developmental Disorder Not Otherwise Specified (PDDNOS)	• Autistic disorder; Asperger disorder; childhood disintegrative disorder; and pervasive developmental disorder, not otherwise specified • No equivalent DSM-IV diagnosis of social communication disorder • Rett disorder • Mental retardation • Reading disorder • Mathematic disorder • Disorder of written expression and learning disorder, not otherwise specified • Expressive language disorder • Mixed receptive-expressive language disorder • Phonological disorder • Stuttering	• Autism spectrum disorder • Social communication disorder • Not included in DSM-5 – Rett disorder • Intellectual disability • Dyslexia • Dyscalculia • Unspecified learning disorder • Language impairment, late language emergence, or specified language impairment • No equivalent DSM-5 diagnosis of mixed receptive-expressive language disorder • Speech sound disorder • Childhood-onset fluency disorder

Table 4.1. Characteristics of DSM versions 3 through 5.

However, ASD, as defined in DSM-5, still describes a neurological disorder of the following characteristics that, in most ways, still mirror those of Kanner in 1943. As noted by Rogers,[41] these include:

A. Persistent deficits in social communication and social interaction across multiple contexts.

B. Restricted, repetitive patterns of behavior, interests, or activities, as manifested by at least two of the following . . .

1. Stereotyped or repetitive motor movements, use of objects, or speech.

2. Insistence on sameness, inflexible adherence to routines, or ritualized patterns or verbal nonverbal behavior.

3. Highly restricted, fixated interests that are abnormal in intensity or focus.

4. Hyper- or hyporeactivity to sensory input or unusual interests in sensory aspects of the environment (American Psychiatric Association, 2013). The definition also states that 'symptoms must be present in the early developmental period', that 'symptoms must cause clinically significant impairment', and that one should make sure to rule out other possible explanations such as 'intellectual developmental disorder or global developmental delay' (American Psychiatric Association, 2013).

Given the above, it is difficult to imagine that any physician would not recognize the essential nature of ASD from Kanner to the present. That aside, while a change in criteria of the inclusion factors *could be* part of the changing prevalence patterns, this could only occur if both were changing in lockstep in the same way. Hence, the change from DSM 3 to 4 might have had part of the increased incidence due to a broadening of ASD criteria. However, with the shrinking of the criteria in DSM-5, the ASD rates should have come down,

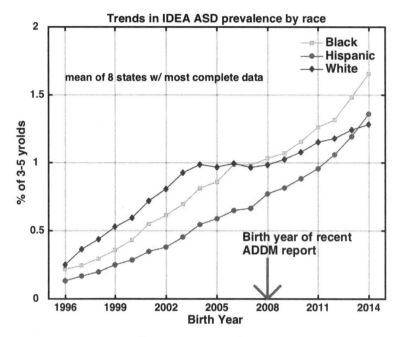

Figure 4.2. Increases in ASD by year. From Cynthia Nevison, as provided to Children's Health Defense. See text for details.

which they did not. However, since this change was relatively recent, the ASD rates may not yet reflect the changed criteria.

Put briefly, this explanation really does not suffice to account for more than a fraction estimated by Rogers to be less than 27 percent of the total increase since the 1980s.

Better Diagnosis/Greater Social and Medical Awareness

One fairly recent idea for the increase in ASD is that there really is none; that is, it is an artifact that stems from physicians and the public being more aware of ASD and thus reporting it more often.[42] This notion follows from a view that if you expect to see something, you are more likely to see it whether there is really more of that something or not.

Reviewing all of the above, the weight of evidence, some to be considered in Chapter 5, is that it is far more likely that the apparent increase in ASD is in fact real. As such, it may arise in a way similar to other multisystem disorders, that is as a result of complex gene-environment (toxin) interactions.[43]

As a side note, another way to sidestep any possible vaccine link is to describe autism and the spectrum itself as a form of "neurodiversity," that is, an expanded range of human brain potential, and thus something to be celebrated rather than be concerned about.[44]

If, however, ASD incidence is indeed increasing, it is far from a trivial problem, morally, socially, or economically. Given the apparent magnitude of the problem, one would assume that the medical profession and governments at all levels would want to understand the disorder, its origins, possible treatments, etc. From this, there might come ways to prevent ASD, or at least diminish its incidence in the future.

Linkage of an Increased Vaccine Schedule with Increases in ASD

Appendix 5 shows the list of vaccines administered according to the CDC schedule in 2020.[45] In comparison, in the United States of the 1950s, children received seven vaccines before age six; in 1983, that number increased to ten. From 1983 to 2013, it grew dramatically to thirty-six.[46] And the vaccine numbers have increased yet again over the last three years. This does not account for booster shots, which can increase the total current number of shots to seventy-two. Of note, the ACIP just recommended Pfizer's mRNA vaccine for children.[47]

One difference that Dr. Toby Rogers notes in relation to a possible vaccine etiology for ASD versus purely an environmental one is that of parental reports: unlike chronic toxicity, the behavioral changes in a child closely following the receipt of some vaccines is not subtle and therefore something most parents would tend to ignore. In this regard, I refer the reader back to the statement made by Dr. Luke Taylor in the previous chapter in the *Epilogue* to his 2014 article.

Much of the discussion on ASD and the possible factors driving the condition will be addressed in more detail in Chapter 5, which deals with aluminum generally as a neurotoxin and more specifically as a vaccine adjuvant. In addition, the role of aluminum in autoimmunity and how it might impact both the CNS and immune system will be presented at that time.

Before doing so, however, let's continue following the thread that links the rise of the "anti-vaccine" movement to autism and the human lighting rod and media whipping boy whose name is Dr. Andrew Wakefield.

The Curious History of the (In)famous Wakefield et al. (1998) Study and Its Aftermath

Being associated with Andrew Wakefield in any way, shape, or form, at least in most medical circles, is sort of the scientific equivalent of the kiss of death. Others, even many scientists, view him quite differently, as very much the lone maverick fighting for those that the mainstream medical system has disbelieved and discarded. The latter, of course, are the parents who hoped, and continue to hope, that Wakefield's work will lead to some relief from their child's autism.

I first met Wakefield in the fall of 2008, about a year after our first paper on aluminum adjuvants in mice came out.[48] Or, more precisely, I met him telephonically when he called my laboratory. Before this, I had only vaguely heard of him and his 1998 study, and then only because one of my graduate students referred to it as we were beginning our work on adjuvants.

At the time, what little I knew about the man was uniformly negative. I knew that he was a gastroenterologist, the specialty that looks at the gastrointestinal (GI) system in health and disease. And I knew, or thought I knew, that he had been held in disgrace after the publication of an article that supposedly claimed a role of the MMR vaccine in the development of autism.

When he called and introduced himself, I had to struggle for a moment to identify who he was. Once I had done so, the conversation was professional.

In brief, his colleagues and he at Thoughtful House,[49] an autism treatment and research center in Austin, Texas, had just completed the behavioral analysis of a series of macaque monkeys given various vaccines in an attempt to see if the vaccines induced anything resembling autistic behaviors in humans. The data seemed, at that time, to suggest that they did. Wakefield, then executive director at Thoughtful House, was calling because he hoped I would agree to do some simple cellular staining of sections from the cerebellum of the brains of the animals, both controls and vaccine-injected.

After giving it some thought, I agreed and the decision made was that he would send us some coded slides from both groups of animals and we would do the histological analysis looking at neuronal cell structure along with the presence of astrocytes and microglia, both glial cell types associated with neuronal degeneration in the brain. I also agreed to look at various markers for types of cell death, one in particular called apoptosis.[50]

In due course, the slides arrived, and shortly afterward we did the stains, packed up the stained slides, and sent them back. Wakefield had not provided us with the codes for which monkeys were in which group, so we had no way to know how to quantify the data. Still, at a glance, the labeling on some of the slides for some monkeys seemed different from those of other monkeys.

Some months went by without hearing anything from Wakefield, and when I contacted him about the data, he told me that the lead investigator on the study had left Thoughtful House and there were no funds to hire a replacement.[51] Soon after this, Wakefield himself left the center.

I didn't give the outcome much more thought, as preliminary projects that go nowhere are typical in science. I kept the protocols and photographs of the labeled slides, however, and while I have these still, to this day I don't know which monkey is which.

About three years later, my laboratory had received funding from what was then called the Dwoskin Family Foundation[52] to pursue our work on aluminum adjuvants in animal models. In one of our early meetings, the head of the Foundation, Claire Dwoskin, and I decided on the spur of the moment to have a miniconference and use as a venue a location outside of Montego Bay, Jamaica, where the Dwoskins had a vacation residence. Dr. Lucija Tomljenovic and I were given the role of selecting conference participants.

We duly created a wish list of those we thought best able to address vaccine safety issues. After some thought, we decided to invite Dr. Chris Exley, an aluminum expert I knew, Drs. Yehuda Shoenfeld and Romain Gherardi, physician-scientists I knew only by reputation,[53] and some others. One of those others was Dr. Andrew Wakefield.

My first inclination was to vote against him given that his reputation in scientific/medical circles was getting worse. However, the others convinced me with the argument that whether he was right or wrong in his 1998 paper, the only way to find out was to hear the man's evidence.

It was at this meeting that I met Wakefield for the first time in the flesh. I'm not sure what I had been expecting, but what I found was a modest, middle-aged man who seemed reserved, informed, and endowed with British good manners. He gave a good talk on his 1998 work and the studies that followed.

It was at this meeting that he received the news that his 1998 article had finally been retracted by the British medical journal *The Lancet*, and it was either while still in Jamaica or shortly after that his media beating by Anderson Cooper on CNN took place.

That was the last meeting until we crossed paths again at a 2015 AutismOne[54] meeting, described later.

While people can be deceptive, Wakefield did not strike me as such, nor as the scientific fraud or charlatan that the media had made him out to be. My conclusion then was what it is now: his negative reputation is, to a great extent, a fiction of the media's own invention. That the latter interpretation is likely to be correct will be documented in the following pages and the chapters that follow.

Before the publication of the paper that started it all, Wakefield and colleagues were faculty clinicians at the Royal Free Hospital in London, an institution that later merged with University College London in 1998.[55] Wakefield and his colleagues reported a case series study in twelve children who had in most cases been referred to them primarily for examination of gastrointestinal problems. Of the twelve children in the study, nine of them presented with autism, three with other neurological features, much of this disputed by British journalist Brian Deer.[56]

Key to this story were Wakefield himself, then a medical doctor holding the title of "senior lecturer," and his established and highly regarded senior colleague, Professor John Walker-Smith.[57] Wakefield's role at the Royal Free was primarily as a researcher, allowed by his contract to address some clinical aspects in the care of the patients, but not treatment. Walker-Smith and most of the other clinicians in the group did the actual treatments.

Let's begin by looking at the actual paper that started the fuss and see what it looks like. Here is the title: "Ileal-lymphoid-nodular hyperplasia, non-specific colitis, and pervasive developmental disorder in children" (AJ Wakefield, SH Murch, A Anthony, J Linnell, DM Casson, M Malik, M Berelowitz, AP Dhillon, MA Thomson, P Harvey, A Valentine, SE Davies, and JA Walker-Smith).[58]

Wakefield and other authors had published an earlier article, also in *The Lancet*, in the same subject area in 1995.[59] The article, "Is Measles vaccination a risk factor for inflammatory bowel disease?", clearly shows that Wakefield was highly interested in exploring the role of measles vaccines in gastrointestinal disorders at that time. Their conclusion, as usual in scientific articles, was pretty carefully worded: "This study shows an association between measles vaccination and inflammatory bowel disease. It does not show a causal relation."

In contrast, as titles go, the 1998 article was extremely tame and something that only specialists in gastroenterology might get worked up about, certainly nothing to capture the media's, and thus the public's, attention. Then again, as I will show, the observation that GI abnormalities might accompany forms of autism was already in the literature. If I had wanted to really call attention to the purported link to the MMR vaccine, I would have given the article a more inflammatory title, something like: "Drawing links between the MMR vaccine and regressive autism with GI abnormalities." Or to be more circumspect but still getting the point across: "Is there a link between the MMR vaccine and regressive autism with GI abnormalities?" as in the 1995 article.

But Wakefield, as lead author, and John Walker-Smith,[60] the senior author, didn't do that. Instead, they gave it a typically boring science title that would likely be seen by only other specialists in the field.

On top of this, if there was a shocking revelation in the article that supposedly demonstrated an MMR link to autism, it was certainly not obvious from the article's discussion. Here is what the authors actually wrote, rather than what they are typically accused of having written:

> We did *not* prove an association between measles, mumps, and rubella vaccine and the syndrome described. Virological studies are underway that may help to resolve this issue.
>
> If there is a causal link between measles, mumps, and rubella vaccine and this syndrome,[61] a rising incidence might be anticipated after the introduction of this vaccine in the UK in 1988. Published evidence is inadequate to

> show whether there is a change in incidence or a link with measles, mumps, and rubella vaccine." [Italics for emphasis, mine]

And, near the end:

> We have identified a chronic enterocolitis in children that may be related to neuropsychiatric dysfunction. In most cases, onset of symptoms was after measles, mumps, and rubella immunisation. Further investigations are needed to examine this syndrome and its possible relation to this vaccine.

All in all, this was pretty standard "sciencespeak" that waffles about whether the data show a link or not and simply notes the need for further studies, the latter phrase used in virtually all biomedical articles.

Next, there were thirteen authors in total. Again, not particularly remarkable for an article in a medical journal. While guidelines for authorship may not have been as rigorous then as they are now, the fact that there are this many coauthors suggests that they all bought into the general findings and conclusions, as vague as the latter were regarding any role for the MMR vaccine in the GI-autism connection.

In other words, there is nothing in this article that should have tweaked anyone not ready to be tweaked.

That being the case, what was the event that happened next to make the article in *The Lancet* the menace to public health that some consider it to be? The answer seems to be the press conference that followed the article's publication, one that featured Wakefield and Dr. Arie Zuckerman, the dean of the Royal Free's associated St. Mary's Medical school.[62]

It is important to realize at this point that universities or hospitals rarely hold press conferences for published articles, and indeed most of the articles produced by their faculty or staff go pretty much unacknowledged. However, when they do hold a press conference, it suggests that there is something worth highlighting, something they consider very significant that will put them in a favorable public light and maybe even lead to more funding. In essence, it is a public affairs stunt.

The fact that the dean was there makes this particular press conference all the more interesting, especially given that links between vaccines and adverse events tend to be pretty much verboten. However, that was back then when the subject of vaccination and its potential harms was far less confrontational than it is today. Regardless, Dr. Zuckerman must have thought the university would get something out of the press conference or he would not have allowed it to go forward, let alone been present there with Wakefield.

During the press conference, Wakefield expanded on the data in the *Lancet* paper, opining that the GI and autism features in the children of the study might have been the result of the nature of the particular MMR vaccine, not just vaccines per se, saying that

> The study has identified a possible link between gut disorders in children and autism. In the majority of cases the onset of symptoms occurred soon after the MMR vaccination. We clearly need further research to examine this new syndrome, and to look into a possible relation to the MMR vaccine.[63]

Wakefield also advocated for using only the separate measles vaccine, then available in the United Kingdom, rather than the combined MMR form.

Notably, neither coauthors Walker-Smith nor Murch, both later accused by the General Medical Council along with Wakefield, supported Wakefield's assessment of the role of the MMR vaccine in the syndrome they had described.

So was this press conference the event that would "Cry 'Havoc!,' and let slip the dogs of war"[64] as far as vaccine controversies go? Maybe. But before deciding, let's review more about the article in question and its short- and long-term outcomes.

The best way to proceed is to consider again just what the Wakefield et al. article actually contained and then see how those elements stand up to scrutiny. The first of these is the purported link between GI abnormalities and some forms of autism. Wakefield et al. cited a number of references that appeared prior to the *Lancet* article, in one case almost twenty years earlier.[65] The 1998 study merely advanced this notion but hardly invented it.

The question then is not whether such a GI-autism correlation exists, but whether the Wakefield et al. paper had accurate data in support of this observation. As will be discussed below, the allegations of British journalist Brian Deer that Wakefield and his coauthors had manipulated or even made up the data may be questioned, but that's as far as dismissing the connection goes.

After the 1998 article, Wakefield continued to publish articles on the same general theme. Of these articles, only one was later retracted[66] as a consequence of the earlier retraction of the 1998 paper, which was based on some of the same data. For example, Uhlmann et al. (2002) reported on the presence of measles virus in seventy-four of ninety-one patients examined. This article, on which Wakefield was a coauthor, did not implicate the vaccine, merely the virus.[67]

In the years after the Wakefield study, a number of other researchers produced evidence in support of a GI-autism link.[68]

In 2004, however, ten of the thirteen authors of the *Lancet* paper "retracted" one interpretation of the article, namely, the suggestion that the data implicated the MMR vaccine.[69] One of these ten was Dr. Simon Murch, whom we will meet again. It was an odd thing to retract what had actually not been more than a fairly vague speculation.

So what's left is the widespread belief, clearly by those who have not read the original 1998 article, that Wakefield et al. claimed that the GI-autism symptoms were caused by the MMR vaccine. As the quotes taken from the paper itself reveal, the authors claimed no such thing. Rather, they noted it as a plausible hypothesis based on the parental reports, a hypothesis that should be examined further. Hypothesis generation is, in fact, one of the primary goals of a case series such as this article was.

Regardless of whether the 1998 *Lancet* article was the causal trigger for all that followed, it certainly remains an enduring feature of the whole *l'affaire* Wakefield that most people recall.

As a sidebar, as long as the singular allegation that Wakefield et al. had claimed the MMR caused autism continues to fester, belief that he did so remains an accurate predictor to which side in the vaccine wars any scientist, journalist, or concerned parent will default.

For the multitudes of supporters around the world who will adore Dr. Wakefield to the very end, he is the first doctor who took seriously their stories of vaccine-induced injury, particularly in regard to autism. For those who despise him, Wakefield will remain a medical charlatan whose "fraudulent," retracted, and "debunked" 1998 study precipitated a dramatic collapse in immunization rates in the United Kingdom and around the world, killed children in the process, and gave birth to the "anti-vaccine" movement. It is indeed rare in science for one figure to be so revered and reviled at the same time, and probably none have achieved the dual status as quickly as Wakefield.

As noted above, I had met Andrew Wakefield on several occasions.[70] The third time was at the AutismOne meeting in 2015, where I had been invited to speak about aluminum toxicity and our work on the subject.

At that meeting I got to see just how much respect Wakefield clearly has in some circles, and AutismOne was one such circle. My first inkling that Wakefield was viewed as an icon was when my talk on aluminum neurotoxicity in animal studies of ASD was scheduled into the same time slot as Wakefield's talk on the MMR vaccine and Somali immigrants: the room I spoke in had about six people in it; Wakefield's lecture room held hundreds.

Later, sitting with Dr. Chris Exley[71] in the hotel bar after the last session of the day, I watched as Wakefield worked the room like a Hollywood celebrity. Wakefield walked through the bar with easily a dozen fans and well-wishers in a cluster around him begging for selfies. He was, in action, the movie star that he would later in actuality become in his various documentaries.[72]

Wakefield stopped at our table for a few seconds to say hello before going on to mingle with more fans.

It was an impressive showing, and Exley was clearly astonished, apparently never having seen anything quite like it. I, in contrast, was born into a Hollywood family, and I grew up knowing my fair share of celebrities. Nonetheless, even I was impressed.

Back in the days after the *Lancet* article came out, pressure on Wakefield was building in his home country. He was dismissed from his position at the Royal Free Hospital in 2001.[73] Whether because of this or due to the search for new horizons, he left the Royal Free and England for Thoughtful House in 2005.

Apart from the random attacks on Wakefield in the next few years, not much else had been made about the 1998 *Lancet* article. That was soon to change. In 2003, a British tabloid writer, Brian Deer,[74] a non-MD/nonscientist, was hired by James Murdock, the managing editor of the *Sunday Times of London* and the son of billionaire owner Rupert Murdock,[75] to investigate Wakefield and the circumstances surrounding the 1998 article and to write a series of articles for the paper about it.[76]

Deer produced a number of apparently damning articles for the paper in 2004, later revisited in another batch of articles for the *British Medical Journal (BMJ)* in 2010/2011. Among other things, the various articles claimed that the GI histology samples taken from the children in the study were fraudulently graded for the pathologies that Wakefield et al. claimed to see. What was especially odd about this was that Deer had no particularly relevant medical training, certainly not in GI pathology, the key issue raised in the *Lancet* article.

In the following, I will highlight some of Deer's specific claims and juxtapose these with the responses of Dr. David Lewis and Mary Holland, the latter a lawyer. Both Lewis and Holland have looked at Deer's accusations and came to very similar conclusions about Deer's veracity.[77]

Here is the summary from Deer's first *BMJ* article,[78] in which he claimed that

- The *Lancet* paper was a case series of 12 child patients; it reported a proposed 'new syndrome' of enterocolitis and regressive autism and associated this with MMR as an 'apparent precipitating event.' But in fact:
- Three of nine children reported with regressive autism did not have autism diagnosed at all. Only one child clearly had regressive autism.
- Despite the paper claiming that all 12 children were 'previously normal,' five had documented pre-existing developmental concerns.
- Some children were reported to have experienced first behavioural symptoms within days of MMR, but the records documented these as starting some months after vaccination.
- In nine cases, unremarkable colonic histopathology results-noting no or minimal fluctuations in inflammatory cell populations-were

changed after a medical school 'research review' to 'non-specific colitis.'

- The parents of eight children were reported as blaming MMR, but 11 families made this allegation at the hospital. The exclusion of three allegations-all giving times to onset of problems in months-helped to create the appearance of a 14-day temporal link.
- Patients were recruited through anti-MMR campaigners, and the study was commissioned and funded for planned litigation.

In the articles, Deer elaborated on the allegation that Wakefield was essentially running a scam for profit.[79] This assertion was based on Wakefield's earlier association with solicitor Richard Barr representing families who claimed that their children had been injured by the MMR vaccine.

In another of the series of articles, Deer went on to blame the editor at *The Lancet*, Richard Horton, for trying to cover up the "fraud" in the 1998 Wakefield et al. paper[80] and followed this up by asserting that Wakefield was claiming that a conspiracy existed to get him.[81]

The Editor of *BMJ*, Fiona Godlee, Supported Deer's Allegations

BMJ Editor Fiona Godlee then wrote several editorials that were widely quoted, serving further to demonize Wakefield and his colleagues.[82]

In response to specific questions on the fraud by Wakefield et al. that Deer alleged, here is what Dr. David Lewis wrote in a detailed rebuttal:

> It's probably worth noting that, in the *BMJ*, Deer said that the proof Wakefield is a fraud lies in the children's biopsy slides which, Deer pointed out, mysteriously disappeared from the UCL [University College London] lab[oratory] where Wakefield worked. However, when I submitted high-resolution color slides of all of the very biopsy slides Wakefield used to write the *Lancet* article, [Fiona] Godlee and other editors at the *BMJ* refused to publish them. They only published Wakefield's grading sheets, which Deer also claimed did not exist. As Deer's own expert of choice said [Dr. Ian Booth; below featured in the Mitting High Court decision on the GMC's investigation] when interviewed in the *Nature* article I sent you, ". . . the grading sheets do not suggest fraud. They were well within the range of simply difference of opinions among experts." [Brackets for addition are mine.]

This last point is key, since any histological examination really comes down to the judgment of experts, is often not quantitative, and thus overall can be extremely subjective. This is as true for much of histological evidence in neurological disease research, and I'm guessing that GI pathology evaluations were at least the same, especially in the late 1990s.

There are two more key points in regard to the possibility, stated by Deer as a certainty, that Wakefield et al. committed deliberate fraud. The first of these is that Wakefield's coauthors would have had to have been complicit in the deception, or at the very least absurdly negligent. The first seems quite unlikely for the simple reason that to get all twelve coauthors, including the very senior Walker-Smith, to go along with a clear fraud would have been quite difficult, if not impossible. Negligence by some authors, hinted at by both Deer and Godlee, is always possible, however, since back in the 1990s, journals tended to be less likely to demand that all authors accept responsibility for *all* aspects of any submitted manuscript. Back then, it was not particularly unusual to have authors on papers who had done little or nothing.[83]

Deer spent a lot of ink in one article in *BMJ* drawing attention to a £435,643 contract fee plus expenses that Wakefield is supposed to have received from Barr. In addition, Deer dwelled at length on Wakefield's other attempts to capitalize on his research on the MMR outcomes by the creation of various companies and applications for patents for alternative means for measles prevention.[84]

In regard to these allegations that the entire episode had been a money-making scheme, Mary Holland wrote this:

> In 1996, an attorney, Solicitor Barr of the law firm Dawbarns, contacted Dr. Wakefield to ask if he would serve as an expert in a legal case on behalf of children injured by vaccines containing the measles virus. The lawyer was bringing the suit on behalf of parents who alleged that vaccines had caused their children's disabilities, including autism. Six months before this, and independent of the litigation effort, parents of children with autism and severe gastrointestinal symptoms began contacting Dr. Wakefield because of his [earlier] publications on the measles vaccine, asking for help for their children's pain and suffering, which they believed was vaccine-induced. Dr. Wakefield made two major, but separate, decisions at about this time - to try to help the families dealing with autism and gastrointestinal problems, and to become an expert in the legal case regarding vaccines and autism. Barr asked Dr. Wakefield to study two questions: (1) whether measles could persist after measles infection or the receipt of the MMR vaccine; and (2) whether the measles virus could lead to complications.[85] [Brackets are mine.]

Another allegation by Deer was that Wakefield was merely attacking the MMR vaccine because he had his own alternative treatment regime that would, presumably, be boosted by a failure in confidence in the MMR. In fact, the aspects of an alternative treatment that Wakefield and others worked really had nothing to do with an alternative vaccine at all.[86]

Further, while it certainly seems damning in 2020 hindsight, such attempts at commercialization back in the 1990s were hardly remarkable. At my own university, dozens of biotech companies were spun off by academic researchers hoping to take often very preliminary, and in many cases erroneous, laboratory discoveries to market. The university was highly active in this regard working through an entity still called the University-Industry Liaison Office, which was even known to approach researchers to find out if they had any research that might be marketable. The overt commercialization of basic science by the pharmaceutical industry will be explored in greater depth in Chapter 12.

Most universities in my experience have something quite similar: all are in the business of making profits for the university on the back of basic research. For this reason, damning Wakefield for doing what most academic or medical researchers were either doing, or hoping to do, is to hold him to a completely different standard. The very facts that the Royal Free participated in such discussions, and the presence of Wakefield's own dean at the press conference after the paper was published, strongly suggest that commercial goals were not solely the design of Wakefield himself.

In medical and media circles, Deer was widely applauded for bringing attention to major flaws, indeed suspected fraud, in the pathology reports that had escaped proper evaluation by experts in the field. Of note, Deer has recently released a book on the whole controversy that recapitulates many of his previous charges against Wakefield and addressed some of the allegations, in turn, made about him.[87]

The articles by Deer in the *Sunday Times of London* had one very pronounced consequence: After receiving a letter from Deer about the *Lancet* article in June 2007, the General Medical Council (GMC)[88] issued a notice of disciplinary hearings against Wakefield, previously a senior lecturer in the Departments of Medicine and Histopathology and a reader in Experimental Gastroenterology at the time the *Lancet* article came out. Additionally, John Walker-Smith, a professor of Paediatric Gastroenterology, was charged, as was Simon Murch, then a senior lecturer in Paediatric Gastroenterology.

GMC's investigation and conclusions, summarized next, eventually came, in turn, to be scrutinized by the British High Court.

The GMC convened a "Fitness to Practice Panel" whose hearings were released on January 28, 2010, after starting in mid-July 2007. According to this 143-page document, it applied the GMC's "Preliminary Proceedings Committee and Professional Conduct (Procedure) Rules 1988."

The Panel was made up of three doctors and two lay members. None were gastroenterologists.

As noted by the GMC in their report:

> . . . the Panel heard evidence and submissions for 148 days over a period of two and a half years. There were 36 witnesses as well as lengthy examination and cross examination of the three doctors. The Panel has deliberated in camera for approximately 45 days.[89]

The panel noted that the case was complex and that "the burden of proof is on the GMC throughout." And that "If the Panel were not sure beyond reasonable doubt, the sub-head of charge was found not proved in favour of the doctor, in accordance with the criminal, as opposed to the civil, standard of proof."[90]

A key point at issue throughout the hearings was which ethics approval the researchers at the Royal Free Hospital had been using, that is, was it an existing project listed as 162–95 or one that had been tendered for future use as a research project, 172–96? This issue was important because which approval was in force spoke to the question of whether the children discussed in the 1998 *Lancet* article were examined for treatment purposes or research purposes, the latter having a treatment component.

The difference, as explained in the various documents put before the panel, was that a research program is intended to benefit others through the outcomes of research, not necessarily the person examined. A treatment program, in contrast, is designed to help the person being treated. It was on this definition that the panel found Wakefield guilty of using all of the children for a research project that did not yet have ethics approval and for which there was little or no clinical benefit. Concerning the last point, the panel noted that the *Lancet* article had stated that approval of the study for research had been obtained when, in fact, this was not the case at the time the article was published.

The panel also took exception to Wakefield not having disclosed in the article his receipt of funds from an ongoing civil action, noted earlier, involving children whose parents alleged that their injuries were the results of the MMR vaccine.

Also at issue, and decided in the negative by the panel, was the notion that the children in the study had not come to the study by a normal referral process upward from other physicians to the Royal Free, but rather were in some cases solicited by Wakefield for the study.

In each instance, the panel termed Wakefield "dishonest" and "irresponsible" and deemed these actions examples of "serious professional misconduct."

The final item was that the panel found Wakefield guilty of the same misconduct relating to a birthday party where Wakefield had obtained blood samples from children and paid them a token fee of five pounds.

To clarify the nature of this event, I asked David Lewis about it. He, in turn, asked Wakefield himself, who wrote:

> The "birthday party" involved a qualified family doctor taking blood samples using standard approved procedures from healthy neuroytypical children at my son's birthday party, to serve as controls in a clinical study of children with autism at the Royal Free Hospital.
>
> Most of the parents of these children were themselves doctors and all samples were taken only following fully informed child and parental consent. The blood taking passed off without incident and the children received 5GBP in their party bag at the conclusion of the event.
>
> These children were not National Health Service patients and therefore they did not fall under the jurisdiction of a Royal Free Hospital or any other IRB [Institutional Research Board].
>
> The General Medical Council was wrong in its conclusions on this latter point.
>
> I made a vaguely humorous reference to the event at a parent conference some months later. This was unfortunate and was later used to characterize me as "callous."
>
> That is pretty much it. A mountain from a mole hill. [Brackets are mine.]

The birthday party blood taking seems like an odd thing for Wakefield to have done, showing perhaps a lack of judgment, but did it really rise to the level of professional misconduct that the GMC was to conclude it did? Probably not, but it seems clear from the GRM's decision that by this point Wakefield was clearly in their crosshairs as an irresponsible liar. Seen in this light, the birthday party incident was merely one more item to hang around Wakefield's neck.

The outcome of the panel investigation for Walker-Smith was much the same, albeit with fewer items to consider. Essentially, the panel focused on whether the investigation of the children was for research or treatment purposes: if the former, did the various tests serve the best interests of each individual child, and was ethics approval obtained? Last, the panel considered whether the *Lancet* article had misrepresented any of the details of these last points.

As with Wakefield, the panel also found him to have been dishonest and to have acted irresponsibly, warranting a judgment of serious professional misconduct.

Murch got off the easiest, and, while the panel suggested that some of the same items applied to him, he was not judged to have engaged in serious professional misconduct.

The final decision of the panel was to "delist" from the registry of physicians in the United Kingdom both Wakefield and Walker-Smith, but not

Murch. What that meant for the first two was that they were no longer able to practice medicine in the United Kingdom.[91]

As a result of the GMC proceeding and punishment, Wakefield was marked as disgraced by his peers in the medical establishment, thus ending his medical career. Following up on this, Thoughtful House, where he had worked as the executive director, severed its affiliation with him.

Walker-Smith appealed the outcome of the GMC to the British High Court,[92] and after several days of hearings involving the lawyers for both Walker-Smith and the GMC, Mr. Justice Mitting released his seventy-six-page decision in December 2012.

Key to Mitting's decision was the question of whether the study that led to the *Lancet* article was for research or treatment.

Mitting stated early in the decision that

> At the heart of the GMC's case against Professor Walker-Smith were two simple propositions: the investigations undertaken under his authority on eleven of the twelve Lancet children were done as part of a research project-Project 172-96-which required, but did not have, Ethics Committee approval; and they were clinically inappropriate. Professor Walker-Smith's case was that the investigations were clinically appropriate attempts at diagnosis of bowel and behavioural disorders in children with broadly similar symptoms and, where possible, treatment of the bowel disorders or alleviation of their symptoms. The GMC's case was that he was conducting research which required Ethics Committee approval. His case was that he was conducting medical practice which did not. Accordingly, an unavoidable and fundamental question which the panel had to answer was: what is the distinction between medical practice and research?[93]

Mitting went on to slam the GMC panel for what he made clear was essentially a sloppy investigation.

Mitting noted that

> These difficulties arose in this case: Dr. Wakefield's purpose was undoubtedly research; Professor Walker-Smith's may have lain anywhere on the spectrum. It was for the panel to determine where it did; but first, it had to determine what his intention in fact was.[94]

And,

> Its conclusion that Professor Walker-Smith was guilty of serious professional misconduct in relation to the *Lancet* children was in part founded upon its

conclusion that the investigations into them were carried out pursuant to Project 172–96. The only explanation given for that conclusion is that it was reached 'in the light of all the available evidence.' On any view, that was an inadequate explanation of the finding. As it may also have been reached upon the basis of *two fundamental errors* - that Professor Walker-Smith's intention was irrelevant and that it was not necessary to determine whether he had lied to the Ethics Committee, it is a determination which cannot stand unless it is justified by the detailed findings made in relation to the eleven relevant *Lancet* children.[95] [Italics are mine.]

Justice Mitting's conclusions were, in polite terms, a severe reprimand to the GMC panel. In terms more familiar to those who might have served in any army, it was an absolute smack down.

His conclusions, which are worth quoting in detail, were that

For the reasons given above, both on general issues and the Lancet paper and in relation to individual children, *the panel's overall conclusion that Professor Walker-Smith was guilty of serious professional misconduct was flawed, in two respects: inadequate and superficial reasoning and, in a number of instances, a wrong conclusion.* Miss Glynn submits that the material which I have been invited to consider would support many of the pane's critical findings; and that I can safely infer that, without saying so, it preferred the evidence of the GMC's experts, principally Professor Booth, to that given by Professor Walker-Smith and Dr. Murch and by Dr. Miller and Dr. Thomas. Even if it were permissible to perform such an exercise, which I doubt, *it would not permit me to rescue the panel's findings.* As I have explained, the medical records provide an equivocal answer to most of the questions which the panel had to decide. The panel had no alternative but to decide whether Professor Walker-Smith had told the truth to it and his colleagues, contemporaneously. The GMC's approach to the fundamental issues in the case led it to believe that that was not necessary-an error from which many of the subsequent weaknesses in the panel's determination flowed. It had to decide what Professor Walker-Smith thought he was doing: if he believed he was undertaking research in the guise of clinical investigation and treatment, he deserved the finding that he had been guilty of serious professional misconduct and the sanction of erasure; if not, he did not, unless, perhaps, his actions fell outside the spectrum of that which would have been considered reasonable medical practice by an academic clinician. *Its failure to address and decide that question is an error which goes to the root of its determination.*

Finally,

> *The panel's determination cannot stand.* I therefore quash it. Miss Glynn, on
> the basis of sensible instructions, does not invite me to remit it to a fresh
> Fitness of Practice panel for redetermination. *The end result is that the finding
> of serious professional misconduct and the sanction of erasure are both quashed.*[96]
> [Italics are mine.]

Translated from legalese, Mitting basically destroyed the GMC's case, going
so far as to note that even the GMC's lawyer, Miss Glynn, did not propose a
reconsideration of the decision. (A shorter summary of the High Court decision can be found at the source in the endnote.)[97]

So Walker-Smith won, and won resoundingly. For him it was more a moral
victory than a financial one. He had retired two years earlier and simply didn't
want the opprobrium of the loss of his licence and the accompanying stain on
his long academic career to be the last things he did in medicine.

Why didn't Wakefield do the same? In actuality, he had also started an
action with the High Court but abandoned it. Why? I suspect there were
multiple reasons. First, taking the GMC to court had to cost a lot of money.
Further, Wakefield was more a researcher than a clinician, regardless of his
training, so maybe he thought that the loss of his medical licence didn't really
matter regarding what he wanted to accomplish in the future in advancing
the hypothesis that the measles vaccine, at least the MMR, was related to
GI-autism outcomes. Then again, maybe he worried that all the add-on charges
would make it hard for him to win.

Many of Wakefield's supporters thought the exoneration of Walker-Smith
suggested that Wakefield would be cleared, as well. This is not necessarily the
case, as the charges he would have had to deal with involved the financial
dealing with solicitor Richard Barr, the birthday party incident, and the extra
responsibility for accuracy that he had as first author of the 1998 article.

Then again, maybe he was just tired of fighting and wanted to get on with
his life.

Whatever it was, whatever his choices at the time and now, the repercus-
sions of the Wakefield et al. study continue to carry a disproportionate weight
in the vaccine wars over twenty years later.

While Wakefield et al. may have been wrong about the MMR vaccine
being the trigger for the GI-autism connection, as Stratton et al. concluded in
Chapter 3, being honestly wrong in science is not a sin.

Scientific fraud is a very different thing, of course, but as discussed above,
the allegations that Wakefield and colleagues manipulated data is simply not
in evidence. This conclusion is not changed by what the Brian Deers of the

world believe. For the GI-autism link alone, one corroborating piece of evidence lies in studies of other neurological disorders. Specifically, it is now becoming clear that the sporadic form of Parkinson's disease also has a clear GI component related to a dysbiosis of the microbiome.[98] Indeed, what seems to be emerging in the newer literature is that such neurological diseases are in reality multisystem disorders whether or not they arise from genetic or environmental factors.[99]

In this sense, Wakefield may have been far ahead of his time, not that he was the first to make a GI-autism link, but that he may have been the first to bring it to widespread attention.

The second thing that continues to have an impact on any vaccine safety discussion is that it influences each and every person who dares challenge the medical orthodoxy or raise even the tamest question about vaccine safety. To do so is a guarantee to be branded as a new "Wakefield," and there is even a verb for what happens to one so described: to be "Wakefielded." In other words, to be stepped on by the medical establishment for transgressions against the established order of things, the orthodoxy, as it were, of the religion of vaccination, described in more detail in Chapter 8.

It is of interest that both Andrew Wakefield and Brian Deer have recent books out about the events described above. In 2010, Wakefield published his own version of the events following the publication of the 1998 *Lancet* paper in a book titled *Callous Disregard*.[100]

While clearly Wakefield is seeking to exonerate himself from all the Deer and GMC allegations, the book itself is actually replete with scientific and legal citations, and quotes from a range of documents. In this regard alone, it far exceeds the Brian Deer product produced ten years later in *The Doctor Who Fooled the World*.[101] Rather than actually refuting Wakefield with the benefit of ten extra years to think about it, Deer contents himself with a series of caricatures of Wakefield, a variety of ad hominem slurs about Wakefield's appearance, his new partner, and various other people whom Deer has come to dislike. Even where Deer is apparently trying for humor, what emerges is not only not funny, but also comes across as the complaints of an aggrieved and bitter man. In brief, Wakefield still writes like a scientist; Deer still writes like one of Rupert Murdoch's tabloid reporters.

Finally, whether Wakefield was right or wrong about the MMR vaccine being linked to autism, both measles itself and concerns about the vaccine continue to percolate with the public, particularly as stoked by the mainstream media. One way to look at this is to scrutinize the responses to so-called measles "outbreaks," considered next.

Measles Outbreaks: The Necessary Bogeyman to Drive Vaccine Mandates (At Least until COVID-19)

Until the 2015 measles "outbreak" at Disneyland in Southern California,[102] measles had mostly disappeared from the vaccine wars, albeit not from the world. To be sure, Wakefield supporters and various investigators still considered that the MMR had a role in autism. Those on the other side had spent a lot of time and money trying to dampen down this speculation, as cited in detail in the previous chapter.

Overall, it would be fair to say that some sort of status quo had evolved. In addition, according to WHO, measles was close to being eradicated.[103] Then came outbreaks of the disease in various places, most famously Disneyland, and then a few years later in other places in North America.[104] Most of the outbreaks were firmly placed on Andrew Wakefield's shoulders by the mainstream media. After all, he was the deemed to be the medical charlatan who had caused MMR vaccine uptake to plummet,[105] so clearly he was the culprit.

The Disneyland outbreak was extensively covered by the media, and the blame, apart from Wakefield, was laid on "anti-vaxxers" in general. This blame game began gathering steam and then reached its zenith with the various measles outbreaks that occurred in New York City and the Tri-Borough area in late 2018 and early 2019. The blame for this last outbreak largely landed on those in the Orthodox (Haredi) Jewish community, some of whom were resistant to vaccine mandates. The resulting governmental and media response had a decidedly bigoted slant, often frankly bordering on anti-Semitism.[106]

The New York measles outbreak was closely followed by others in other states, including Washington and California, and in due course in British Columbia, where I live. In the following, I will focus on the latter, since I know this situation the best. However, I believe that what happened in British Columbia was symptomatic of what occurred in the various American states, albeit to an extent that was not as extreme in terms of human rights violations.

In the following, I want to consider a number of facets of the outbreaks and the panic that ensued. In some ways, what happened in 2018/19 was the harbingers of what was to come with COVID-19 less than one year later. In both cases, the spin of the media and various levels of government drove the public perception of the dangers of each disease and what the required response should be. Those inclined to so-called "conspiracy theories" noted that the scenario resembled the almost classic form of "problem-reaction-solution."[107]

Let's start with the questions of just how hazardous the measles is and, returning to the events of the previous chapter, what level of harm, if any, the MMR vaccine poses. Finally, how should we view the media coverage, and what role does it play in how politicians and the general public respond?

To answer the first, measles is a *Paramyxoviridae* virus,[108] which has certainly proven lethal in the past to populations that had not acquired natural immunity, for example, the Native population of the "Americas"[109] prior to European contact. At present, however, for most of those infected with measles in North America, the disease is typically neither life-threatening nor of long duration. In a small fraction of cases, the outcomes can be more severe leading to hospitalization; in a smaller fraction, disease complications can lead to death.[110] Thus, the disease is not necessarily trivial, but nowhere near the Black Death (bubonic plague) caused by the bacterium *Yersinia pestis*, or the life-threatening Ebola virus, either.[111]

During the course of the outbreak in British Columbia, the provincial health authority reported some thirty-one cases. There were no deaths.[112]

BC's Provincial Health Officer Dr. Bonnie Henry, who we will meet later in Chapter 13 concerning COVID-19, became a new provincial heroine with her frequent press conferences to keep the public informed about the spread of the disease. In spite of this, neither she nor the BC ministry of Health/ more local health authorities was able to say what percentage of those infected had been previously vaccinated against the MMR. When I tried to find out this information, I got passed from Dr. Henry to the Health Ministry to the regional health authorities, then to the BC CDC, which declined to tell me because of "confidentiality concerns."

Although the measles outbreak simply went away on its own, the government of the day was pressured by some panicked citizens and some in the health establishment to adopt more stringent measures for the future. One of the measures approved by Minister of Health Adrian Dix and Henry's office was for a mandatory registration of the vaccination status of all children in public schools. Vaccine mandates were not proposed in BC, unlike elsewhere,[113] but the requirement for reporting vaccine status did come with a provision for "education" sessions for "refusenik" parents with a nurse or other health-care provider. These requirements never came to pass, or at least did not do so in the first full school year after the outbreak. This was in marked contrast to other provinces, which went steaming ahead with proposed mandates. As of this writing, New Brunswick's mandatory vaccine legislation failed, and that of Ontario and other provinces is still pending.

Juxtaposed to Henry's statement about the dangers of measles was her take on adverse effects of the vaccine: "There are very rare serious adverse events following MMR and we have no deaths recorded; in contrast to the much higher risk of serious illness or death with measles infection."[114]

That clearly wasn't true, at least in British Columbia.

The numbers in the United States for young children tell a somewhat less comforting story about MMR vaccine safety: from 2015 to 2018, there were eighteen deaths in the MMR vaccinated group in an estimated number of 21.5 million zero-to-five-year-olds in that time frame. In the same period, in the MMR-unvaccinated population of zero-to-five-year-olds, there was one death in 1.99 million from measles. This puts the numbers of deaths at just under one per million in the vaccinated group and about one per two million in the unvaccinated group.

The BC outbreak mirrored larger ones in Washington State and in New York, New Jersey, and Connecticut among the Haredi Jewish communities. In all cases, there were no deaths, and the numbers of partially or fully vaccinated people who still caught measles was significant.[115]

All US states then had, and continue to have, vaccine mandates in place for public school entry, but some still allowed at that time for exemptions based on medical, religious, or philosophical reasons. In Washington, the measles outbreak in Clark Country in the southwest of the state led to the declaration of a state of emergency by the governor and to the passage of a law by the state legislature removing philosophical (personal) exemptions for vaccines.

In New York, the state legislature earlier passed a similar law. Other laws in other states followed suit.

Going back to the Disneyland measles outbreak in 2014–2015, the statistics are that it affected 147 people, 5 percent of whom were fully vaccinated.[116] The California state legislature passed a law barring vaccine exemptions for all but medical reasons.

Two things emerge from a consideration of the numbers. The first is that the disease numbers are not large, certainly not in proportion to the overall populations in these areas. Proponents of the measles vaccine say the numbers are low thanks to the fact that most people are already vaccinated against the disease.

Accepting this, however, brings us to the second point: what should one make of the numbers who are vaccinated and yet still develop the disease? The CDC claims that the MMR vaccine is highly effective at preventing measles overall, but individual levels of protection may vary, and hence some of those fully vaccinated will still get the disease if exposed to an infected person. During the period in which people are contagious, they can transmit the disease to those who are not vaccinated, including those who cannot be vaccinated due to allergies to vaccine ingredients or a compromised immune status.

In an earlier request by me to the BC CDC about the percentage of the latter, the response was that children who cannot be vaccinated comprise 0.03 percent of the population. Including adults, the number is 0.3 percent; in other words, three in ten thousand and three in one thousand, respectively.

One of the major talking points in much of the media coverage in British Columbia and elsewhere concerned the severity of measles for health consequences. In 1962, just a year before the first measles vaccine was put on the market, the US government's Vital Statistics of the United States, Vol. 2-Mortality, Part A, published data showing the overall causes of mortality in the United States by disease.[117] Measles deaths in the United States were listed as occurring at a ratio of one in five hundred thousand. This number is misleading, as the total fatalities in that year were about five hundred in about four million people infected, thus giving a ratio of 1.25 in ten thousand fatalities. This latter number closely matches a later estimate of one in one thousand from the CDC.[118]

The Public Health Agency of Canada's online report on measles does not reveal any recent deaths from measles.

In spite of this, a commonly quoted number for fatalities is one in one thousand, which is frequently repeated in the media with some variations.

Some researchers maintain that agencies such as the CDC may be conflating cases that show up in hospital versus the bulk of measles cases that do not by including the numbers of serious complications and/or by giving the measles-related fatalities worldwide. The latter is problematic due to the overall health of people in much of the Third World. In addition, the inflation of disease harms and deaths in relation to the overall number infected is a problem that has cropped up in relation to COVID-19 (see Chapter 13).

Dr. Bob Sears, a California pediatrician, addressed this issue when speaking to senior rabbis in the Haredi Jewish community in Staten Island in December, 2019:

> Measles is a routine childhood illness for 99.9% of people who catch it. I know you have likely been told that measles has a very high complication rate (20 to 30%) and can kill 1 in 1000 children who catch it. These concerning numbers come from US measles outbreaks from 30 years ago in which ONLY people who came to a hospital or saw a doctor were counted in the data analysis.[119]

What then is the true measles fatality rate in North America? The VAERS surveillance system introduced in Chapter 2 gives some clues. The CDC lists 111 deaths due to measles complications from 1990 to 2018. In the same time period, VAERS data show 148 deaths attributed to the measles vaccines.[120] However, as noted previously, the 1 percent actual reporting of adverse effects may suggest that the real numbers of MMR deaths are quite a bit higher.

Critical pieces of information usually left out of mainstream media reports include a Finnish study that showed that the MMR vaccine's long-term ability to prevent the disease may be compromised in some individuals by average declining antibody levels of about 50 percent at eight years after the recommended second MMR shot and by 60 percent at fifteen years.[121] This is termed secondary vaccination failure and has been discussed earlier. In this particular study, some 95 percent of those tested had some seropositivity remaining, a finding that has led some medical authorities to conclude that the immunity conferred by two doses of the vaccines, even with declining antibody levels, is effective for life. Dr. Henry, for example, tends to believe that this is correct.

This view is actually not generally true, as various studies attest, as measured either by antibody levels or even by measuring T helper or T killer cells.[122]

In the United States, one estimate is that less than 30 percent of adults receive booster shots for measles, which, with declining antibody levels in some individuals, may make them as likely as the unvaccinated to acquire and transmit the disease.

In the face of the measles outbreaks, the calls for vaccine mandates and the elimination of exemptions, and the pushback against them due to safety concerns, seem to reflect very different narratives about the quality and independence of the existing research that was examined in detail in Chapter 3.

Alan Cassels, a Victoria-based journalist who writes on medical issues, cited a 2012 meta-analysis from the Cochrane Collaboration that examined fifty-seven studies from clinical trials involving approximately 14.7 million children who had received the MMR vaccine. The Cochrane's conclusion stated that "the design and reporting of safety outcomes in MMR vaccine studies, both pre- and post-marketing, are largely inadequate."[123]

A similar point was made repeatedly in Chapter 3.

The late Dr. Toni Bark, a Chicago MD and emergency medical planner, agreed with Sears about the quality of the evidence on MMR safety and noted a crucial point in regard to vaccine mandates and possible adverse reactions: "If you eliminate the exemptions you are making a large minority of people susceptible to very serious risks including death."[124]

The waning immunity of some non-vaccine-boosted adults may further compromise the widespread belief in so-called "herd immunity," that is, the concept that when a high enough percentage of all persons naturally acquire the disease, or are vaccinated, the disease dies out before it can spread to those who can't be vaccinated.

Models of herd immunity require vaccine coverage for measles of well over 90 percent to be effective. In the United States, children comprise something like 24 percent of the population and, assuming all are vaccinated, the problem may lie with the adult population, who did not acquire natural immunity from the disease or do not get booster shots later if their seropositivity to the vaccine has significantly declined. Of the latter, as noted above, only about 30 percent had received the booster shots. Taking these numbers together suggests a percentage too low to achieve herd immunity based on standard herd immunity models, as discussed in the next section of this chapter.

For measles, and more so with pertussis (whooping cough), achieving herd immunity would likely require mandatory vaccination of adults with low antibody levels in addition to school-aged children. Legislators in some US states are apparently contemplating just such laws for the adult population.

Who would benefit from such mandates? Cui bono? This issue is addressed in Chapters 13 and 14.

The measles issue is thus not without ongoing controversy and seems a clear remnant of the initial vaccine wars that may have begun with the MMR but are nowhere near ended, as we see with COVID-19. This is a gift for the media that keeps on giving and, at least until COVID-19, served as the hinge on which to pillory those deemed to be "anti-vaxxers." And, of course, measles outbreaks allow the media to drag out, again, their portrayal of the whole issue arising from the Wakefield et al. study of 1998.

Some of the ongoing issues that should be mentioned before moving on to the last section of this chapter are the following:

First, there is some evidence in the literature to suggest that those who get the measles suffer from prolonged overall immune suppression that makes them more susceptible to other diseases, particularly those that might induce secondary respiratory or digestive system infections. The key idea here is that there is an "immunological amnesia" that occurs when the measles virus attacks memory T-lymphocytes, thus weakening the immune system response overall to other pathogens. How long does this immune suppression last? It varies, but some data suggest a few weeks to some months[125] and even as much as two to three years.[126]

How likely is this to be generally true, and does immune system suppression occur for naturally acquired diseases such as measles? If true for measles, this might serve to support the MMR vaccine as a general boost to an overall effective immune response.[127]

Another question to ask is whether the MMR vaccine, designed to elicit a protective immune response, also triggers such immune suppression. It may well be that it does not, since the real virus attaches to host cells by way of a receptor site on the cell, while the attenuated virus does not. It would be interesting to know if anyone has looked at this in more detail.

Another observation in relation to the MMR vaccine concerns antibodies passed on from the mother to a newborn that are deemed to be lesser in number than those passed from an unvaccinated mother who had the actual disease. In both cases, the antibody levels in newborns decline with age, but in children with vaccinated mothers, the decline was considerably faster, observations that again suggest that the notion of secondary vaccine failure needs to be revisited.[128]

These outcomes lead to some authors suggesting an earlier vaccination of infants with the MMR to combat maternal antibody decline. In turn, there have been suggestions that natural measles antibodies in breast milk may inhibit the antibodies generated by the vaccine given to infants. I mention this last point because it highlights yet one more aspect of the increased pace of overmedicalization, in which vaccine-derived immunity is considered to be superior to naturally-acquired immunity.

And finally, there is the Dr. William Thompson case, in which Thompson, a CDC epidemiologist, turned whistleblower and, in recorded telephone conversations with vaccines skeptic Dr. Brian Hooker,[129] revealed that a group at the CDC had deliberately removed, indeed destroyed, data suggesting that Black children had a significantly higher probability of developing ASD following the MMR vaccine. Thompson has never testified publicly and continues to work at the CDC, so there is currently no way to know if these allegations are correct.

Herd Immunity: What It Is, and What It Isn't

A topic that gets trotted out quite frequently when considering vaccine mandates is the need to maintain, or establish, "herd" or "community" immunity. As discussed in the section above, this is frequently used as an article of faith when it comes to mandates for the MMR vaccine and is now widely bandied about in relation to COVID-19 and the various vaccines being developed or even already deployed.

The term "community immunity" is the more politically correct version that came into use some years ago, perhaps driven by umbrage and the resulting pushback from some people, mostly on the vaccine-skeptical side, who felt demeaned about being considered a member of a herd.

The concept of herd/community immunity itself is fairly simple at first glance and even makes some intuitive sense: if enough people (or animals in a herd) are immune from a particular pathogen, then, depending on the characteristics of that pathogen, it won't be able to spread to infect others who may not yet have immunity. In the broad view of the term, immunity will arise through prior exposure to the pathogen and thus natural immunity involving both innate and adaptive immune systems, or maybe because of immunity acquired through vaccination that typically only involves the latter.

In this view, if one knows some key variables, one can calculate the number needed to vaccinate successfully to achieve this overall level of herd immunity. These factors include the type of disease transmission; the Basic Reproduction ratio (Ro, called R "nought"), e.g., how many persons an infected person can, in turn, infect; the serial interval, that is, the time between successive cases in a chain of transmission; and the H* (the herd immunity range). Taking a disease like measles, in 1993 the Ro was 12–18, making it very infectious. The calculated H* at the time was 83–94 percent needing to be vaccinated to achieve herd immunity.[130] That changed and by 2018, Hussain et al. were citing and calculating the percentage for herd immunity at 96 to 99 percent.[131]

Herd immunity as a concept can be modeled by creating a schematic program that shows a given number of people (1500 in the example provided soon) and the overall number who cannot be vaccinated for various reasons, the latter usually taken as about 0.3 percent based in this model on data from the BC CDC.[132] Finally, a disease vector, or initially infected subject, sometimes called "Patient Zero," is added to the population. The program is then allowed to run for a set period of time, allowing one to visualize disease progression in the overall population.

The number who cannot be vaccinated is estimated based on those known or suspected of having some level of immune compromise. For example, immune compromise might be due to treatments that limit immune responses such as radiation therapy in cancer treatment, or due to allergies to some vaccine component. In British Columbia, this number is estimated to be 0.03 percent in children under eighteen and 0.3 percent for the population overall (BC Centres for Disease Control). Put another way, this comes out to be three persons in ten thousand or three in one thousand, as noted previously. Patient Zero is considered to be a disease-infected person who comes into contact with the model population and starts the spread of the disease.

In Fig. 4.3, found in the photo insert in the center of the book, the vector is the red dot with a black center, those who cannot be vaccinated are in purple, those immune because of prior infection or vaccination are in green, and those not immune due to either no vaccination or prior disease-induced immunity are in yellow.

For measles, the estimated effectiveness with two doses of the MMR vaccine is considered to be about 97 percent. Or, put another way, some 3 percent of the vaccinated population has "primary vaccination failure" and is comprised of those for whom the vaccine does not provide any initial immunity as measured by surrogate markers such as antibodies.

The table below the various graphs shows the values for measles and some other infectious diseases.[133]

In the upper panels of Fig. 4.3, we have set the numbers who have disease immunity at 0 percent, 50 percent, 75 percent, and 97 percent. When we allow the program to run for a set time, the model provides a demonstration of what herd immunity should look like: In the 0 percent graph, the disease spreads to all persons, at 50 percent and 75 percent it spreads to most people, but at 97 percent disease spread is rapidly contained.

So far, so good. But there are significant problems with this ideal model. The first problem is termed "primary vaccine failure," as noted previously. The second, and larger, problem is called "secondary vaccine failure" and includes those in whom immunity has worn off such that antibody and T cell levels have dropped below some critical level for protection.

For measles, there are various assumptions or measurements of secondary failure. One widespread assumption is that the number is zero, but according to the available scientific literature, this seems to be largely incorrect. The actual rate estimation can be quite variable from less than 1 percent to almost 80 percent over the course of twenty some years after vaccination.[134] Here is where it gets interesting: In a 2004 article, Naniche et al. measure antibody levels for measles as well as CD4 and 8 cells (see Chapter 2) producing either interferon or tumor necrosis factor measured twenty-three to forty-seven years postvaccination.[135] The authors claim that only CD4 cells decline significantly at times greater than twenty-one years since vaccination in the tested population, so this should reassure us that immunity from measles following the MMR is pretty much for life. This may be true at a population level, but the huge standard deviations also mean that a large number of people have lost both antibodies and T cells and B cells. Insofar as this is true, it means that a number of people are not protected and, in turn, are infectious. How big is the problem? Big enough to make it highly unlikely for herd immunity to occur without revaccinating much of this age grouping. The whooping cough (pertussis) part of the DPT vaccine has a secondary failure of close to 100 percent after eight years.[136]

The actual level of secondary failure is crucial to a consideration of herd immunity, particularly as it applies to school mandates, not to mention mandates in general. For example, let's assume that school mandates in a state or province, or even country, are in place and 100 percent of those who can be vaccinated are in fact vaccinated to achieve the required overall percentage for herd immunity. In North America, the percentage of the population under nineteen is about 23 percent,[137] of whom some 3.03 percent cannot be vaccinated and have primary vaccine failure.

The rest of the population now needs to be considered. Those over the age of fifty-six, born before the MMR vaccine became available, possibly have had the actual measles and now have natural, lifelong immunity. For the purposes of this illustration, let's assume that this is correct. This is about 23 percent of the overall population.[138] The rest of the population, between the ages of nineteen and fifty-six, become the wild cards. Some, in North America most likely, were vaccinated with the MMR vaccine. Some had the recommended two doses; some had one dose (and some had none). Now, depending on the amount of secondary vaccine failure in the vaccinated group, the model shows an increasing failure to achieve herd immunity and the disease spreads. This outcome is shown in the lower panels of Fig. 4.3. The model can be run with pertussis as well, and the results are just as predictable (see Fig. 4.3).

What would happen if we did the same thing with any of the COVID-19 vaccines? Given that no coronavirus, such as those that cause the common cold, has yet succumbed to a vaccine, the outcome would presumably show a higher level of both primary and secondary vaccine failure in a much shorter time frame. This has led to predictions of having to give boosters for any COVID-19 vaccine yearly or even more frequently.[139]

In brief, herd immunity as proposed by mainstream medical authorities and as envisioned by legislators voting on any vaccine mandates for diseases such as measles, pertussis, or COVID-19 cannot be achieved by simply going after school-aged children, the low hanging fruit in typical herd immunity considerations. Rather, in order for vaccine mandates to work in such cases, secondary vaccine failure has to be considered in those previously vaccinated. The only way to do so, however, is to arbitrarily revaccinate all of those in this previously vaccinated demographic, or do so more selectively based on antibody levels. In other words, to get herd immunity by mandatory vaccination in school-aged children, a much larger part of the population needs to be vaccinated, as well.

All of this leads to several quite predicable conclusions, the key one being that, for vaccine mandates to work, all have to be vaccinated either by choice or by compulsion. Based on the emerging data coming from companies seeking COVID-19 vaccines, mandatory vaccination for almost all of the population at large would be necessary. Indeed, there have already been opinion pieces in scientific journals and newspapers calling for just such mandates at large.[140]

With this, the key concern then is for the preservation of basic human rights, not to mention the possibility of greater adverse vaccine effects in a larger fraction of the population. These last points will be considered in Chapters 11 and 13 in more detail.

Moving the "Goal Posts" in the Search for ASD Etiology (Part 1)

Any plot of ASD incidence shows a dramatic increase since 1986, as previously shown in Fig. 4.2, with the increase seemingly accelerating over time.[141] Even mainstream medical organizations, including the CDC, agree with this upward trend, without necessarily agreeing on why the increase is happening, or even if it is real.

As I've tried to illustrate previously, genes or SNPs are not likely to be causal, at least not alone. Further, notions that the apparent increase in ASD over time is due to changing diagnostic criteria, better detection, and/or greater awareness are also likely to be incorrect.

What's left of the obvious etiologies is the environment. This factor, passed over briefly before, needs a bit more evaluation, and I will provide this in the following.

Environment, now more concretely and broadly termed the "exposome,"[142] encompasses a huge range of toxic substances to which humans are exposed in an increasingly polluted world. These can include various toxic molecules in the environment such as pesticides and heavy metals like lead, cadmium, and mercury, and other metals and toxins in "biosolids,"[143] aluminum, various sources of radiation, and a huge variety of infectious pathogens, to mention just the most obvious. Much of the human exposure to such toxic compounds has increased dramatically since when our grandparents were born.

Any of these toxicants, in principle, could disrupt the nervous system in early development. Further, we can be reasonably sure that the environmental factors that could play a role in the initiation of the events leading to an autistic brain work in some sort of synergy, creating the equivalent of a "perfect storm," discussed previously. Add on genetic susceptibility or enhancer regions to interact with the exosome, and you have all the ingredients you need for ASD, not to mention other neurological disorders across the lifespan.[144]

Is it therefore possible that vaccines with some of their ingredients might be involved at some level? This question is at the heart of Chapter 5.

One of the most prominent bloggers on the subject of vaccines, Dr. David Gorski, frequently points out that no matter what, those he describes as "anti-vaxxers" (pretty much anyone he disagrees with) always come back to vaccines as the causal factor in ASD. He may be right about that, but wrong to assume that it is simply a mental tic that those who blame vaccines display.

In fact, the reasons that those concerned about autism return to vaccines as part of the pathological process is twofold. First, vaccines may contain substances known to be neurotoxins; second, the temporal sequence of a child receiving a vaccine and the resultant autism is clearly evident in a number of cases. Unless all the parents who report a vaccine injury to their child are delusional or lying, a seemingly unlikely proposition, then the notion of vaccines serving some role in autism cannot be simply discounted.

Maybe the vaccine factor involved is not Thimerosal or even the MMR vaccine, as considered in detail in the last chapter. But it has to be recalled that of the myriad studies cited in Chapter 3, none compared any vaccine alone or in combination to true control subjects.

For this very reason, vaccines will continue to be under observation, at least by some parents and scientists, until such interactions are explored by scientists with no ties to the pharmaceutical industry.

The solution is actually pretty simple: If we truly want the hypothesis that vaccines are somehow involved in ASD to be nullified (and remember that any collection of null results can be overturned by one positive outcome), someone has to do the detailed research.

Would negative answers derived from well-designed studies that looked at vaccine interactions convince everyone? For certain, they would not.

To begin with, there are many in the mainstream medical professions who would continue to claim that the necessary studies have already been done and that the science is now "settled," as if somehow the weight of dead trees comprising the bulk of the scientific literature, cited in Chapter 3, is more than sufficient to resolve the issue.

In this regard, statements such as "We don't know what causes autism, but it is not vaccines" are simply ill-informed and not actually very scientific: if one does not know what causes event "X," it is sort of ludicrous to state that it can't be factor "Y" for the simple reason that not all combinations of Y have been tested, and those that have been have not been tested very well. In particular, up until now there have been no studies of whole vaccine schedules, nor of vaccines-gene interactions in either humans or animals.[145]

Does this mean that vaccines are responsible for the increases in ASD over the last forty years? No, it certainly does not. But is it still plausible that vaccines, in ever increasing amounts, play some role? The answer is very much yes. Denying this reality without the critical experiments being done doesn't make the problem go away, it just fosters cynicism and contempt for medicine in general and in modern medicine's ability to achieve some level of immune protection from infectious diseases.

Would a null result to a well-designed study convince enough of the vaccine-hesitant to accept a more mainstream medical view and thus lead them to focus on other factors that may actually be involved in autism? Apart from doctrinaire true anti-vaccine advocates, the answer is "maybe."

So, are the vaccine-hesitant simply moving the goalposts following some bizarre obsession aiming to keep the focus on vaccines? Maybe some people are, but in my opinion the majority are simply looking for a clear, rather than pat, solution to ASD.

With the above in mind, in Chapter 5, I will deliberately move the goalposts twice more to consider the role(s) that several metals, mercury and aluminum, might play in ASD and other neurological disorders. In this regard, it is important to always remember that in science, goalposts are not necessarily fixed, unlike in sports.

CHAPTER 5

On Mercury and Aluminum: General Aspects of Neurotoxicity and the Role of Aluminum Adjuvants

Aluminum (Al) is present in very small amounts in living organisms but is abundant in the environment. In no case has Al^{3+} been shown to have a definite biological function. Taken together, this suggests that Al^{3+} possesses properties incompatible with fundamental life processes.

—P. O. Ganrot[1]

Moving the "Goalposts" in the Search for ASD Causality (Part 2)

As noted in Chapter 4, a common theme in the mainstream media and blogosphere is that, for so-called "anti-vaxxers," the cause of autism is always going to be vaccines no matter what. The notion here is that when one hypothesis about vaccines and autism is discarded or shown to be unlikely, for example Thimerosal or the MMR, another vaccine bogeyman will immediately take its place. From this perspective, those who harbor an innate hatred or fear of vaccines will always seek and find another vaccine-related bad actor responsible for autism. That bad actor might be aluminum, or too many antigens at once, or any of a rather large shopping list of other vaccine ingredients alone or in combination ad infinitum, essentially an endless game of vaccines = autism "whack-a-mole."

Those who like this argument seem to like it a lot. But is such goalpost shifting, if that is what it is, unique to vaccines and any controversy surrounding them?

The answer is that it is not.

I will give one example from my other main field of study, the Guamanian neurological disease spectrum, ALS-parkinsonism dementia complex.[2]

Long before I became interested in this topic, indeed before I was even born, a series of epidemiologists and neurologists were looking into what appeared to be a real cluster of neurological disorders[3] that seemed suddenly to spike on Guam at the end of World War II. Some of those affected seemed to have a rather conventional form of Lou Gehrig's disease (ALS). Others were found to have a form of Parkinsonism with dementia. Some of those affected had both disorders. Incidence levels were vastly higher than for ALS or Parkinsonism (including conventional Parkinson's disease) in North America or Europe.

The lead epidemiologist studying ALS-PDC, Dr. Leonard Kurland of the Mayo Clinic, saw this disease spectrum with its very high numbers as the neurological equivalent to the Rosetta Stone.[4] And, as a result, a number of scientists joined the hunt for the causal factor(s).

The hunt started with hypotheses about genetic causality, then with a lot less understanding of genetics, since the structure of DNA had not been described. No clear Mendelian dominant inheritance pattern was seen. Those involved in the hunt then moved the "goalposts" to look at various toxic substances on Guam, some naturally occurring such as aluminum, some not, and from there to possible toxins in the local diet. From this last came a focused inquiry on a particular foodstuff—the seeds of the cycad tree—which led, in turn, to intense efforts to identify every toxic molecule that cycad seeds might contain.[5]

Of the latter, a series of studies first focused on a compound called methyl-azoxymethanol (MAM). When that didn't generate a clear neurological outcome resembling ALS-PDC in experimental animals, researchers moved on to look at several nonprotein amino acids, BOAA and BMAA.[6] When none of these seemed to give the features of ALS-PDC, more genetics studies followed. Then the work went around yet again, back to the amino acid toxins. For researchers trying to understand the origins of ALS-PDC, the goalposts were always in motion.

Eventually, our group proposed still another toxin isolated from cycad seeds, BSSG, and other steryl glucosides[7] just at the same time that BMAA returned as a favorite toxin for the disease, thus moving the goalposts twice in a short period. It seems almost certain that genetic factors will come back yet again, focusing on enhancer regions of genes of those affected, or some combination of genetic susceptibility factors that act synergistically with certain environmental factors.

Is all this goalpost moving the result of a bunch of neurotic neurologists and neuroscientists desperately looking for scapegoats for the disease?

Actually, no. This is actually how science works: You start with an observation, in this case the apparently sudden explosion of ALS-PDC on a small island. You then hypothesize some plausible causal factor based on observation. If this factor turns out not to be involved, it doesn't mean your original observation was wrong, merely that one hypothesis was wrong. That happens in science all the time. You then try another hypothesis, and this fails, too. The observation remains, and the reasons for the disease are still not known. Another hypothesis arises, etc. And on and on it goes until one of the hypotheses turns out to be correct, if it ever does.

With autism, you start with the nature of the disease itself and what seems to be a changing temporal incidence, as discussed in the last chapter. So, in this case you have a developmental disorder with certain characteristic behavioral features, named autism, an observation that has stood the test of time from Kanner through to DSM-5. A corollary observation is that the incidence of autism and related disorders seem to have increased enormously since the 1980s. Is incidence really changing, or is it consequence of the methods used? There are quite different opinions on this, as already discussed.

Still, there is this disorder. What causes it? Is it genes, noncoding DNA regions, toxins/toxicants in the environment, social factors, gene-toxin interactions, etc.? How do we find out?

If we suspect genes, we screen for mutant genes or regions of DNA that might be affected differently in those with the disease versus those without. If genetic explanations seem to come up short, then we move along to look at environmental factors. For example, are there toxic substances early in life—fetal or early postnatal—that might cause neuronal changes to the developing brain? Are these toxic substances something that only those who later become autistic encountered, or were they ubiquitous? An example of the former would be that of the birth defects linked to giving some pregnant women the sedative thalidomide to control nausea and morning sickness: not all pregnant women took thalidomide, and only the ones who did had children with the characteristic birth defects. This sort of cause-and-effect outcome is pretty simple, at least in hindsight.

A more ubiquitous etiology would be something like drinking water contaminated with some heavy metal or pesticide. If in high enough concentrations, the impact will be more obvious, as it will affect more children, and even adults. If lower concentrations are responsible, fewer children and adults will fall victim. At this point, the challenge becomes one of figuring out why some people are susceptible to this level of toxin while others are not.

In the case of autism, what might be the equivalent of thalidomide? So far, nothing like this has come up in the literature. What about the ubiquitous factors that might be variable in amount, such as particular types of food, certain known or unknown toxic substances in food or water, particular kinds of drugs taken by most children, or particular health measures, such as vaccination? Which of these could be evaluated using the Hill Criteria discussed in Chapter 1 to give clues? All of the above, but none would be as likely to satisfy as many of the Hill Criteria as the childhood vaccination schedule. In particular, the criteria of biological plausibility and temporality may be the most relevant.

If you don't find any environmental factors that seem clearly to be causal, the next step is to look at interactions of genes and environment that might make some individuals susceptible to the disorder while others are not. The observation of the existence of a neurological disorder called autism is the bedrock to which you return time and time again as each hypothesis is tested and discarded.

At the end of the process, you might discover that for some children, vaccines alone or in combination are the key causal factor in early development leading to ASD. For other children, it may not be vaccines at all, but some other toxin acting alone or in combination with some genetic susceptibility to that toxin. For some children, autism might be the end product of a "perfect storm" of interacting factors.

Some call this goalpost moving. Others call it the process by which science advances.

Overall, the point of the above exercise is simple: once you have an observation that you want to understand, the framing and reframing of hypotheses does not end until you have found a likely causal factor. This is true whether the observation is neurological diseases on Guam or autism around the world. This is not a case of shifting goalposts, to use the dismissive term, but rather the normal process by which science zeros in on the real causal factor(s).

For this reason, statements often made that "we don't know what causes autism, but we know it is not vaccines" are both unscientific and meaningless. Put as concisely as possible, if you don't know what really causes autism, you cannot say that vaccines are not part of the set of factors involved unless and until you have investigated all of the permutations of those factors. And, as demonstrated in the other chapters of this book, the CDC and others haven't done anything of the sort. Indeed, as shown in the previous chapter, the CDC hasn't even clearly demonstrated that Thimerosal or the MMR vaccine are not partially responsible given that most of the work they cite as proof is so compromised that it makes any null conclusion highly questionable. The exception to this might well be the Stratton et al. report of the IOC in 2012.

The reality is that if the CDC, FDA, and NIH really wanted to answer the question and absolve vaccines, they would spend the money to do a lot more searching for alternative hypotheses, environmental and genetic, and combinations of the two. That they don't suggests either that these agencies are so tightly locked into dogmas of their own design that they can no longer be objective. Or that they fear the answer that might arise, an answer that might upset a corporate agenda and the careers of a lot of people who have built reputations denying any possibility of vaccines having any role whatsoever in autism. Further, it hardly helped for physicians and the medical establishment to belittle parents by telling them that the onset of autism in their children was merely coincidence or a figment of their imaginations.

But none of this happened, so desperate parents latched onto the one thing they could see with their own eyes: the often abrupt change in their children following some vaccine. Moving goalposts? Not really, more like trying to get governments, the medical establishment, and the pharmaceutical industry to stop playing "Calvin ball" with constantly changing rules.[8]

The enormous irony here is that while some of the mainstream medical community, the above agencies, and the trolls accuse the vaccine-hesitant of being unscientific "woo meisters," their own behaviors in regard to basic observation and inquiry into the possible connection between autism and vaccines is anything besides scientific. The term in psychiatry/psychology for attributing to others what you yourself exhibit is called "projection."[9]

I will examine some of the psychology at play in Chapter 8, and while this examination is hardly definitive, it is potentially revealing.

All of the aspects mentioned being the case, parents and some researchers kept looking for explanations for autism, and the next hypothesis involved mercury (Hg) in vaccines.

Thimerosal and ASD

As discussed in the previous chapter, the Wakefield et al. paper of 1998 led to a fairly widespread concern about MMR as a causal factor. It was perhaps this linkage, correct or not, and its aftermath in the Wakefield drama that triggered a growing concern about the potential for adverse effects from vaccines more generally. The plethora of scientific studies that came out after Wakefield et al. was intended to stem these concerns. To a large measure, these studies succeeded in doing so, and the issue was for all intents and purposes laid to rest with the release of the Statton et al. (2012) report by the IOC, as described in Chapter 3.

But continuing concerns about vaccines quickly focused on the presence of a mercury compound, Thimerosal,[10] added to multiuser vaccine vials to control bacterial growth. The historical and experimental observations that mercury (Hg) could be a neurotoxin were already well established leading to a general feeling that Hg in vaccines, in any formulation, could be a problem, particularly perhaps on a developing nervous system.

The Neural Toxicity of Hg

Ionic mercury is a known toxin, particularly a neurotoxin, with observations that go back easily more than hundreds of years. Terms such as "mad as a hatter" reflected the common observation that those who made hats, hatters, and other workers exposed to mercury often had severe neurological consequences.[11] In fact, mercury is classified as the most toxic nonradioactive element on Earth. It has an atomic number of 80, making it one of the heavier metals, and carries a charge of +1 or +2, meaning that it can have either one or two fewer electrons than protons. Hg is a liquid at room temperature.

In spite of this background knowledge of the toxicity of mercury, exposure to mercury in any of its forms was treated pretty casually, at least until fairly recently. For example, I remember as a kid playing with mercury that would come out of a broken thermometer, and no adult in my world seemed unduly perturbed.

Part of this lack of concern involves the route of exposure. Playing with metallic mercury is not safe, more so if one has skin lesions, but even then such exposure was not then viewed as particularly harmful, but actually sometimes beneficial for its antibacterial properties. After all, this was still a period in which minor cuts and scrapes were routinely treated with an antiseptic mercury ointment called Mercurochrome,[12] which was widely used until at least 1998.[13]

Mercury in diet, however, has been shown to be extremely harmful. One of the better-known cases of a neurological disease cluster involved a severe case of industrial pollution at Minimata Bay, Japan, in the late 1950s. Health authorities had noted that some inhabitants of Minimata showed a range of neurological signs, including seizures, intermittent loss of consciousness, and repeated periods of perturbed mental state. Many of those affected progressed into a permanent comatose condition. Birth defects, including severe mental retardation, rose to epidemic levels.

The impact on so many people in such a relatively short period of time led to the identification of the likely causal factor: an organic form of the element, methyl mercury, coming from a nearby chemical facility that had been released into the bay. In turn, the mercury found its way up the food chain into fish, and from the fish into humans.[14]

All of the above should have been a reason for some caution when considering injecting mercury in any form into the body or in application to areas of high absorption. Nevertheless, this was done with a range of vaccines using a second organic form, ethyl mercury, which was used in North America and elsewhere to prevent bacterial contamination in multiuser vials or in solutions used repeatedly, such as eyedrops.[15] Perhaps the perspective here was like mine as a kid: if it's safe enough to play with, then how bad can it be in vaccines or eyedrops?

Thimerosal was removed from both eyedrops and most vaccines in 2000, in the latter case only remaining in the influenza vaccine. The "official" reason was that Thimerosal was being removed, not because it posed any serious threat to human health, but due to *public concerns* about its potential link to ASD.[16]

The mainstream medical view about ethyl mercury, especially in regard to Thimerosal, then and now, is that it is not toxic compared to the methyl form and thus "safe," is only found in tiny amounts in vaccines, and is rapidly removed from the body after injection. In reality, the toxicity of mercury lies in its ionic form, so the discussion really boils down to how the actual element is released into the environment or in the body.

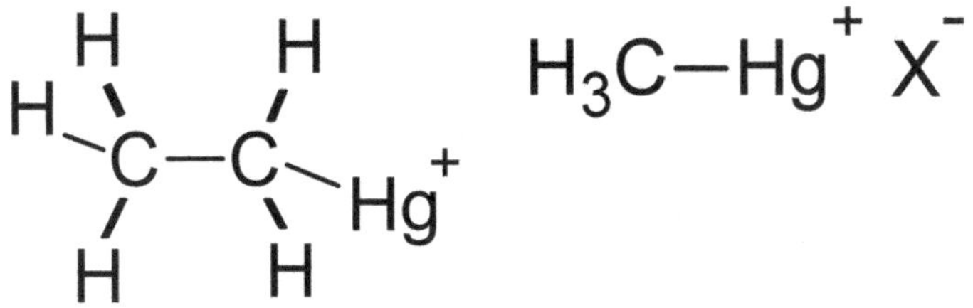

Figure 5.1. Chemical structures of ethyl and methyl mercury compounds.

Scientific Reviews of Thimerosal Toxicity

In the early 2000s, the CDC took the concerns about Thimerosal and any potential link to ASD seriously enough to hold a small conference at the Simpsonwood Retreat Center outside of Atlanta, Georgia. Not long afterward, the Institute of Medicine issued a report on their findings on the subject in 2001. With these, the CDC and FDA considered the matter closed, particularly since Thimerosal had been removed from the list ingredients of vaccines in the United Staetes in 2001, apart from some influenza vaccines.[17] In 2015, Robert F. Kennedy Jr. published a book, *Thimerosal, Let the Science Speak,*[18] that contained an extensive review of the scientific evidence in support of the hypothesis that Thimerosal had a decisive role in the increase in ASD.

In the following, I will consider each of these documents in turn.

Simpsonwood Meeting on Thimerosal

On June 7 and 8, 2000, the CDC, facing growing concerns about the presence of Thimerosal in some vaccines, hosted a miniconference in Norcross, Georgia. At the meeting, fifty-three participants listened to a presentation by Dr. Thomas Verstraeten, then a CDC scientist.

Verstraeten's data were drawn from the VSD, a vaccines surveillance system controlled by the CDC as described in Chapter 2. Titled "A Scientific Review of Vaccine Safety Datalink Information," the material was debated over the two days of the meeting. The 259 transcript pages from the meeting are available online[19] and are certainly revealing for what the conference transcripts do, and don't, contain.

The meeting was called to order by Dr. Walter Orenstein, then the director of the National Immunization Program at the CDC. Orenstein, in his introduction, termed the meeting a "simultaneous individual consultation," noting that each of the formal participants, termed consultants, would be asked for their opinion on a series of questions. The focus of the meeting was to be the Verstraeten report concerning Thimerosal and its presence in some vaccines, including the Hep B, DPT, H. influenza, and others, and the possibility that the compound might have a role in triggering adverse effects, in particular ASD.

Part of the introduction to the participants delivered by one of Orenstein's colleagues summarized the goal of the meeting:

> Those who don't know, initial concerns were raised last summer that mercury, as methylmercury in vaccines, might exceed safe levels. As a result of these concerns, CDC undertook, in collaboration with investigators in the Vaccine Safety Datalink, an effort to evaluate whether there were any health risks from mercury in any of these vaccines.
>
> Analysis to date raise [sic] some concerns of a possible dose-response effect of increasing levels of methylmercury in vaccines and certain neurologic diagnoses. Therefore, the purpose of this meeting is to have a careful scientific review of the data. [pg. 2–3]

And,

> . . . I believe in October of 1999 the ACIP looked this situation over again and did not express a preference for any of the vaccines that were Thimerosal free. They said the vaccines could be continued to be used, but reiterated the importance of the long term goal to try to remove Thimerosal as soon as possible. [pg. 13]

In other words, the organizers didn't want to commit to Thimerosal being a problem but wanted it out of vaccines "as soon as possible" anyway.

Of those attending the meeting, twelve were from the CDC itself, including Verstraeten, Orenstein, and Dr. Frank DeStefano, the latter the project director of the VSD, along with a number of participants from the ACIP. Several international observers were also present, notably Dr. John Clemens, a member of the WHO's Global Advisory Committee on Vaccine Safety (GAVCS). (GACVS will be discussed in context to WHO in Chapter 10.) Most of the participants were physicians; few basic scientists were present.

The summary notes included some details about Thimerosal, used since 1930 as a bacteriostatic (not an antibiotic) in multiuser vaccine vials, which consists of 49.6 percent elemental Hg. Thimerosal as an organomercury compound breaks down into its main constituent parts: ethyl mercury and the sulfur compound thiosalicylate.

After Orenstein did the introductions , Dr. Johnston noted that

> There is a very limited pharmacokinetic data concerning ethylmercury. There is very limited data on its blood levels. There is no data on its excretion. It is recognized to both cross placenta and the blood-brain barrier. The data on its toxicity, ethylmercury, is sparse. It is primarily recognized as a cause of hypersensitivity. Acutely it can cause neurologic and renal toxicity, including death, from overdose. [pg. 14]

Johnston went on to refer to a meeting held a year earlier convened by the National Vaccine Advisory Committee and the Inter-Agency Working Group on Vaccines on the same subject as that at Simpsonwood. Presumably, the Simpsonwood meeting was meant to reach a more definitive resolution to whatever the previous meeting had concluded.

Johnson went on to say some interesting things about vaccinologists with regard to other health professionals at the previous year's meeting:

> As an aside, we found a cultural difference between vaccinologists and environmental health people in that *many of us in the vaccine arena have never thought about uncertainty factors before*. We tend to be relatively concrete in our thinking. Probably one of the big cultural events in that meeting, at least for me, was when Dr. Clarkson repetitively pointed out to us that we just didn't get it about uncertainty, and he was actually quite right. It took us a couple of days to understand the factor of uncertainty in assessing environmental exposure, particularly to metals. [pg. 16; Italics for emphasis are mine.]

Continuing, Johnston said that

> There were a number of things that we got a consensus on in that meeting. First is that there was no evidence of a problem, only a theoretical concern that young infants' developing brains were being exposed to an organomercurial. We agreed that while there was no evidence of a problem, the increasing number of vaccine injections given to infants was increasing the theoretical mercury exposure risk. We agreed that the greatest risk for mercury exposure from vaccines would be to low birth weight infants and to infants born prematurely. [pg. 17]

But best of all, Johnston revealed very clearly what the various vaccinologists and others didn't know at the time, in spite of the then decades-long use of Thimerosal in vaccines:

However, we also learned that there is absolutely no data, including animal data, about the potential for synergy, additivity or antagonism, all of which can occur in binary metal mixtures that relate and allow us to draw any conclusions from the simultaneous exposure to these two salts in vaccines. [pg. 20]

A Dr. Weil then joined in the discussion about the previous year's meeting to say that:

The one thing that was a take away from that meeting was that if there were an increased risk, it would be in the low birth rate and preterm infants. I wanted to ask an unrelated question, and this has to do with potentially look-ing at confounding as we go through this. You mentioned the issue of alumi-num salts. I know it's an issue, but I don't know the specifics of it. *I wonder is there a particular health outcome that has been of concern that is related to the aluminum salts that may have anything to do with what we are looking at here today?* No, I don't believe there are any particular health concern that was raised. It was raised as an issue, and clearly it's a confounding issue in that exposure to vaccine includes exposure to things other than Thimerosal. Two things. One, up until this last discussion we have been talking about chronic exposure. I think it's clear to me anyway that we are talking about a problem that is probably more related to bolus acute exposures, and we also need to know that the migration problem and some of the other developmental prob-lems in the central nervous system go on for quite a period after birth. But from all the other studies of other toxic substances, *the earlier you work with the central nervous system, the more likely you are to run into a sensitive period for one of these effects, so that moving from one month or one day of birth to six months of birth changes enormously the potential for toxicity. There are just a host of neurodevelopmental data that would suggest that we've got a serious problem. The earlier we go, the more serious the problem.* [p. 24; Italics are mine.]

This was a pertinent observation and, sadly, one that is still unresolved over twenty years later. The question that must be addressed is *why?* The answer is that in neurodevelopment there are "critical periods" in which certain aspects of brain structure and ultimate function are laid down.[20] Interfering with any of these critical periods usually has lifelong consequences, a point my colleague Dr. Lucija Tomljenovic and I made in regard to aluminum adjuvants and brain development in a peer-reviewed publication in 2011.[21]

After these preliminaries, Verstraeten presented on the epidemiology and statistics of the purported link between Thimerosal-containing vaccines and ASD. The data were drawn from a survey of two health maintenance organizations (HMO) databases.[22] The data suggested to Verstraeten that there was, indeed, a "signal" in the data linking the two.

In the transcripts, Verstraeten is quoted as saying:

> From those risk analysis [sic], excluding those dichotomized for EPA [presumably the Environmental Protection Agency], we have found statistically significant relationships between the exposure and the outcome for these different exposures and outcomes. First, for two months of age, an unspecified developmental delay which has its own specific ICD9 code.
>
> Exposure at three months of age, Tics. Exposure at six months of age, an attention deficit disorder. Exposure at one, three and six months of age, language and speech delays which are two separate ICD9 codes. Exposure at one, three and six months of age, the entire category of neurodevelopmental delays, which includes all of these plus a number of other disorders." [pg. 40-41; Brackets for addition are mine.]

In other, more genteel, words, most of the features of ASD.

Verstaeten notes, mirroring Weil above, that he couldn't be certain that it was the Thimerosal in the vaccines that is doing this, as it could just as well be vaccine antigens or aluminum adjuvants.

Verstraeten finished by noting that

> In conclusion, the screening analysis suggests a possible association between certain neurologic developmental disorders. Namely Tics, attention deficit disorder, speech and language disorders and exposure to mercury from Thimerosal containing vaccines before the age of six months. No such association was found for renal disorders. [pg. 50]

What followed Verstraeten's presentation was then a long, widely ranging discussion over the next day and a half with some of the participants concerned that the signal *was* real and others trying to knock down this notion. During the discussions, the participants discussed other possible factors for the results, including a host of methodological issues, different sources of bias, the statistical evaluation, as well as the reliability of the two HMO sources of the data given variabilities in symptom coding. The participants also considered other possible confounds from the vaccines themselves such as the antigens present and any aluminum adjuvants, issues that Verstaeten himself had recognized, as well as the possible impact of dosage of Thimerosal and the scheduling of the vaccines. One participant even stressed the need for genetic data.

Much of the initial discussion following Verstraeten's talk examined the epidemiological data about the human health impacts of Hg in fish contaminated with methyl mercury in the Seychelles and the Faroe Islands. These citations, although completely irrelevant to the topic of a very different Hg compound in vaccines, nevertheless alerted the participants that Hg might be a neurotoxin, the evidence lending a degree of "biological plausibility" to the idea that Hg in vaccines might have CNS impact.

One Dr. Koller, a veterinarian working with Hg, noted that

> . . . I have always considered the neurological effects of ethylmercury and methylmercury to be somewhat similar at a similar dose. Now ethylmercury has thought to cause maybe some of the other organ abnormalities. Maybe more so than methylmercury, but I have considered the responses, the toxic effects to the nervous system to be similar at a similar dose. [pg. 122]

This was a very pertinent comment and highlighted one reason why animal preclinical data, then and now, are likely to be predictive of human outcomes in vaccine phase trials.

Questioned about the biological plausibility of his data, Verstraeten stated that

> You are asking for biological plausibility? Well, yes. When I saw this, and I went back through the literature, *I was actually stunned by what I saw because I thought it is plausible.* First of all there is the Faeroe study, which I think people have dismissed too easily, and there is a new article in the same Journal that was presented here, the Journal of Pediatrics, where they have looked at PCB. They have looked at other contaminants in seafood and they have adjusted for that, and still mercury comes out. That is one point. Another point is that in many of the studies with animals, it turned out there is quite a different result depending on the dose of mercury. Depending on the route of exposure and depending on the age at which the animals were exposed. *Now I don't know how much you can extrapolate that from animals to humans, but that tells me that mercury at one month of age is not the same as mercury at three months, at 12 months, prenatal mercury, later mercury. There is a whole range of plausible outcomes from mercury.* [pg. 161–162; Italics are mine.]

Dr. Weil again, one of the more astute members at the conference:

> Although the data presents a number of uncertainties, there is adequate consistency, biological plausibility, a lack of relationship with phenomenon not expected to be related, and a potential causal role that is as good as any other hypothesized etiology of explanation of the noted associations. In addition, *the possibility that the associations could be causal has major significance for*

public and professional acceptance of Thimerosal containing vaccines. I think that is a critical issue. [pg. 187; Italics are mine.]

Dr. Johnston, who was cited above, came back to say that

. . . This association leads me to favor a recommendation that infants up to two years old not be immunized with Thimerosal containing vaccines if suitable alternative preparations are available. I do not believe the diagnoses justifies compensation in the Vaccine Compensation Program at this point. I deal with causality; it seems pretty clear to me that the data are not sufficient one way or the other. My gut feeling? It worries me enough. *Forgive this personal comment, but I got called out at eight o'clock for an emergency call and my daughter-in-law delivered a son by C-section. Our first male in the line of the next generation, and I do not want that grandson to get a Thimerosal containing vaccine until we know better what is going on. It will probably take a long time.* In the meantime, and I know there are probably implications for this internationally, *but in the meanwhile I think I want that grandson to only be given Thimerosal-free vaccines.* [pg. 199; Italics are mine.]

A Dr. Caserta commented that

One of the things I learned at the Aluminum Conference in Puerto Rico that was tied into the metal lines in biology and medicine that I never really understood before, is the interactive effect of different ions and different metals when they are together in the same organism. It is not the same as when they are alone, and I think it would be foolish for us not to include aluminum as part of our thinking with this.

Again, aluminum in vaccines comes up in the Puerto Rico conference along with concerns about toxic metal synergies, topics that I will address in the next section.

And then, to finally put the baby to bed, one of the last commentators was WHO's John Clemens, who summed up what was really at stake. What was at stake, as the comments clearly reveal, was this: it was not the health of children, but the health of organizations like the WHO.

In the following, I will quote his comments in their entirety as they reveal precisely the dimensions of the overall problem when it comes to vaccine safety evaluations, at least by the CDC and WHO:

Thank you, Mr. Chairman, I will stand so you can see me. First of all I want to thank the organizers for allowing me to sit quietly at the back. It has been a great privilege to listen to the debate and to hear everybody work through with enormous detail, and I want to congratulate, as others have done, the work that has been done by the team. *Then comes the but. I am really concerned that we have taken off like a boat going down one arm of the mangrove swamp at high speed, when in fact there was no enough discussion really early on about which way the boat should go at all. And I really want to risk offending everyone in the room by saying that perhaps this study should not have been done at all,* because the outcome of it could have, to some extent, been predicted and we have all reached this point now where we are left hanging, even though I hear the majority of the consultants say to the Board that they are not convinced there is a causality direct link between Thimerosal and various neurological outcomes. *I know how we handle it from here is extremely problematic.* The ACIP is going to depend on comments from this group in order to move forward into policy, and I have been advised that whatever I say should not move into the policy area because that is not the point of this meeting. But nonetheless, we know from many experiences in history that the pure scientist has done research because of pure science. But that pure science has resulted in splitting the atom or some other process which is completely beyond the power of the scientists who did the research to control it. And what we have here is [*sic*] people who have, for every best reason in the world, pursued a direction of research. But there is now the point at which the research results have to be handled, and even if this committee decides that there is no association and that information gets out, the work has been done and through freedom of information that will be taken by others and will be used in other ways beyond the control of this group. And I am very concerned about that as I suspect it is already too late to do anything regardless of any professional body and what they say. *My mandate as I sit here in this group is to make sure at the end of the day that 100,000,000 are immunized with DTP, Hepatitis B and if possible Hib, this year, next year and for many years to come, and that will have to be with Thimerosal containing vaccines* unless a miracle occurs and an alternative is found quickly and is tried and found to be safe.

So I leave you with the challenge that I am very concerned that this has gotten this far, and that having got this far, how you present in a concerted voice the information to the ACIP in a way they will be able to handle it and not get exposed to the traps which are out there in public relations. My message would be that any other study, and *I like the study that has just been described here very much. I think it makes a lot of sense, but it has to be thought through. What are the potential outcomes and how will you handle it?* How will it be presented to a public and a media that is hungry for selecting the information they want to use for whatever means they have in store for them? I thank you for that moment to speak, Mr. Chairman, and I am sorry if I have offended you. *I have the deepest respect for the work that has been done and the deepest respect for the analysis that has been done, but I wonder how on earth you are going to handle it from here.* [pg. 247–249; Italics are mine.]

And that was that.

Clemens had expressed the WHO/corporate bottom line that safety data be damned if larger agendas were in play. And with this, the very realistic concerns that members of the group had voiced, including the author of the presentation, Thomas Verstraeten, simply vanished.

Verstraeten went on to reanalyze the data and statistics over the next few years, including in a presentation to the IOM, cited next. When he finally published his paper in 2003, the significant associations between Thimerosal and autism-like features had simply evaporated like morning fog off a Georgia lake.

To see the next stage in the Thimerosal story, I now turn to the IOC's 2001 report.

The IOC's Contribution to the Question of Thimerosal and Autism

A sequel to the Simpsonwood conference was a volume put out by the IOC in 2001: *Immunization Safety Review: Thimerosal Containing Vaccines and Neurodevelopmental Disorders.*[23] The editors are listed as K. Stratton, A. Gable, and M. C. McCormack, members of the Immunization Safety Review Committee, Board of Health Promotion and Disease Prevention. Stratton was the editor as well for the IOC's book on the MMR vaccine and autism, as discussed in the previous chapter.[24]

Predictably, as in the earlier MMR book from the IOC, the editors felt that they were required to reaffirm the basic vaccine creed. They wrote:

Vaccines are among the greatest public health accomplishments of the past century. In recent years, however, a number of concerns have been raised about the safety of, and need for, certain immunizations. Indeed, immunization safety is a contentious area of public health policy, with discourse around it having become increasingly polarized and exceedingly difficult. The numerous controversies and allegations surrounding immunization safety signify an erosion of public trust in those responsible for vaccine research, development, licensure, schedules, and policymaking. Because vaccines are so widely used—and because state laws require that children be vaccinated to enter daycare and school, in part to protect others - immunization safety concerns should be vigorously pursued in order to restore this trust." [pg. ix]

Having thus ticked off the required central statement of faith, the editors were then ready to proceed with their evaluation.

The editors, statement of faith aside, still seemed quite aware of the need to at least assure readers of some semblance of rigor in the evaluations and further to try to make sure that no blatant conflicts of interest might arise. They wrote:

> Given the sensitive nature of the present immunization safety review study, the IOM felt it was especially critical to establish strict criteria for committee membership. These criteria prevented participation by anyone with financial ties to vaccine manufacturers or their parent companies, previous service on major vaccine-advisory committees, or prior expert testimony or publications on issues of vaccine safety.

The IOC thus established a valuable guide on how to do unbiased evaluations based on actual science, apart from the catechism that opens their efforts, that is. That other organizations do not establish similar rules, such as ACIP, has been noted elsewhere in the book.

After evaluating the scientific literature and viewing a revised presentation from Thomas Verstraeten that did not convince them, the committee realized that some things were undoubtedly true: (1) not much was known about ethyl mercury in terms of pharmacokinetics or neurotoxicity; (2) nevertheless, the general toxicity of Hg was known, including that of methyl mercury; (3) it remained biologically plausible that Thimerosal could be involved in neurodevelopmental disorders, although still not conclusively demonstrated.

Based on these outcomes, the committee decided that

> . . . the evidence is inadequate to accept or reject a causal relationship between thimerosal exposures from childhood vaccines and the neurodevelopmental disorders of autism, ADHD, and speech or language delay . . . [pg. 13]

This conclusion differed from that of the IOC committee, which dealt with the MMR and rejected any link to autism.

The committee's final summary statement was that

> Mercury is a known neurotoxicant. Little is known about ethylmercury (the active component in thimerosal) compared to methylmercury. However, the committee believes that the effort to remove thimerosal from vaccines was a prudent measure in support of the public health goal to reduce the mercury exposure of infants and children as much as possible. The committee urges, in fact, that full consideration be given to removing thimerosal from any biological or pharmaceutical product to which infants, children, and pregnant women are exposed. [pg. 13]

Recognizing that the biological plausibility of Thimerosal/ethyl mercury toxicity needed to be demonstrated, the committee suggested that human or animal studies be performed, particularly in relation to dose. The first step was to clearly establish that Thimerosal's Hg would show up after injection in the brain.

This is precisely what a study with infant macaque monkeys by Burbacker et al. later showed in 2005.[25] In this study, the authors used forty-one monkeys of unknown sex divided into three groups: seventeen given methyl mercury [MeHg] by gavage, seventeen injected *i.m* with Thimerosal, and seven not treated. Hg blood levels were measured two, four, seven, or twenty-eight days after the last treatment. Although blood and brain concentrations of the compounds were considerably lower with Thimerosal, the authors also observed that

> There was a much higher proportion of inorganic Hg in the brain of thimerosal monkeys than in the brains of MeHg monkeys (up to 71% vs. 10%). Absolute inorganic Hg concentrations in the brains of the thimerosal-exposed monkeys were approximately twice that of the MeHg monkeys.

No behavioral or histological measurements were made, but the key finding remains the above, namely, that inorganic Hg, the known neurotoxin, can be found in the brains of Thimerosal-treated infant monkeys, and thus potentially in human infant brains, as well.

Robert F. Kennedy Jr.'s Book

Thimerosal: Let the Science Speak came out in 2015 at a time when the earlier efforts to address a possible connection between Thimerosal and autism had long been thought to be resolved in the negative by most of mainstream medicine and the media. In some ways, Kennedy's book was an elaboration of the earlier 2006 book by David Kirby called *Evidence of Harm: Mercury in Vaccines and the Autism Epidemic: A Medical Controversy.*[26]

What Kennedy did in his 443-page tome was to summarize all of the evidence in favor of a Thimerosal-autism link, including most of the material from the Simpsonwood meeting, the IOC report of 2001 cited above, and a later IOC report in 2004 that claimed that there was no link. Kennedy also cited a huge number of other papers from the peer-reviewed literature demonstrating convincingly the neurotoxicity of Thimerosal.

Several of these deserving mention include two studies by Dr. Mark Geier and his son, David. Although much criticized by bloggers such as Dr. David Gorski (see Chapter 9), the first of these articles makes the case that autism and speech disorders are dose-dependent with Thimerosal amount.[27] A second, more recent study[28] was a case control study looking at the years 1995–1999 of 3486 cases drawn from the VAERS database for adverse effects following the Hib vaccine. Geier et al. compared two groups: One included children receiving Hib with Thimerosal, and the second group receiving a mercury-free version. The authors report that there were significantly more cases of autism, developmental delays, and psychomotor disorders in the mercury group. To their credit, the authors acknowledge a host of caveats to these results, and one could add that quantification of VAERS data in both studies is somewhat problematic in general. This last, however, does not stop the CDC from listing their own vaccine safety studies also drawn from the VAERS database (see Chapter 3).

It is worth noting the trajectory of Verstraeten's work from the original presentation at Simpsonwood to what he and his colleagues finally published in 2003.[29]

The key element in this article is that the authors had added another HMO in their evaluations to the two that had been presented at Simpsonwood. HMO A still showed a significant risk of tics; HMO B showed an increased risk of language delay at two time points. But analysis of HMO C found no significant associations in respect to either autism or attention-deficit disorder. The final conclusion of the article tried to have the outcome support the original Simpsonwood presentation by making the results appear to vanish at the same time:

> Conclusions. No consistent significant associations were found between TCVs [thimerosal-containing vaccines] and neurodevelopmental out-comes. Conflicting results were found at different HMOs for certain outcomes. [pg. 1039; Brackets are mine.]

The take-home message was clearly intended to be that there was no causal relationship between Thimerosal and autism, and yet the authors had clearly shown that there was.

In his book, Kennedy also addressed the machinations that made Verstraeten's initial conclusions about the link of Thimerosal to autism go away from the time of his presentation at Simpsonwood until his 2003 article. In regard to this last point, it is possibly more than a casual coincidence that the Verstraeten data modifications while he was still at the CDC were eerily predictive of the same sorts of events that would later happen in regard to CDC studies on the MMR vaccine. These latter events were disclosed in telephone conversations between CDC researcher William Thompson and independent vaccine researcher Dr. Brian Hooker in 2013[30] and became the subject for a movie made by Hollywood producer Del Bigtree and Andrew Wakefield, *Vaxxed*, in 2016.[31]

In both cases, Verstraeten's immediate superior was Dr. Frank DeStefano, described above in the Simpsonwood conference's discussion.

In his book, Kennedy also critiques various studies in the literature for not using the actual doses of Thimerosal that might be relevant, his critiques in many ways mirroring some of my own as cited in Chapter 3's review of the CDC list of vaccine safety publications.

Kennedy discussed his own experience with the retraction of an article, *Deadly Immunity,* about Thimerosal published in, and then promptly retracted from, *Rolling Stone/Salon.*[32] This sort of media censorship turned out to be highly predictive of the wave of retractions that would soon afflict scientists publishing in the peer-reviewed literature, as cited in Chapter 9.

Finally, Kennedy explored the issue of conflicts of interest in the scientific journals, particularly *Pediatrics,* the house journal of the AAP and the source of many articles reviewed in Chapter 3 that allege the absolute safety of vaccines. The role that pharmaceutical industry funding plays in academic publishing in the biomedical field is a topic to which I will return in Chapter 12.

Thimerosal and Autism: Conclusions and Speculations

Those who had linked ASD to Thimerosal in vaccines were thought to be on the defensive after the compound came out of US vaccines in 2000 while the rates of ASD continued to climb year after year. The CDC, which tends to dismiss any actual increase in ASD in spite of their own numbers, championed the notion that Thimerosal was not the culprit, as indeed it may not have been. However, the data reanalysis that followed the initial Verstraeten presentation at Simpsonwood would tend to raise concerns about their credibility in reaching this conclusion. Similar data manipulation has also been alleged in regard to the MMR vaccine, as detailed in Chapter 4.

In addition, as Kennedy demonstrated in his book, the notion that all Thimerosal had been removed from all vaccines in the United States early in 2000 is not correct. In fact, Thimerosal remains in most influenza vaccines with perhaps trace amounts in other vaccines, as well. It is important in this regard to remember that the ACIP recommended expanding influenza vaccination to both children and pregnant women, a recommendation that found its way into the CDC's vaccine schedule, where it remains.

A key piece of evidence that Thimerosal was not involved in ASD might be seen in graphs of ASD incidence levels, particularly if the numbers appeared to drop, or flatten, following Thimerosal. There is a hint of this occurring in graphs from Dr. Cynthia Nevison and provided to Children's Health Defense and to me. Other graphs of ASD before and after 2000 do not appear to show a significant decrease, such as one provided by Nevison in a publication from 2014.[33] This paper shows a continuing increase in ASD over time and also shows that it occurs in much the same way across various racial groups in the United States.

Do these data suggest that there really is no validity to the Thimerosal-autism hypothesis? Actually, they do not, rather possibly revealing something far more interesting: Thimerosal was not the sole or even main driver of the ASD increase, leading to speculation that another ingredient in vaccines may have been the key accelerator. One possibility, explored in detail in the next section, is that this ingredient is aluminum in the form of various adjuvants.

In this interpretation, aluminum may serve as the primary driver of the overall increase in ASD numbers, with Thimerosal being the compound that determines how severe ASD will be.

Nevison writes that

> . . . this commentary[34] illustrates what I would suggest might be the signature of the thimerosal phaseout, i.e., the proportion of ASD cases with co-occurring intellectual disability (ID) dropped markedly between birth year 1998 and 2002. My personal view or hypothesis regarding ASD is that some aspect (likely, but not necessarily limited to, the aluminum adjuvant) of the forced immune system activation associated with today's vaccine schedule is responsible for the core features of ASD. Thimerosal comes in as a synergistic agent that increases the severity of ASD. This hypothesis seems consistent with how ASD rates have increased but lessened in severity as thimerosal was phased out while the vaccine schedule itself and accompanying adjuvant load continued to expand. The media misinterprets this as, 'we're just expanding the diagnosis to milder cases.' I'll attach a paper from Australia that I think illustrates quite clearly the phenomenon of increasing overall ASD with milder presentation.[35]

The article Nevison cites, Whitehouse et al. (2017), makes it quite clear that ASD continued to increase after 2000, but in examining ASD cases from 2000 to 2006, the authors showed that the cases had a significant reduction in two of twelve criteria and a reduction in the number of cases with an "extreme severity" in six of the criteria.[36]

It is interesting to note also that various participants at Simpsonwood considered the very same sort of synergistic neurotoxicity due to aluminum and mercury, as cited in the previous chapter. Also, as noted above, they even referenced a conference held in Puerto Rico that discussed aluminum adjuvants and what was known about aluminum's potential toxicity.

The meeting report was published in *Vaccine* in 2002.[37] The conference had hosted two panels, one to discuss what participants thought they knew about aluminum adjuvants, the second to discuss what they felt they still didn't know. The first was full of assumptions of safety based on the history of aluminum adjuvant usage. The second panel's conclusions were far more revealing.

The list of things that were not known included:

> 1. Toxicology and pharmacokinetics of aluminum adjuvants. Specifically, the processing of aluminum by infants and children. 2. Mechanisms by which aluminum adjuvants interact with the immune system. 3. Necessity of adjuvants in booster doses. 4. Definition of frequency and duration of the MMF lesion in normal people. 5. Role of aluminum in the pathophysiology of the MMF lesion. 6. Human control studies to assess the relationship between the 'symptom complex' identified by Dr. Gherardi in patients who have the MMF lesion. 7. New adjuvant development. 8. Expanded trials of IM rather than the SQ route of injection for anthrax vaccine and non-needle vaccine administration technologies.

In other words, quite a lot was clearly not known, and some of these gaps will be addressed in the following section, including the work of the Gherardi group on macrophagic myofasciitis (MMF).

My inclination, like Nevison's, is to take the view that Thimerosal did not drive the rise of ASD, although such a conclusion should not blind us to the longer-term impacts on neurological health across the lifespan of repeatedly injecting any mercury compound. Whether or not aluminum can provide the necessary toxicity in place of Thimerosal is the subject for the next section.

In partial summary so far, up to this point in the book I've considered a range of possible triggers for ASD, from genetics to environmental toxins, then more specifically to some vaccines and/or vaccine ingredients.

The first of these discussed in the previous chapter was the MMR vaccine. The evidence reviewed does not suggest a strong link at a population level with autism. In this chapter so far, I've considered the evidence for and against Thimerosal, which was, as suggested by the initial IOC report in 2001, inconclusive.

I now want to do more than move the goalpost; I want to change the playing field to look at a ubiquitous element that may be hiding in plain sight in food, water, pharmaceutical products . . . and in vaccines. It is an element that plays a major role in our industrial civilization, and it is none other than that amazing metal: aluminum (Al).

Aluminum Chemistry and Place in the Biosphere

Al is the third most common element after oxygen and silicon on Earth and the most abundant metal in the Earth's crust.[38] That this statement is correct is demonstrable. It is, however, often used by both mainstream medical people and the vaccine-hesitant, albeit for reasons that are polar opposites. In the first case, the statement is meant to suggest that because Al is so common on Earth it must be benign, or even beneficial. As we will see, this interpretation is pretty much as far from the truth as one can get. The vaccine-hesitant use the same facts to explain that while Al is ubiquitous in our environment and thus finds its way into a range of processes and materials that humans use, including vaccines, it is not always benign.

Al was first isolated from various minerals, many of these silicates, by Hans Christian Oersted in 1825,[39] pretty much near the end of the first Industrial Revolution. This was a period from approximately 1750 to 1840, which saw the widespread transition from a rural to an urban environment in England, together with the growth of widespread industry and manufacturing. Then, as now, most pure Al came from refining it from various molecular complexes, such as the mineral bauxite.

Al has an atomic number of 13 and a charge of $+3$, meaning three fewer electrons, a fact that provides the basis for much of its chemical activity, and it can accept electrons from other elements, enabling binding to them.

In terms of physical properties, Al is light, malleable, durable, conducts heat, and is corrosion resistant, the latter due to binding to oxygen and thus making an aluminum oxide surface layer.[40]

As with oxygen, Al is highly reactive with a range of elements essential for life, avidly binding to carbon, phosphorous, and sulphur, all key elements in biological systems. It is this aspect that provides the potential to significantly impact such systems.

In spite of claims often made in regard to Al's medical applications, for example on the website of the Vaccine Division of the Children's Hospital of Philadelphia (CHOP), Al is certainly not inert nor, as will be shown soon, is it harmless to life, as noted in the quote that opens this chapter, perhaps particularly in relation to the nervous system (see Fig. 6.2).[41] It is also manifestly not an "essential" element, as the same website once maintained.

Al was not widely bioavailable until the early nineteenth century, as noted earlier, with most sources of ionic Al found near areas of volcanic activity, which tend to be acidic. Due to this lack of bioavailability, Al seems to have been "selected out" of a role in terrestrial biochemistry.[42] It was only during the extraction of Al during the first Industrial Revolution that Al and its later myriad materials applications made it a major feature of the biosphere, particularly for humans. It is this observation that has led Exley and others to opine that we now live in the "Age of Aluminum."[43]

Sources of Aluminum Exposure in Humans

Al in the biosphere, particularly that which may affect humans, arises from various sources.[44] Sadly, in most of the cases to be discussed, the amounts are not as small as in Ganrot's quote that opens this chapter. Rather, the overall amounts to which humans in particular are exposed has been rising continuously since Ganrot's 1986 paper.

Al has a significant presence in processed foods due to its deliberate addition because of its chemical properties and due to contamination during manufacturing processes. Salts of aluminum[45] also show up in a great variety of medicinal products. Included in the latter are antacids, various coatings for pills, and some vaccines. In regard to the latter, Al salts serve as adjuvants to improve the immunogenicity of antigens, as will soon be discussed. Al salts are also used as mordants to fix colored dyes in textiles, in cosmetics, and in antiperspirants.

Food remains one of the key sources of human exposure to this element. The second most common appears to be from Al vaccine adjuvants.[46] In both cases, Al can readily enter the body by way of the soluble salts. In the case of food, Al is absorbed through the gastrointestinal system with an average daily human range of 3 to 10 mg. Intestinal absorption is influenced by other compounds that increase absorption (e.g., citrate and fluoride compounds) or is decreased by substances such as milk.[47]

Given normal patent kidney function, most dietary/waterborne Al ions will be excreted through the kidneys relatively rapidly. Young children in whom kidney function is not yet mature (two and under) and the elderly in whom kidney function may have declined are less able to excrete Al. An additional major means of Al excretion is through sweat.[48] Notably, the same is not true for Al bound up in fluoride complexes or, most relevant for the following discussion, for Al that has a different route of administration, as, for example, by injection into muscle or skin.

Al in drinking water can arise through the use of Al sulphate as a flocculant, but its overall impact seems to be low (0.3 percent) [49] except in unusual circumstances such as the large Al sulphate spill into the water supply in Camelford, England, in 1988. High concentrations can also arise naturally in well water near volcanic or acidified soils.

The addition of fluoride to drinking water as part of a campaign against dental caries[50] has been a source of concern in some circles for years from a health perspective. Fluoride promotes gastrointestinal disorders according to some studies.[51] Further, the joint presence of Al and fluoride can form aluminofluoride complexes, which can be extremely toxic to humans, since these can act as phosphate analogues.[52]

Al can also enter the body by inhalation with an estimated daily uptake of 4.4 µg in industrialized areas. Al metal workers may show higher levels in blood, urine, and bone. Health outcomes of inhaled Al can include respiratory tract infections with asthma-like symptoms, as well as cognitive disorders, the latter implicating uptake into, and damage to, the CNS.[53] The impacts of Al by inhalation are discussed below in more detail.

Table 5.1 summarizes some of the above sources of Al exposure for humans; while food is the largest source of aluminum, the second most common is from medicinal products, including from vaccines that use Al salts as adjuvants.[54]

Major sources of Al exposure in humans	Daily Al intake (mg/day)	Weekly Al intake (mg/day)	PTWI (1 mg/kg/bw; for an average 70 kg human PTWI = 70 mg)	Amount delivered daily into systemic circulation (µg; at 0.25% absorption rate)
Natural Food	1-10	7-70	0.1-1	2.5-25
Food with Al additives	1-20 (individual intake can exceed 100)	7-140 (700)	0.1-2 (10)	2.5-50 (250)
Water	0.08-0.224	0.56-1.56	0.008-0.02	0.2-0.56
Pharmaceuticals (antacids, buffered analgesics, anti-ulceratives, anti-diarrheal drugs)	126-5000	882-35,000	12.6-500	315-12,500
Vaccines (HepB, Hib, Td, DTP)	0.51-4.56	NA	NA	510-4560
Cosmetics, skin-care products, and antiperspirants	70	490	NA	8.4 (at 0.012% absorption rate)
Cooking utensils and food packaging	0-2	0-14	0-0.2	0-5

Table 5.1. Aluminum from all sources in the human biosphere.[55]

Aluminum and Human Health

As far back as 1911, William Gies, quoting even earlier work, noted the potential for harm from Al in the human biosphere, in particular, from food. Gies said this in his paper:

> These studies have convinced me that the use in food of aluminum or any other aluminum compound is a dangerous practice. That the aluminum ion is very toxic is well known. That aluminized any other aluminum food yields soluble aluminum compounds to gastric juice (and stomach contents) has been demonstrated. That such soluble aluminum is in part absorbed and carried to all parts of the body by the blood can no longer be doubted. That the organism can 'tolerate' such treatment without suffering harmful consequences has not been shown. It is believed that the facts in this paper will give emphasis to my conviction that aluminum should be excluded from food.[56]

Gies was right about Al exposure from food. What he did not know was that Al would also be used as an inhalant in a deliberate exposure by industries such as those involved in hard-rock mining.

As reported by Janice Martell, workers in a number of mines in Northern Ontario were exposed to Al dust from what was termed "McIntyre powder."

McIntyre Power, according to the McIntyre Research Foundation,[57] was composed of 15 percent elemental Al and 85 percent aluminum oxide. McIntyre began human trials of the powder in the early 1940s at the "Silicosis Research Clinic" in Timmins, Ontario, using local miners deemed to be "silicotic or presilicotic" based on chest X-rays. The official first use of the powder was at the McIntyre Porcupine Mine in Timmins in 1943. The use of McIntyre powder continued until 1979 in at least fifty-two mines in Ontario, but maybe even longer in some cases. The ostensible goal was to prevent silicosis[58] by binding the powder to silicates inhaled from working in the mines. The rationale for the use of McIntyre powder was that the aluminum would bind to silicate particles and together enable their excretion. This strategy did not work but did leave the miners exposed to longer-term impacts of Al, including those of the nervous system.

Based on a study in a voluntary registry compiled by Martell, some two-thirds of those on her registry have respiratory symptoms; one-third have neurological symptoms. Of the latter, fifty have Parkinson's disease and seven have ALS, a percentage that may be an underestimate.[59] These numbers, if correct, put the rate of these neurological diseases in the McIntyre powder victims at many times higher than in the general population.[60]

McIntyre licensed the use of McIntyre powder extensively in the United States, and some reports suggest it was also used in South Africa and Australia.

The miners so exposed never gave informed consent for being given McIntyre powder. I asked Martell if there was any attempt to do so. Her response:

> Never. Most of them [the miners] were given little to no information about the aluminum dust. They were forced to do it as a condition of employment (sometimes in a locked room - like [Martell's] Dad), and under threat of discipline/job loss if they did not do it. I have spoken to hundreds of miners. Not one of them was allowed the courtesy of informed consent. This was a massive human experiment. [Parentheses are hers; brackets are mine.]

In addition to this evidence for Al's toxic actions on the CNS following deliberate exposure, accidental exposure to intravenous solutions containing Al were reported in the 1970s, the most famous example termed "dialysis associated encephalopathy"(DAE), which occurred when kidney dialysis patients were accidentally given dialysis fluids containing high levels of the element.[61] The outcomes were typically of relatively rapid onset and severity. The resulting neurological signs included cognitive dysfunctions resembling Alzheimer's disease, as well as epileptic seizures. Postmortem histology showed some of the hallmark pathological features of Alzheimer's disease including neurofibrillary tangles (NFT) and amyloid β (Aβ) plaques. It is likely that the mechanism by which Al ions were transported into the brain involved one or more of various carrier proteins, such as ferritin and transferrin.

The above two examples highlight a key feature of Al exposure, namely, impacts on the CNS. These observations raise the question of just how much Al enters the brain following exposure. The amount of Al in the normal adult human brain is less than 2 µg/g, with the distribution reflecting higher concentrations in gray compared to white matter.[62] Along with bone, the brain has the highest potential to accumulate Al.[63] Postmortem brain samples of individuals exposed during the Camelford incident[64] in the United Kingdom showed an Al concentration from 0.75 µg/g in frontal white matter to 49 µg/g in the choroid plexus.[65] In addition, an association of Al with the hallmark abnormal proteins entities Alzheimer's disease, Aβ plaques, and NFT has been well documented.[66]

Clearly, Al can enter the brain, but how much is removed? There is disagreement about this issue, perhaps reflecting the route of exposure, but it is now clear that retained Al seems to be stored in five main CNS compartments including the blood brain barrier, the brain interstitial fluid, neurons, glia, and pathological inclusions such as Lewy bodies (found in Parkinson's disease), and NFT, and in Aβ plaques in Alzheimer's disease.[67]

The outcomes in humans from the Camelford incident were reflected in a series of studies by Dr. Judie Walton in which old rats given Al in drinking water showed significant cognitive decline along with the presence in the brain of the abnormal proteins associated with Alzheimer's disease.[68]

Perhaps most telling from a general perspective on Al neurotoxicity was aluminum's clear role in DAE that served, in principle, to highlight the possible link to Alzheimer's disease. In this regard, the notion of such a role in Alzheimer's has been considered at various times, one of the most contentious of these when McLachlan and colleagues linked Al in water and in Al-based materials such as in cooking utensils to Alzheimer's disease, at the very least as a part of the overall etiology.[69]

Water treated with Al-compounds remains a possible link, but Al pots and pans are far less likely to be a major culprit. While nonanodized[70] Al utensils with acidic foods can indeed release Al ions into the food, the risk was never all that great, since in those with patent kidney function, most all dietary Al is excreted. However, the purported link between Alzheimer's and Al cookware made for a good media story.

In due course, the notion that Al from cookware could be harmful faded away but led to the unfortunate conclusion that Al had no role at all in Alzheimer's disease. This last notion was not correct, as amply demonstrated in the seminal and heavily referenced review by Tomljenovic in 2010.[71]

Tomljenovic's work cited the most likely sources of Al for humans (see Table 5.1 earlier).

In the years after the various articles by McLachlan, converging lines of evidence from a number of studies, cited next, were beginning to consider the role of Al in vaccines where this element was used as an adjuvant to further boost the immune response to some antigen. So extensive has the use of Al adjuvants become that Table 5.1 shows that it can be one of the main sources of Al exposure in humans, particularly in the young.

There are a variety of Al adjuvant preparations, but the two most common are variants on Al hydroxide and Al phosphate.[72] Each vaccine may contain only a relatively small amount of whichever compound (usually less than 0.5mg), and actual elemental Al is only part of this. Al adjuvants, however, may cumulatively constitute an important source of an overall Al body burden.[73] For example, the administration of twenty or more vaccines containing Al as adjuvants containing 0.5 mg of Al compound could add up to 10 mg Al compound to the body burden. The latter would be the equivalent of a dietary intake of Al of over 4000 mg/day based on excretion.[74]

The Rationale for Using Aluminum in Vaccines

By the 1920s, vaccines as injectable substances had been used for over seventy years, the most common being for smallpox. But as researchers began to develop vaccines for other infectious diseases, they found a significant problem: if using only parts of viral or bacterial proteins, antibody production was often too low to allow for a sustained immune response, primarily by seroconversion, the latter concept discussed in Chapter 2. The answer, at least to researchers at the Wellcome Physiological Research Laboratories in the United Kingdom, was to try to find an additive that would enhance antibody production and duration. (We will meet the offspring of the Wellcome laboratories later in Chapter 12).

In response to this problem, Glenny and colleagues at the Wellcome laboratories seem to have rummaged through the chemical cabinets at the company before finding that various Al salts[75] could do the job of increasing immunity very well. In a paper published in 1926, Glenny et al. used aluminum potassium with a diphtheria toxoid to try to enhance an antibody response in guinea pigs.[76]

The rest, as they say, is history, and the use of various types of adjuvants, mostly based on Al, has been a staple of vaccinology and vaccine manufacturing ever since. In most instances, whatever the antigen used, usually a piece of coat protein, is "adsorbed" (chemically adhered) onto the Al nanoparticles in the compound. The creative use of Al compounds to create ever-greater adjuvant abilities is now a key aspect of vaccine research and is considered essential to delivering effective vaccines. Just how safe this practice is will be soon discussed.

How do Al adjuvants actually work to enhance the immune response? We don't know for sure, and as the current website of the British Society for Immunology states, "Sometimes in medical science it is not always possible to explain how something works, but only to accept that it does."[77]

Maybe, but if that "something" *might* be harmful, it might be very useful to have a bit more information in order to figure out why. Take for example the study by Dr. S. H. Lee showing that the Gardasil HPV vaccine contained residual HPV DNA for the HPV L1 capsid protein for strains 11 and 18 that was bound to the proprietary aluminum adjuvant.[78] Viral RNA and DNA is not normally likely to be a problem given that endonucleases in the body should take the DNA apart, and indeed the CDC and FDA consider that these are merely the residual by-products of the manufacturing process for this vaccine. However, if the DNA in the vaccine is bound to the Al adjuvant that gets moved around the body by immune cells, could such a novel molecule become a problem, especially in the CNS? Maybe not, but how would we know unless we did some experiments? Almost needless to say, such experiments were never performed.

As for what Al adjuvants actually do, researchers used to think that the adjuvant and antigen just sat where they had been injected, in what was called the "depot effect." Once parked, the idea was that the adjuvant released the antigen gradually, thus allowing it to stimulate immune cells and generating the hoped-for immunity.

What we now think is a more likely explanation based on newer studies: the newer version is that injected Al adjuvants induce local irritation of the surrounding tissues, and various immune cells infiltrate into the area and, in the process, pick up the adjuvant/antigen. The immune cells, mainly macrophages, then transport the Al complex to other regions of the body, beginning with the draining lymph nodes.[79] The extensive work by the Gherardi group in Paris has further amplified our understanding of Al transport from muscle into other systems such as the brain.[80]

Before considering the potential adverse actions that Al adjuvants may have, particularly in the CNS, it is worth noting that while Al adjuvants are the most common vaccine adjuvants currently licenced in the United States and Canada, other adjuvant formulations such as squalene are used in some influenza vaccines elsewhere in the world. Squalene, a natural lipid found in both plants and animals, is used in a formulation termed M59 by vaccine maker Seqiris.[81]

Squalene itself is not an adjuvant in the same way that Al salts are but, when combined with various molecules in M59, is an immune stimulant, although why is not clearly known. The M59 adjuvant was reportedly used along with Al hydroxide in the anthrax vaccine made by BioPort and linked by various investigators to Gulf War Syndrome.[82]

In addition, GlaxoSmithKline (GSK), a major UK pharmaceutical firm, has a series of "adjuvant system" (AS) variants, numbered AS01 to AS04. AS01 through 03 use a combination of extracts of various plants and lipids; AS04 adsorbs the other components onto either Al hydroxide or Al phosphate.[83]

In spite of this and other still-experimental vaccine adjuvants such as toll-like receptor agonists, the key ones in use in North America and much of the world remain those based on Al: Al oxyhydroxide and Al potassium or Al potassium sulfate. It is worth noting here that the still-experimental COVID-19 vaccines used in Europe and North America do not use aluminum adjuvants, since the claim is that they are not needed (see Chapter 13).

In Vivo and Human Studies of Aluminum Adjuvant Neurotoxicity

The general notion that Al may be harmful to living things has an extensive history, as partially cited earlier. Some of this literature clearly implicates Al harms to the CNS in particular. In regard to CNS damage, apart from the studies on Al compounds in water by Walton and colleagues (cited earlier), most of the relevant work done in the last fifteen years really arises in relation to Al adjuvants with the data collected by three independent laboratories: mine, that of Dr. Lluis Luján in Spain, and that of Dr. Romain Gherardi in France.

These studies, particularly from the last two laboratories, show some overlap with considerations of Al-induced autoimmune dysfunction.

Some of the first work on Al adjuvant neurotoxicity came from my laboratory with experiments using young adult male mice given *s.c.* injections of Al hydroxide.[84] The first of these studies was an attempt to explore the hypothesis that this adjuvant, linked to some of the neurological outcomes of Gulf War Syndrome (GWS), would induce behavioral and pathological CNS outcomes resembling those seen in ALS. Indeed, the study results showed precisely such an outcome: neuronal death in motor neurons in the spinal cord and motor cortex accompanied by degraded motor function that became progressively more pronounced. The second study from my group used slightly older mice and found the same general outcomes. This latter study also showed clear evidence for Al having moved inside motor neurons along with evidence for an involvement of the resident immune cells of the brain, microglia, somewhere in the process leading to motor neuron degeneration. (See Fig. 5.2 and Fig 5.3a-c in the color insert.)

While these studies seemed to support the view that Al hydroxide delivered by injection can damage the CNS, this was a provisional view. These studies, although peer-reviewed and published in decent journals, were nevertheless critiqued by both bloggers and entities such as the WHO.[85]

Were those critiques justified? Not in my opinion, but to quote from the Stoic philosopher Epictetus, "If anyone tells you that a certain person speaks ill of you, do not make excuses about what is said of you but answer, 'He was ignorant of my other faults, else he would not have mentioned these alone.'"[86]

This was because both bloggers and WHO missed the more valid critiques I would have raised myself. These were: (1) Even though the appropriate number of animals satisfying statistical requirements was used, the number of animals in each group was still relatively small. This limitation left open the possibility that the outcomes were a statistical anomaly. The fact that we were able to duplicate many of the initial findings in a second study, somewhat diminishes this concern, but that too was a smallish study, and the number problem might simply have been repeated; and (2) Were the histological stains the best that could be used, that is, the most selective and quantitative? No, but at the time they were.

Both critiques serve to make the general point that it is always easy to critique *any* work after it is done, but harder to anticipate which advances the future may bring in techniques and interpretations based on other studies in the field.

WHO's critiques were simply a joke: they were vague to the point that one had to doubt they had even read the study and had merely reacted to defend aluminum adjuvants. A more detailed return critique of the ability of WHO to understand, let alone do, actual science will be provided in Chapter 10.

In addition to the above, we also did some primary behavioral work looking at anxiety behaviors in mice given *s.c.* injections of Al hydroxide early in postnatal life. Both studies demonstrated increased anxiety and lower levels of social interactions in the Al-treated animals.[87]

Finally, in 2020, we published an article that reported a range of behavioral abnormalities in mice following injection with vaccines from the CDC's current pediatric schedule for children up to eighteen months of age. The injected mice were age-adjusted to the best guess for human equivalence, and the vaccine dose was adjusted for weight.[88] Because we were using whole vaccines and not just the adjuvant, these outcomes from a purely Al perspective are more complicated to interpret, but they are also more realistic in attempting to understand how Al-adjuvanted vaccines might lead to neurological disorders.

A fellow Al researcher Dr. Lluis Luján and colleagues have published several studies on the impact of Al adjuvants in sheep in 2013 and later in 2019.[89] This work originated in the observation that commercial sheep in Spain seemed to have an adverse CNS outcome as a result of a vaccine directed against "blue tongue," a nonfatal ovine disease. Following blue-tongue vaccination, chronic adverse effects had been observed in 50–70 percent of flocks, and up to 100 percent of animals within a flock showed behavioral disturbances such as restlessness, compulsive wool biting, generalized weakness, muscle tremors, the loss of response to external stimuli, various movement disorders, and stupor. Coma and death could follow. On histological examination, inflammatory lesions in the brain and spinal cord were found to be associated with aluminum (see Fig. 5.4).

The pathologies observed on examination included a number of key features including the loss of the myelin covering of long axons and motor neuron loss in the spinal cord. The disorder was made worse by cold weather conditions, perhaps suggesting some synergy with other environmental factors. These initial observations were successfully reproduced under experimental conditions following the experimental administration of aluminum-containing vaccines.

Figure 5.4. Severely emaciated sheep following vaccination against the ovine disease "blue tongue." Luján and colleagues attribute these features of extreme muscle and fat wasting (cachexia)(a) and evidence of "barbering" or wool/hair biting (b) to the aluminum hydroxide adjuvant in the vaccine. (Photos courtesy of Prof. L. Luján.)

The similarities in behavioral outcomes and neuropathology that the Luján group found in sheep seemed to be broadly similar to our own mouse studies, thus adding to evidence linking Al hydroxide injection with CNS disorders and pathology.

Overall, the converging data from studies across species may be of obvious relevance to humans similarly exposed to Al from various sources. Further, the noted CNS pathologies are worth considering with regard to the development of age-related neurological diseases in humans.

A key question to raise is how Al might be transported from the site of injection into the CNS, where it is able to inflict harm.

The answer has been provided by the work of Dr. Gherardi's group, which showed that Al hydroxide administered intramuscularly in mice does not stay localized in the muscle, but rather migrates to different organs, including the brain. The path by which it does so is now clear from various tracking experiments with fluorescent markers, notably rhodamine- or nano-diamond-labelled Al hydroxide.

These studies demonstrated that a significant proportion of the nanoparticles escape the injected muscle within macrophages, travel to regional draining lymph nodes, and then exit the lymphatic system to reach the bloodstream, eventually gaining access to distant organs including the brain. Such a "Trojan-horse"-like transport mechanism in which Al-containing macrophages enter the brain may result in the gradual accumulation of Al due to lack of recirculation of the adjuvant-antigen compound.[90] These studies clearly refute previous notions that injected Al adjuvant nanoparticles remain localized at the injection site.

Modeling Studies of Aluminum Pharmacokinetics

It is important in this next section to remember the words of the physicist Richard Feynman: "It doesn't matter how beautiful your theory is. It doesn't matter how smart you are. If it doesn't agree with experiment, it's wrong."[91]

With the Feynman quote in mind, let's consider the evidence from the other side that aluminum from vaccination does not enter the CNS in sufficient amount to do harm because it is supposedly rapidly excreted. The basis of these arguments is based on modeled pharmacokinetics, pharmacokinetics being basically what the body does to a substance and how it moves into and out of the body. In this, aspects considered are bioavailability, absorption, how the substance is metabolized (if it is), and how it is distributed into different bodily compartments.[92] As we will see, not only do some of these studies not agree with actual experiments that show CNS damage and the presence of Al inside the brain, but the calculations are based on unwarranted assumptions, not to mention rather fundamental mathematical errors as in the case of the best-known one, that of Mitkus et al. in 2011. The Mitkus study was written as a counterweight to concerns about Al adjuvants, particularly in infants and children.

Let's consider the scientific value of these articles in order of chronological appearance.

Before I do, however, I want to introduce some terms and definitions. The Agency for Toxic Substances and Disease Registry (ATSDR) had previously calculated an overall safe level, termed "No observed adverse effect level" (NOAEL) of 2mg/kg/day for aluminum.[93]

The next term to define is that of Minimal Risk Level (MRL) or Minimum Safe Level (MSL) used in some of the papers considered below. The MRL/MSL is basically an estimate of the daily human exposure to a toxic substance that is without appreciable risk.

Earlier, Priest et al. in 1995[94] calculated the pharmacokinetics of aluminum excretion using a radiolabeled (Al^{26}) Al-citrate solution injected intravenously in *one* male volunteer. Yes, one. The results showed a drop of the radiolabel of more than 50 percent in fifteen minutes postinjection and over 99 percent in two days. The problems with this study for understanding potential Al toxicity, especially for Al adjuvants in children, should be readily apparent: one subject, an adult, using an Al compound not used in vaccines and thus one with questionable relevance to actual adjuvant toxicity that is not intravenous. In addition, Priest et al. were modeling serum clearance, not clearance from organs like the brain. In fairness, this was not why Priest et al. were doing their study. The fact is they had simpler goals: trying to see what happened to Al in the body.

Next up was Flarend et al. in 1997,[95] who used six female New Zealand white rabbits of unknown age, two each receiving intramuscular injections of radiolabeled Al hydroxide or Al phosphate. Of the other rabbits, one had a radiolabeled Al citrate injected, one just had the radiolabel. These researchers collected blood after injection for twenty-eight days, then sacrificed the animals and examined Al location in various organs. The results showed that Al was present at one hour after injection, peaked at ten hours, and then declined over time. The Al phosphate levels were about three times higher than those of the hydroxide. Looking at the tissues, the highest Al levels remaining were found in the kidney, followed by the spleen, liver, lymph nodes, and finally the brain.

What can we conclude from this study? Apart from the number of animals per condition being too low to do reliable statistics, the fact that after twenty-eight days there was any Al at all in brain should have been a cause for concern, not celebration. Note, however, that in this study, like Priest et al. before, the researchers were not trying to assess the potential adverse effects of Al adjuvants in infants or children.

The next up was Keith et al. in 2002.[96] Keith et al. used the ATSDR data and the NOAEL to compare the MRL for Al in two vaccines, DPT and Hep B, administered to children, versus the body burden of Al in milk formula and breast milk. The MRL was first calculated from the above based on weight. The resulting graphs showed that the vaccine value crossed the MRL at approximately fifty days following some of the vaccines and rested just below this for most of the remaining time, except at the over-three hundred day time point. Is there a concern that the MRL was crossed at all? We will see soon.

The most influential article written in an attempt to address concerns about Al adjuvants in vaccines was that of Mitkus et al. 2011,[97] cited earlier. Curiously, this article is not cited by the CDC in their hit parade of articles demonstrating the safety of vaccines discussed in Chapter 3, although it is cited elsewhere.

Mitkus et al. remodeled the work of Keith et al. and, like the latter, found mostly the same results, namely, that the MRL is only transiently exceeded after each injection, after which the levels of Al trailed off. These data allowed the authors to conclude that Al adjuvants were unlikely to pose any health risk to children. These toxicokinetic studies have been critiqued by various researchers in recent years. For example, Masson et al. (2018) revisited the above studies and found them inaccurate and unable to account for the movement of aluminum adjuvants in the body postinjection.[98] A far more detailed study was provided by McFarland et al. (2020).[99] MacFarland et al. calculated the aluminum body burden with a focus on the first postnatal months of life up to eighteen months, comparing the CDC's schedule with an alternative schedule using the lowest Al levels in the DTaP vaccine and no Al at all in the Hip vaccines, and finally with a modified schedule devised by Dr. Paul Thomas (the "vaccine-friendly plan"), one of the authors.[100] This latter plan contains several Al-containing vaccines and eliminates others. Their table showing the vaccines and cumulative Al levels is shown here as Table 5.2:

Age (Months)	CDC 2019 Schedule		Modified CDC Schedule		Vaccine Friendly Plan	
	Vaccine (Al µg)	Total Dose (µg)	Vaccine (Al µg)	Total Dose (µg)	Vaccine (Al µg)	Total Dose (µg)
Birth	HepB (250)	250	HepB (250)	250	None	0
2	HepB (250) DTaP (625) Hib (225) PVC13 (125)	1225	HepB (250) Low Al DTaP (330) ActHib (0) PVC13 (125)	705	Low Al DTaP (330) ActHib (0)	330
3	None	0	None	0	PVC13 (125)	125
4	DTaP (625) Hib (225) PVC13 (125)	975	Low Al DTaP (330) ActHib (0) PVC13 (125)	455	Low Al DTaP (330) ActHib (0)	330
5	None	0	None	0	PVC13 (125)	125
6	HepB (250) DTaP (625) PVC13 (125)	1000	HepB (250) Low Al DTaP (330) PVC13 (125)	705	Low Al DTaP (330) ActHib (0)	330
7	None	0	None	0	PVC13 (125)	125
12	Hib (225) PVC13 (125) HepA (250)	600	ActHib (0) PVC13 (125) HepA (250)	375	ActHib (0) PVC13 (125)	125
18	DTaP (625) HepA (250)	875	Low Al DTaP (330) HepA (250)	580	Low Al DTaP (330)	330
	Total (µg)	4925	Total (µg)	3070	Total (µg)	1820

Table 5.2. Amounts of aluminum hydroxide from pediatric vaccines exceeding official "safe" levels by age. (Derived from McFarland et al., 2019, as cited in the text.)

In brief, the CDC schedule gives a total Al level (µg) of 4925, versus 3070 for the modified schedule and 1820 for the Thomas plan.

Based on the FDA's MSL of 850ug/dose for adults, McFarland et al. calculated the amount of Al as a function of weight in the children using a value of 14.2ug/kg. None of the vaccine schedules exceed this in adults, but the problem arises in children with multiple vaccines administered at the two-, four-, and six-month "Well Child" visits with Al numbers far exceeding the FDA MSL.

Although critical of various aspects of the Priest et al. study, using Priest's equations, McFarland et al. calculate both short and long retention models scaled to weight, and thus age. (Not considered here is the stage of neuronal development, which may be a major concern for the impact of any level of Al.)[101]

Of the three schedules, that of the CDC for short-term retention shows the highest overshoot by 15.9 times "safe" levels at two months of age when four Al-adjuvanted vaccines are typically given at the "Well Child" visit. Overall, safe levels are exceeded 70 percent of the time in this age range. The modified schedule exceeds the safe level 26 percent of the same time period. In contrast, the Thomas schedule only goes over by 5 percent of the time. A comparison of these schedules is shown in Fig. 5.5 in the color photo insert.

McFarland et al. also note the strong potential for genetic variance in different populations to impact the amount of time that the various schedules will exceed the MSL.

Taking these numbers into account, it will be apparent that at certain time points, the level of Al from vaccines exceeds that taken in from food, a point also made by Dorea and Marques and in an appendix to McFarland et al.,[102] the latter cited earlier. Dorea and Marques also make the very valid point that, as of yet, there has been no actual experimental data provided by the CDC validating their pediatric vaccine schedule in relation to Al.

To finish this section, I want to turn to a commentary made by Dr. Shira Miller of the Physicians for Informed Consent.[103] In their reevaluation of Mitkus et al.'s determination of the MSL for Al adjuvants, the key observation was that Mitkus et al. based their calculations on a value of 0.78 percent of oral aluminum being absorbed into the bloodstream from diet. In actuality, the ATSDR value for the same was 0.1 percent, thus making Mitkus et al.'s MSL almost eightfold too high. Combined with the evaluation by MacFarland et al., these are fatal errors in the Mitkus article that render their conclusions meaningless.

Questions about Aluminum and Immunotherapy in relation to Pediatric Vaccine Schedules

The discipline of immunotherapy employs various substances designed to trigger an immune response. The idea here is that such activation will provide a more general chronic immunity. One of the techniques often used involves the *s.c.* injection of Al hydroxide in various doses and with various schedules, mostly in adults.

A comment our work with Al adjuvants has sometimes received is that such injections do not cause any adverse effects and that therefore our critiques of the use of Al adjuvants is misplaced. In response to this, we have maintained that the notion that there are no adverse effects to immunotherapy with Al hydroxide is simply incorrect.[104]

On top of this, while it is true that the impact of *s.c.* injection may differ from that of *i.m.* injection in some regards, what is not taken into account by the critics is that weight is a critical variable in any calculation of Al body burden, as is the stage of neural development.

For a fast comparison of the amounts given over the first eighteen months of life, we have taken the vaccine schedule in my home province of British Columbia (Table 5.3A), calculated the amount of Al in this schedule based on the vaccines administered, and compared it to Al hydroxide regimes in immunotherapy based on amount given and how often. The results are shown in Table 5.3A through D.

In brief, considering only the amount of Al hydroxide given over eighteen months, a child in British Columbia will have a total of 2,087µg/kg of Al injected solely from Al hydroxide vaccines versus a total of 964µg/kg with the most aggressive immunotherapy schedule, the latter rarely applied. It is also well to remember that these numbers only apply to µg/kg of Al hydroxide and do not take into consideration stages of brain development and how Al, in any form, may affect it.

Immunization Schedule for B.C. Infants and Children

Vaccine	2 Months	4 Months	6 Months	12 Months	18 Months	Starting at 4 Years of Age (Kindergarten Entry)
Chickenpox (Varicella) Vaccine (#44b)[1]				✓		
Diphtheria, Tetanus, Pertussis, Hepatitis B, Polio, and *Haemophilus influenzae* type b (DTaP-HB-IPV-Hib) Vaccine (#105)	✓	✓	✓			
Diphtheria, Tetanus, Pertussis, Polio, *Haemophilus influenzae* Type b (DTaP-IPV-Hib) Vaccine (#15b)					✓	
Hepatitis A Vaccine (#33) Indigenous children only			✓		✓	
Inactivated Influenza (Flu) Vaccine (#12d)[2]					✓	
					Annually for children 6 months to 4 years of age	
Measles, Mumps, Rubella (MMR) Vaccine (#14a)				✓		
Measles, Mumps, Rubella and Varicella (MMRV) Vaccine (#14e)[1]						✓
Meningococcal C Conjugate (Men-C) Vaccine (#23a)	✓			✓		
Pneumococcal Conjugate (PCV 13) Vaccine (#62a)	✓	✓		✓		
Rotavirus Vaccine (RotaTeq®)(#104)	✓	✓	✓			
Tetanus, Diphtheria, Pertussis, Polio (Tdap-IPV) Vaccine (#15a)						✓

Figure 5.6. British Columbia immunization schedule from age two months to four years.[105]

Vaccines in BC Immunization schedule for 2, 4, and 6 months old with Al content

Vaccine	2 months	4 months	6 months	Dose	Al form	Al amount	Net amount
DTaP-IPV-Hib-HB vaccine	Y	Y	Y	0.5 mL	Aluminum salts	0.82mg (.5mg aluminium hydroxide + 0.32 mg aluminium phosphate)	1640ug/mL
Pneumococcal Conjugate (PCV 13)	Y	Y		0.5 mL	Aluminum phosphate	125µg	250ug/mL
Meningococcal C Conjugate (Men-C)	Y			0.5 mL	Aluminium hydroxide	1 mg	2000ug/mL
Hepatitis A			Y	0.5 mL	Aluminum hydroxyphosphate sulfate	0.225 mg	450ug/mL

Table 5.3A.

Al calculation

Name	Adjuvant	MW (g/mol)	Adjusted (mg)/ml of Vaccine	Vaccine (ml)	Adjusted (g)	mol Adjusted	Al (µg)	Al(OH)3 (µg)	Al (mol)
DTaP-IPV-Hib-HB vaccine	Al(OH)3	78.0036	1	0.5	0.0005	6.41E-06	172.9407	1445.582	1.85E-05
	AlPO4	121.9529	0.64	0.5	0.00032	2.62E-06	70.79454	925.1724	1.19E-05
Pneumococcal Conjugate (PCV 13)	AlPO4	121.9529	0.25	0.5	0.000125	1.02E-06	27.65412	361.3955	4.63E-06
Meningococcal C Conjugate (Men-C)	Al(OH)3	78.0036	2	0.5	0.001	1.28E-05	345.8815	2891.164	3.71E-05
Hepatitis A	AlHO9PS-3	235.03	0.45	0.5	0.000225	9.57E-07	25.82862	650.5119	8.34E-06

Table 5.3B.

Al amount in BC Immunization Pediatric Vaccines

Number of vaccines	Infants age (Months)	Infants weight (kg)	Al(OH)3 (µg)	Al(OH)3 (µg/kg)
DTaP-IPV-Hib-HB + Pneumococcal Conjugate + Meningococcal C Conjugate	2	5	5623.314	1124.663
DTaP-IPV-Hib-HB + Pneumococcal Conjugate	4	6.5	2732.15	420.3307
DTaP-IPV-Hib-HB + Hepatitis A	6	8	4336.746	542.0932
Total: 7 Injections			12692.21	2087.087

Table 5.3C.

Al amount in sub-cutaneous allergy immunotherapy

Length of treatment	Total number of injections	Al(OH)3 amount			Al(OH)3 amount in adult human (weight 70 kg)		
		Low (0.1 mg) µg	Medium (0.6 mg) µg	High (1.25 mg) µg	Al(OH)3 Low µg/kg	Al(OH)3 Medium µg/kg	Al(OH)3 High µg/kg
Starting	1	100	600	1250	1.428571	8.571429	17.85714
1st year	15	1500	9000	18750	21.42857	128.5714	267.8571
2nd year	34	3400	20400	42500	48.57143	291.4286	607.1429
3rd year	54	5400	32400	67500	77.14286	462.8571	964.2857

Table 5.3D.
Tables 5.3A-D. Aluminum hydroxide amounts in the British Columbia pediatric schedule from birth to eighteen months of age compared to the amounts of aluminum hydroxide delivered in "immunotherapy" treatments. Numbers are adjusted for approximate weight in kg.

The Unlikely Assertions of Dr. Paul Offit and CHOP

Dr. Paul Offit is a Philadelphia pediatrician, a faculty member at the University of Pennsylvania, and a vaccine inventor whose rotavirus RotaTeq made both him, and the university, a lot of money in royalties when the vaccine was sold to Merck. Offit also heads up the Vaccine Education Center at the Children's Hospital of Philadelphia (CHOP).[106]

Offit and CHOP, both discussed in more detail in the next chapter, are perhaps best known as staunch proponents of vaccination and equally strong antagonists to vaccine hesitancy and those considered to be anti-vaxxers. Let's leave the last aside until Chapter 6, where I plan to focus in much more detail on the misstatements made by CHOP's Vaccine Education Center in relation to Al adjuvants.

Aluminum and Autoimmunity

In Chapter 3, I touched on the immune system in health and disease. In the following, I want to briefly introduce the concept of autoimmunity before considering the evidence that Al adjuvants may have a role in triggering some autoimmune conditions.

First, the word autoimmunity simply means that one's own immune system is triggered to attack one's own body. The causes for such a reaction, in many cases, are not known. However, the subject is the focus of intense interest in medicine and at meetings of various professional bodies, such as the International Congress of Autoimmunity at its biannual meetings.[107]

Currently, some eighty autoimmune conditions are recognized with more discovered yearly. These include various forms of bowel disease, including inflammatory bowel disease (IBD), Crohn's disease, and ulcerative colitis; Type 1 diabetes; psoriasis and psoriatic arthritis; systemic lupus erythematosus (SLE); and a number of syndromes impacting various glands (Addison's,

Graves', Sjogren's, and Hashimoto's). Finally, there are autoimmune diseases of the nervous system including multiple sclerosis[108] and myasthenia gravis.[109]

In regard to vaccines, one of the earliest possible examples of an autoimmune reaction in the literature was to Guillain-Barré, a neurological disorder in which the Schwann cells that provide the myelination (insulation) to long axons in the peripheral nervous system are attacked and destroyed by one's own antibodies. Since effective nerve conduction from spinal cord motor neurons to the muscles they control and from sensory cells in the spinal cord to the sensory receptors in skin both depend on the axons being myelinated, the loss of this covering leads to loss of both motor and sensory function.

The initial link between vaccines and Guillain-Barré was in connection to the swine flu vaccine in 1976.[110] The idea that there was some potential linkage was not set in stone, but even conventional medical schools, such as mine, teach that the possibility exists.

In this view, the vaccine triggers some level of what is termed "molecular mimicry," in which some amino acid sequence in the viral/bacterial protein antigen in the vaccine resembles some amino acid sequence in the myelin sheath and marks it for destruction by immune cells.

A variety of other CNS disorders of an autoimmune nature have also been associated with Al adjuvants in vaccines. These are discussed in more detail in Shaw 2017[111] but are presented in brief in the following.

Of those neurological disorders associated with Al hydroxide in vaccines, a major one is macrophagic myofasciitis (MMF),[112] a deteriorating neuromuscular disorder that follows intramuscular injections of Al hydroxide in standard vaccines.

Patients diagnosed with MMF tend to be female (70 percent) and middle-aged at time of muscle biopsy (median age forty-five years). The patients have received up to seventeen Al-adjuvanted vaccines in a ten-year period prior to diagnosis (mean delay 5.3 years).[113]

Clinical manifestations of MMF patients include diffuse myalgia (muscle pain), arthralgia (joint pain), chronic fatigue, muscle weakness, and cognitive dysfunction with mild cognitive impairment (MCI). The latter is viewed as a key observation,[114] sometimes considered in other circumstances a precursor to Alzheimer's disease. MMF also features a variety of disturbances in interhemispheric functioning. Further, overt cognitive alterations affecting memory and attention are manifested in over half of all cases.

In addition to the other symptoms, 15 to 20 percent of patients with MMF concurrently develop some type of autoimmune disease, the most frequent being a multiple sclerosis-like demyelinating disorder.

The clinical significance of the MMF lesion at the site of injection was not fully understood until the work of Khan et al., 2013, cited earlier. These studies used Al tracers in mice following Al intramuscular injection to follow Al from muscle to other sites in the body, making the pathway quite clear: from muscle, Al nanoparticles are transported to the draining lymph nodes by circulating macrophages and then into the brain.

In addition, Al that made it into brain did not seem to be excreted during the approximately two-year time frame of these experiments.

There are obvious implications of these data when considering the impact of Al vaccines in children. While an adult MMF patient may have received up to seventeen vaccines over ten years, the average child in the United States following the CDC's vaccination schedule will receive the same number of aluminum-adjuvanted vaccines in their first eighteen months of life.[115] (The full CDC recommended vaccine schedule as of this writing is shown in Appendix 5.) As noted earlier, early postnatal life in humans and other mammals is a period of intense neurological development during which the CNS is extremely vulnerable to neurotoxic and immunotoxic insults.

More broadly, MMF is considered by Gherardi to be part of the "autoimmune/inflammatory syndrome induced by adjuvants" (ASIA), to be discussed in more detail below. The WHO, which we will consider in a later chapter, does not agree that MMF actually exists as a defined syndrome.[116]

ASIA was first described by investigators in Tel Aviv, all members of the clinical research group of Dr. Yehuda Shoenfeld, one of the leading figures in autoimmunity research and the head of the International Congress of Autoimmunity.

ASIA as described by Shoenfeld and colleagues comprises a wide spectrum of adjuvant-induced conditions characterized by a misregulated immune response.[117] As with MMF, various authors have questioned the existence of ASIA, of which MMF may be a subtype, as too vague in definition to actually describe a real syndrome.[118] Both the Gherardi and Shoenfeld groups have disputed these contentions.[119]

The studies by Luján and colleagues on commercial sheep and the response to either the blue-tongue vaccine or the injection of Al hydroxide are also considered by these authors to fall into the ASIA category.

What Do We Know about Aluminum Adjuvants and ASD?

As we've seen earlier in this chapter, ASD is not likely to have a simple etiology. As noted, it is not likely to be a genetic disorder per se, since no gene deletions or gain or loss of function mutations have been described, certainly none that would comprise the majority of cases. If not genetics as ardently promoted by much of mainstream medicine, if not changing diagnostic criteria or more general awareness, what is left? What's left is the exposome, that is, some external factors that may play out in context to some inherent (genetic) susceptibility.

So, let's consider what we do know.

At the beginning of this chapter, I reviewed the literature demonstrating the innate connections between the immune and nervous systems across the life span, but perhaps particularly so in early neural development. Especially at these early stages of development, things that strongly stimulate the immune system will necessarily have a neural impact, most commonly a negative one.[120] These negative outcomes may even be more pronounced if the stimuli are provided during fetal development as in the case of the literature on maternal inflammatory activation (MIA) brought on by various cytokines, notably IL6. These results from the work of the late Dr. Paul Patterson may actually be crucial to understanding how immune modulation shapes neural development.[121]

In regard to this last point, cytokines produced in peripheral tissues by immune stimulation can enter the brain by way of the circumventricular organs (CVO).[122] CVOs are structures in the brain with an extensive vasculature and are among the few sites in the brain devoid of the blood-brain barrier, the collection of membranes that function to keep many molecules out of the brain.

CVOs provide one link between the CNS and peripheral blood flow and thus are an integral part of neuroendocrine function. The absence of a blood brain barrier to CVO molecule release allows the CVO to provide an alternate means for the release of hormones and various peptides from the CNS into peripheral circulation. In addition, structural connections now demonstrated between the lymphatic system and the CNS add to the potential for immune-CNS bidirectional ingress of molecules that affect both,[123] in normal and abnormal brain development.[124]

In the above sections, we have also shown that Al can negatively impact neurons in the CNS; and Al adjuvants by their very purpose chronically stimulate an immune response. Taken together, it would seem highly unlikely that Al from any source, perhaps particularly from vaccines that contribute so heavily to the Al body burden in young children, fail to have a role in abnormal brain development.

For this reason, the ability of Al to adversely affect both the immune and the nervous system in an interactive manner makes it a strong candidate risk factor for triggering developmental disorders such as ASD. It is in ASD that the two principal features of the disorder are precisely those of neurological and immune system signaling dysfunctions.

There are various caveats to the above, especially considering Al adjuvants, but one in particular is worth considering in more detail. This is that not all forms of Al adjuvants have the same properties in terms of distributions and longevity in the body.[125] In addition, it must be considered that the studies that have implicated Al adjuvants in either immune or CNS disorders do not use the actual form of Al used by the various companies in their adjuvant formulations. The obvious reason for this is that these are proprietary inventions, not generally available to noncompany scientists. It is certainly possible, if unlikely, that these forms of the Al adjuvants may not provoke the same sorts of response that the commercial forms have been shown to do.

One finding that makes this last possibility unlikely is that the behavioral and neurological pathologies associated with aluminum-adjuvanted vaccines administered to commercial sheep by the Luján group, compared to the available forms of the adjuvants alone, are largely the same.[126]

Aluminum and Biosemiosis

All of the issues surrounding Al toxicity in the nervous system and the effects on the immune response, as cited earlier, are potentially surmountable, at least in the short term. Yes, Al adjuvants have the potential to cause harm, but a facile solution would be to switch to less potentially toxic adjuvants. These already exist. One example is a calcium phosphate developed by the Pasteur Institute in Paris[127] and the calcium fluoride adjuvant tested by the pharmaceutical company GSK.[128]

The simple reality is that the pharmaceutical industry knows this perfectly well and won't make the switch until they are forced to do so. Switching adjuvants, doing the effectiveness and safety trials again, as minimal as the latter are, would cost money. From a dollars and cents perspective, this won't happen, as these highly profitable corporations maintain their focus, as always, on the corporate bottom line.

Would either alternative adjuvants be safer than Al adjuvants? Probably, but that still doesn't get to the fatal flaw in the use of any adjuvant, although the severity of the problem really remains highest in relation to Al-adjuvanted vaccines.

To explain this, I need to go into the nature of biological signaling, a discipline called "biosemiosis," where "bio" indicates living things and "semiosis" reflects what the symbols mean. Biosemiosis is basically how living things communicate at all levels within themselves and at much higher levels with other organisms.

Some simple examples will suffice to make the nature of biologically signaling clear. Most are familiar with what happens when there are gene deletions or mutations. In the first case, a deleted gene simply won't make a protein on which the survival of the cell, and maybe the organism, may depend. As there is enormous redundancy in coverage for crucial functions, a deletion may not matter much, but then again, it can be crucial. Gene deletion therapies are currently in development for various medical conditions.[129] This is termed a "loss of function" mutation.

However, a "gain of function" mutation makes the gene do something it would not normally do, and this change can lead to fatal consequences. An example is the gain of function mutation in the gene coding for superoxide dismutase (SOD), often considered one of the major genes associated with familial ALS.[130] In both cases, an essential bit of information from the DNA to the cell's protein manufacturing organelles has been changed. In other words, there has been a failure to communicate a usable signal to make the correct protein needed for normal cellular and systems functions. At a much higher level, let's say we were talking in English, which we both understood well, then I suddenly switched to the Kurmanci dialect of Kurdish. Unless you also understand this language, the message I was trying to send you won't work. Worst, if the message is crucial to your survival, it will fail to alert you to any danger.

Dr. John Oller, a mathematician and linguistic scholar, has studied biosemiosis in great detail and has elaborated on the basic theory by defining what he terms the "true narrative representation" (TNR), which is the information that passes from a sender through a translation to a receiver as a concrete concept.

The basic idea is that if the message sent is corrupted, the receiver will have no idea what to do with it. Take as an example of this the children's game "broken telephone." I say to the person next to me, "The house is blue." What may come back, determined by how many people are passing the message sequentially, may be a complete biosemiotics mess that makes no sense at all. If there are no real consequences to this, it remains a silly party game. What if, however, the message was "The house is on fire!"?

Here is where the problem with an adjuvant like Al becomes acute: it is not a real message.[131] Aluminum is like an alien. No life form on Earth evolved with aluminum and hence knows what to do with it. Al sends a nonsense signal that your cells, particularly those in your immune system, can't interpret. Instead, what they get is a garbled message that they react to, but incorrectly.

Al adjuvants viewed through the lens of TNR are not just wrong, but a biosemiotics "lie."[132] They not only don't convey any real information, they actually prevent it.

Al as an adjuvant thus presents an intentional deception when it inexplicably appears in various organs and the lymphatic system along with partial viral or bacterial fragments that are supposed to be the informative antigenic components of one or more vaccines. These pathogenic components are also deceptive insofar as they are not the actual immune system stimulant, that is an actual pathogen that will accurately trigger an appropriate immune response.

The consequence, if repeated often enough, is increasingly likely to be dysregulation, which is almost a classical definition of autoimmunity. In turn, when the immune system is undergoing rapid development in infancy and early childhood, it is axiomatic that biosemiotics confusion will impact the CNS, which is where the greatest damage occurs.

If the organism survives the initial presentation of Al or other bad signals, the resulting errors in biosemiosis and thus neural development are likely to become magnified through successive stages of development.

This is the core message of the paper that Oller and I wrote in 2019. Although this message is contained within a large theoretical construct defining how biosemiosis works,[133] the lie of Al adjuvants is primarily that it presents an imaginary signal to the immune system. From this point onward, it is almost a certainty that since antigens in vaccines are not a complete TNR either, such incorrect signals alone, not to mention when amplified by Al, may limit the effectiveness of multiple vaccine schedules by devolving into serious rates of adverse effects.

In other words, the more you try to fool the immune system with incomplete TNRs, the greater the likelihood of failure and biological pushback. You may get away with it for a while: in some individuals for a long time with multiple vaccines, in others for a much shorter time or none at all.

Nevertheless, these considerations may set a finite limit on the number of incomplete/erroneous TNRs that can be sent to the immune system. Beyond that limit, autoimmune reactions are all but inevitable.

What Do the Leading US Health Organizations Know about Aluminum Adjuvants in Vaccines?

In the United States, there are three main entities involved in the surveillance of human health: the CDC, the FDA, and the NIH. The first, as per its name, is responsible for understanding and preventing diseases that may affect the American population. The second agency monitors the safety of food and drugs used by Americans. The NIH, one institute of which is the National Institute of Allergy and Infectious Diseases headed by Dr. Anthony Fauci, is supposed to sponsor, or even do the research on, any biomedical topic involved with infectious diseases and allergies. We will meet Fauci again in Chapter 13.

The work of these three agencies intersects in a number of ways, but none more than in regard to the evaluation, licensing, and surveillance of vaccines. Given this, it would seem part of their respective mandates to know a great deal about the subject of Al in general and about Al in vaccines more specifically. All have made statements on their respective websites that Al in vaccines is not harmful.

Aaron Siri, a New York City lawyer, decided to find out what scientific literature each agency used as the basis for these conclusions.

On behalf of Mary Holland[134] and me, Siri wrote a Freedom of Information Act, or FOIA, request to each agency requesting a list of the scientific evidence each relied on for this evaluation of Al adjuvant safety. An example of this letter, in this case the one sent to the NIH, is shown in Appendix 7.

Regarding this example, this was not a complex request. In due course, the NIH replied that they had found "no records responsive to your request." CDC said much the same, which is odd given the number of papers they feature on their website supposedly demonstrating vaccine safety, as also highlighted in Chapter 3. The FDA seemed to have difficulties understanding the question, but they too seemed to have nothing to report.

The reply from the NIH is shown in Appendix 8.

One would think that the FOIA person at each agency would at least cite Mitkus et al., maybe even some of the other Al kinetics studies cited in the section above. Since these FOIA respondents didn't even cite such studies, we are left with the difficult conclusion that none of these agencies is capable of backing up their assertions on Al adjuvant safety with any scientific literature.

In reality, it should not have been a difficult problem for any of these agencies to satisfy the FOIA: all they had to do was to cite any scientific literature, not only the peer-reviewed part. It seems, based on this outcome, that they all believe that it is far simpler to make blanket pronouncements about safety, while hoping that no one notices that this particular emperor has no clothes on.

And, mostly, they seem to have gotten away with this strategy . . . at least so far.

Summary of the Effects of Aluminum in Vaccines

One way to view Al toxicity in the CNS, generally and in relation to the impact of Al adjuvants in vaccines, may be to consider it as a generally neurotoxic element with spatial variations in CNS subsystem impacts that are extremely diverse. In this view, the precise outcome may depend on a variety of intrinsic and extrinsic factors such as age, sex, and individual genetic polymorphisms and biochemistries, individual microbiomes, etc. Extrinsic factors include the type of the Al compound, the amount of exposure, the route of exposure (for example, by food, water, intramuscular versus other types of injections, inhalation), etc.

For all of these reasons, the toxicity of Al in the CNS appears to depend on a number of variables that include both direct and indirect cellular mechanisms. In both cases, factors include the form of Al complex, the size of the adjuvant particles, route of administration, and dose. Dose itself may not be the major consideration. Further, in animal models, species and even strain may influence the level of Al neurotoxicity.[135]

Does this suggest that I am convinced that Al adjuvants are causal to ASD? Far from it. My studies of various neurological diseases do not lead me to think that any of them, from ALS to ASD, have a single causal factor. Rather, I think that in each case, they all require a constellation of collaborating effects triggered before the disease can be expressed.

In this view, Al adjuvants are merely another such element. They may perhaps be a crucial element and one without which ASD would decrease. However, I don't think they are the only ones.

This outline of the toxicity of Al, especially in relation to neurological disorders like ASD, opens many windows into the status quo about vaccines and neurological adverse effects. The reasons why some cannot/won't see this, both inside the field and among lay people, thus becomes the subject of some of the following chapters.

Before going there, however, I want to close this chapter by acknowledging the seminal work of Prof. Chris Exley, discussed in various places in this book. Prof. Exley has just completed his own book, *Imagine You Are an Aluminum Atom: Discussions With Mr. Aluminum*.[136] In his online blog,[137] Exley writes that

> The book is a warning. When Rachel Carson wrote *Silent Spring* it was not as a reference guide to pollution of the environment by pesticides. It was a warning. As I wrote my book, I unravelled forty years of research and revealed, at least to myself, that human exposure to aluminium is the, yet unrecognised, unprecedented threat to the future of mankind.

As in any subject in science, not everyone sees all of the literature on a subject, in this case Al, in the same way. For example, a few authors don't admit to seeing Al as the same sort of serious toxicant that Exley and various others, including this author, do, like a 2014 systematic review by Willhite et al.[138] To their credit, Willhite and colleagues review the impact of different forms of aluminum in humans and in animal studies in this review. Here is what they conclude:

> The results of the present review demonstrate that *health risks posed by exposure to inorganic Al depend on its physical and chemical forms and that the response varies with route of administration, magnitude, duration and frequency of exposure.*
>
> These results support previous conclusions that there is little evidence that exposure to metallic Al, the Al oxides or its salts increases risk for AD, genetic damage or cancer. . .
>
> The results of the present review support previous conclusions that there were no clear associations between vaccinations using Al adjuvants and serious adverse events (GACVS 2012, Kelso et al. 2012). [Italics are mine.]

The italicized comments are precisely the point that I was making earlier: the impacts of Al are variable, depending on all of the factors that Willhite et al.

mention. This does not, however, translate into a valid conclusion that Al is harmless to humans or animals. In fact, a detailed reading of this meta-analysis makes it quite clear that the opposite is true. In a review of this nature, the authors apparently decide that one demonstrated harm is negated by null results in other studies.

It's an odd way to do science in my view and one that might possibly have arisen due to the links and funding of some of the authors to the International Aluminum Institute, the Aluminum Research Consortium, and Risk Science International.

A more recent article from 2019 by Corkins holds that in general aluminum is not a problem for human health.[139] However, this article is largely focused on aluminum in infant formula and tends to make the same general assertions that dietary aluminum has the same pharmacokinetics in the body as that of injected aluminum. That this is not correct has been amply demonstrated by the data from the Gherardi group, cited earlier. Finally, a recent article by Boretti clearly supports the notion that adjuvant aluminum is associated with ASD.[140]

If one has to choose a credible narrative, I'd be inclined to trust a scientist like Prof. Chris Exley with forty-plus years of bench science behind him over those with ties to the aluminum industry or those who confuse routes of administration. As in many things, the "devil is in the details" here, as I will show when I turn next to considering the pro-vaccine thought leaders, highlighted in the next chapter.

CHAPTER 6

The Vaccine Wars and the Pro-Vaccine "Thought Leaders"

One certain effect of war is to diminish freedom of expression. Patriotism becomes the order of the day, and those who question the war are seen as traitors, to be silenced and imprisoned.

—Howard Zinn[1]

On War Metaphors

It is far too common to use war metaphors for nonwar topics. Indeed, it has become typical in current politics to do so, at least in certain countries such as the United States, especially given the close relationship between failed diplomacy and the use of military force. War metaphors are also widely used elsewhere for a variety of topics, perhaps the best example being medicine and social policy, the latter in relation to what are actually issues related to health. We speak, for example, about wars on poverty, on crime, on drugs, or on disease.

The language of war is used extensively in regard to the latter: conquering cancer, battling neurological diseases, or defeating COVID-19 come to mind. In addition, with COVID-19 we talk in pseudopatriotic slogans such as "We're all in this together."

In brief, we are surrounded by warlike messaging about us as human beings fighting against concepts, social situations, or medical conditions. To a great extent, the (mis)use of war terminology makes little real sense, no matter how evocative it may be.

Where this war metaphor may not be stretched too far, however, is how those of us involved in the discussions about vaccines speak about the controversies that have arisen in the last twenty some years. As the title of this book attests, I am guilty of using the war metaphor myself.

Some might say that this metaphor for vaccines is still overblown because it doesn't actually involve killing, but those critics would be wrong. At least in the minds of some of the participants on the opposing sides, it very much involves killing. Those on the mainstream medical side quite often make statements to the effect that by not vaccinating your children you are, in effect, risking— maybe even taking—their lives, or the lives of those who can't be vaccinated. In contrast, those who have lost a child to a vaccine adverse reaction may very much see the medical establishment as a killer enemy. Marry these views up to statements made in both directions that their opponents on the other side are somehow lesser beings, and you have the very fertile ground in which real violence could grow. We will come back to this last point near the end of the book, where I will indulge my speculations and concerns for the future.

Wars, chaotic as they are, have structure, mostly in the form of hierarchical ranks of the participants. At the low end are the "grunts," the foot soldiers, who do most of the actual fighting. Higher up are the senior noncommissioned officers and the junior officers who are also down there in the weeds slugging it out with an enemy. Further up the food chain are the field rank officers who only rarely get their hands bloody. Above them are the general officers. And at the top of that pyramid are those who ultimately call the shots, literally and figuratively, and who usually have something in particular to gain. Whether that something is money or power, or both, varies.

One thing to keep in mind throughout the remainder of this book is a key question: who are the people at the top? Intimately related to this question is another, and that is cui bono, who benefits? The answer is inevitably going to be that those at the top are the ones who do.

Some readers may wonder if I am starting to slip into "conspiracy theory" mode and will soon be evoking the illuminati and shape-shifting lizards.[2] No, I won't, not because the idea of lizards controlling human destiny lacks for entertainment, but because I think the answer is far more mundane. As we will see in later chapters dealing with the pharmaceutical industry and their allies, those calling the shots (again, figuratively and literally) have more the characteristics of mob families than aliens.

But for now, I want to consider who the middle-ranking officers are, the military equivalent of the majors and lieutenant colonels controlling battalions and regiments. In other words, who are the individuals driving the vaccine campaigns, but not necessarily giving the orders overall? We will look at who holds these ranks on the vaccine-hesitant side in the next chapter, but here I want to focus on those sometimes described by the media as the "thought leaders," those whose pronouncements on vaccine issues get widespread media coverage and serve to mold public opinion.

The "Thought Leaders" of the Pro-Vaccine Camp

At the top of virtually everyone's list of the most influential, "go to" spokespersons for vaccines, you will find the name of Dr. Paul Offit. Whether the media wants to talk about measles outbreaks, damn a publication that questions any aspect of vaccine safety, get an update on the development of vaccines against COVID-19, or discuss pretty much anything vaccine-related, Offit is the man they seek out first, at least in the United States.

In Canada, that person has become University of Alberta law professor, and Offit wannabe, Timothy Caulfield. Caulfield has the disadvantage that he has neither the professional chops of Offit combined with the rather glaring problem that he often doesn't seem to know what he is talking about. However, he can be counted on to give snappy quotes that skirt the border of defamation, a plus from a media perspective. If your news organization is the public Canadian Broadcasting Corporation (CBC), an entity that many Canadians worship but one whose reporting has become sloppy in the extreme, Caulfield is about the best you can get. I will return to both Caulfield and the CBC in a later chapter.

So who is Paul Offit, MD, or "Dr. Proffit," as he is sometimes referred to by his detractors? The epithet arises from royalties received from his work as a coinventor of the rotavirus vaccine, RotaTeq®, made by Merck.[3]

Offit received his undergraduate degree at Tufts University in 1973 before going on to do his medical degree at the University of Maryland. From there, he took his internship and residency in pediatrics at the Children's Hospital of Philadelphia (CHOP) and basically never left. His board certification is in pediatrics, not infectious disease.

Here are some of the titles he holds, all notionally quite impressive:

> Offit is currently a professor of Pediatrics in the Division of Infectious Diseases at the Children's Hospital of Philadelphia; director of the Vaccine Education Center; Maurice R. Hilleman and professor of Vaccinology at the Perelman School of Medicine at the University of Pennsylvania.[4] He holds a $1.5 million research chair at CHOP funded by the large pharmaceutical company Merck. He is a member of the Institute of Medicine.[5]

In addition, from 1998 to 2003, he served on the Advisory Committee on Immunization Practices (ACIP)[6] for the CDC. He advises the FDA in his role (2018 to the time of writing) on the Vaccine-Related Biologicals Approvals Committee. From 2005 until the time of writing, he has been on the Allergy/Asthma Data Safety Monitoring Board of the National Institute for Allergy and Infectious Disease, part of the National Institutes of Health. This is the same institute that has had Dr. Anthony Fauci as director since 1984, a man we will meet again in Chapter 13. Offit is also codirector (2008 to the time of writing) of the Center for Vaccine Ethics and Policy (CVEP).[7] CVEP is a program of the GE2P2 Global Foundation[8] and has primary affiliations with the division of Medical Ethics at New York University School of Medicine and the Vaccine Education Center of CHOP. CVEP's website states that they receive support from the Bill and Melinda Gates Foundation, PATH,[9] the International Vaccine Institute, and various pharmaceutical companies including Crucell (a lesser-known vaccine company), Janssen, Johnson and Johnson, Pfizer, Sanofi Pasteur US, Takeda (the largest Japanese pharmaceutical company), Valera (part of Moderna, which we will review in Chapter 13,[10] and something called the Developing Countries Vaccine Manufacturers Network. CVEP on their website also states that ". . . regardless of the source, form or scale of support we receive, we do not accept and will not operate under any conditions which interfere with our independence . . ."[11]

Good to know. And at least they admit that there could be a perceived conflict of interest in such extensive ties to the pharmaceutical industry, unlike Offit, who only occasionally acknowledges his own conflicts of interest.[12]

From 2011 to now, Offit has been involved with the Foundation for Vaccine Research,[13] about which little information is available online apart from the fact that it received a $23,000 grant from the Gates Foundation. Also, from 2009 to the time of writing, Offit was on the Founding Advisor Board of the Autism Science Foundation (ASF),[14] the latter created in a split from an organization called Autism Speaks, which had considered the possibility of vaccination having a role in autism. ASF disavows any such link. Offit apparently donated the royalties of his books *Deadly Choices* and *Autism's False Prophets* to ASF. Other than this, ASF claims that their funding comes from donations and fundraising events.

In addition, Offit has served from 2006 to the time of writing on the American Council on Science and Health (ASCH) as an advisory board member. ACSH is a controversial "nonprofit" advocacy organization, founded in 1978. It describes itself as a "consumer education consortium" focusing on food, nutrition, chemicals, pharmaceuticals, lifestyle, the environment, and health, but others have accused it of being biased in favor of industry. ACSH has defended DDT, asbestos, and Agent Orange, as well as common pesticides; and they are known to be funded by big agribusinesses and big companies such as Kellogg, General Mills, Pepsico, and others.[15]

Adding to the above is this: From 2006 to the present, Offit has served on the advisory board of Every Child By Two (ECBT), the latter criticized in a CBS News feature called "How independent are vaccine defenders?" which questioned the group's links to big pharma and thus had a conflict of interest. A spokesperson for ECBT told CBS that ". . . there are simply no conflicts to be unearthed."[16]

ECBT is now called "Vaccinate Your Family." The 2019 sources of funding for this organization include the CDC, GSK, Merck, Novavax, Novolex
Holdings, Pfizer, Sanofi Pasteur, and STC Health.[17] No conflicts to be
unearthed, indeed.

Of interest, particularly in relation to the COVID-19 pandemic, Offit sits
on the Board of Directors for the GAVI Campaign, in turn part of the Global
Alliance for Vaccines and Immunisation (GAVI), the self-described "vaccine
alliance." GAVI describes itself as a "public-private global health partnership
with the goal of increasing access to immunization in poor countries."[18]

GAVI has been criticized for giving private donors more unilateral power
to decide on global health goals, prioritizing new, expensive vaccines while
putting less money and effort into expanding coverage of older, cheaper ones,
harming local healthcare systems, and spending too much on subsidies to
large, profitable pharmaceutical companies (mostly to GSK and Pfizer) without reducing the prices of some vaccines. In regard to its conflict of interest,
it acknowledges having vaccine manufacturers (Susan Silbermann from Pfizer
Vaccines) on its governance board, along with members of the Bill and Melinda
Gates Foundation. These latter members include Orin Levine, director of the
Vaccine Delivery Global Development Program, and his deputy director,
Violaine Mitchell. In addition, there are WHO members on the board, as well
as members of the World Bank.[19]

GAVI sponsors include: The Bill and Melinda Gates Foundation, which
had donated $1.56 billion as of March 2019, followed in June 2020 with a pledge
for another $1.6 billion. The largest donor is the United Kingdom; other coun-
try donors are Norway, Germany, and the United States. In the case of the
United States, when President Trump announced during late 2020 that he was
pulling US funding out of WHO, he suggested that would put some of these
funds into GAVI instead.[20]

Is any of the above a mortal sin?[21] Maybe not mortal, but likely one of a
moral nature.

All of these relationships show quite clearly the highly interconnected
nature of the various regulatory bodies of the US government, the extensive
reach of the pharmaceutical industry into a range of organizations, and the
role played by someone like Paul Offit, a willing hired hand whose conflicts of
interest would sink anyone else, or at least anyone else who had not brought an
endless royalty stream into his institution.

The underlying role of the pharmaceutical industry will be highlighted in
Chapter 12.

Offit has won numerous awards and has authored, at least as of 2018,
150 scientific papers.[22] Offit has also written at least seven popular books on
medicine and more specifically on vaccines. In some of these, Offit spends a

lot of his effort slamming what he considers to be bad science (*Autism's False Prophets: Bad Science, Risky Medicine, and the Search for a Cure*),[23] a peculiar position to be taking given some of his own publications, to be discussed in detail soon. Other titles are simply bombastic, such as *Deadly Choices: How the Anti-Vaccine Movement Threatens Us All*.[24] Then there is this: *Bad Faith: When Religious Belief Clashes with Modern Medicine*.[25] The irony of this last title will become clearer when I discuss the essentially religious and authoritarian nature of many ardent vaccine advocates as typified by Offit. (The religious aspect will be explored in greater detail in Chapter 8 and Offit's apparently authoritarian instincts in Chapter 11, which deals with what is termed "state of exception.")

Figure 6.1. Pictures of well-known proponents of mass vaccination: Drs. Paul Offit, Stanley Plotkin, and Peter Hotez.[26]

Dr. Offit has impressive credentials for sure, but before we get too starry-eyed about his formal accomplishments, let's explore a few more items in detail. These will include the often casual nature of the scholarship that Offit exhibits, the track record of RotaTeq, and the often undisclosed conflicts of interest that seem to be present in relation to his royalties from this vaccine.

Before going into these items, I should mention that I have had one direct interaction with Offit, albeit via television link. This was on *Stossel*, John Stossel's show on the Fox Business Channel, in 2010.

In regard to this show, I am not sure what I had been expecting, but what I had been *told* to expect by the show's producer was a civil, academic discussion on the merits of Al adjuvants in vaccines. I quickly realized that this was not to be when the show began with the sound of a child's labored breathing and Stossel asking me right out of the gate why I wanted children with pertussis (whooping cough) to suffer and perhaps die.

I had been led to believe that Stossel was a libertarian, at least in leaning, so was surprised by this opening. Offit, for his part, was true to form and started off complaining that he didn't like having to go on shows to discuss the obvious fact that, using the typical cliché, "the science was settled" about vaccines and their safety, as it only provided false balance between those in authority, such as himself, and those who were simply raising fake concerns, i.e., me.

I was perhaps naive then, not fully understanding how the entire scientific debate on vaccine safety was really no longer an academic one, but rather one of corporate interests protecting their products. I digress, but I will come back to the latter observation soon.

After the show, I made a note to myself to have a ready counter sound track to a kid suffering with pertussis. How about this: a kid on the spectrum having a raging meltdown followed by an epileptic seizure? There are certainly enough examples of something like this out there. Next time, I am going to be ready.

I don't recall whether or not Offit stated the typical talking point that Al can't possibly be harmful because it is only found in tiny quantities in vaccines, less than in a cup of tea, and because it is rapidly excreted. If he did not use these arguments at that time, he and others certainly have at other times. For this reason, this is the opportune moment to refute them.

The first part of the "amount" argument follows along the lines of the famous statement attributed to Paracelsus that "The dose makes the poison."[27] This is true but is also totally irrelevant unless one is also going to claim that the route of administration makes no difference to the pharmacokinetics of Al in the body. In the case of dietary Al, the Al goes from the stomach/gastrointestinal system into the blood and is then, in most cases, effectively and rapidly excreted. This view would follow from the studies of Mitkus et al., shown in the previous chapter to be incorrect.

In distinct contrast, injected Al, either subcutaneously or intramuscularly, is highly concentrated in a vastly smaller space where immune cells find and transport it to various organs. Rather than rapidly leaving the body, we have reason to believe from the work of McFarland et al. and others that, at least in small children, Al remains for considerable periods in the body. To confuse the two routes and their outcomes for Al levels in the body is a gross distortion of the science, done either through ignorance or a deliberate attempt to deceive.

The second part of this argument about the amount of Al in any vaccine is that it is so small that regardless of where it goes it can't possibly do any harm. Again, such an argument can only be made out of ignorance or malice, since, as demonstrated in Chapter 5, that same small amount is perfectly able to highly stimulate the immune system. Further, as already noted, the interactions between the immune system and the CNS ensure that what stimulates one will necessarily stimulate the other.

In addition to the above, as director of the Vaccine Education Center, Offit is responsible for the content of the information and papers cited by the Center, if for no other reason than "ministerial responsibility." In other words, this is the notion that the person at the top very much "owns" any mistakes made by subordinates.

In spite of this, Offit does not seem troubled about the content of the Center's website. In regard to vaccine adjuvants, the website previously made some rather egregious errors of fact, one of which appeared on the website concerning aluminum and pregnancy.

This was, in fact, the first thing I originally noted about the Center's website—the material contained in a screen shot about Aluminum and Pregnancy (see Fig. 6.2). This rather remarkable collection of statements included the notions that Al is an *essential* element, a blatantly incorrect assertion, and the even more bizarre notion that it was part of a healthy pregnancy. Here is what CHOP had posted:

> Aluminum is considered to be an *essential* metal with quantities fluctuating naturally during normal cellular activity. *It is found in all tissues and is also believed to play an important role in the development of a healthy fetus.* This is supported by several findings: During healthy pregnancies the amount of aluminum in a woman's blood increases; the amount of aluminum in the blood of the fetus increases between four and a half and six months gestation and again at eight months gestation; at delivery, the blood of full-term infants contains more aluminum than the mother's, but it decreases shortly after delivery; the blood of premature infants has more aluminum than that

Aluminum and pregnancy

Aluminum is considered to be an essential metal with quantities fluctuating naturally during normal cellular activity. It is found in all tissues and is also believed to play an important role in the development of a healthy fetus. This is supported by several findings:

- During healthy pregnancies the amount of aluminum in a woman's blood increases.

- The amount of aluminum in the blood of the fetus increases between four and a half and six months gestation and again at eight months gestation.

- At delivery, the blood of full-term infants contains more aluminum than the mother's, but it decreases shortly after delivery.

- The blood of premature infants has more aluminum than that of full-term infants.

- The concentrations of aluminum in brain tissue are high during gestation and highest immediately after birth.

- The breast milk of moms with premature infants contains more aluminum than that of moms who carried their babies to term.

Figure 6.2. Screen shots from the Vaccine Education Center. Screen shot from the CHOP Vaccine Education Center (VEC) website, later deleted by VEC, to remove the claim that aluminum is an essential element during development, including during fetal development. The page was taken down not long after I gave a talk at George Washington University in Washington, DC, in 2013 pointing out the falsity of these claims.

> of full-term infants; the concentrations of aluminum in brain tissue are high during gestation and highest immediately after birth; the breast milk of moms with premature infants contains more aluminum than that of moms who carried their babies to term. [Italics for emphasis are mine.]

None of these comments are referenced to the scientific literature, hence it is difficult to know if these Center "facts" were taken from actual papers or were simply made up. A key point to note is the suggestion that aluminum was an "essential metal" came down from the site not long after I gave a talk at George Washington University, noting that such a claim was without any scientific foundation. Maybe the Vaccine Education Center had been planning to change this claim anyway, or perhaps they felt embarrassed by being caught peddling nonsense. It is not clear which motive had the higher value. Whichever the case, the fact that such a statement was even posted suggests that Offit and the Center have a fairly meager understanding of aluminum in the body and what it does and does not do.

This interpretation is supported by a reference they do cite in general support of their view about aluminum. The reference in question is to a massive 1986 paper by Ganrot that contains 959 references. The opening paragraph of this paper states that

> Aluminum (Al) is present in very small amounts in living organisms but is abundant in the environment. *In no case has Al³⁺ been shown to have a definite biological function. Taken together, this suggests that Al³⁺ possesses properties incompatible with fundamental life processes.* Despite this, Al³⁺ has generally been regarded as virtually biologically inert and the interest shown for its biochemistry and metabolism has been very limited. However, during recent years, an increasing number of toxic effects have been established. Interest in Al³⁺ has therefore increased, but many basic questions still remain unanswered. [Italics are mine.]

This paragraph, indeed the entire paper, fits well within the body of scientific studies that have typically characterized aluminum as a toxic substance, in particular a neurotoxic one, as discussed in detail in Chapter 5. Indeed, it would seemingly be impossible to use Ganrot as a reference for aluminum being beneficial to a fetus or mother, not to mention to anyone else. Further, in more general terms, it is rather hard to reconcile this paragraph being used to support a benign, let alone a beneficial, role for aluminum in biological systems at all.

Nevertheless, what persisted was the reference to Ganrot, of which part of the first paragraph is shown in the previous quote.

The Center cites this article as somehow supporting the view in the screenshot that Al is beneficial.

How one gets from Ganrot to such a conclusion is not at all clear. The most obvious interpretation is that Offit and colleagues at the Center simply hadn't read any of the article, in which case they are guilty of unbelievably careless scholarship. Or, the alternative is that they had read the article, knew that it stated the opposite of what they claimed, and cited it as support for their views anyway. Such behavior is simply a major misrepresentation of the existing data and in a scientific context would certainly cross into the zone of academic malfeasance at any university that even pretends to care about research ethics.

Let's, however, give the Center the benefit of the doubt. Errors in citations can happen for sure, and one might forgive the Center for letting the wrong article slip through onto their website.

It is more difficult to reconcile the evaluation of an Al study cited in an article by Offit and Jew in 2003,[28] which reveals yet another example of the tendency of Offit and his colleagues to grossly misrepresent the actual scientific literature. On page 1396 of this article, Offit and Jew state that

> For determining the quantity of aluminum below which safety is likely, data were generated in mice that were inoculated orally with various quantities of aluminum lactate.[42] No adverse reactions were observed when mice were fed quantities of aluminum as high as 62 mg/kg/day.

The endnote 42 in the article refers to a paper by Golub et al. from 1989 in which mice were given different levels of Al lactate in water, observed for six weeks, then had various organs harvested and assayed for Al levels. Here is what the Golub article actually says in the final sentence of the abstract: "These data demonstrate that short term feeding of aluminum at levels within an order of magnitude of estimated human intake can influence neurobehavioral function as indexed by motor activity."[29]

The paper noted more abnormalities such as fur loss in the high Al group, a 20 percent decline in activity levels, less activity during the day, and an over threefold increase in Al in the brain.

How does one possibly square what Offit and Jew claim with what was actually reported by Golub et al.? One can't. It's simply a grotesque misrepresentation.

In addition, Golub's laboratory carried out studies leading to a whole series of papers, all published before Offit and Jew's article was published in 2003. In Golub et al.'s 1994 article, the authors demonstrated that mice given Al lactate from conception showed a reduced auditory startle reflex at an early time point.[30] A 1995 paper by the same group added to the earlier findings and noted decreased fore- and hindlimb grip strength and increased cage mate aggres-sion,[31] the latter resembling the data from sheep in Lluis Luján's study, cited in Chapter 5. Later, Golub et al. 2001 showed that higher concentrations of alumi-num exposure in male and female mice led to lower body weights, but greater brain weight,[32] the latter observation similar to that seen in human ASD.[33] *All concentrations* of aluminum led to longer latencies on a Morris water maze, a test of spatial learning. This effect was dose-dependent with the amount of alu-minum given such that the latency increased in lockstep with increased dose. Hindlimb grip strength, performance on the rotarod, and wire suspension tests also significantly decreased in the highest aluminum concentration group.

Admittedly, Golub and colleagues used Al lactate in water whose pharmaco-kinetics are quite different than that of injected Al adjuvants. Offit presumably knows this, or at least should. So even in the best case, Offit and Jew used the wrong model in an attempt to demonstrate Al safety and then misrepresented the data so as to ensure that the outcomes backed their preconceived notions.

As in any field of religious morality, there are sins of commission and sins of omission.[34] The problems mentioned with the work of the Vaccine Evaluation Center and Offit specifically concern the former.

The sins of omission are perhaps subtler and involve what can only be con-sidered cases of ignoring data that don't fit into a clear and biased narrative such as with the Golub papers that came after 1989.

In this regard, it should be mentioned that there are also two articles on Al and human infant development, the first by Bishop et al. 1997.[35] These authors examined 182 surviving premature babies of less than thirty-four weeks dura-tion who were fed by intravenous infusion a standard formula that delivered 45µg Al/kg/day versus one in which most of the Al had been deleted (4-5µg Al/kg/day). No other aspects between the groups differed. Infant development was measured by the Scales of Infant Development at eighteen months of age. In the first study, the differences were not significant between the groups. However, in a subgroup of about half of the infants based on elimination of those with neuromotor dysfunctions, the outcomes were highly significant between the groups: those with the higher Al levels had a lower mental development index.

Given all of this, one has to question both the honesty of Offit and the Center, or at least their dedication to any reasonable standards of scholarship. And, of course, an error of this dimension should lead to follow-up questions. One such question might be how many of their other statements on vaccines are accurate and if these are actually from the scientific literature at all, or are merely propaganda talking points cobbled together to sway parents into accepting vaccines without much respect for scientific accuracy.[36]

One final note about the Vaccine Education Center's website: as of February 18, 2020, the reference to Ganrot was still on this website[37] and was still used to support the notion that aluminum is not harmful in vaccines.

The RotaTeq® vaccine is a live oral vaccine, manufactured by Merck and licensed by the FDA in early 2006. It was designed to target the rotavirus, which can cause severe diarrhea in infants.

RotaTeq® and its competitor, Rotarix®, manufactured by GSK and licenced in 2008, replaced a previous rotovirus vaccine, RotaShield,® made by Wyeth-Lederle. RotaShield® was licensed in 1998 but withdrawn in 1999 due to concerns about a complication called intussusception, a condition in which one segment of intestine "telescopes" inside of another, causing an intestinal blockage. This condition can have serious consequences. RotaTeq® can also cause intussusception, but some reports suggest this occurs at a lower rate.[39]

The assignees on the patent for RotaTeq® were the financial arm of CHOP, the Children's Hospital Foundation, and the Wistar Institute.[40]

Offit's RotaTeq® made him a very wealthy man with one estimate of his total profits to date from direct payments and royalties of somewhere between thirteen to thirty-five million dollars. *Age of Autism* writers Dan Olmsted and Mark Blaxill came up with this range after a detailed forensic financial evaluation, with similar amounts noted by other researchers.[41]

In regard to disclosing his conflicts of interest due to his sale of RotaTeq®, Offit often appears remarkably lax, often failing to do so in a way that would have serious consequences for lesser persons or those at other medical centers, as noted previously. Not only does Offit rarely note his profits from RotaTeq® or his extensive connections to the pharmaceutical industry in his papers, but he also did not apparently recuse himself as a member of ACIP from the organization's discussions about recommending the vaccine.[42]

So, what can one say in summary about the highly lauded Dr. Offit? Would it be fair to say the following: many awards, lots of good press, but scientific honesty apparently badly lacking and ties to the pharmaceutical industry directly and indirectly quite extensive? Maybe the first cancels out the latter? Or, maybe those who give the awards and do the book reviews don't actually read the scientific literature or care what it actually says? Nor, apparently, do they delve too deeply into all of the alliances with industry that Offit has. And maybe organizations like CHOP don't care that much about scientific and academic integrity when a lot of royalty money from RotaTeq® rolls in each year.

Is making a lot of money for an invention a bad thing? Not necessarily, at least not in the United States, where the glorification of private enterprise and the wealth it can bring reigns supreme. There is, however, a minor note of irony here in that previous vaccine inventors such as Edward Jenner (smallpox) and Jonas Salk and his competitor Albert Sabin (polio) chose not to profit from their inventions. One might suspect that Offit admires these men for their achievements, just not their lack of greed.

Finally, it is worthwhile looking at some of the press interviews with Offit, particularly a 2020 interview by Priyanka Boghani from *FRONTLINE*.[43] In these interviews, Offit can come across as quite genial, even charming. He seems to exude calm and scientific wisdom in the face of the raging danger he describes in his book as coming from "anti-vaxxers."

The portrait painted in this interview was that of a man driven by his desire to help children avoid horrible diseases, carrying out his lonely, but necessary, vigil on behalf of "us all." He can be stern when he needs to be, but charitable, as well. How knavish must one be to not respect the man?

Then again, since the journalist in the above case was lobbing the softest pitches in the world and since she never once thought to question Offit's expertise on things like aluminum, nor even look very hard at his massive conflicts of interest, what's not to be charming about?

In regard to the last, one point Boghani does cover near the end of the interview is about Offit's remuneration from RotaTeq®.

Boghani lobs out the softball, noting that "You also benefit from that vaccine." Offit replies,

> I know this isn't going to sell, but it doesn't matter. It doesn't matter whether I financially have benefited or not. The only thing that mattered is, did the vaccine we created at Children's Hospital in Philadelphia do what it was claimed to do? And the answer to that question is yes.

And:

> What isn't OK is that the profit motive gets in the way of explaining what vaccines are and how they work and how they're made; that the profit motive obscures real information about vaccine . . . I think this whole discussion is about conflict of interest, profit motive, who's saying what, is irrelevant. The question is, what do the data show, and what has been the impact of vaccines, and have vaccines been as safe and effective as they've been claimed to be? And frankly, they consistently have been . . .

These statements are very much like how politicians, particularly those who are dissembling, answer questions by not answering them, instead shifting the focus to something else, like some benefit, real or imagined.[44] And, of note, in another recent interview, when asked, Offit doesn't mention any financial conflicts at all.[45] One presumes, as in the *FRONTLINE* piece, that such things as conflicts of interest are "irrelevant."

Sometimes, just as in his articles, Offit seems to be speaking through his hat. In the same interview with Boghani, in replying to a question about modified vaccine schedules, such as those provided by Dr. Bob Sears, Offit stated that

> I think that people should be reassured by something called *concomitant use studies*, which is to say, every time a vaccine comes into the schedule-a new vaccines- the FDA makes the company prove that that vaccine doesn't interfere with the safety profile or the immunogenicity profile of the existing vaccine and vice versa. So the schedule is very well tested. [Italics are mine.]

That would be lovely if that were true, but searching with the term "concomitant use studies of vaccines" and FDA brings up a website,[46] which, in turn, directs you to another website with a PowerPoint presentation,[47] which actually says nothing about what Offit thought he was talking about.

While it completely makes sense to do such studies, postlicensure or ideally before, it is not at all clear that FDA actually monitors that this is actually done by the companies that make vaccines (see Chapter 2). Maybe what Offit meant was that concomitant use studies are actually aspirational goals, rather than a real deliverable in the process of approving vaccines. Or, maybe some eagle-eyed reader will see where I have missed the information. If so, let me know. However, based on Offit's previous apparent lapses in providing correct information, I'd hardly be surprised if nothing comes up.

In one telling quote, Boghani has Offit noting that "Well this webpage: there're certain things about vaccines that are *absolute truths*. One is that they work." [Italics are mine.]

Throughout the article, Offit also uses the term "frankly," or "to be frank," rather a lot. Is this just a verbal tic? Maybe, or maybe it is a sign that maybe we are not getting the whole truth in some statements . . . and that Offit knows this.[48]

Although Boghani made the interview come out as laudatory as humanly possible, a more thorough inspection of Offit's words provides some real insight into Offit's character and core beliefs. We have Offit's view that his profits from vaccines—which seems to extend to the pharmaceutical manufacturers, as well—are "irrelevant." We have dissembling answers to the easiest questions, and we have his use of some telltale phrases. And best of all, we have his version of "absolute truth." I will come back to the notion of absolute truth in Chapter 8 when I consider the ideology of vaccination and its resemblance to organized religion.

As noted above, Offit holds a Maurice Ralph Hilleman professorship and clearly crossed paths with Hilleman when the latter was an adjunct professor of Pediatrics at the University of Philadelphia.

It seems abundantly clear that the path Hilleman followed in his career influenced Paul Offit, so it is worthwhile briefly reviewing who Hilleman (1919–2005) was and what he did.

In brief, Hilleman was a microbiologist/vaccinologist, credited with a range of vaccines that he helped to develop. These included measles, mumps, hepatitis A, hepatitis B, meningitis, pneumonia, Haemophilus influenzae bacteria, and rubella, some of them still in use in current vaccine schedules. For the most part, Hilleman worked within the pharmaceutical industry: first at the precursor to Bristol-Meyers Squibb and then at Merck, where most of the vaccines he was involved with were created.

Hilleman worked in the Merck Research Labs until 1984 and then moved to the Merck Institute for Vaccinology for the next twenty years. Curiously, he was apparently one of the first scientists to express concerns about viral contamination in vaccines, notably concerning the monkey virus, SV40, which contaminated the Salk polio vaccine, as well documented elsewhere.[49]

Hilleman served on the National Institutes of Health's Office of AIDS Research Program Evaluation and the Advisory Committee on Immunization Practices of the National Immunization Program and also served as an advisor for WHO. All in all, a notable career, but one that has come under increasing attention in recent years, particularly in regard to the origins of AIDS.[50]

Another vaccinologist who had a major influence on Offit, as well as on the general public above vaccine-related issues, is Stanley A. Plotkin (1932–).

Plotkin received an MD from the State University of New York in 1956 and completed more medical education at the University of Pennsylvania in 1963. He was part of the research faculty from 1960 to 1991 at Wistar Institute in Philadelphia, where he invented the rubella vaccine, the only one that we examined that seems not to have done much to diminish the incidence of an infectious disease (see Chapter 2).

For the most part, Plotkin has been a corporate man from his years at Wistar Institute and together with his ongoing consultancy for a range of vaccine makers, including Sanofi-Pasteur, GlaxoSmithKline, Merck, Pfizer, Inovio Pharmaceuticals, Variations Bio, Takeda Pharmaceutical Company, Dynavax Technologies, the Serum Institute of India, CureVac, Valneva SE, Hookipa Pharma, and NTxBio, in most cases as a principle of what appears to be his own firm, Vaxconsult, LLC.

In 1991, Plotkin left the Wistar Institute to take the lead as medical and scientific director for vaccine manufacturer Pasteur-Mérieux-Connaught. He held this role for seven years. The same company is now called Sanofi-Pasteur.

Plotkin is listed as cofounder of, and continuing advisor to, the Coalition for Epidemic Preparedness (CEPI), which will be discussed again in Chapter 12.

It is difficult to accept the notion that the links between Hilleman, Plotkin, and Offit are accidental. In fact, the extent of the interconnections of these three men to one another and to a huge piece of the vaccine industry makes it quite clear that we are dealing with the equivalent of a vaccine scientific dynasty under the firm grip of corporate interests. And the scientific line of succession is quite clear: Hilleman to Plotkin, Plotkin to Offit.

It gets better: A frequent critic of so-called anti-vaxxers, and on record calling for people opposed to vaccines to be penalized for refusal to vaccinate their children, is Prof. Arthur L. Caplan, described as a "bioethicist." In an article in the *Washington Post (Feb. 6, 2015)*, Caplan wrote about denying parents the right to spread "misinformation" by speaking out against "the truths of vaccination."[51] Caplan also opined that physicians who agreed with such misinformation should lose their licenses.

Caplan is the current Drs. William F. and Virginia Connolly Mitty professor and the founder of the Division of Medical Ethics at the New York University (NYU) School of Medicine in New York City. What did Caplan do before this? Well, before NYU, Caplan was the Sidney D. Caplan professor of Bioethics at the University of Pennsylvania's Perelman School of Medicine in Philadelphia, where he created the Center for Bioethics and the Department of Medical Ethics. To no great surprise, this is the same medical school where Offit holds an appointment in Pediatrics.

Bioethicists, like Caplan, have come under increased scrutiny. Dianne Irving[52] finds the very subject an odd collection of principles that don't really mesh and certainly provide nothing resembling real medical ethics. In this regard, and it clearly applies to Caplan in spades, Irving states that "Using blatantly incorrect science in the design, protocol, or analysis of an experiment is *per se* unethical, as well as unscientific." [pg. 37]

Of interest from the bioethics perspective that Caplan is so fond of touting is a deposition conducted by New York attorney Aaron Siri on Stanley Plotkin in 2018. In the hours of the deposition, Siri shows that Plotkin's own ethics have been highly questionable in regard to how he helped develop vaccines.[53]

Peter Hotez

A newer "thought leader," one who was often trotted out pre-COVID-19 to address the issue of vaccines and autism, is Dr. Peter J. Hotez. Hotez holds both PhD and MD degrees.

Hotez does not seem to be part of the Philadelphia clique; indeed, his work seems vastly more honorable: Hotez's key focus is on global health, including tropical disease control, and of course, in these contexts, vaccines.

His list of affiliations is significant: founding dean of the National School of Tropical Medicine and professor of Pediatrics and Molecular Virology and Microbiology at Baylor College of Medicine, the latter where he serves as director of the Texas Children's Hospital Center for Vaccine Development and as the Texas Children's Hospital endowed chair in Tropical Pediatrics. His main work appears to be developing vaccines for various diseases including leishmaniasis, Chagas disease, and for two coronavirus vaccines for SARS and MERS.

It was in regard to the last two diseases that the media sought his comments on a future COVID-19 vaccine. In these interviews, Hotez expressed concern about the rush for a such a vaccine, a view that seemed to go against media stories hyping the same to let the world go back to "normal."[54]

Prior to this, Hotez was often in the media expressing his opinion that autism is a genetic disorder and thus not related in any way to vaccines. In this regard, much of Hotez's appeal has to do with the fact that he has an autistic daughter, and his recent book about her, *Vaccines Did Not Cause Rachel's Autism: My Journey as a Vaccine Scientist, Pediatrician, and Autism Dad*,[55] initially drew considerable media attention for this reason. In the book, and in a variety of interviews, the one with *Vox* being very much like the others,[56] Hotez holds forth his reasons for denying that his now-twenty-plus-year-old daughter did not develop her autism from vaccines. First, he claims that a genome profile for Rachel may have identified a novel gene that might be involved in autism. That's fine, and I will be happy to see the published, peer-reviewed article, because if there is not one, then all we have is his opinion as a vaccine scientist, much like Taylor cited in Chapter 3. Hotez, presumably, had his child fully vaccinated.

Another thing that stands out about Hotez is his fierce dislike for those he considers to be anti-vaxxers. In this he resembles Paul Offit, and, although it is certainly his right to like or dislike those he disagrees with, his rhetoric sounds more like a frustrated priest railing against the heathens.

Harassment of Pro-Vaccine Advocates

Much tends to be made by various bloggers about harassment and even the threats of violence that those who are pro-vaccine are allegedly exposed to from some anti-vaxxers. Those who are prominent defenders of vaccine schedules and mandates, such as Paul Offit, sometimes report both harassment and outright death threats.

At least as far as the threats go, since most tend to be anonymous, it is difficult to judge just how serious such threats really are, or were. Which is not to say that they are not concerning, even frightening, to the recipients. And, in the highly polarized world of vaccine controversies, they might be very real.

Various bloggers, such as David Gorski, whom we will meet in more detail later in the book, report both physical threats as well as attempts to get them fired for some alleged malfeasance. Offit has complained about the same, and it may well be that such attacks on scientific opponents of mass vaccination and/or mandates occur with some regularity. Dr. Richard Pan, a California state senator, has also reported repeated threats and even an assault.[57]

Obviously, none of this is acceptable in any possible way. Those who engage in threats are clearly taking the war metaphor too far if they are seriously moving from words to potential actions.

Yet, some of us on the vaccine skeptical side have had both harassment and threats, as well. I've had anonymous email writers suggest that my death would be a good thing for the world and had several others post that they hoped my children died of some disease.

As for harassment at the work place, yes, I've had this, too. The same applies to Prof. Chris Exley for his work on aluminum toxicity, Prof. Romain Gherardi for his work on MMF, and Prof. Yehuda Shoenfeld for work on the ASIA syndrome.

Thus, whatever potentially violent or intolerant nature lurks in the hearts of some "anti-vaxxers," the same can be said as well about the threats directed at some vaccine researchers by the other side of the house. Further, a major difference between the two situations is this: those harassing vaccine critics are likely backed, if not sponsored, by the vaccine industry. These points will be elaborated below when we consider the manner in which any industry that feels threatened for the continued success of their products responds.

It may be that all of the harassment and threats are symptomatic of the level of discourse found in relation to so many contentious issues in modern society. This seems to be particularly true on social media, where anonymity and the use of aliases are often the norm. It is difficult not to notice just how prevalent this is. For example, most of Gorski's acolytes (sycophants might be a better term) are wont to use aliases, as indeed do many who write into various media websites on a whole host of issues.

In most cases, the threats of violence seem mere bluster. Then again, America is a country in which a significant proportion of the population has ready access to firearms of all descriptions, where many see an absolute right to these weapons guaranteed by the US Constitution's Second Amendment.

It hasn't yet come to this, but in the future the will to use these weapons may in turn be driven by fear. For example, if you fear that someone on the other side of the vaccine issue is going to hurt your children, then the seed is planted to take measures for your children's protection. Far-fetched? Perhaps, but remember that back in the heyday of the abortion debates, abortion-providing doctors were sometimes killed.[58]

Whichever side the threats and harassment come from, this is certainly neither a moral nor even particularly effective way to go about changing minds.

What all the talk, and reality, of threats may reveal is the very human impulse to suppress, by whatever means necessary, ideas and people we don't like, forcing a zero-sum game outcome on what, in reality, are different but perhaps complementary sources of knowledge. Then again, such behaviors may simply reflect the growing lack of civility in social discourse that the Internet seems to allow.

Trust Us, We're Experts

As we've seen, Offit and the other so-called thought leaders have attempted, in many cases successfully, to push forward a one-sided narrative on vaccines. Especially in regard to safety, this narrative takes a zero-sum game approach: vaccines are safe and effective, the science is settled, and any other view is simply the spawn of blatant ignorance and an antiscience agenda to boot.

So who then are thought leaders, and where do they come from? One definition is that thought leaders are those who can shape public opinion.[59] Sometimes they take on such a voluntary role due to a burning desire to educate their fellow citizens. Sometimes, however, they are the products of a different agenda that is not seeking to educate, but to defeat any alternative narrative that may be anathema to special interests.

Let's look further into this last possibility.

In 2002, journalists Sheldon Rampton and Jon Stauber published a remarkable book called *Trust Us, We're Experts*.[60] In the book, written long before the current vaccine wars began, Rampton and Stauber examined a number of controversies about various products, from lead in paint, to biosolids and pesticides, to climate change, and more.

In each case examined, what they found was that as lay people and scientists raised questions about the safety of some products, the manufacturers of these products fought back. Various industries developed methods to diminish the potential impact about the public perception of their products, and thus the corporate bottom line, by a comprehensive collection of actions. These actions were identical regardless of the industry under threat. For example, lay critics were accused of being ignorant, or even generally antiscience. Credentialed scientists who joined the critics were described as "quacks" peddling "pseudoscience."

The so-called thought leaders in these cases, much like Offit today with vaccines, provided critiques of both types of critics, wrote books to hammer the message home, and provided ample media fodder designed to diminish any concerns in a welter of ad hominem attacks. To a great extent, each concerned industry succeeded in dampening down opposition, at least for a while.

It is of more than passing interest that in each of the cases cited in the book, the critics were shown to be correct in the end. This observation did not always save the careers of the critics at the time, but the vindication that came later must have been some solace, as it was for Prof. Walker-Smith, whose case was highlighted in Chapter 4.

Of course, the bad news part of this is that those who critique vaccines are at the same place as those who questioned pesticides were in the 1950s. For pesticides, in the ensuing decades there was an inevitable shift of public opinion that led to changes in the regulation and mass use of these chemicals.

Will the same happen with vaccine safety issues and their critics? This is hard to tell in advance. While pesticides were widely used at the time, and still are in some quarters, the industry marketing these chemicals was more a creature of the United States and lacked the reach that globalization would bring to the pharmaceutical industry. Further, the Internet now allows attacks on vaccine critics to be coordinated almost instantly in ways that would have been impossible up until the 1990s. The Internet also allows trolls,[61] or just interested individuals, operating in a number of theaters to see critical articles as soon as they come out, if not before. This, in turn, allows them the opportunity to strike proactively, sometimes within days of when a publication that threatens their industry's masters comes out.

It is ridiculously simple: just create an alert message on *Google Alerts* for a particular subject, for example, human papilloma virus (HPV) vaccines, and Google will let you know each time a paper on this is published or if something comes up on the web. It's easy, and anyone with Internet access can do such ongoing monitoring for certain subjects, or names. A few dedicated trolls, paid by the pharma, can cover it off completely, each one assigned a few key subjects/names.

I have experience this personally in the case of a small-scale study that looked at how the VAERS database categorized some reactions to Gardasil, one of the vaccines against strains of the human papilloma virus (HPV).

While reports are voluntary and thus largely underreported,[62] anyone can enter a report, each screened by medical doctors working for these agencies.

In our paper, we asked a very simple question: if we take a random selection of VAERS reports about Gardasil adverse events evaluated by the medical doctors working for the CDC/FDA, would the severity of the score given by them differ from that of medical doctors chosen independently? The answer that came back, highly statistically significant, was that it did indeed matter who did the evaluation. Routinely, the doctors we had used in the study tended to report the adverse events as more serious compared to the official doctors.

This was all well and good, but the real point here is not the paper or its data, nor even its quality, but what happened to it after it was accepted for publication.

In brief, after revisions of the manuscript, it was finally accepted by a rather obscure journal, at least from a biomedical perspective: the *Indiana Law Review*. In part, this choice was due to the fact that one of the authors, Mary Holland, is a lawyer.

The paper was duly reviewed, revised as per standard peer review practices (see Chapter 9 on the problems with the peer review system), accepted, and posted to the journal's web page.

Within days, we were notified that the paper needed to be rereviewed due to critiques of the methods. This alone was odd and not in keeping with conventional editorial practice in which reviewer critiques come before acceptance, not afterward. Whose critiques, we asked? The editor provided no answer to this query. Further weeks of correspondence with him never elicited any details of why the paper needed a rereview.

Since we could not get any clarity, we refused the editor's decision to have the paper rereviewed and withdrew the paper instead.

The question we were left with at the time was this: how did a minor article, published in a relatively unknown law journal, get flagged by the editor who is a lawyer, not a scientist, as needing rereview?

Was this a case where the editor, secretly a statistician or epidemiologist interested in HPV vaccines, suddenly had a change of heart? Possible, but rather unlikely. Or was it a case, as will be documented in the section of the book on peer review, that letters to the editor had come in with complaints leading the editor to get cold feet about this article? This possibility seems more likely. But, again, it begged the question of how the paper was so quickly found by the complainant(s). Were we dealing with people who have nothing else to do besides troll the Internet for any new articles on vaccine adverse events? How thorough would one have to be to find the *Indiana Law Review* of all possible publications?

The answer is likely that one does not have to be thorough at all; a search engine, such as *Google Alerts,* does that instead. All that one has to do is set up an ongoing search of all new articles anywhere on the web that have any relation to vaccine adverse events, or even to authors known to be critics of any aspect of vaccines. Once one such new article is found, hostile letters to the editor who allowed publication will surely follow.

Who has the time, energy, and resources to do this? Of course, it could be anyone on a crusade seeking to block vaccine-critical articles from remaining in the literature. Or, it could be that a company like Merck, which makes the Gardasil vaccine, has trolls who do this for pay. Occam's Razor applies here: which is the simpler explanation that matches the facts? I think it is the latter.

As it happens, we know precisely what happened in this case thanks to a Freedom of Information request filed to the University of Indiana by coauthor Mary Holland.

What came back from the university was very revealing. Appendix 10 contains a copy of the emails received between Nicolas Terry, the editor of the journal, and Dorit Reiss, a law professor in California.[63] From these emails it seems that Reiss approached Terry with concerns about the article, and this led to the plan to rereview it. The back-and-forth between the two is fascinating to watch as they get onto a first-name basis while attempting to derail the normal peer review process.

Such backroom maneuvering seems to be the new "normal," as we will see in Chapter 9, with similar types of exchanges between the blogger David Gorski and Dr. Catherine Roe, an academic who had been invited to attend that rarest of events, namely, a two-sided debate on vaccine safety. As with Roe, I wrote to Terry giving him the opportunity to explain the emails with Reiss. As with Roe, Terry never responded.[64]

It is in this way that the attacks on vaccine critics markedly differ from what happened to the critics of other industries in the past, namely, the speed of the attacks that seek to remove dangerous (to them) articles before almost anyone has seen them. It seems rather likely that some eighteen years after Rampton and Stauber's book came out, those who seek to protect their industries have harnessed the Internet on their behalf to turn it into a vehicle for censorship. Of course, having editors who don't understand the peer review system and who also lack a backbone helps, too.

Eating Their Own: What Happens to Pro-Vaccine Scientists and Physicians Who Step out of Line?

As it happens, so-called thought leaders like Offit and the others don't just go after those they consider to be on the anti-vaxxer fringe, but even take cudgels to their own.

In 2018, science journalist Melinda Moyers, writing for the *New York Times*, wrote an article titled "Anti-vaccine activists have taken vaccines science hostage."[65] The title, however, was misleading, as titles often are in magazines and newspapers where someone who is not the article writer provides it. The print version, published the next day, had the more textually accurate title, "The Censorship of Vaccine Science."

The original title had clearly suggested that it was "anti-vaxxers" who were harming vaccine science. In reality, the point Moyers was trying to make, as per the second title, was that in the highly polarized and politicized realm of vaccine research, any deviation from what could only be considered the official narrative was "punished" by those still 100 percent in the fold. How the punishment was accomplished was in many ways reminiscent of how some religions treat any apostasy within their own ranks. This topic will be explored in much more detail in Chapter 8.

It's an interesting article that raises a number of important topics that are relevant to not only vaccine science, but any science in which scientists fear reprisals for honestly reporting their findings.

Here is what Moyers wrote as she introduced the article:

> As a science journalist, I've written several articles to quell vaccine angst and encourage immunization. But lately, I've noticed that the cloud of fear surrounding vaccines is having another nefarious effect: It is eroding the integrity of vaccine science.

And,

> When I tried to report on unexpected or controversial aspects of vaccine efficacy or safety, scientists often didn't want to talk with me. When I did get them on the phone, a worrying theme emerged: Scientists are so terrified of the public's vaccine hesitancy that they are censoring themselves, playing down undesirable findings and perhaps even avoiding undertaking studies that could show unwanted effects. *Those who break these unwritten rules are criticized.* [Italics are mine.]

Moyers gives various examples, but one that struck me the most forcefully, and poignantly, was the story of Dr. Danuta Skowronski, listed in the article as the lead epidemiologist in the division of Influenza and Emerging Respiratory Pathogens at the British Columbia Centre for Disease Control. Skowronski's building is literally around the corner from my own laboratory.

What was Skowronski's sin? She and her many colleagues from across Canada had published a 2010 analysis of data from four epidemiological studies dealing with observations comparing those who had previously received an influenza vaccine (trivalent inactivated vaccine, or TIV) versus those who had not, and both in relation to the 2009 H1N1 strain of influenza. In these cases, they were primarily dealing with a TIV made by GlaxoSmithKline (GSK), Fluviral. The odds ratio[66] for these outcomes was a 1.4 to 2.4 times greater serious H1N1 outcome than those naive to the TIV. At the same time, Skowronski et al. noted that the TIV itself had provided a 56 percent reduction in the risk of serious illness.[67]

These results seemed to the authors to be paradoxical. This study was therefore an attempt to see if the various outcomes were valid and could be explained by some known mechanism. One such mechanism is known as "original antigenic sin," that is, how the first exposure to a virus, like in this case one of the influenza strains, influences the later immune response to related strains, in this case H1N1.[68] The concept has been known since the 1960s and may be highly relevant to the future of COVID-19 vaccines.

Skowronski and colleagues also considered the possibility of "antibody dependent enhancement" (ADE), a much more significant negative outcome to a new vaccine after a previous one. This concept is emerging as an important one to consider also in relation to COVID-19. This concern will be revisited in Chapter 13.

Skowronski followed up the epidemiological work with an in vivo study in 2014 using ferrets, a standard animal model to study influenza, and found the same sort of outcome.[69] In this study of thirty-two male ferrets, half were given the Fluviral vaccine, and half, the controls, had a saline injection. The injections were given twice, at days zero and twenty-eight, and the animals monitored for an additional forty-nine days. When a challenge experiment was done using the H1NI virus, the Fluviral-injected animals showed significantly lowered appetite levels, weight loss, higher viral lung loads, and some increased cytokine expression compared to the untreated ferrets. As in the epidemiological study, Skowronski and colleagues were *beyond* careful in discussing alternative interpretations. Even though the human study and the ferret outcomes seemed to match, the authors went out of their way to note that the ferret results *did not* mean that human responses would be the same.

Skowronski and her colleagues, we should remember, were trying to honestly report anything that might serve to make vaccines better, safer, and more widely acceptable to the general public. In their article, these authors skated very carefully around the possible implications of these findings and went on to stress that these were preliminary. From a vaccine promotion perspective, what could possibly be better?

Well, lots, apparently.

Here is what Moyers quotes Skowronski as saying: "'There was tremendous pushback' … and some questioned whether 'the findings were appropriate for publication.'" And, "'I believed I had no right to not publish those findings. They were too important.'"[70]

Skowonski's situation simply mirrored that of other investigators interviewed by Moyers who had experienced similar dramas when they reported things that did not fit neatly into the official narrative.[71]

Moyers interviewed Paul Offit for the article as well, noting that

> Dr. Offit says that researchers should handle findings differently when there's a chance they might frighten the public. He thinks that small, inconclusive, worrying studies *should not be published because they could do more harm than good.* [Italics are mine.]

Moyers then quoted Offit as saying: "'Knowing that you're going to scare people, I think you have to have far more data.'"

Where have we heard this sort of thing before? From fundamentalist religion, of course, which is fearful that any questioning established dogma might simply confuse and frighten the faithful. Could there be a clearer declaration of vaccine orthodoxy? Hardly, but then such is even more telling coming from a person who seems completely capable of deliberately and grotesquely misstating the work of others in order to keep the faithful in line.

Moyers asked the question:

> If a study scares parents away from vaccines, people could die. That's a big risk to take to protect the sanctity of scientific discourse. I was warned several times that covering this issue could leave me with 'blood on my hands,' too. But in the long run, isn't stifling scientific inquiry even more dangerous?
>
>   It's a good question.
>
>   Moyers finished with this:
>
> One thing vaccine scientists and vaccine-wary parents have in common is a desire for the safest and most effective vaccines possible - but vaccines can't be refined if researchers ignore inconvenient data. Moreover, vaccine scientists will earn a lot more public trust, and overcome a lot more unfounded fear, if they choose transparency over censorship.

It's hard to argue with this, as it is a perfectly rational approach to science's role in gaining knowledge for the benefit of humanity. What it doesn't do, however, is consider that rationality and what is actually for the good of us all, which may not neatly mesh with what has become essentially a statement of religious faith in the church of vaccination, a church whose foundations are built on money.

Before I get to that, however, I want to cite some of the thoughts of Dr. Mateja Cernic that seem particularly apropos at this juncture:

> *. . . discourses are simultaneously inclusionary and exclusionary because they define what may be said or not, or thought and done. [An] example is allopathic medicine in general by denying those not part of the community to think or talk about anything related to allopathic medicine in terms of health and illness; does not answer critique with arguments, rather engages in 'personal and professional discreditation.'*[72] [Brackets for addition are mine.]

The above applies to outsiders. Cernic easily anticipated what happened to insiders such as Skowronski and others when she wrote:

> *It makes no difference if the critics come from within their ranks - as soon as they begin to express doubts about the sacred truths of allopathic medicine, the process of discretization kicks in: their scientific titles and education suddenly lose importance or they are not 'good enough,' and they are characterized as irrational, unscientific, fraudulent - in short, as people who have really lost it and whose words no longer bear any weight.*[73]

Cui Bono?

When I began this book, I made a pledge to myself to look at all issues and all of the data, no matter whose feathers got ruffled. I pretty much anticipated annoying, once again, the various trolls/bloggers and the official arms of the various international and national agencies such as WHO and the CDC.

In order to keep my pledge, I looked extensively at the CDC's list of articles that they provide on their website to demonstrate that vaccines are safe, as discussed in Chapter 3. It was a long process. Reading many of them was an even longer endeavor. I also read the original Jenner papers (Chapter 2) and came away impressed by his achievements, his willingness to think outside the box, and his general concern for humanity.

At the same time, I felt it important to "go down the rabbit hole" so to speak and look at aspects of the vaccine story that will seem to many people on the official narrative side to be wild "conspiracy theories." Indeed, many of them are. But are all of them simply the result of people with too much time on their hands being creative with their use of historical accounts, current news, and various articles on the web? In many cases, yes. But all? Almost certainly no.

A conspiracy in law is simply defined as two or more people acting together to do (usually) an illegal act. Conspiracies do exist, although they are often hard to prove. And it is important to remember that the very term "conspiracy theory" was created by the CIA to deflect any suspicion of their possible (probable) role in the assassination of John F. Kennedy.[74]

But as *Trust Us, We're Experts* makes clear from studies of other industries, there is almost certainly a conspiracy to control the narrative on vaccine safety.

This conspiracy has multiple moving parts that include people like Offit and other so-called thought leaders, along with agencies such as the CDC, FDA, and NIH, which apparently don't know, or care to know, the actual data on which their case for vaccine safety depends. It includes the various bloggers and trolls, some volunteers, some paid, who attack scientists who question the vaccine orthodoxy. Then there is the system of medical education entrusted with educating the next generation of physicians, doctors who will likely go forth with the one-sided information received from their years in medical school and pontificate to an increasingly restive collection of parents. And let's not forget the importance of the media, which plays a major role in the process of indoctrination.

None of this is to say that all of the above are conscious "card-carrying" members of the vaccine orthodoxy, nor do they necessarily know one another or what each is doing. What I imagine is happening is that all of these entities are drawn together by a common faith in vaccines, a faith that self-identifies with one another as members of the same brotherhood/sisterhood of the faithful. Would this be unlikely? Not at all. Members of formal religious denominations seek one another out and offer support, as do "social" organizations such as the Masons. Nor is it necessarily bad that the vaccine faithful cleave together, but it certainly explains a lot of the behaviors when this grouping is confronted by those who doubt or actively reject the faith.

The system is thus clearly designed to enforce the catechism about vaccines holding an elite and unique status in medicine, one that allows no official doubts about safety. For those who stray, at each level the system has mechanisms in place to punish dissenters, no matter how mild that dissent may be.

How does such an enterprise arise? Is it simply the case that the official truth is so obvious for all of these participants that they see no need to revisit the articles of their faith? Or, just perhaps there is a guiding principle here, one that serves an overwatch function to keep the machine running smoothly?

I expect that many who read this book already know who shifts the gears in this vehicle. I further expect that those who don't know won't be particularly surprised when they find out.

Cui bono? At the top of the heap sits the pharmaceutical industry, the topic of Chapter 12.

CHAPTER 7

The Resistance to Vaccination Policies: Vaccine Hesitancy to Outright Refusal

People get used to anything. The less you think about your oppression, the more your tolerance for it grows. After a while, people just think oppression is the normal state of things. But to become free, you have to be acutely aware of being a slave.

—Assata Shakur[1]

As already discussed in Chapter 1, the basic controversy for some people about vaccines lies particularly in the most triggering of issues, namely, the push for school and other vaccine mandates.

The Spectrum of Vaccine Resistance

Just as ASD describes a spectrum of neurological dysfunctions, the suitcase word "anti-vaxxer" brings in a spectrum of beliefs about vaccines. This can include those who really are against any form of vaccination. Inside this general grouping are those who simply don't think that vaccines work at all, to those who think that regardless of whether they work or not, they can only cause harm to people or animals. And, of course, some of those in these subgroups believe both. Some examples of this end of the spectrum include the views of Dr. Tetyana Obukahanych, a true apostate to the vaccine orthodoxy who holds a PhD in immunology from Rockefeller University.

A corollary position in this, held by some, is that viruses actually don't exist as independent entities, but are rather by-products of one's own toxic state of health. This view holds that the things that look like viruses are actually something called exosomes.[2] I will leave it to others to debate the merits, or lack thereof, of this view. Here, I merely mention it to show the so-called "hard shoulder" of vaccine resistance. Part of this hard shoulder, or at least an adjacent part, has a political wing that I will discuss below.

Views such as these allow mainstream medicine, various bloggers, and especially the mainstream media to dismiss all vaccine concerns of any sort as the handiwork of a wing-nut, tin foil (or these days, aluminum foil) hat-wearing collective that also believes in lizard people and aliens underground at the South Pole.[3] If all of those with vaccine concerns are of the same ilk, why listen to them at all? It would be like me looking at Paul Offit, whom we met in the last chapter, and concluding that all who favor vaccination are just like him, and definitely not in a good way.

The danger with religious extremism is that it can be very serious. Sadly, for those in medicine and some branches of science who don't seem 100 percent orthodox enough to the prevailing dogma, being labeled a heretic has consequences. On the vaccine front it may not be dangerous to life and limb but can still come with significant consequences for friendships, professional standings, and even careers, as cited in Chapter 6.

The groups that are truly against vaccination might be considered the fringe element in what has become a far larger and more nuanced vaccine opposition. For example, a newer descriptor among some of those considered by the mainstream to be anti-vax are those who self-describe as "vaccine-hesitant." Once again, there is quite a range here from those who reject most vaccines for themselves or their children to those who draw a line at some vaccines or some of the recommended schedules for vaccines.

Those in this fairly broad vaccine-hesitant category can still be lumped into the "anti-vaxxer" category by the various pro-vaccine shills and some of the media. Indeed, they often are, and one has only to peruse almost any posting on the *Respectful Insolence* website (see Chapter 9) to see how this sentiment plays out.

As much as some might prefer to see the world only in black and white, the reality is that there is a lot of gray, and vaccine hesitancy, even resistance, is a pallet of gray, not to mention a virtual rainbow of colors, as well. This includes those who favor some vaccines, just not all, and not all the time no matter what the ACIP and the CDC say. There are some who have had a family member, usually a child, with a vaccine injury that has pushed them to question the entire vaccine paradigm to control infectious diseases. This initial push can become a hard shove when the same parents face the sometimes arrogant and dismissive behaviors of some physicians. Finally, there are quite a lot of the so-called vaccine-hesitant who support vaccines personally but find mandates abhorrent from a human rights perspective.

In the following, I will look at the demographics of the vaccine resistance in more detail.

Vaccine Resistance Demographics

One characterization of those who are hesitant to accept mainstream vaccination recommendations, as cited previously, is that such persons are called "anti-vaxxers," or simply "anti-vax," names that are increasingly used as descriptors. In the earliest incarnations, the names simply meant that someone was against vaccines in general. In current parlance, at least in much of the media and the medical community, an *anti-vaxxer* is simply anyone who does not follow the full recommended vaccine schedule as laid out by the CDC or other health agencies at national or state levels. For example, if you and your children are "up to date" on most vaccinations but have declined the human papilloma vaccine (HPV) or the yearly influenza vaccine, these rejections are more than enough in some circles to get the anti-vaxxer epithet applied to you.

It's as if your personal decision about how you choose to use vaccine products is not about choice as a consumer or one who simply chooses to make a free choice about one's own health. Rather, such choices are now open to critique in a way that other life choices are not. The equivalent would be if you decline to use a car generally out of concerns for the environment, or simply don't like one particular car company and refuse to buy their products, or don't live in a city where car ownership is feasible. As others have noted, only the first *might* make you "anti-car," but to assume that everyone who does not use a car is a tin foil hat-wearing anti-car activist is simply illogical, even ludicrous.

What the use of the *anti-vaxxer* term really shows is that it has become a "suitcase" word, which, as described by Marvin Minsky,[4] is a word that contains within itself multiple meanings. Thus, *anti-vaxxer* is meant to describe someone who not only chooses to reject some aspect of vaccine promotion, but also that such a person is scientifically illiterate, believes in a range of wild conspiracy theories, and harbors a serious distain for the health of others. In other words, it is meant to describe a person who is ignorant and selfish, maybe even criminal.

Further, *anti-vaxxer* is tossed out like a pejorative epithet, like those slurs so often used against Jews,[5] Blacks, Arabs, Mexicans, Chinese people, and so on. Of course, such epithets often are more than just insults, as they carry with them the weight of an often brutal history: anti-Semitism and the Holocaust for Jews, the centuries of slavery and continued oppression for Black people in the United States, the genocide of the Native peoples of the Americas, etc.

Clearly, this is not true for the use of *anti-vaxxer* as a term of denigration. However, the latter is surely used in much the same way that "dog whistle" politics and politicians demonstrate that a particular leader or political entity is aiming for a visceral resonance with some social base by using words or terms whose underlying meaning only the base hears. *Anti-vaxxer* is one such word, used precisely the same way to evoke the same sort of distaste for those so categorized, people who can now be dismissed at will. It is also used to stifle debate, since why even bother to talk science to someone who seems by definition to reject science in general?

In this view, the deductive flow would seem to be that vaccines are part of the science of vaccinology (the theory and practice of vaccines) and are a pure example of science generally. Hence, to reject one is to reject the other in its entirety.

For this reason, the term *anti-vaxxer* is used to demonize people and put them outside the boundaries of scientific, even civilized, discourse by intention, as shown in the previous chapter. This form of "shunning" will be discussed further in Chapter 8.

To some great measure the intention of using *anti-vaxxer* as a slur has succeeded, at least judging by the media's increasing use of the term. This intention has even served to categorize as anti-vaccine those who actually believe in, and use, vaccines as somehow untrustworthy if they don't fully comply. I will show some examples of this later in the chapter.

In many ways, the use of the term *anti-vaxxer* to write off any person critiquing vaccines is reminiscent of calling those fighting for civil rights or against the Vietnam War (or more recent causes or wars) "Commies" as if that label simply ended the matter. Back in the day, those tossing around the "Commie" epithet really had no inkling what those they were denigrating actually believed, merely that it was something the speaker didn't like.

A "Commie," of course, is short for communist, and while there were certainly people who identified as communists involved in civil rights and anti-war activities, the bulk of the protesters in both cases were simply opposed to either racism or the war(s). To many of us involved in any of these protests, the name-calling simply reinforced the view that those doing the name-calling really didn't know what a communist actually was (or believed), compared, say, to an anarchist. Or, more likely they didn't care, since the point was not to discuss economic/social theory, but rather to discredit the target without considering the arguments being made.[6] Ironically, name-calling is not restricted to the right wing side of the political spectrum, but now routinely shows up on the left, as well. It is now common to find that anyone opposed to COVID-19 lockdowns is described as "alt-right" by leftists. Typically, at least in my recent experience, the irony of this situation tends to be lost on those doing the name-calling.

There are many more examples of how the anti-vaxxer label has been applied as a blanket denunciation, often incorrectly applied to those who actually fully believe in vaccination. The sin of such doctors, scientists, and lay people is not that they don't believe, but that don't believe 100 percent in the holy scripture of vaccination. This 1 percent unwillingness to simply take on faith the clear gaps in vaccine theory and practice is what does them in if their fellows get wind of such apostasy. The imperative dictated by the official "vaccine church" to apply vaccines to any and all persons under virtually all circumstances cannot show any weakness, because God alone knows what might happen next.

Perfect examples of this categorization have occurred in the recent past to Drs. Bob Sears and Prof. Yehuda Shoenfeld, now routinely cast as "anti-vax" in the media and elsewhere. Sears is a working Southern California pediatrician who comes from a family of professionals who all work in pediatric medicine.[7] Sears gives vaccines to kids every working day. Where he suddenly became an anti-vaxxer was when he simply proposed in his *Vaccine Book*[8] that the CDC's rigid vaccine schedule did not need to be applied in precisely the same way for each child. Rather, Sears suggested that vaccine delivery could be accommodated to the wishes and concerns of the parents. Dr. Paul Thomas, discussed previously, has run smack into the same problem and even had his license to practice medicine suspended following a publication with Dr. James Lyons-Weiler.[9]

Shoenfeld runs an autoimmunity clinic, the Zabludowicz Center for Autoimmune Diseases, and holds an appointment at the Sheba Medical Center/ Tel Hashomer Hospital, both affiliated with Tel-Aviv University. He routinely begins each talk or seminar that I've heard him give—probably at least a dozen—with a statement to the effect that "Vaccines are the greatest medical invention of all time and have saved millions of lives." Are these the words of a true vaccine opponent? Hardly; nonetheless, Shoenfeld's work on suspected adjuvant-induced autoimmune reactions, the ASIA syndrome described in Chapter 5, was enough to negate his routine defense of vaccination. He has also come under attack by other doctors in Israel who have sought to strip him of some well-deserved awards.[10]

What is being discussed with the anti-vaxxer label is thus a symptom of something else, something far bigger than what it seems. That something else, to be described in Chapter 12, is the unholy alliance of the pharmaceutical industry and the medical establishment.

Demonizing those who are not "vaccine pure enough" tends to have a very predicable consequence, namely, pushback. In this sense, the daily drumbeat of calling out all of those who resist some aspect of vaccine orthodoxy can often have the opposite consequences from those that the CDC and other pro-vaccine entities hoped to achieve. Calls to penalize parents who don't fully vaccinate their children are now common,[11] going hand in hand with vaccine mandates for school-aged children in the United States and around the world. Further, various social media entities have succumbed to calls to ban "anti-vax misinformation,"[12] and some commentators have even speculated on the need for civil and criminal penalties for the vaccine-hesitant. All of this has only intensified in the COVID-19 fear campaign of 2020/21 where social media sites now routinely block "COVID-19 misinformation."[13]

Pre-COVID-19, if the intention had been to frighten the relatively mild portion of the vaccine-hesitant population back into the fold, it failed spectacularly. Indeed, such an outcome was completely predictable from any historical, social, or political perspective. People in notionally free societies simply don't like being coerced and like it even less if it involves their social and religious rights as parents vis-à-vis their children. In turn, this led to the rise of what could legitimately have been described as a vaccine "resistance movement," one that had taken on the trappings and language of many previous civil or human rights groupings.

In brief, what had been some isolated outposts of vaccine hesitancy were driven by pressure from the medical establishment and pharmaceutical industry to create a very vocal and public resistance. And in the creation of this resistance, members found a form of unity that for a time seemed to cross conventional political lines. Even now with COVID-19 hysteria, as various branches of government strive to impose their will on the resisters, it is this heavy-handedness that tends to make even those who fully support vaccination, and might well take a COVID-19 vaccine, wonder what the hell is going on.

As with the suppression of other forms of dissent, the more those in power coerce and use bullying tactics, the more those being coerced push back, often for the first time asking themselves why they are being forced to do something that seems to be agenda-driven. That alone, even without knowing much about the science, perhaps fuels their suspicions.

With the questions raised, many then go on to ask the awkward questions: Why are we being forced to follow the CDC vaccine schedule? Why is a COVID-19 vaccine the *only* way to get back to normal? Is it really for the greater good of us all, or are there other, baser motives? Or, maybe is it for someone else's greater good? What role does the pharmaceutical industry have in all of this? And who in hell made Bill Gates the health czar of planet Earth?

A perfect example of how a vaccine mandate can fail due to overreach by those in power comes from the province of New Brunswick in Canada. In this case, legislation to mandate the full range of vaccines for school children initially failed in 2019. The minister pushing this mandate, Minister of Education Dominic Cardy, thought he had the needed votes to put the legislation in the bag. Where there had been some isolated opposition, Cardy's high-handed statements increasingly riled New Brunswick legislators and led previously supportive members to vote against his bill.

Then with the COVID-19 pandemic, Cardy thought he would try again, anticipating a slam-dunk victory due to COVID-19 panic.[14] In order to ensure that such legislation could not be overturned by federal courts, Cardy tried to tack on the "Notwithstanding" clause of the Canadian Constitution. This clause essentially says that any province can contest any national legislation they don't like and create the equivalent to a Monopoly "get out of jail free" card from being overturned by higher courts. It was a fatal mistake. The other parties in the legislature were already jumpy enough about the legislation from its earlier iteration. Trying to make the law immune from any higher court oversight went too far, and the bill lost, twenty-two against to twenty for.

Perhaps the clearest example of how the mainstream medical establishment has failed in its ostensible mission to have full vaccine compliance is how unprepared the various entities have been to respond to what might normally seem reasonable questions. Instead, this last year has witnessed half-baked answers that simply repeat previous talking points and make declarative statements such as "the science is settled" and "vaccines are safe and effective" and a host of others without any real scientific citations to back up such assertions.

The notion that science is ever settled, let alone in the case of vaccines, tends to strike many working scientists, as well as educated lay people, as an absurd thing to say. Such statements then tend to diminish whatever credibility the tellers of the official story may have had before. Similarly, endlessly repeating the mantra about "safe and effective" when the only citation many can give is to the retracted the Wakefield et al. study is simply lame.

The numbers of the vaccine-hesitant, however defined, may not (yet) be that great, but they are surely growing as the pro-vaccine side fails to make their case and then defaults into repressive legislation to do it for them.

All the unanswered questions and deflections ultimately trigger more resistance and lock it into a feedback loop such that coercion breeds resistance, resistance breeds more repression, which breeds still more resistance, etc. The COVID pandemic has accelerated the process.

Whether it was the CDC with their vested interests or simply the pharmaceutical industry with its need to follow the logic of capitalism and endlessly expand its market, the plan to force greater vaccine compliance has simply failed and created the very thing that these same bodies least wanted. The push for mandates and the use of lockdowns will likely turn out to have been fatal mistakes.

More Demographics Arising from Vaccine Controversies

Trying to decide if a particular social, political, religious, or racial class supports, opposes, is hesitant, or is indifferent about vaccines in the United States or Canada is largely a fool's errand. The CBC Marketplace report about "anti-vaxxers" suggested that almost half of Canadians "have some concern about vaccine safety."[15] CBC directed the readers of this article to the British nonprofit entity the Wellcome Trust, whose 2018 "Global Monitor" report[16] addressed attitudes around the world about vaccine safety and effectiveness.

In the Wellcome Trust report, the numbers cited are curious. In North America, for example, only 48 percent of respondents strongly agree that vaccines are safe, with another 24 percent somewhat agreeing. In Western Europe, the numbers were 36 percent and 23 percent, respectively, with the French apparently showing the greatest skepticism about vaccine safety. In Northern Europe, the numbers are 44 perecnt and 29 percent, respectively. In terms of effectiveness, North America scores 59 percent for strongly agreeing that vaccines are effective and 24 percent somewhat agreeing. Western Europe shows low scores of 44 percent and 33 percent; Northern Europe comes in at 58 percent and 26 percent.

North America and Western and Northern Europe are cumulatively the richest parts of the world at present and have been for hundreds of years. With such relative wealth, educational levels might be expected to be higher than elsewhere. For this very reason, the above results beg the question of why richer and more educated people are less inclined to agree that vaccines are safe and effective. This will seem to some to be paradoxical based on the mainstream narrative. Indeed, one might have expected the opposite.

One argument floated to explain this paradox is that the poorer, less well educated countries show higher levels of agreement with the benefits of vaccination for precisely that reason; that is, they do so because they have seen the terrible impacts of vaccine-preventable diseases. Certainly, that is one explanation. Another might be that in more authoritarian countries, most people do what they are told.

Returning to North America, the majority of all races in both countries support, or think they do, vaccination under most circumstances, but the numbers may be trending to the vaccine hesitancy side, particularly in those with higher educational achievements, as noted.

In the United States, the concern about vaccines and mandates seems most prominent among those who are pro-Republican party, although it may be circular in that those supporters favor the Republicans because the Republicans have more legislators who seem to oppose vaccine mandates. This latter result may arise because Republicans in many cases have closer philosophical links to libertarian politics, and, at least for some of them, there is no jarring disconnect between support for both vaccines and vaccine choice.

In contrast, the Democratic Party in the United States keeps itself far from recognizing, let alone encouraging, in its own ranks the left version of libertarianism, e.g., any anarchist-like philosophy. Similarly, in Canada, it tends to be the more conservative political parties that more often resist the siren song for mandates.

It should be mentioned, however, that anarchist philosophy is quite nuanced. Yet in my experience, most of those who self-describe as anarchists are bizarrely bipolar on the issue of vaccination. This is really odd, at least to me, because this is a political grouping that notionally should have little trust in any branch of government in the United States or Canada (or anywhere) and routinely believes they are being lied to by governments across the board. And, just as routinely, they are correct: They are. But when it comes to vaccines, the CDC, any official health agency, and any government, all are virtually always deemed to be correct. Are these people better versed in science than those on the right? In my experience this is not the case at all.

These apparent disconnects highlight the lack of a clear pro- or anti-vaccine demographic based on political leanings. I will return again to the issues of political persuasion and vaccine issues when I consider in Chapter 14 the response of both the left and right during the COVID-19 pandemic.

In terms of religious groupings, most major religions in North America are on record supporting vaccination, but within the various religions there is quite a range of opinion about mandates. Among Christians, the impression I get is that some of the most vociferous opponents of both vaccination and mandates are those of a fundamentalist bent, but, like the political groupings, this is far from universal. The same applies to Orthodox Jews with major Orthodox organizations taking polar opposite positions based on their interpretation of "Halacha," or Jewish religious law.

Racially, the outcomes are likely to be mixed as well, but it is pretty obvious at first glance that most of those who show up at anti-vaccine/mandate events are white. This outcome may more reflect a position of privilege rather than that of race per se, since one thing that seems to distinguish pro- from hesitant/anti grouping is relative wealth and level of general education.

The latter two often go together. What this means, if it is even correct, is not at all clear and indeed has stymied pro-vaccine commentators. The most obvious interpretation from the hesitant side is that *because* of relative wealth, those with the highest levels of education (not counting those in the medical professions) are better able to interpret the wildly divergent claims of both camps and feel better able to use their education to make personal choices rather than default to the medical "experts." As noted, this does not apply to those in the mainstream medical professions who, at least officially, all sing from the same song sheet.

The same outlook may or may not apply to those in various sciences, since these fields tend to be highly specialized and do not necessarily include in their training any conventional courses on vaccination theory or practice. Some of those medical and nonmedical professionals who do diverge from the mainstream narrative on vaccine safety and effectiveness often find themselves rapidly branded as quacks and pseudoscientists by the media if they disclose any such leanings. For this reason, most probably don't get into any controversies, as there are no career advantages in so doing.

Coming back to the level of education, those on the pro side and most of the media seem to struggle with what seems a glaring anomaly: those who do not agree that vaccines are safe and effective tend to be better educated. One way around this anomaly has been provided by the whole "anti-vaxx" suitcase of meanings. If the vaccine-hesitant are better educated, then they must be deficient is some other way. This can only be some personality flaw such as privilege and/or selfishness and the lack of any commitment to a larger community.

One supposes this might be true, but to me taking this tack feels like the desperate attempt by the media and pro-vaccine bloggers to find something, anything, that makes the anomaly go away.

Maybe something will emerge in the future that distinguishes the groupings. But until then, the bottom line is that it may not be simple. There are likely many factors that determine an individual's choice of vaccine narratives.

As long as the medical authorities, CDC, mainstream media, the pharma, et al. confined themselves to the usual glowing statements about the unadorned benefits of vaccines, they might get away with keeping the official narrative alive and uncontested for many of the otherwise nonengaged. This may change in unpredictable ways with COVID-19, as I will discuss later in the book.

The push for mandates, and all of the issues arising from COVID-19 lockdowns, however, has triggered a counter push and one not just about COVID-19 issues. That is, attempts to suppress alternative narratives seem to have led a lot of people to examine in detail vaccine issues for the first time, often with the aid of an Internet that remains, for now, *partially* free. Such easy access to information, correct and fake, has generated calls to ban online vaccine "misinformation" that companies like Facebook, Amazon, Pinterest, and others seem to have acquiesced to, or at least paid lip service to.[17]

The reality for these companies is that they are stuck in a capitalist economic system where they cannot really afford to alienate any major potential sources of revenue. The growing vaccine-hesitant community represents one such grouping. If indeed this group is wealthier than most, then all the more reason for the Internet companies to be somewhat careful lest they drive them all away to some of the newer competitors.

Another key reason, that by now might be obvious to even the dimmest politician contemplating censorship, is that bans only push people underground. For Facebook's Mark Zuckerberg and other CEOs from other online companies, unless they are true believers in all things vaccine and are willing to take the possible financial consequences, driving people and their wealth to MeWe[18] or the dark web[19] is hardly a smart financial move. And make no mistake, if the online companies really try such suppression of information, this is where a significant portion of their audience will go.

The Rise of the Vaccine-Hesitant Movement:
What Are the Reasons?

Any movement has distinct phases in its evolution, and the emerging vaccine-resistance movement is no different. The initial phases are typically composed of individuals or small groups who have been impacted negatively, in this case, by vaccines. This doesn't mean the grievances such small groupings hold are valid, merely that they are perceived by the individuals involved to be valid. Most such groups fade away, unless (a) the grievances turn out indeed to be valid and increasing numbers of people begin to see the same things; or (b) connections between like-minded groups sprout. Analogies to the prerevolutionary period in American colonial history are probably not misplaced.

There is little doubt that for the vaccine-hesitant the multiplying effect of social media has enormously helped individuals and groups to coalesce and further spread their message at a rate that would have been vastly more difficult pre-Internet. Indeed, the very facility that the Internet allows those concerned about vaccines to reach out to one another and organize has not escaped the notice of either the State or the owners of the Internet services. This growing awareness spurred California Congressman Adam Schiff to write to Facebook founder Mark Zuckerberg in 2019[20] urging him to shut down "anti-vax" groups and their purported "mis"-information. Initially, when Schiff sent his letter, there was concern among some that full-on censorship was at hand for views anathema to Schiff and others holding pro-vaccine viewpoints.

Quite rapidly, however, the concerns abated. There were two main reasons for this. The first is contained in a comment attributed to Vladimir Lenin: "The Capitalists will sell us the rope with which we will hang them." In other words, Zuckerberg is running a very profitable business, and whatever he may or may not believe about vaccines, good, bad, or indifferent, he is rather unlikely to do much that will hurt his profit margin. The simple reality is that it would take a lot for him to deliberately alienate potential consumers of the various products that Facebook draws revenue from.

One presumes that Schiff, and others like him in Congress, could in the future move from friendly requests to social media owners to outright legislation to enforce censorship. Doing so, however, opens up a Pandora's Box of nasty outcomes, not least of these being raising the question of which views need to be suppressed in a notionally free country for "the greater good." I rather doubt that Schiff thought much about longer-term consequences after his initial knee jerk attempt at social media censorship.

Much the same problem bedevils attempts by various governments, including Canada's, to suppress the Boycott, Divestment, and Sanctions (BDS) movement that is aimed at the Israeli economy.[21] While the attempts to demonize BDS have gained support by some political groupings and with politicians of various stripes, they also run the risk of exposing a lot more people to nuances of the ongoing conflict between Israel and the Palestinians. Doing so also leads to the primary question of why a nonviolent movement that simply asks people to exercise free consumer choice about which products they buy is such a threat and to whom.

Simply equating BDS with anti-Semitism is not cutting it either for most concerned about Israeli policies in Gaza and the West Bank. In fact, branding those who support, or at least lean in the direction of, BDS as anti-Semites is about as effective as branding the vaccine-hesitant as anti-vaxxers. While there may indeed be anti-Semites in the ranks of BDS, just as there may be true anti-vaccine people among the vaccine-hesitant, suppression is usually not the best long-term strategy, since it mostly calls attention to the very things the powers that be want people *not* to think about.

One superb example of such unintended publicity came about a few years back when Dr. David Gorski, writing under the pseudonym "Orac," decided it was part of his mission in life to suppress a pending showing of the documentary *Vaxxed* at the 2016 Tribeca Film Festival in New York City. He did so by blogging endlessly about the film—which he had not seen—and then followed up his blogs by writing to the festival's organizers. Gorski got his wish: Tribeca canceled the showing of *Vaxxed* that had originally been scheduled to run one time in the middle of a Sunday afternoon.

Had the film been shown at Tribeca, perhaps a few hundred people might have seen it there, and it might have occasioned some media commentary. Then it would have gone to the place most documentaries go to die. Instead, the negative publicity that Gorski helped orchestrate led to the film being kicked out of the festival, which in turn created the opposite effect from what he and others had hoped. Instead of going off quietly into relative oblivion, banning the film effectively gave it a huge publicity boost, fostering, in turn, a cross-country tour with massive publicity for the film's basic premise that the CDC had covered up data on adverse effects of the MMR vaccine.

In addition, the attempted suppression of *Vaxxed* led to a sequel, *Vaxxed 2*. Neither film was about the retraction of the famous (or infamous) 1998 article in the journal *The Lancet* by Dr. Andrew Wakefield and colleagues, discussed in Chapter 4. Of note, *Vaxxed 2* has now been followed by a newer effort by Wakefield including *The Pathological Optimist*, after a book titled *Callous Disregard*, which documents his own story.[22] The most recent documentary is *1986: The Act,* which focuses on the 1986 National Vaccine Compensation Act.[23]

Much of the mainstream media picked up on the false narrative, in most cases without actually having seen *Vaxxed,* either. Regardless of what Gorski and others thought the film to be about, their vociferous attempts to crush it gave it a vast audience that would not likely have occurred otherwise.

This situation recalls a basic rule in politics: if you are far ahead against a relatively unknown opponent, don't give that opponent any publicity, as it simply increases their name recognition and perhaps, eventually, votes.

This is not a new concept, but Gorski fell into the "It is difficult to get a man to understand something, when his salary depends on his not understanding it"[24] trap and then, oddly, did it again when several vaccine skeptic activists, Dr. Shannon Kroner and Britney Valas, tried to host a civil scientific debate between skeptical medical doctors and scientists and their more pro-vaccine counterparts.

Once again, pressure from Gorski and others scared the pro-vaccine people off, and the event, called *One Conversation*, which would have likely faded from view in a day or two, got a huge boost. Gorski blogged about the event a number of times, easily exceeding the minor notoriety that the event might have had on its own.

In addition, Gorski wrote personally to the various people who were supposed to be on the pro side, effectively scaring them off. One such example of this obtained through a Freedom of Information (FOI) request is quoted in Chapter 9.

As an aside, once I saw the FOIA outcome, I wrote to one person who had dropped out, Dr. Catherine Roe from Washington University, to get her reason(s) for not attending. In my letter, shown in Appendix 14, I simply stated that I had obtained a copy of the FOIA request and, as per standard journalistic practice, wanted to give her the chance to comment. She chose not to respond, her right of course, but then went running to Gorski for help. In turn, he blogged about it, making the story all about him, which it wasn't, and essentially tossing Dr. Roe under the bus by making her look like a hapless child.[25]

We will meet Gorski again in Chapter 9, but these two examples demonstrate that he still does not get the difference between effective versus ineffective ways to prevent the public from being influenced by ideas he hates.

Key Players in the Resistance to Mandatory Vaccination: Who Are They and Why Do They Believe What They Believe?

The next stage of any movement is when the various elements have started to come together and a form of leadership starts to emerge.

The leadership that started to come together in the vaccine-hesitant world centered on celebrities, as in the past, but with a host of new faces. The initial star of the disparate vaccine-hesitant groups had been Dr. Andrew Wakefield, discussed in detail in Chapter 4. Wakefield is still lionized by the parents of vaccine-injured kids as the man who stood up for their kids and against the vaccine industry.

In turn, Wakefield was preceded by vaccine safety advocates who had children they thought had been injured by vaccines. Included in this grouping were Barbara Loe Fisher, founder of the National Vaccine Information Center,[26] who began her advocacy in the early 1980s by looking into adverse effects of the whole cell pertussis vaccine.[27] Some celebrities also got involved in early vaccine-hesitancy work, notably former *Playboy* bunny Jenny McCarthy and her then-partner, Jim Carrey.

Later the philanthropist Claire Dwoskin, cofounder of the Dwoskin Family Foundation and later the Children's Safety Medical Research Institute (CMSRI), became involved in vaccine safety issues. The Dwoskin advocacy for vaccine safety was backstopped by considerable private funds that enabled various scientists, often those not particularly interested in vaccines or vaccine safety per se, to take a walk on the wild side and actually do research in this area. Some of those researchers included this author and a variety of scientists from various countries, notably Dr. Yehuda Shoenfeld of Israel, Dr. Romain Gherardi and his colleague Dr. J-P Authier in France, Dr. Chris Exley in the United Kingdom, and some others.

As the wave of attempted mandates in various American states followed the measles outbreaks of 2018/19, newer faces came to the fore. These included Robert F. Kennedy Jr., a very successful environmental lawyer,[28] and Del Bigtree, a television producer.

In the pages to follow, I will highlight these two prominent vaccine critics of the last several years, both still involved in this effort; another one seems to have been a faux opponent for whom interest appears to have faded.

Figure 7.1. Robert F. Kennedy Jr.[29]

Figure 7.2. Del Bigtree. (Courtesy of Del Bigtree.)

Robert F. Kennedy Jr.

Robert F. Kennedy Jr. (1954) comes from the incredibly connected and influential Massachusetts Kennedy family whose mark on twentieth-century American politics has been vast. The most famous of the Kennedys were his uncle, President John F. Kennedy (JFK), and his father, Robert F. Kennedy, who was first JFK's attorney general and later a US senator. As most readers will know, both JFK and Robert F. Kennedy were assassinated, both in the official narratives by "lone" gunmen.

Over the last several years, Kennedy has become a highly sought-after speaker at vaccine safety events and protests. Indeed, his passionate speeches about vaccines choice are often the highlight of such events. But how did he come into this scene and what drives him?

With this family background in Democratic politics, it was probably inevitable that Robert Jr., called Bobby by most people who know him, would become committed to public issues. One of the Kennedy family initiatives was the Special Olympics, a cause that Kennedy was involved with from an early age.

Kennedy attended Harvard and studied at the London School of Economics, earning his BA from the former in 1976. He received his law degree from the University of Virginia and devoted his practice to environmental issues from an early stage in his legal career, helping to found a number of groups.

Kennedy started the Waterkeeper Alliance in 1984. The Alliance was a blue-collar group whose goal was, and still is, to protect rivers and lakes across North America from pollution. The Alliance's first efforts in this regard began with trying to clean up the Hudson River in New York, where a primary source of pollution was mercury compounds from industrial activities. Mercury, as we saw in Chapter 5, is a major neurotoxin in water, moving up the food chain to humans from contaminated fish. And, as described, mercury in the form of Thimerosal appeared in various vaccines prior to 2000. As a lawyer with the Alliance, much of Kennedy's work involved suing plants and utilities for polluting waterways and lakes.

It is often confusing for people to understand how Kennedy as an environmental crusader against toxic pollution became a vaccine critic and activist. In personal correspondence with this author, Kennedy described how his journey into vaccine issues originated: he began to notice at his talks on environmental pollution the same group of women who would often come to hear him speak. These women would typically sit near the front of the room. It turned out that these were women who believed their children had been harmed by Thimerosal in vaccines.

Kennedy related that initially he didn't want to get involved, as he has described himself as "fiercely pro vaccine" with all of his children fully vaccinated. With this, Kennedy perhaps thought that the women at his talks were just part of another kooky anti-vaccine fringe group.

What changed Kennedy's mind, certainly about the potential for Thimerosal to cause neural injury to some children, was when one of the women who frequented his pollution talks, Sarah Bridges, showed up at his doorstep in Hyannis Port in 2005. Bridges refused to leave until Kennedy agreed to look at an eighteen-inch-high stack of articles and papers on mercury and Thimerosal. Bridges believed that her son, Porter, had been injured by Thimerosal, and she had recently been awarded a twenty-million-dollar Vaccine Court settlement.

Kennedy had been well trained to read scientific articles though his litigation efforts on behalf of the Waterkeeper Alliance. He was immediately struck by the apparent gap between what the articles showed and what official sources were claiming in regard to Thimerosal.

Kennedy was used to going straight to the top in his environmental work, so he cold-called Dr. Michael Collins, then head of the NIH. Kennedy asked Collins to explain the difference in toxicity between methyl mercury, such as that found in contaminated fish, and the ethyl mercury in Thimerosal. Oddly, Collins didn't seem to know but referred him to Paul Offit, discussed in detail in Chapter 6.

Kennedy fairly quickly realized that he was being given the runaround when Offit claimed that there was "a whole mosaic of studies" about the difference between mercury compounds but didn't seem aware of the Burbacker et al. study.

From that point onward, Kennedy began to doubt the official narrative. His work on environmental issues had trained him to spot industries and politicians who were lying. In looking into vaccine safety issues, he found plenty of liars and a pharmaceutical industry, one of the most powerful in the United States and one that was, and is, largely unregulated, at least as far as vaccines go. He writes in Children's Health Defense:[30]

> Loads of you who have been involved in the past in the battle to protect our children from poorly made vaccines or toxic chemicals in our food or in our water know the power of these industries and how they've undermined every institution in our democracy that is supposed to protect little children from powerful, greedy corporations. Even the pharmaceutical companies have been able to purchase Congress. They're the largest lobbying entity in Washington DC. They have more lobbyists in Washington DC than there are congressmen and senators combined. They give twice to Congress what the next largest lobbying entity is, which is oil and gas, and four times what defense and aerospace. Imagine the power that they exercise over both Republicans and Democrats on Capitol Hill.

And,

> They've [the pharma] turned the regulatory agencies that are supposed to be protecting Americans, and particularly our children, from toxic exposure.

They've captured them and turn them into sock puppets for their own industries. They've compromised the press. Any of you who watch evening news and see the pharmaceutical ads one after the other, that's a way of buying the news in this country. They've undermined the medical establishment. They've cowed the scientists and defunded the ones who are trying to tell the truth, and they destroy the publications that publish real science. [Brackets for addition are mine.]

It's a damning charge and yet one that is completely in synchrony with research from other investigators like Marcia Angell and Alan Cassals and Pat Moynihan, to be described in detail in Chapter 12.

The power that the pharmaceutical industry wields leads to predictable consequences. According to Kennedy,

The greatest crisis that America faces today, as big as any of our environmental crises is the chronic disease epidemic in America's children. The generation of American children born after 1998 are arguably the sickest generation in the history of our country. With chronic ailments and pediatric rates of some chronic conditions that are the highest in the world. For example, according to CDC, one in every six American children has been diagnosed with a developmental disability such as autism, ADD, ADHD or learning disabilities. And according to HHS, an astonishing 54% of American children has a chronic disease like rheumatoid arthritis, or juvenile diabetes.

Much of this is summarized in Kennedy's book, cited in the previous chapter.

In 2016, Kennedy started an organization called Children's Health Defense (CHD), where he is the chairman of the board and senior prosecuting attorney. CHD arose out of an earlier initiative called the World Mercury Project. CHD currently has several lawsuits ongoing on behalf of those injured by vaccines.[31] CHD has also taken a lead role in the fight against COVID-19 restrictions. It would be safe to say that anytime vaccine safety is discussed or mandates are proposed, Kennedy and CHD are there fighting for those who may be impacted.

CHD has become one of the few organizations taking an in-depth look at a range of COVID-19 issues that go far beyond any health considerations of this one virus. The interlocking themes explored in great detail by CHD include the massive economic dislocations and transfers of wealth from the middle class around the world into the pockets of the 0.1 percent, along with some of the frankly Orwellian social changes now being test-driven by the World Economic Forum and its supporters. CHD's recent report, just published in the *International Journal of Vaccine Theory, Practise, and Research,* discusses in great detail the web of links involved in the COVID-19 pandemic.[32] This analysis will be highlighted in Chapter 13.

One last comment on Kennedy: Those who have heard him speak note that his voice seems perpetually hoarse. The condition is described by Kennedy as spasmodic dystonia,[33] which Kennedy believes was the result of an influenza vaccine he received.

Del Bigtree

Del Bigtree (1970) was not known in vaccine safety circles until the movie *Vaxxed* came out in 2016. Indeed, his trajectory into this movement seems more a matter of odd coincidences than deliberate planning.

Bigtree, who uses his mother's Mohawk family name, was born in New York City to parents who were both entertainers working in the theater and also social activists.

Bigtree grew up performing and dancing, and then followed these passions to a Vancouver film school. From there, his path took him to Los Angeles and then back to New York to work on short films, documentaries, and videos. In turn, these efforts led to helping with interviews for the Dr. Phil show.[34] This last gig led to work on *The Doctors*, a syndicated television talk show about medicine. Eventually, Bigtree became one of seven producers for the show. This period, from roughly 2009 to 2015, kindled in him an interest in medicine and with this he learned to read and evaluate the scientific literature.

One of his episodes featured a critic of Monsanto and a Monsanto executive. In this period, Bigtree had been only vaguely aware of the emerging story of the CDC whistleblower, Dr. William Thompson, and the revelations about corruption of the scientific process at the CDC (see Chapter 3).

One episode during a production for *The Doctors* stood out to him when he had California State Senator Richard Pan on the show. Pan is widely hated in the vaccine-hesitant world for pushing through the requirement for school vaccine mandates with his bill SB77.[35] The arrogance and the way Pan talked about SB277 and his disdain for his opponents was the spark for Bigtree to dig deeper, in turn kindling a desire to pursue documentary work on vaccine mandates, as well as to explore in more detail the Thompson case.

As Bigtree described it, fate literally intervened in the next step in his trajectory: a publicist at the show called to ask if Bigtree had heard about Andrew Wakefield and did he want to meet him? Three days later, Bigtree met Wakefield in person.

Wakefield had already finished filming *Vaxxed* and showed Bigtree the rough cut. This version of the movie was heavily driven by the science, and Bigtree suspected that his own previous work in television could help make the film accessible to a lay public, much the same way that *The Doctors* had made medicine more available to nonspecialists. That crucial decision led Bigtree to take on an associate producer role for *Vaxxed*. Luck once again intervened: the show, as cited earlier, was kicked out of the Tribeca Film Festival. From the perspective of widespread publicity, this was one of the best things that could have happened to it.

From *Vaxxed* and its overall success, Bigtree was offered a studio at half price to start his widely viewed new show, *The Highwire*, which explores a range of issues involving vaccines and now COVID-19.

With Kennedy and Bigtree, the similarities between the two are striking: both share a background in social activism. And, like Kennedy, Bigtree denies the frequent critique that he is "anti-vaccine." In regard to this charge, Bigtree replies that "When people say I'm anti vaccine, it's actually not true. I am pro-science. I approach everything with curiosity and skepticism."[36]

The Others

And, as is usual with any growing movements, some of those drawn to what was coming to be known, at least internally, as the vaccine resistance comprise a mixed bag of personalities and aims. Many of these aims were legitimate, but some were sketchier.

For the most part, the movement seems to be grassroots-based, often led by mothers who have witnessed adverse effects to their own children.

Inevitably, those from various backgrounds given to grandstanding have also found it easy to do so in this new incarnation of the vaccine safety constellation of groups; those with products to sell see opportunities in the growing recognition that such a movement really exists, and some simply find in the movement a means to increase their own publicity.

In all of these ways, the rapid explosion of vaccine resistance groups has created a new on-the-ground reality as the movement begins to mature.[37]

One of the newcomers is Dr. Shiva Ayyadurai, called Dr. Shiva by his fans. Shiva was born Vellayappa Ayyadurai Shiva, in Mumbai, India, in 1963. Whatever else his fans or detractors may say about him, none can really deny that he is a colorful character. His academic credentials include a BSc, two master's degrees, and a PhD from the Massachusetts Institute of Technology.

Shiva's claims include that he was the inventor of email, a claim that his critics claim is highly unlikely. One counterclaim is that what he actually did was name an interoffice computer system, where he worked as a teenager, "email." This is obviously not the same as inventing the method used by people around the world for Internet communication.

To take a deeper look at his academic credentials, I examined his various publications, only one of which seemed to be a peer-reviewed journal article on which he is listed as the senior author. His publications also include approximately nineteen abstracts of presentations at scientific meetings. His website lists twelve of his own books for sale.[38]

One thing that becomes very clear when looking at Shiva's website or listening to his podcasts is that he appears to be heavily into self-promotion. Some of this promotion is political: Shiva ran for the US Senate in 2016 as an independent and attempted, unsuccessfully, to capture the Republican nomination to run against the incumbent Elizabeth Warren in 2020. Shiva also plays the race card heavily and seems to have a tendency to sue those he thinks have offended him.

Shiva was virtually unknown to me until last year, when he seemed to have suddenly appeared with various comments at events organized to highlight the issues of vaccine safety and mandates. Since then, he has established a podcast and has gone on the attack against Kennedy, Bigtree, and others, accusing all of selling out the movement by being a "controlled opposition" put in place to pretend to question vaccines while actually trying to divide the movement. Although he has gathered a number of adherents, his increasingly shrill accusations border on farce. For his part, Kennedy tried to ignore the attacks at first and then, finally fed up, published a reply.[39] In his reply, Kennedy pointed out clear evidence linking Shiva to vaccine companies, to the Bill and Melinda Gates Foundation, and to a host of other entities that Shiva routinely attacks. In other words, the attacks seem to be a screen to promote a very different agenda. It's a trick as old as the hills and one that only serves to aid those who want to destroy, or at least neutralize, the vaccine awareness movement.

The attacks against other vaccine safety advocates is problematic in an obvious way in that it carries with it the potential to fragment the movement and make it more vulnerable overall. To those who study the politics of existing and emerging movements, this is nothing new. In politics, a classical example in the early US Revolutionary Period was the attempts by some soldiers and civilians to undercut the authority of George Washington.[40] There are many more examples from history and even today.

In regard to those attempting to fracture a movement, the question is always who benefits from achieving such a goal. One clear answer can be the person who is doing the fracturing. The other answer is that such a person is acting as an agent provocateur and doing this at the behest of a larger power, in the current case perhaps the pharmaceutical industry. In the case of other political movements from anti-Vietnam protests to Occupy Wall Street and its offshoots, to Black Live Matter, agents provocateurs were invariably government agents.

The Emerging Vaccine Skeptic's Literature

In addition to Kennedy's 2015 book, a number of others have written critical evaluations of vaccines in recent years. In addition to some already cited above, these include Ted Kuntz's *Dare to Question: One Parent to Another*.[41] Each and every one of these faces an uphill battle against an entrenched orthodoxy, as described in the next chapter.

CHAPTER 8

Vaccine Ideology and Religion

Persecution is not an original feature in any religion, but it is always the strongly marked feature of all religions established by law.

—Thomas Paine[1]

The unconditional right to decide about our own body and the bodies of those whom we are responsible for (children, animals) is the most fundamental 'sacred' and inalienable right which must never be trodden on by a society, state, or system. Therefore, no one but ourselves can or should have the 'jurisdiction' over our body and decide on our behalf, let alone against our wishes, that procedures we should undergo.

—Mateja Cernic[2]

Religion versus Science

At first glance, it might seem that religion and science are the very antitheses of each other. Religion deals with the largely unseen presence of some deity (or deities). In the large monotheistic religions, those that arose in the Middle East, predominantly Judaism, Christianity, and Islam, the followers of each believe in their own, and sometimes shared, collection of stories, fables, aphorisms, and what they consider to be sacred revelations, teachings, and sayings of their founders.

In the following, I will stay with these religions as a lead in to the central thesis of this chapter, namely, that aspects of science, medicine in particular, and most particularly that subset of medicine, vaccinology, have more in common with formal religion—some would even call it a cult—than is commonly imagined.[3]

With that said, let me return to considering the three main religions of Middle Eastern origin. Each of these acknowledges the others, which is not to say that they accept the doctrines of one another. For example, both Christianity and Islam see Judaism as the originator of their own creeds, but supplanted by later revelations of later prophets. Thus, for believing Christians, the Jews came first, and their book, the Old Testament (to the Jews the Torah, which also includes the Tanakh), contains much that is true, but much that was replaced by the teachings of Jesus and later the New Testament. Muslims accept both Jews and Christians as "people of the book" but see Islam as completing the processes with their own holy book, the Koran, which Moses and Jesus began. In the Muslim view, the Koran supersedes both Old and New Testaments.

Both Christians and Muslims accept many Old Testament figures as prophets in their own religions and have more or less "grandfathered" them into their own core beliefs. Jews, at least Orthodox Jews, certainly do not feel that their religion needs modifying by either Jesus or Mohammed and hence do not accept that the holy books of the others have any particular meaning for them, at least in a religious sense.

Wikipedia defines a religion in this way (and this definition is just as accurate as others I have found):

> Religion is a social-cultural system of designated behaviors and practices, morals, worldviews, texts, sanctified places, prophecies, ethics, or organizations, that relates humanity to supernatural, transcendental, or spiritual elements. However, there is no scholarly consensus over what precisely constitutes a religion.[4]

The last part is certainly true. And, to make it even more complex, religion and science, particularly medical science, actually have a number of characteristics in common.

To probe this in more detail, I asked two individuals I know who are people of faith in their own religious traditions to comment on how they see the differences and similarities of religion and science. I further asked them to comment on cultlike behaviors.

The first response was from Rabbi Zev Epstein, a very wise man described elsewhere in the book. Epstein is not only a Torah scholar, but also someone highly knowledgeable about vaccine issues. Here is what Rabbi Epstein had to say, starting with his definition of religion:

> I define religion as a program for serving G-d, the Creator of the universe. Thus, religion begins after the scientific question of 'Is there a G-d, a Creator, or did the world get here some other way?' is answered-on the side of there is a G-d. This answer can be arrived at through philosophical inquiry, or through faith/tradition.
>
> Different religions have different beliefs as to what G-d's revelation to mankind is (was) so the program can vary but they are all based on the elementary belief that G-d exists.

Zeroing in on deviations from this definition of religion, Epstein made a key point when he wrote:

> Idolatry moves away from G-d in the sense that it worships physical forces. It grants the sun, or some other facet of the physical universe some sort of independence or intelligence and worships it so that it will be good to man-good crops, etc. Some forms of idol worship were simply occult practices that used black magic (which most Rabbinic sources believe exists) to manipulate the physical world for one's benefit. Idol worship is infamous for serving idols-meaning for granting divinity to an actual physical object. This is what G-d abhors. Idolatry is a rejection of G-d, and therefore one of the three cardinal sins, and also, therefore not a 'religion.'

As a follow-up, Epstein wrote this about cults: "As for cults, I think these are forms of an extreme sort of social structure, almost like a commune where you not only surrender your money, but your mind as well . . ."

In relation to science, Epstein stated:

> Science, the scientific method of inquiry and of gathering and garnering knowledge etc. are fine tools to help us become informed about the world around us. But some view 'science' as the only effective means of ascertaining truth. And since science cannot see G-d or observe Him, or prove His existence, He does not exist. So that type of science cannot be called a religion, it is a philosophy—a way of looking at the world and of processing reality—which perhaps can be called a form of idolatry, ascribing not intelligence but at least independence to the laws of nature—which cannot be broken, and which can be harnessed and even manipulated to serve our needs if we will understand them properly using our science tools.
>
> I believe that the term cult cannot be applied here [in general in relation to science overall] because the scientific community does not fit the bill for that type of social structure. The mindlessness—blindly following intellectual leaders is 'cult-like' but this does not a cult make." [Brackets for addition, mine]

I then asked him how the above ties into the "science" of vaccines. Epstein commented that

> So I think that the vaccinologists . . . are guilty of a 'cult-like' surrender of their independent thought, against a backdrop of the idolatry of science which grants exclusive primacy to 'nature' and seeks to bend it to mankind's needs.

To get a Christian perspective, I asked Dr. David Lewis, the former EPA scientist we've met before in Chapter 6. Lewis had this to say about religion, science, and vaccinology:

> Briefly summarized in the Ten Commandments, the Law of Moses, which Moses delivered to the Jewish people when they escaped Egypt, has served as a template for civil laws throughout the world for thousands of years. World peace, particularly in the Mideast, can only exist when nations that follow the Law agree upon what it was intended to achieve. It has nothing to do with following any particular religion, and everything to do with treating everyone the same regardless of their religion. As Jesus taught: In all things, do unto others as you would have done unto you. This is the Law and the Prophet [Matthew 7:12].[5]

And,

> In my mind, there is no difference between religious cults deceiving their followers for personal gain, and the CDC and pharmaceutical companies downplaying the risks associated with vaccines. This is especially true with regard to live vaccines, which disproportionately harm economically and educationally disadvantaged communities via environmental transmission. I am reminded of the similarities between vaccinology and cults whenever I recall serving on the Board of Directors of the National Whistleblower Center in Washington, DC. When I asked the Executive Director about taking on cases involving vaccine safety, he answered 'no way.' He explained that massive financial, legal and political forces protecting the vaccine industry make it impossible to win such a case.
>
> I am also reminded of the cult-like forces involved with vaccinology every time I recall a scientist who worked for a leading vaccine manufacturer showing me some of the company's internal records illustrating how it was manipulating data to cover up adverse health effects of the MMR vaccine. As we wrapped up, I asked: 'How many vaccines on the pediatric vaccine schedule is (this manufacturer) doing this with?' He replied without hesitation: 'All of them.'[6]

Paul Bloom, writing in *The Atlantic*, noted that:

> Many religious narratives are believed without even being understood. People will often assert religious claims with confidence - there exists a God, he listens to my prayers, I will go to Heaven when I die - but with little understanding, or even interest, in the details. The sociologist Alan Wolfe observes

that 'evangelical believers are sometimes hard pressed to explain exactly what, doctrinally speaking, their faith is,' and goes on to note that 'These are people who believe, often passionately, in God, even if they cannot tell others all that much about the God in which they believe.'[7]

Bloom went on to say that

People defer to authorities not just to the truth of the religious beliefs, but their meaning as well. In a recent article, the philosopher Neil Van Leeuwen calls these sorts of mental states "credences," and he notes that they have a moral component. We believe that we *should* accept them, and that others - at least those who belong to our family and community - should accept them as well.[8]

Although Bloom's article was meant to distinguish religion from science, the notion of "credences" is key to what follows. Credences in any religion are beliefs that need to be accepted by the faithful whether they are understood or not. And because they have a "moral" component, then others should accept them as well, at least as long as those others want to remain a part of the same religious community.

In Chapter 1, I discussed what science is, and isn't. As mentioned, mostly science is a method for understanding the natural world. While some have made a pseudoreligion out of it or have had their work inter-preted in such a fashion,[9] there is nothing per se sacred in either the methodology of science or in the interpretation of the results generated. Certainly, in most branches of science, today's results and interpretation can be modified, even negated, by tomorrow's data. Science, as described, is not about proof, rather about the probability that some observation is correct. Science does not set itself up as the final judge of what is eternally true in the natural world.

That, however, is not how some scientists and science followers see it, and, as noted above, this is particularly true in medicine.

It would be fair to say that the general public holds scientists in high regard, and one continually hears from various voices that we "need to trust the science" as a credence, while precisely what the speaker means by this is not always clear.

Medical doctors tend to receive even greater levels of esteem, even though, as noted in Chapter 1, not all are really all that well versed in the scientific method or even fully understand all of the emerging knowledge in specialties other than their own. What many have instead of this knowledge are, like the general public, credences.

Dr. Mateja Cernic, writing about the set of core beliefs in modern medical practices, states that truth and reality are, in fact, social constructs, established at particular points in time. She wrote that

> It is the person who accepts the established truths and structure of medical knowledge and thus the medical license. But to keep it, he/she must renew it. He/she must confirm that he/she is loyal to the paradigm . . .[10]

For paradigm, substitute the word ideology: "The task of ideology is to protect certain practices as universal and self-evident, so its role in complementing economic and political power is priceless and irreplaceable."[11]

The latter points are broadly similar to those of Dr. Toby Rogers when considering the interplay between ideology and economics in medicine.[12]

It is also clear that most people hold religious figures in high regard, whether they are priests, imams, rabbis, or the faith leaders of other religions.

It is precisely this level of esteem that typically leads to obedience to the advice of religious leaders *and* physicians, although the previous dominance that religious/medical authorities have conventionally wielded over the laity seems to have seriously diminished in our Internet-based world.

Does Vaccinology Behave Like a Cult?

With the above in mind, let's take a deeper look at vaccinology. Vaccinology, discussed in Chapter 2, is defined as: "The branch of medicine concerned with the development of vaccines."[13]

To recapitulate, vaccinology as a discipline is focused on vaccines, the development of them, their characteristics and actions inside the body, the kinetics and dynamics of vaccine stimulation of the immune system, and finally and particularly the materials science that makes things like adjuvants possible and effective.

Nothing too strange about this, so far. Basically, vaccinology lays claim to being a scientific discipline like any other. As such, it is expected to adhere to the core principles of observation, hypothesis generation, experimentation, data analysis, and interpretation of results based on other research in the field. To be effective in vaccinology, a solid grounding in immunology is essential. Hypotheses that don't generate an expected result are expected to be modified or discarded.

But what if vaccinology doesn't strictly play by these basic rules but has another element, one more closely resembling the dogma of formal religion, or even cults, in which it cannot allow an opportunity for any aspect of its core assumptions to be falsified and replaced? What if vaccinology believes that it has created such a mighty edifice within medicine that to question any aspect of that edifice risks sacrificing the entire structure? In other words, what if vaccinology and some related disciplines in medical science, such as parts of epidemiology and public health[14] as they relate to vaccines, have now more fully embraced similarities to religion?

I will explore the similarities between these branches of medicine with religion in more detail later in this chapter. First, however, I want to return to my own journey into vaccine research, a subject that I came upon by chance. It was this chance encounter that first began to educate me on the overlapping core world views of the medical disciplines that deal with vaccination and how these core views are more akin to actual religion than most scientific areas of study.

On Ideology

It would be safe to say that my path into vaccine safety research was largely accidental, as detailed in the Preface. To recapitulate, I was previously purely an ALS researcher, with ALS still a key area of investigation in my laboratory to this day. As part of this work, I was simply following a thread that led from putative ALS clusters to aluminum, then to vaccines with aluminum.

If my laboratory's first experiments on aluminum adjuvants and their impacts in the CNS hadn't found something credible, and statistically significant, we'd have followed the "nothing to see here, move along" perspective. That is to say, if the Gulf War Syndrome[15] link to vaccines and aluminum had not borne fruit, we'd have been long gone searching for other ALS clusters with some resemblance to ALS-PDC.

In fact, since in my view there really weren't other clusters, nor are there now, I'd have simply dropped the aluminum aspect and continued looking for other environmental factors on Guam and elsewhere.

Alas, it didn't turn out that way, and we *did* see negative aluminum adjuvant effects in the spinal cords and brains of the treated mice. And, being naive, I then saw no problem continuing to explore what was then an unexpected finding.

Maybe I should have paid more attention to conversations I'd had with several prominent ALS researchers. Two of these scientists had told me about having had disturbing interactions with people they thought were representatives of the aluminum industry. Both were very clear in warning me that it would probably be wiser to simply leave the whole issue of aluminum in general alone. In particular, they advised dropping any thought that aluminum might have a role in neurological disease.[16] In hindsight, that would likely have been a happier career decision for me had I followed it.

The second bit of my naïveté was that I simply assumed that our results would trigger some concern in the official bodies such as the CDC and NIH that aluminum in vaccines might be a problem for nervous system health.

In fact, there was no response to our work at all, at least initially. Eventually, both the WHO quoting themselves (without references, as usual) and the CDC, citing WHO, simply dismissed our study, as cited previously.

Back then, I still believed in the quaint notion of "speaking truth to power." In other words, if someone points out a serious problem in any realm of human affairs, the entity responsible for that problem should say something like, "Aha. Now that we know there might be a problem, let's go take a look and if we need to, go fix this."

The speaking truth to power notion crops up in politics all the time, for example, when Black Lives Matter details the chronic discrimination and murder of Black people, when antiwar activists point out the sheer folly of wars such as that in Afghanistan, and many more. And, of course, it equally applies in a range of scientific issues from climate change to the increasing toxification of the biosphere. Needless to say, it applies to vaccines.

It was at this point that I began to suspect that in our investigations of aluminum in vaccines, I was not merely dealing with a scientific dispute where data could be presented, reviewed, critiqued, and discussed. Rather, that something more fundamental was going on, something that was more like a worldview that had a very clear right side and wrong side. The critiques of vaccine adjuvants were definitely on the wrong side from a mainstream medical perspective.

Put another way, I began to suspect that, for the powers that be, having me talking about the possible dangers of vaccine adjuvants was speaking a language that the other side simply did not speak, nor care to understand.

In this way, the power was acting almost like a religion.

While I have great respect for most religious traditions, arguing about matters of faith is a fool's errand and quite unlikely to advance the debate. For example, if you are a religious Christian, then there is little point to try to counter a belief that Jesus died on the cross, rose from the dead, etc. Or for Muslims, that Mohammed received the Koran from God, or for Jews that Moses did the same with the Ten Commandments.

Such efforts go nowhere, as perhaps they shouldn't, because they are matters of faith, rather than science, in which there is no middle path: You believe in these articles of the religion or you don't. And at least for Christians or Muslims, if you don't, you are no longer a member of the faithful.[17]

What was gradually dawning on me was that the whole subject of vaccinology and the application of this to medicine was very much like a religion, or worse, a cult,[18] in that one had to accept all of the catechism or be cast out.

Continuing to study aluminum adjuvants and reporting on the outcomes was rapidly putting me on the path to excommunication, and I would soon join other scientists who, in the words of one hostile blogger, "used to be pretty decent scientists" until all of us went over to the dark side.

I just didn't know it at the time. And by the time I found out, it was too late to honorably turn back.

The idealized version of science is not supposed to work like this, although some would dispute the notion that scientific belief systems differ that much from religious ones. Indeed, in these days of COVID-19, a typical response to any critique or question about various State policies is to provide the shorthand reply, "because of the science." It is as if simply evoking "science" without necessarily knowing what science does generally, or in relation to COVID-19 specifically, were considered a sufficient reply. This sort of statement alone provides a clear insight that far too many people have come to see science as a secular religion in which, if you disagree, you must "hate science."

A Brief Excursion into North American Archeology

Insights often come from unexpected sources. At first glance, none may seem odder than to ask why mastodons aren't like vaccines. Simple, right?

Mastodons are an extinct branch of mammals distantly related to the elephant, of the species *Mammut americanum*; vaccines are inventions of medical science designed to prevent infectious diseases. Mastodons no longer exist, except as bones, sometimes fossilized. In contrast, vaccines are alive and well in medicine and continue to be a boon for the pharmaceutical industry. Increasingly well, in fact, in that the development and application of existing and newer vaccines continues to proliferate wildly, providing an increasingly profitable revenue stream for the pharma, as detailed in Chapter 12. With a COVID-19 vaccine on the horizon, the market may be doing a lot better in the near future, especially if these vaccines are seen as the savior from the disease or, even better, can be mandated for everyone.

The whole question, why aren't mastodons like vaccines, feels like the set-up for a punchline to a joke, which it is, just not in the way most might expect.

Delving into North American prehistory may help make the joke clearer.

Mastodons became extinct about ten thousand years ago just as the ice sheets that once covered most of North America began to recede. There has been abundant speculation that humans migrating into the continent had something to do with the demise of mastodons and other megafauna.

Conventional wisdom in archeology has long held that humans first crossed the Bering Straits to begin the long process of settling North and South America about ten thousand years ago. A few anomalous dates have cropped up over the last few decades potentially pushing human habitation in the Americas further back in time.[19] Human habitation in coastal British Columbia has been dated to at least fifteen thousand years ago.[20]

Where some of this conflicts with notions of the land bridge between Asia and North America and with changing sea levels, archeologists have adjusted their hypotheses accordingly, just as scientists are supposed to do with new data in any field.

One notion now gaining credibility is that early migrants did not come by land, but by boat, and did so by hugging the coastline along what has been called the "kelp highway."[21] This last notion inevitably leads to the obvious conclusion that, if true, humans have been building and using boats for far longer than most anthropologists used to think. The incidence for boat use by humans may indeed be quite ancient, not only in North America, but also in Australia, Indonesia, and elsewhere.

However, what really threw the archeology community on its collective head, or might have been expected to, was a discovery from San Diego, California, in 1992/3, later published in *Nature* in 2017 by Holen et al.[22] In brief, during the construction of a freeway, workers found bones that turned out to have come from a mastodon. No great surprise here, since mastodon and related mammoth bones have been found across North America. Nor was it a particular surprise for the archeologists who examined the bones to find evidence that the remains had been altered, leaving behind characteristic fracture marks where the latter had been deliberately smashed to extract the marrow. Implements called anvil stones were found at the site that could have been used for this purpose. The signs on the bones suggested predation by humans, again similar to other animal-butchering sites found in Europe and Africa. Overall, nothing too shocking for the prehistory of archeology in North America: mastodons carved up by the first North Americans.

Instead, the shock lay in the uranium dating of the bones, which gave a date of 130,000 years plus/minus some ten thousand years. This was completely unexpected by almost a factor of ten.

First, if correct, it put humans, or some other tool-using hominin, in what would become San Diego, a lot further back in time than any previous chronology had even hinted at.

In turn, a date as long ago as this was clearly going to cause a minirevolution in terms of when our species first emerged. This latter number is currently pegged at about three hundred thousand plus years ago[23] but if correct would in turn mean that modern humans had emerged and spread very rapidly around the world. It would certainly suggest that First Nations claims that their ancestors had been on the North American continent ("Turtle Island") from time immemorial was, in fact, essentially correct and not just a legend.

Or, if it were not modern humans responsible for the smashed bones in San Diego, then it would mean that other hominins had crossed into North America long ago.

We are left with four choices that are the most likely. First, the report is correct, and humans of the species *Homo sapiens* arrived long ago, either relatively soon after emerging from the Old World, or that the origin of the entire species goes back even further than we have thought and our current timeline for our species is simply wrong.

Second, some unknown hominins were wandering around North America killing mastodons, an idea for which there is, as yet, no archeological or paleontological evidence.

Third, maybe the mastodon bone discovery and the resulting date were simply a methodological error in the dating techniques. Finally, fourth, maybe the marks on the bones had nothing to do with humans/other hominins at all, but rather were the results of geological forces or the tractor that first uncovered the bones causing rocks to collide with the bones giving the *appearance* of being the result of actions by ancient humans/hominins.

As fascinating as this is, at least to me, it doesn't really matter to our inquiry into vaccines which of the archeological outcomes turns out to be correct. Rather, the point of the story is to show how even earthshaking scientific discoveries, as the mastodon story surely is, are treated within their own scientific disciplines and by the media.

In this regard, it would likely be safe to say that most people have never heard about the 130,000-year-old mastodon bones and the conclusions that arose from this discovery. The media may have mentioned it in passing but certainly didn't dwell on the finding or even speculate all that much on what the implications for human prehistory might be.

Instead, the article flashed through the literature and then vanished into academic obscurity as far as the general public was concerned. Several research groups wrote letters to *Nature* to discuss or dispute the findings,[24] but no trolls have apparently attacked the central thesis or data, or wrote to *Nature* demanding a retraction. In addition, the various universities whose researchers made and reported the discovery were not critiqued by their university heads.

Much the same lack of concern attends most new discoveries in physics, astrophysics, paleontology, oceanography, neuroscience, and many other disciplines.

Clearly, as documented throughout this book, the same cannot be said to apply to discoveries about vaccines, unless, of course, they are laudatory. As journalists Jon Rampson and Sheldon Stauber point out in their 2002 book, *Trust Us, We're Experts*, the negative spotlight on studies critiquing tobacco, biosolids (or sludge), pesticides, and, more recently, genetically modified organisms (GMOs) is now shared by vaccine-critical articles.

What these other topics share with vaccine studies is primarily one thing: vaccines are commercial products and make a lot of money for companies and their shareholders. In contrast, little or no money changes hands when contemplating the mechanisms of mastodon death, however scientifically explosive the results may be. No one is going to make money from the mastodon study, apart maybe from Disneyland if they one day make a ride called "130,000 years BC."

Essentially, the attention given to a particular discovery by the media is the result of money to be spent trying to sway public opinion away from any critique of corporate products. It has become very clear that the level of critique and the slander visited upon the skeptics is primarily about money, who makes it, and how those who make it try to keep their product's money stream flowing.

There is another dimension to this, however: while those critiquing GMOs can also find themselves taking a lot of flack for their work and views, very few on either side of the question of GMO safety likely view GMOs as of divine origin. Thus, while being pro- or anti-GMOs may polarize the conversation around a Thanksgiving dinner table, it does not evoke the same level of intensity, the same "you are with us, or with the terrorists" mentality that vaccine disputes seem to generate.

One also does not hear in any of these fields, even those that challenge other industries, a phrase commonly used about vaccines: "the science is settled." It is an odd thing for anyone to say about any field of study, but more so when pro-vaccine advocates say it, particularly when it is blindingly clear that the science of how vaccines work and their effects is not at all "settled," as the previous chapters have demonstrated.

This aspect of "settled-ness" puts vaccines and their proponents in a quite unique status, one that more than anything else resembles a religion with its devoted adherents. Vaccinology has achieved a status usually reserved for religions and acts of faith therein. As such, it is less and less about what science actually shows, and more about how rigorously, and how much, any person actually "believes," as noted in the Bloom article. Indeed, it is not just a religion, but a *fundamentalist* religion, bordering on a cult, that can tolerate no deviation from the true faith.

I will consider this last point later in this chapter.

Ideological Constructs of Vaccination

A book by this title was published in 2018 by Dr. Mateja Cernic. Cernic is a Slovenian social scientist who decided to explore the similarities she saw between some aspects of religion and vaccination theory and practice.[25]

I was given a copy of this book about a year later, and the central thesis of the book has stayed with me since. The COVID-19 crisis has only deepened my evolving concerns about the religious-like nature of the various proponents of vaccination, and particularly of mass vaccination programs. These include the stated rationales for school and other mandates to enforce vaccine compliance.

As we saw with the measles vaccine issues of 2019, discussed in Chapter 4, the issue often turns on the notion of herd immunity. Herd immunity, to recapitulate, has been demonstrated in natural infections in animal populations, just not in humans through vaccination. This doesn't mean that it can't be true, but stating it as an established fact is simply wrong: an assumption cannot be considered to be a validated experimental finding. It may be worth mentioning at this point that WHO has now modified their prior definition of herd immunity from one arising from natural infections *and* vaccination to one purely from vaccination. This was, according to reports,[26] in response to the COVID-19 pandemic.

At several points in this book, I have mentioned that the scientific method does not prove anything, but rather deals in probabilities. The stronger the probability, the more likely it is that some scientific hypothesis or theory is probably correct.[27] The level of this probability may be very high, but it can never be 100 percent for the simple reason that to be scientific, any hypothesis or theory must be falsifiable, and all it takes to falsify it is one experiment that does not give the expected answer. In contrast, mathematics does provide proof through a formal process. Religion does the same, not through mathematical and logical means, but through faith.

Faith in the tenets of a religion is a self-contained world view: one accepts the core theses of that religion as innately true, thus not needing any formal proof or research. One cannot prove with the scientific method that Jesus rose from the dead, nor does one need to. If you have been raised in the religion, or even came to it later, it is true because you and your faith community assert that it is true as a foundation on which your belief is anchored.

Cernic describes religion in her book and in a detailed email response to me that is much like the formal definition at the beginning of this chapter:

> The essential characteristic of religion, its essential and fundamental role/ function is **to explain the world**. To explain what the world is, how it functions, what are its rules and laws, what is truth/lie, black/white, food/poison, how we should behave, how we should function, what to expect, how to cope, etc . . . It creates the order out of chaos, gives us maps and directions, categorizes/molds/ organizes the 'great unknown' into something manageable. **This is also the fundamental role/function of science.**[28] [Bolding is hers.]

Cernic goes on to write that, regarding similarities between religion and science, there are a great many:

> Here we can most clearly see how the two are truly just two sides of the same coin (there are also of course many differences between science and religion

as well, but these differences are 'superficial'; in their core, they are more or less the same).

Another of these similarities is that each system wields enormous power over their respective communities, although the relative amount of power has changed. At least in the West, science now holds more clout in shaping how people view themselves, their society, and their future. As Cernic describes, much of this shift arose with historical events, the Renaissance and the Industrial Revolution, for example, not to mention that capitalist economics played a role in shifting the balance between the two. Before, simply stating a doctrine of any religion was often sufficient to sway behavior; now, for many who are less inclined to listen to religion, arguments are framed as being because of "the science."

It's the same behavior.

Cernic wrote in the email: "So, this explanation of the world (and consequently, creation/legitimization of social order) is where religion and science (science as a whole) most closely match."

Cernic also made note of the hierarchies that religion and science both create:

> Both science and religion distinguish between the 'lay' and 'professional/ ordained' public, with the same positions in both. The professional/ordained public is made up of the individuals who have attained the high social position of expert researchers and explicators of the world or truth by fulfilling specific conditions and carrying out specific activities (e.g. many years of education in formal scientific or religious institutions). In this relationship the lay public is distinctly subordinated. It must not doubt the expert knowledge, the claims of the scientist or the priest.

Although Cernic believes that this applies to science generally, the distinction between the laity and the priesthood most clearly finds its full expression in medicine:

> All of the above is even more profoundly, more clearly expressed in the medicine. Medicine became secularized religion in a very literal sense of the word. Domaradzki explains that beautifully: 'Although medicine presents itself as rational, i.e., scientific, objective and neutral, its organization and functioning are typical of religion. Thus, while defining itself as a secular enterprise, medicine is deeply waterlogged with the spirit of the old religion. Even more, for many, medicine becomes a new, secularized religion and takes up its social functions. It is present in people's life from the womb to the tomb, provides

a response to the same fears and angsts of humanity as the Church, and the pursuit of 'eternal' health, youth and beauty has substituted the religious zeal for salvation. Medicine's war on diseases and death is similar to a religious war against sin, as viruses and bacteria have replaced devils and demons, and the structure and functioning of the World Health Organization (WHO) is similar to that of the Church. Physicians have replaced priests and old, religious morality is being substituted by a new moral code: healthism; even though the object of faith and its expression are different, their religious nature persists.'

And,

'Moreover, a physician, like a priest combines the roles of the judge and moralist, ethicist and politician and thus becomes the agent of the State . . . the Medieval principle of Dictatus Papae has changed and now integrates the State with medicine . . . Consequently, a confessional state is replaced by a therapeutic state and medicine becomes a modern inquisition.'

The reader will remember some of the things that I quoted Paul Offit as saying in Chapter 6. From that chapter to here, it becomes obvious that if anyone typifies this mentality it is he, and pretty much everything he writes or says is expected to be taken not only as truth, but absolute truth, as well. Others like Plotkin are much the same. So, for that matter, is Dr. Anthony Fauci in his role of medical overlord of all things COVID-19-like in the United States, regardless of whether or not his statements about the disease or measures to control it on one day radically conflict with things he said a few days previously, or will say a few days in the future.

And, of all the subdisciplines in medicine, vaccinology takes the cake for just how far this field has descended willingly and completely into a role as a new religion for the twenty-first century.

Cernic wrote,

Simply put, one does not question vaccination. Ever, in any way whatsoever. If one does question it, one is sanctioned and expelled immediately. Questioning vaccination is the greatest sin, the greatest taboo in modern medicine. Vaccination is medicine's holy cow and no one can ever doubt it in any way. In a way, vaccination is like a Holy Grail of modern medicine, one of its foundations. If this foundation falls, everything falls (or at least cracks, is at least severely threatened). And when I say "foundation" I don't mean it health wise, I mean from the ideological aspect.

Punishing the Apostates

The mindset of apparently many in the mainstream medical community is pretty clear, namely, that it would be better to censor and self-censor data about vaccine safety that is inconvenient for several reasons. First, to avoid frightening a public perceived as too ignorant to understand the literature, or at least that part which supports vaccination. Second, to avoid the opprobrium of colleagues that far too easily can be career-ending. A perfect case in point was provided in Chapter 6 when I discussed the demonization of even the very pro-vaccine when they do the wrong experiments or even dare to ask the wrong question.

The punishment inflicted on almost complete believers is accomplished through shunning, in removing them from speaking positions, trying to prevent funding for their research, or even getting them fired from their jobs. Of note, for an academic scientist, the loss of funding can lead to the end of their job in due course for faculty members who have only "grant tenure."[29]

The response of some scientists so targeted is to give up, or at least try to return to the fold. But what about the laity who have suffered a crisis of faith, perhaps because of a vaccine injury to a loved one, particularly to a child?

Cernic stated that

> The foundations of one's world, one's core beliefs, shake. This is one of the most profound betrayals a person experiences in today's world. They lied about vaccines . . . vaccines for God's sake! And so, the journey down the rabbit hole begins. What was previously a good, trusting, law-abiding citizen [in regard to vaccination] now becomes a rebel, a fighter, a cynic, a heretic. The enchantment, the veil, the trust, the belief . . . all gone. Because the betrayal was so deep (and so deep was precisely because vaccination was put on the highest pedestal). When person truly researches vaccines, the state/authorities/doctors, etc. lose [the] majority of their power over that person. This person then starts to research, truly research other things. Learns other truths, tears other veils. But usually no other awakening is as brutal as awakening from the vaccine-faith. [Brackets are mine.]

And there we have summarized in a single paragraph how most of those who were once pro-vaccine came to be vaccine resisters. The alliance of much of the medical profession with the State and the pharmaceutical industry has created rebels, not only about the merits of vaccination, but about a range of medical, political, and social issues, as well.

This is clearly what is now occurring in various countries around the world, most notably the United States. This is also why WHO, the Wellcome Foundation, and the Vaccine Confidence Project consider vaccine hesitancy among the greatest current health threats. An example of the latter group's webinars is shown in Appendix 12, as found on dispatchesfromthevaccinewars.com.[30]

This is further why subservient governments try to suppress any vaccine resistance, why they impose mandates, and why they provide civil penalties against those fighting back.

Can the powers that be in medicine, government, and the pharmaceutical industry, not to mention power brokers like Bill Gates, relent? No, they cannot because they are far too invested, literally and figuratively, in keeping the faith intact. Instead, like authoritarian regimes in the political sphere, they seek to crush dissent and rebellion by whatever means necessary.

As Cernic noted,

> So, it is really not surprising that state, system, science, medicine, etc. battle vaccine-hesitancy with such force, with everything at their disposal. This ideological/religious aspect of vaccination is a very important one and is part of the reason why antivaxxer(s) are attacked like rabid dogs (there are others reasons as well, of course, through mass vaccination everything can be injected into general public, anything can be done to the public), but these ideological reasons are in my opinion a very, very important part of the picture. With vaccination our liberties, our freedom, stand or fall. If they can forcibly vaccinate us, they can do anything to us. But this is perhaps a topic for another discussion . . .

One thus sees in Cernic's lucid explanation the fundamental nature of the conflict between the establishment and the growing ranks of rebels. The panic surrounding COVID-19 may dampen this conflict for a time but can only really suppress it finally by taking government and society to a place far beyond what most of us can imagine.

I will return to the above in the final chapter of the book.

A Consideration of Medical Ethics in Light of Vaccine Ideology

From attacking those who critique vaccine safety or effectiveness, to trying to stifle any whiff of apostasy and even shunning their own if not deemed pure enough, the vaccine orthodoxy resembles all other religious orthodoxies. Further, just as in some proselytizing religions,[31] the vaccine orthodoxy is always seeking to expand its hegemony over medical affairs, even if this means trampling on human rights in the process.

This last aspect reaches its apogee in considerations of vaccine mandates, whether aimed at school-aged children, or directed more broadly at the entire population.

To understand the ethical implications of mandates in full, it is necessary to delve into the codified protections that have arisen to prevent unethical experiments, particularly the strict guidelines for informed consent, and see how these might apply to vaccine mandates of any kind.

Guidelines for work on human subjects were initially developed to address the medical experiments conducted by Nazi scientists in the 1930s and '40s.[32] In the context of the United States, the revelations about the so-called Tuskegee syphilis study with its violations of human rights led to vastly more stringent regulations. In brief, the Tuskegee study[33] began in 1932, when the US Public Health Service working with the Tuskegee Institute decided to examine the course of syphilis in black adult males. Some six hundred men participated, of whom almost two-thirds already had the disease. In essence, the study was intended to watch the progression of syphilis over time and thus understand the impacts on various organ systems leading eventually to death.

The failings of the study on ethical grounds consisted of a number of issues, including a lack of informed consent, as the men did not really know what they were being treated for, nor were they allowed effective treatment if they were in the syphilis group.

The study was finally halted in 1973 after media attention led to the (then-) department of Health and Scientific Affairs to appoint an advisory panel to examine the experiments.

The panel concluded that the study was "ethically unjustified." In addition, the panel concluded that a primer on the types of clinical trials that might be used to answer medical questions involving human subjects needed to be introduced.

Oddly, some ethical guidelines for clinical research were already available and had been since the end of World War II.

Of these, one document, the Nuremberg Code of 1947, was already in place. The other four documents came after the Tuskegee study had wound down.

In the following, I will examine these five key documents, although there are various others, and all medical organizations and universities doing any human research have their own variations all largely based on those discussed below. These are:

- the Nuremberg Code of 1947;
- the Belmont Report of 1979;
- the US Common Rule, 1991;
- the Declaration of Helsinki of 2000; and
- the Council for International Organizations of Medical Sciences (CIOMS), International Ethical Guidelines for Biomedical Research Involving Human Subjects, 2002.

These are briefly discussed in the chronological order of their creation.

The Nuremberg Code

The oldest of these modern documents, the Nuremberg Code, arose after World War II and the trials of Nazi war criminals, including those accused of conducting medical experiments in the death camps.[34]

The Nuremberg Code document contains ten items. Of particular relevance to the subject of consent in vaccination are the following items, which I quote verbatim from the source document, as these are key to determining the extent to which such codes are supposed to be followed in medicine:

1. *The voluntary consent of the human subject is absolutely essential.* This means that the person involved should have the legal capacity to give consent; should be so situated as to be able to exercise free power of choice, *without the intervention of any element of force, fraud, deceit, duress, over-reaching, or other ulterior form of constraint or coercion*; and should have sufficient knowledge and comprehension of the elements of the subject matter involved, as to enable him to make an understanding and enlightened decision. This latter element requires that, before the acceptance of an affirmative decision by the experimental subject, there should be made known to him the nature, duration, and purpose of the experiment; the method and means by which it is to be conducted; all inconveniences and hazards reasonably to be expected; and the effects upon his health or person, which may possibly come from is participation in the experiment.

3. The experiment should be so designed and based upon on the results of animal experimentation and a knowledge of the natural history of the disease or other problem under study, that the anticipated results will justify the performance of the experiment.

7. Proper preparations should be made and adequate facilities provided to protect the experimental subject against even *remote possibilities of injury, disability, or death.*

9. During the course of the experiment, the human subject should be at liberty to bring the experiment to an end, if he has reached the physical or mental state, where continuation of the experiment seemed to him to be impossible. [Italics for emphasis are mine.]

Of note, the Universal Declaration of Human Rights, adopted by the General Assembly of the United Nations in 1948,[35] provided a prelude and backstop to the Nuremburg Code. The Universal Declaration of Human Rights also provided the legal basis for the adoption for the later 1966 International Covenant on Civil and Political Rights (1966), adopted by the General Assembly. Article 7 of this document states that

No one shall be subjected to torture or cruel, inhuman, or degrading treatment or punishment. *In particular, no one shall be subjected without his free consent to medical or scientific experimentation.* [Italics are mine.]

Commentary on Item 1: Considering that the first sentence in the Nuremberg Code makes the consent of the subject "absolutely essential," one has to question just what medical practitioners, or governments, think they are complying with in this code when they then browbeat individuals into getting certain vaccines or legislate mandates for vaccination for children (or adults, if this comes to pass) in the first place.

Concerning Item 3, if animal experiments are deemed to be required, we have a major discrepancy: while the vaccine manufacturers have likely done pilot studies in animals, few that I have seen of these would be likely to pass a master's thesis exam, let alone get published in the normal peer-reviewed literature. In other words, the animal experimental work often done by the pharmaceutical industry is often so casually done as to be worthless, and this shows up in the lack of published results in peer-reviewed journals. On top of this, trusting a company, any company, to do honest due diligence on their own products is likely an overly optimistic take on human behavior, especially corporate behavior in a capitalist society. It is of concern, to be discussed in more detail in Chapter 13, that in the search for vaccines against the current COVID-19 pandemic, animal trials have almost been uniformly jettisoned.

In the case of aluminum adjuvants, consider this: all of the agencies I contacted about the safety of these compounds (FDA, CDC, NIH) [see Appendix 8] wrote back to say that they "had no records responsive to your request." From this, should we conclude that the absence of any citation of experimental results means that the medical practitioner delivering the vaccines, or the company making them, is in breach of the Nuremburg Code? Maybe.

The Belmont Report

The Belmont Report[36] builds on the Nuremberg document by restating a number of basic ethical principles, including "Respect for Persons" and "Beneficence," in which "Persons are treated in an ethical manner not only by respecting their decisions and protecting them from harm, but also by making efforts to secure their well-being."[37]

In Belmont, this last is viewed not simply as a guideline, but as an absolute obligation. The last principle is "Justice," which attempts to answer the question of who bears the burden of the research and who receives the benefits.

In a section labeled "Applications," the authors of the Belmont Report consider the practical aspects of the stated principles, these being "informed consent," "comprehension," or the ability of the patient or a caregiver as third party. The last of these applications is "voluntariness," i.e., the consent must be given voluntarily, free from coercion and "undue influence." In this regard, the authors note that "coercion occurs when an overt threat of harm is intentionally presented by one person to another in order to obtain compliance."[38]

And,

> Unjustifiable pressures usually occur when persons in positions of authority or commanding influence-especially where possible sanctions are involved-urge a course of action for a subject. A continuum of such influencing factors exists, however, and it is impossible to state precisely where justifiable persuasion ends and undue influence begins. But undue influence would include actions such as manipulating a person's choice through the controlling influence of a close relative and threatening to withdraw health services to which an individual would otherwise be entitled.[39]

Commentary: The panel distinguishes between research and practice, without clearly noting that practice arises from research, or should do so in evidence-based medicine, and thus does not stand alone. For this reason, behaviors that are unethical in research cannot be magically transformed into being ethical just because the application of the research has moved into human practice. It is in this sense that withholding health care by physicians who won't tolerate the vaccine-hesitant as patients, or the frankly punitive actions taken by governments against those who do not accept vaccine mandates, appear to be in direct violation of Belmont. If various rights are withheld for failing to comply with any mandate, how is this not a violation?

Of note, the Belmont Report has come under significant critique, in particular to the branch of ethics from which it derives, "bioethics."[40] One major critique by Diane Irving is that while Belmont pretends to be about the rights of individuals, it defers these rights to the decision of the State for the "greater good," thus in contradiction to other ethical treatises such as Nuremberg and the Declaration of Helsinki (discussed later). As Irving writes,

> At any rate, after all is said and done, bioethics is ultimately reduced to more or less to some form of utilitarianism or relativism, where 'the good of society' is the morally relevant principle, and 'the good of the individual person' is clearly *not* top priority.[41]

The Common Rule for the Protection of Human Subjects [42]

The National Science Foundation (NSF) in the United States, like other funding organizations, has adopted a series of guidelines, or rules, governing the nature of research with human beings and defining what is allowable and how to determine this status through the review by any Institutional Review Board (IRB). The Common Rule also goes into some detail on the "General requirements for informed consent," covering many of the same key points as the other documents cited previously.

Commentary: Once again, we have a document that makes it quite clear that truly informed consent is paramount before using human subjects in experiments.

World Medical Association Declaration of Helsinki, Ethical Principles for Medical Research Involving Human Subjects[43]

This document mirrors many of the same guidelines and concerns of those previously cited. The Declaration of Helsinki is divided into three sections, and although all of it is relevant to the guidance of human experimentation, the following are most pertinent to the current discussion on vaccine mandates:[44]

Part A, Introduction
Items:

4. *Medical progress is based on research which ultimately must rest in part on experimentation involving human subjects.*
5. In medical research on human subjects, considerations related to *the well-being of the human subject should take precedence over the interests of science and society.*
6. The primary purpose of medical research involving human subjects is to improve prophylactic, diagnostic and therapeutic procedures and the understanding of the aetiology and pathogenesis of disease. Even the best proven prophylactic, diagnostic, and therapeutic *methods must continuously be challenged through research for their effectiveness, efficiency, accessibility and quality.*
9. Research Investigators should be aware of the ethical, legal and regulatory requirements for research on human subjects in their own countries as well as applicable international requirements. *No national ethical, legal or regulatory requirement should be allowed to reduce or eliminate any of the protections for human subjects set forth in this Declaration.* [Italics are mine.]

And from Part B, Basic Principles for All Medical Research:

11. *Medical research involving human subjects must* conform to generally accepted scientific principles, *be based on a thorough knowledge of the scientific literature, other relevant sources of information, and on adequate laboratory and, where appropriate, animal experimentation.*[Italics, mine]

Commentary: The Helsinki Declaration reinforces the documents I have considered so far and makes it crystal clear that advances in medical treatment arise from research and that such research must be based on the best interests of the human subject. My view is that one would seek in vain to find anything in this declaration that would support vaccine mandates without evidence that the mandates themselves were based on definitive science.

CIOMS (2002)[45]

CIOMS is an international nongovernmental organization (NGO) with an official relationship to WHO, as it was founded by the latter and by the United Nations Educational, Scientific and Cultural Organization (UNESCO). This document has come under the same level of criticism by Irving, as cited previously for the Belmont Report, particularly in regard to the level of expertise applied and the precedence of the "greater good" over individual choice and freedom.

The stated initial goal was to bring biomedical research involving human subjects into line with the Declaration of Helsinki with an appreciation of the differing burdens of biomedical research in rich versus poor countries and hence with differing ethical concerns.

The 2002 version of CIOMS followed two previous versions in 1982 and 1993. The intent of these documents was to

> . . . be of use to countries in defining national policies on the ethics of biomedical research involving human subjects, applying ethical standards in local circumstances, and establishing or improving ethical review mechanisms.

This extensive document provides three general ethical principles for research involving human subjects:

> Respect for persons, including respect for autonomy; and protection of persons with impaired or diminished autonomy.
> Beneficence, the 'ethical obligation to maximize benefits and reduce harms.'
> Justice, specifically distributive justice requiring the equitable distribution of burdens and benefits of the research.

There are in total twenty-one guidelines. Most of these resemble those seen in the documents cited above, although often with a lot more detail and commentary. Of these, the key guidelines relevant to our discussion are the following:

4. Individual informed consent;

11. Choice of control in clinical trials;

19. Right of injured subjects to treatment and compensation.

Commentary: Guideline 4 is much the same as the requirement for informed consent seen in the other documents cited. Guideline 11 is interesting in that it may provide the basis for the assertion, sometimes heard in response to calls for a vaccinated versus unvaccinated study in which the latter are essentially control subjects, that to do the study is "unethical." In the guideline in this document, the primary concern is that controls are not treated. Here is the description about what this guideline actually says:

> As a general rule, research subjects in the control group of a trial of a diagnostic, therapeutic, or preventative intervention should receive an established effective intervention. In some circumstances it may be ethically acceptable to use an alternative comparator, such as placebo or 'no treatment.'

Relevant to the issue of what constitutes a control, the document notes that

> A clinical trial cannot be justified ethically unless it is capable of producing scientifically reliable results. When the objective is to establish the effectiveness and safety of an investigational intervention, the use of a placebo control is often much more likely than that of an active control to produce a scientifically reliable result. In many cases the ability of a trial to distinguish effective from ineffective interventions . . . cannot be assured *unless the control is a placebo.* If, however, an effect of using a placebo would be to deprive a subject in the control arm of an established effective intervention, and thereby to expose them to serious harm, particularly if it is irreversible, it would obviously be unethical to use a placebo. [Italics are mine.]

The use of this guideline as the basis to reject such a vaccinated/unvaccinated study was previously discussed in Chapter 2. This will be further explored in more detail next, but it is clear that such a distinction hinges on the fundamental belief that vaccines are vastly more likely to provide benefits than any harms. Given that this is a document written by the medical establishment, such a conclusion was predictable.

The extent to which this latter belief is accurate has been explored in previous chapters, especially Chapters 3 and 5. In addition, this issue will come up in Chapter 13, which considers the impact of the COVID-19 pandemic and the possibility expressed by various medical and political figures that mandating the vaccine(s) may be required.

One key factor to consider here is how the FDA classifies vaccines as "biologics,"[46] that is, a process arising from biological systems. Calling vaccines biologics distinguishes them, at least according to the FDA, from synthetic drugs. If a substance to be tested is considered a biologic, then the regulatory guidelines change accordingly.

Concerning Guideline 19, the consideration is relevant in relation to whether a vaccine mandate is an experiment or not. Insofar as it is, then who is responsible for any harms linked to the vaccines?

All of this raises some basic questions. First, can the State compel one to accept a medical treatment against one's will? If so, what are the justifications for doing so, and how does the State propose to get past the Universal Declaration of Human Rights or past similar documents from various countries that have enshrined the notion of a basic natural right termed "security of the person"?

The short answer is that to compel people to receive some medical treatment against their will, any State has to resort to a "state of exception" doctrine. This latter doctrine will be further considered in Chapter 11.

Next, when is an experiment an experiment, not merely a standard of medical practice, and thus needing clear, informed consent?

There are two clear examples. The first would entail passing a mandate based on the "need" to create herd immunity. As discussed in Chapter 2, vaccine-induced herd immunity has not yet been demonstrated to exist, unlike natural herd immunity. If herd immunity hasn't been clearly demonstrated, any mandate for vaccination against any disease is thus a clear violation of all of the above documents, since no State has the *moral* authority to mandate that people participate in a medical experiment. That they have the *power* to do so under a state of exception is, however, clear.

The current calls by politicians of various stripes, by Bill Gates whom we will meet later on, and by physicians for mandates for any possible COVID-19 vaccine is thus a concern from the perspective of these documents.

Of interest for a possible COVID-19 vaccine is the apparent lack of most safety concerns, especially given the broadly shortened timeline from early phase trials to market. Essentially, people have been asked to take the safety aspect of a vaccine on faith, much like a formal religion may ask individuals to take some aspect of the dogma on faith. Further, the oft-stated view that we will not get "back to normal" until there is a vaccine highlights the religious nature of the entire enterprise, as enterprise indeed it is. "Normal" does not return until we are back in a post-COVID-19 state of grace with a vaccine, one or more, for everyone. The coming of the vaccine for those so inclined to believe was the religious equivalent of waiting for the messiah.

One thing to note here in terms of possible penalties for noncompliance with vaccine mandates is that the penalties might be different from what some might think. A common Facebook comment from vaccine-skeptical people is that the powers that be will (a) pass COVID-19 mandates to be delivered directly into your body by armed agents of the State and (b) these State thugs will deliver the vaccine with or without your consent.

As scary as it sounds having jackbooted men come to your door to force a vaccine on you, this is not likely to be how the whole mandatory enforcement would occur. Rather than a boot on your neck in the middle of the night, the mandate would likely be enforced on the recalcitrant by penalties of various degrees. Do you need to renew your driver's license? How about getting on an airplane or a train? Or how about losing your right to go shopping in your local market? Or what about your pension? Or signing a rental contract or a house purchase agreement? And so on.

In other words, it doesn't have to be a sort of head-slamming "big brother," but only your kindly old Uncle Joe in the United States (or Uncle Justin in Canada) withholding your popsicle until you've cleaned up your toys and done what you have been told to do. Chapter 14 illustrates just how some champions of mandates envision compliance occurring.

How will you be compensated if any adverse reactions occur? You won't, at least in the United States, where the federal government is already promising liability protection for companies making the vaccines.[47] If this is the case, will the government automatically step in and assist you financially, or will you have to fight your way through the Vaccine Court to get recompense? It may be a very safe bet that it will be the latter, particularly as governments in the United States and elsewhere feel the economic bite of the pandemic. (Canada has just promised legislation for compensation for adverse effects to COVID-19 vaccines, but as of this date, details are sketchy.)[48]

As we will see in Chapter 13, mandates for any health procedure for the "greater good," perhaps vaccine mandates especially, carry more than a whiff of what true fascism is actually about. And while definitions of what fascism is abound, that of the Italian dictator Benito Mussolini (who should know) suffices: "Fascism should more appropriately be called Corporatism because it is a merger of state and corporate power."[49]

The religious nature of vaccines comes out yet again in the frequent pronouncements from various sources in politics and media that we are not "getting back to normal" until a vaccine is widely received. Take, for example, the doctrine of the major monotheistic religions that the messiah will be coming when the crises plaguing mankind are severe enough. Now, replace the word messiah with vaccine.

We see this view played out on various stages and in art imitating life imitating art. Consider, for example, the 2011 movie *Contagion*,[50] in which a virus named MEV-1 kills a large part of the population. MEV-1 is much like COVID-19 as a respiratory virus, just a lot more lethal.

In the movie, Matt Damon plays a husband and father caught up in the rolling pandemic after his wife, "patient zero," dies.

The movie manages to reinforce two religious themes. The first is that the only real salvation can come from a vaccine that the CDC miraculously provides at the last minute.

The second theme of the movie seemed to have been intended to provide a lesson in morality: good (CDC) against evil (anti-vaxxers). The evil character was played by Jude Law, named Alan Krumwiede in the movie, an unscrupulous journalist and peddler of fake medicines to control the virus. Law's character disparages the safety of vaccines and manages to get in some lines about conflicts of interest at the CDC in relation to the pharmaceutical industry, comments that are dismissed as just conspiracy bunk by the director of the CDC, played by Laurence Fishburn.

In hindsight, Law's character was more prescient than the movie's screenwriter might have imagined. Interestingly enough, in the movie the CDC's director does not refute the charges, merely kicks them sideways and talks instead about how we all have to pull together to end the pandemic. Art imitating life indeed, just nine years in advance of COVID-19.

The movie is essentially presented as a religious morality play. It features the dangers of hell (the virus) versus trust in the divine (the vaccine) delivered by the messiah (the CDC and authorities), who will save us all from havoc and death if only we believe and resist the temptations of the devilish anti-vaxxers. In the movie, the heroes prevail and the world is returned to "normal," a bit shaken, but normal, thanks to the vaccine.

One more note: the term "social distancing" was used in the movie, something I had not heard before COVID-19.

With Hollywood on their side, can there be any doubt that the CDC, Bill Gates, and the pharmaceutical industry only want what's best for us?

We will see if this is so in Chapter 12.

CHAPTER 9

Attack of the Bloggers

Asking a working writer what he thinks about critics is like asking a lamp-post what it feels about dogs.

—John Osborne[1]

On Critics

The hoary aphorism that "Those who can, do. Those who can't, teach. Those who can't do or teach, criticize" is apt. The quote that opens this chapter pretty much summarizes, in Osborne's acerbic style, how writers and others tend to feel about critics in general.

While I share this view, my take on most critics is that they are more like mosquitoes: annoying bloodsuckers, but hardly a great danger in most circumstances.

Where I see critics having a useful role is when criticisms are actually intended to be constructive, as in good peer reviews of papers or grant applications. Contrast this to reviews designed to tear down others to make the critic look good. The same applies to all types of narcissistic critics, particularly those who engage in the online blood sport known as blogging.

I used to do a fair bit of blogging about political issues back in the early 2000s, back when it was still pretty new as a means of social commentary. Even then, however, it was pretty clear that blogging versus political column writing or opinion pieces in magazines and journals was going to be different. First, the "rules" about what one could write were a lot laxer. For example, in an opinion piece, if you were to make a claim that someone had done something nefarious, an editor would always ask for verification of that statement. A blog, on the other hand, had far fewer checks on validity, particularly if it was your own blog and not subject to editorial scrutiny. Today, this lack of oversight is even worse. Further, back in the day, any commentary by readers on any post had a more reasonable tone of civility, and people rarely used pseudonyms.

None of that is true today. Blogs in the political realm and, as I will show, even on matters of science and medicine have become attack vehicles that offer little in the way of original commentary on any subject, nor verification of any claims provided in the blog.

In addition, bloggers often seem to relish the praise they receive from their followers, as well as often ad hominem comments by the same followers directed at the target of the blog. Some such blogs are essentially ego-driven efforts in which the blogger is playing to an adoring audience by using all of the dog-whistle phrases that his/her followers lap up with their morning coffee.

I suspect we've all seen the decreasing lack of civility in many blogs. Oftentimes, this is also clearly visible in reader responses to articles in online news organizations. All one has to do to verify this is to go onto the Canadian Broadcasting Corporation's (CBC) website,[2] for example, and check any article to see the comments. Many are full of backbiting rage against the author if the respondent didn't like the article or slurs against other commentators if they did. CBC is not unique in this regard, and I expect readers will have their own examples.

The general lack of civility and accountability on social media is now widely obvious to most. Perhaps the lack of civility arises from the blogger him/herself with, in many cases, pretty hostile and uninformed commentary on whatever issue is at hand. And, sadly, this general problem far too often spills over into supposedly "scientific" blogs and commentaries, as well.

One Conversation, the Event and the Outcomes

I mentioned in Chapter 2 the event that billed itself as *One Conversation*,[3] held in Atlanta, Georgia, in October 2018. In brief, Dr. Shannon Kroner and her colleague Britany Valas decided to have a forum on vaccine safety in the broad format of a debate. The key concept was to have opposing views on this topic, including very pro-vaccine advocates and some critical voices, with roughly six on each side. It seemed like a good way to bridge an obvious scientific and social divide in an attempt to find common ground.

Kroner and Valas wanted to make the event as fair as possible and therefore set up some basic ground rules: First, participants should have the necessary credentials and expertise to address the issues and, secondly, would be required to keep the debate civil, no matter how much they might disagree with those on the other side. Kroner and Valas argued that since you could put a debate together with Israelis and Palestinians to discuss that Middle East quagmire, why couldn't you do the same sort of thing with vaccine safety?

Why not, indeed?

Kroner approached me a few months prior to the scheduled date, wanting to know if I would be interested in participating. I said I would be but expressed my doubts that the event would actually come to pass. This skepticism was based on my own failed attempt to do something similar in a lecture series at my university.[4] I told her that in my opinion the pro-vaccine side simply wouldn't show up, giving instead the typical response that to debate "anti-vaxxers" would be to provide false equivalence between "real" science and "pseudoscience." This view holds that featuring both sides in a debate format would somehow imply that the sides were equal in scientific weight and thus potentially equally valid. Since many pro-vaccine people don't see the vaccine-hesitant side as equal by any means, framing it as such would only confuse the general public.

Kroner assured me that I was wrong and that she already had received acceptances from scientists who clearly identified on the pro side. Awesome. With this, I agreed to participate, made some suggestions for others to invite on the skeptical side, and started preparing my talk.

All the while, a fairly well-known blogger, Dr. David Gorski, a.k.a. "Orac" on his blog site[5] knew about the proposed debate, as he had been invited, as well. He declined Kroner's invitation and then wrote a series of blogs about the meeting, effectively drawing attention to an event that otherwise would not have likely had that much publicity.

And then, one by one, the pro-vaccine people started dropping out. I had noted earlier when I looked at their biographies that many were not yet full professors, making them quite vulnerable to pressure from an academic promotions perspective.

The excuses for withdrawing varied from having scheduling problems to fearing reprisals from higher ups, both reasonable concerns that might lead one to not attend, particularly the latter for junior faculty hoping for future promotions.

It turned out that there was another reason that had spooked them that had less to do with not properly checking their daybooks for scheduling conflicts and more to do with secret email lobbying by the same David Gorski, whom we will consider in more detail now.

Dr. David Gorski, Here to Save Us All from Pseudoscience in Medicine

Dr. David Gorski (MD and PhD) is a member of the Barbara Ann Karmanos Cancer Institute specializing in breast cancer surgery. At this institute, Gorski serves as the medical director of the Alexander J. Walt Comprehensive Breast Center. He holds the rank of professor of Surgery and Oncology at the Wayne State University School of Medicine and is on the faculty of the Graduate Program in Cancer Biology. Wayne State is one of the largest medical schools, perhaps the largest, in the United States. It ranks at 351 of 400 universities worldwide in terms of academic status.[6]

Many of Gorski's official associations can be found online.[7]

Gorksi's degrees, in order, are from the University of Michigan with a BSc in chemistry in 1984 and a PhD in cellular physiology from Case Western Reserve University in 1994. In between, he did his MD training at the University of Michigan Medical School, graduating in 1988 and then interning at Case Western University's medical school.

Some of his current official positions include co-medical director, Michigan Breast Oncology Quality Initiative (MiBOQI); cochair, Cancer Committee, Barbara Ann Karmanos Cancer Center; and the American College of Surgeons Committee on Cancer (ACS CoC). His details, as cited in the references provided, note that Gorski is the "managing editor" of the Science-Based Medicine (SBM) weblog,[8] for which he writes, as well. In regard to this, Gorski states that "SBM exists to take a skeptical, science-based view of medicine in general and in particular the infiltration of pseudoscientific practices into medicine, even in academic medical centers."

He doesn't comment here on Respectful Insolence, his other blog.

Gorski has held various grants, and these are listed on the Karmanos website as of 2015.[9] Most of these seem to deal with clinical breast cancer studies. A few of his publications are also listed on the Wayne State website.[10]

In terms of publications, the website lists eight publications from 2015 to 2018. This is a respectable number from someone who is also an active physician, but there is a notable lack of any of these publications with him as first author and only two listing him as the senior author. Both of the latter are on an area of interest for him, that is, the role of the drug Riluzole to inhibit cancer cells.[11]

Going to the scholar.google site fills in more details about Gorski and also raises more questions.[12] The listing of 162 entries is very much a mixed bag, comprised of research articles, abstracts to scientific meetings, patent applications, and, of course, his various opinion pieces and blogs.

The trend noted in the 2015–2018 holds, at least after about 2005, from which point Gorski was most often listed in the middle of a pack of authors. This is precisely the sort of thing that grant review panels note, and tend to worry about. In other words, the first author is usually the person who did the bulk of the work, wrote the paper, and may have been the one who initiated the study in the first place. The senior author is usually the laboratory director who probably had a major role in the study design and likely paid for it out of grant funds. The folks in the middle, well, they did something for the article, but that something is sometimes harder to define precisely.

As for grants Gorski has held since being at Wayne State: of eight grants listed, three are from a research branch of the US Army in the Department of Defense-Congressionally Directed Medical Research Programs. From this program from 2001 to 2017 or so, Gorski captured about $1,261,000. He also held a National Cancer Institute (NIH) grant from 2005 to 2010 for $1,316,000 and one from the American Society for Clinical Oncology for $450,000. There are a few other items in the list, but these tend to be smaller granting organizations, including one from his own clinical center. All in all, he has managed to pull in pretty close to three million dollars over the years.[13] This is a decent amount of money for someone like Gorski, and therefore what is striking is the overall lack of output in terms of actual research publications. Were I a grant reviewer and saw something like this in a curriculum vitae attached to a grant application, I'd be inclined to wonder why this was so. And, although I might not say it in a review panel, I'd probably wonder if the time he spends chasing anti-vaxxers and others who he disagrees with on his blog might be better used in service to his laboratory research projects.

Does any of this mean that Gorksi is a terrible scientist and doctor? Not at all. But it may suggest that his extensive self-aggrandizement might just be overblown, a conclusion I will return to below.

Next question: How is Gorski rated as a physician by his patients? Not great, at least according to outcomes from two evaluation sites. It is important to keep in mind with this, however, that like VAERS, not all patients may have written reviews, and they may be biased for one reason or another.[14]

Overall, at a first glance of the Karmanos and Wayne State websites, Gorski's curriculum vitae appears to be reasonable enough, but the details are

Figure 9.1. Dr. David Gorski, a.k.a. "Orac," author of the blog Respectful Insolence.[15]

revealing if one does a bit more digging. Is he a major figure in oncology, for example? No, apparently not. How about in any other field of medicine? Also, not. In brief, is he as accomplished as his presumed hero Paul Offit? Clearly not. Nor is he even as accomplished as those other scientists or physicians he loves to trash, such as Profs. Chris Exley, Romain Gherardi, or Yehuda Shoenfeld.

In brief, David Gorski is a solid middle-of-the-road researcher at a middle-of-the-road university with a satisfactory level of funding, and a less than satisfactory level of productivity in his field of research. Nothing to be ashamed of for sure; nothing to be so arrogant about, either.

What instead makes Gorski stand out then? Answer: his blogs.

Thus, when not doing his official day job, Gorski is one of the main biomedical bloggers who spends a lot of his free time debunking, in his view, any research that suggests that vaccines might, in any way, have a role in autism or other disorders. However, Gorski is in some ways an "equal opportunity" skeptic going after virtually any branch of what he considers "alternative" medicine, which he variously describes as "woo," "quackery," "pseudoscience," etc. His skepticism, however, does not extend to mainstream articles on vaccination or other branches of allopathic medicine.

According to some of the biographical tidbits he sometimes shares, he first became interested in such topics in 2000, at about the time his own productivity as a basic scientist began to drop off.

Gorksi's blogs combine ridicule for research he considers shoddy by his standards, ridicule that is particularly "insolent" against those scientists (and lay persons) he has decided to include in what he tends to call a "heaping helping of insolence." Ad hominem attacks, the latter almost one of his trademarks, complete the picture.

Gorksi's loyal followers, who mostly use pseudonyms, number at least several dozen on a routine basis. These followers comment on his pronouncements in a manner that might be considered as fawning. It also seems from Gorksi's comments in response that he basks in the adoration he receives.

Gorski gets his nom de guerre, Orac, from the late television show *Blake's 7*, in which the character Orac was a computer.[16]

The name of his blog itself is interesting. *Insolence*, according to the Merriman-Webster dictionary, is defined as: "insultingly contemptuous in speech or conduct: overbearing. 2: exhibiting boldness or effrontery: impudent."

How one gets from this definition to Respectful Insolence is likely part of the secret only he knows, apart from maybe some of his most devoted sycophants. True to the name of the blog and the above definition, Gorski comes across as contemptuous and overbearing. Occasionally, his blogs are humorous, if one likes one's humor spiked with ad hominem slurs, that is. His followers appear to like this very much.

As noted above, Gorski has another blog called Science-Based Medicine. It contains much the same sort of material and level of analysis. And like Respectful Insolence, much of it is ad hominem.

One of Gorski's typical lines of attack on individuals is to fault their lack of formal training or expertise in a particular field if they deign to publish, or speak, in such an area. While on the face of it this might be a reasonable critique, Gorski's critiques themselves are often in areas in which he also has no formal training.

Gorski is frequently wont to cite the Dunning-Kruger effect. The Dunning-Kruger effect is described as one where a person's level of expertise on a given subject is *inversely* proportional to that person's confidence of their *actual* knowledge in that subject.[17]

His own tendency to lapse into Dunning-Kruger is quite apparent in regard to his comments about subjects that, as seen in his training, are outside his own limited areas of expertise. For example, in regard to anything to do with the neurosciences, he seems to have only the level of sophistication in the field that an average student might get in a neuroscience first-year graduate school program. About aluminum in general and as a neurotoxin, he seems to know relatively little, but this hardly stops him from slamming people who do know.

And so it goes. It seems that Orac is engaged in some rather striking demonstrations of the psychological feature of what is termed "projection," that is, attributing to others the same characteristics that they themselves possess.

One particularly funny blog, although not in the way Gorski likely intended, had him mocking Kent Heckenlively, the latter a writer for Age of Autism.[18]

In Heckenlively's piece, the author tried to rally those opposed to vaccine mandates by reminding them of a dramatic scene in the last movie in the *Lord of the Rings* cycle, the *Return of the King*. In this final movie, the hero, Aragorn, rallies his motley army by exhorting them to stand firm and fight against the army of orcs streaming out of the Black Gate:

Hold your ground! Hold your ground!

Sons of Gondor, of Rohan, my brothers, I see in your eyes the same fear that would take the heart of me. A day may come when the courage of men fails, when we forsake our friends and break all bonds of fellowship, but it is not this day.

An hour of wolves and shattered shields, when the age of men comes crashing down, but it is not this day! This day we fight!

By all that you hold dear on this good Earth, I bid you stand, Men of the West!

Aragon's army goes charging into the fray in a glorious effort to win or die to save all of Middle Earth.

It's lovely, heroic, dramatic fantasy stuff, and I would imagine that many of those who saw the movie imagine, in some way, themselves fighting the good fight against apparently insurmountable odds. With his blog, Gorski mocks the idea that Heckenlively projects, that fighting against seemingly impossible odds when you have no choice is not only necessary, but honorable. And yet, it is hard to escape the impression that Gorski gives in all of his blogs that he sees himself like Aragorn at the Black Gate, the lonely hero trying to stem the advance of pseudoscience and medical darkness: Gorksi-Aragorn will save us all, if only we listen to his words of inspiration and reason: Hold your ground, indeed.

The various levels of projection, the blatant hostility he brings to his critiques, and his written verbal "tics" that he repeats frequently remind me of people I have known in various armies in which I have served. In brief, the ones who seem to have had the worst time in their own training due to bullying became the harshest corporals and sergeants. In this, Gorksi seems to fit the mold. If one had to speculate, and that's all it is on my end, I'd guess that Gorski was a nerdy kid who got bullied a lot. And with this, Respectful Insolence became his revenge.

To look into the psychology a bit more professionally with a view to understanding the nature of bloggers like Gorski who circulate in the vaccine world, I looked for an outside opinion. I asked a clinical psychologist I know well to look at some of Gorski's blogs and give me a candid, albeit not necessarily a definitive, opinion about what drives the man. (Note: my speculations were written before I contacted the psychologist.)

Below is what the psychologist, a PhD, wrote in a snap analysis. (It is important to remember in reading this précis that this is not a formal analysis. The latter would normally involve interviews with the subject. Since this was not done, all of the following are merely impressions and opinions by the psychologist):

> Preliminary impression on Gorski: based on his self-description in *Science-Based Medicine*, is wow, is he ever an aggrieved victim. All he does is debunk harmful sham pseudo-science, he only wants to save humanity from evildoers, and he is persecuted, slandered, etc., Joan of Arc comes to mind, or Salem. We should all sympathize with him, the unfairness . . .

In Appendix 11, I provide the more complete analysis in which the psychologist goes on at more length, although acknowledging that this is not a formal report. I should also note that I did not initially direct the psychologist to the blogs where Gorski attacks me, and indeed I am not even sure the psychologist saw any of the latter in forming his initial opinion.

I think the analysis speaks for itself. The reader will decide for him/herself.

It becomes ever clearer that Gorski feels he is a man on a mission to save us all from pseudoscience in general and, most particularly, vaccine pseudoscience.

Gorski is a man who usually blogs five days a week, and not short blogs, either. He has been doing this for years. And he has to be a busy guy with a laboratory to run, grants to write, patients to see and treat, not to mention, presumably, a family to be with.

Gorski is so dedicated that he even found it within his calling to try to derail the planned meeting of pro and skeptical scientists and lay people on the subject of vaccine safety, *One Conversation*.

As mentioned, Gorski blogged about this event a number of times.

One by one, the pro scientists dropped out as the date approached.

Aaron Siri, a lawyer with a New York City practice, submitted a Freedom of Information Act request (FOIA) to the universities of those same scientists to see what had led them to drop out. Surprise: it turned out, at least in part, to be letters from Gorski to each individual that led them to abandon their participation. Gorski made sure to cc Dorit Reiss,[19] also a frequent opponent of anti-vaxxers and one of Gorski's usual acolytes.

One of these letters is shown next. Although it is very long, I quote it verbatim for two reasons: First, for what it reveals; second, because I didn't want Gorski to ever be able to deny his role about what he wrote to the various individuals. As it turns out, he had no desire to hide his role; indeed, he was rather proud of it. Since this is the case, I am only too happy to share it more widely.

Gorski wrote:

From: David Gorski
Sent: Thursday, September 13, 2018
1:21 PM
To: Roe, Catherine
Cc: Dorit Rubinstein Reiss
Subject: One Conversation

Dear Dr. Roe,
I'm writing to you because your name came up as a participant in One Conversation, which is billed on its website as seeking to "break down and clear the barriers of confusion with scientific data, critical thought and engaging conversation" regarding vaccines. This concerns me because you are a legitimate scientific researcher and academic and I want to make sure that you know what you are likely to be in for on this panel.

How do I know? Shannon Kroner and Britney Valas reached out to me a month ago to be on the panel. The reason was that my main "extracurricular activities" involve refuting medical misinformation and combatting quackery on social media. To that end, I edit the Science-Based

Medicine Blog {sciencebasedmedicine.org), write my own blog under a pseudonym {respectfulinsolence.com, although my true identity is a poorly kept secret given that it's on the blog), and engage on Twitter {@gorskon, over 19K followers). I have nearly 20 years of experience combatting antivaccine misinformation, nearly 14 of them running my own blog. After a prolonged back-and-forth email exchange between Ms. Kroner, Ms. Valas, and myself, I politely declined their invitation. The reason was that it quickly became very clear to me in my interactions with the organizers that they are antivaccine activists and that the purpose of One Conversation was not education but propaganda. {For instance, Ms. Kroner has spoken at at [sic] least one antivaccine rally, a rally that Ms. Valas helped organize.)

I described my experience being asked to be part of this panel here: https://respectfulinsolence.com/2018/08/10/shannon-kroner-invited-me-vaccine-panel/.

Upon learning who is going to be on the panel, I wrote a followup post last night: https://respectfulinsolence.com/2018/09/13/one-conversation-medical-authority-antivaccine-trap/.

I'll give you the short version, though, in case you understandably don't want to be bothered reading a few thousand words about this {although, of course, I'd be happy and honored if you would). I'm also including additional links just in case you are interested in more information. Basically, there eight panelists, of whom five can definitely be described as antivaccine. Of the antivaxers, Del Bigtree is probably the most vocal and famous. He is the producer of the antivaccine propaganda "documentary" VAXXED, a film by the guru of the antivaccine movement, Andrew Wakefield. It's a film that I once described as so over- the-top and unsubtle that even Leni Reifenstahl, were she still alive to see it, would say it was too much. Mr. Bigtree is a master propagandist prone to flights of hyperbole. For instance, in a video at the end of the this post, he speaks to a Michigan antivaccine group about how he's willing to fight and die for "vaccine freedom": https://respectfulinsolence.com/2016/10/28/nobody-promotes-antivaccine-nonsense-in-my-statewithout-receiving-some-insolence-2016-election-edition/.

In any event, VAXXED was so full of antivaccine misinformation and conspiracies {particularly the "CDC whistleblower" conspiracy) as to have caused me pain to watch it and try to count and deconstruct all the lies and bits of misinformation: https://sciencebasedmedicine.org/andrew-wakefields-vaxxed-antivaccine-propaganda-at-its-most-pernicious/.

Also, I note that there is no mention of Andrew Wakefield in One Conversation's bio for Mr. Bigtree, and, although I could be wrong, I'd be willing to bet that Kroner and Valas probably didn't mention Wakefield when trying to persuade you to be on the panel and that you probably have no idea just how closely Bigtree and Wakefield work together.

Moving on, two of the physicians on the panel, Drs. Tenpenny and Bark, are "holistic" doctors practicing alternative medicine. They are very antivaccine.

Bark practices naturopathy and homeopathy, while Dr. Tenpenny is a fairly big name in the antivaccine movement. Both of them were featured speakers with Andrew Wakefield on the "Conspira-Sea Cruise," which featured all manner of cranks, from antivaccine cranks, to quacks, to 9/11 Truthers to sovereign citizens:

https://sciencebasedmedicine.org/the-woo-boat-or-how-far-andrew-wakefield-has-fallen/.

If you want a flavor of the sort of nonsense that Dr. Tenpenny believes in, consider that she thinks that vaccines contaminate our DNA in the name of transhumanism:

https://respectfulinsolence.com/2017/04/28/quoth-an-antivaxer-dna-vaccines-are-contaminating-our-dna-in-the-name-of-transhumanism.

Next, Christopher Shaw is a scientist in the Department of Ophthalmology and Visual Sciences at the University of British Columbia, However, over the last few years he's become antivaccine and has become known for doing very bad studies funded by the Dwoskin Foundation, which funds antivaccine studies and activities. His "hypothesis"{if you can call it that) is that aluminum in vaccines causes brain damage through immune activation. He's tried to convince New Zealand authorities that Gardasil killed a young woman and has had at least one paper I know of retracted for image manipulation. I've written about hims [sic] so many times that he has his own tag: https://respectfulinsolence.com/tag/christopher-shaw.

Finally, there is Mahin Khatami. I must admit that when I first perused the list of participants I didn't recognize the name and didn't suspect that Dr. Khatami was antivaccine given her background at the NCI. However, it didn't take me long to find that she's very hostile to conventional medicine, refers to cancer treatment as a scam designed to make money for big pharma, and is very, very anti-HPV vaccine. To give you an idea, one of the books she's written/edited is Cancer Research and Therapy: Scam of Century- Promote Immunity.

I don't know what Dr. Kroner and Ms. Valas told you about the panel, although I can probably make a reasonable guess based on my interactions with her. I also don't know how familiar you are with antivaccine pseudoscience, misinformation, and lies. I, unfortunately, am all too familiar with them. Because I'm so familiar with the tactics and tropes of the antivaccine movement, I hope you will strongly consider what I say when I urge you to back out of this event.The [*sic*] odds are stacked against you and your fellow provaccine advocates Drs. Brown and Stringer, and just by appearing on the same stage with them you will elevate them so that they win, no matter how the panel discussion goes.

Please believe that I am not disparaging you in any way. It has nothing to do with how knowledgeable you are about vaccines, how smart you are, or how good a speaker you are. You might be great on all three scores for all I know. However, it's what you're knowledgeable about that matters. If you aren't familiar with the deceptive tactics of antivaccine activists, they will be able to do what we like to refer to as the Gish gallop, in which they bury you in dubious studies, bogus "criticisms" of studies showing vaccines to be safe and effective, and various other distortions, misinformation, and distractions. If you are not intimately familiar with these tropes, you will almost certainly be overwhelmed and unable to answer.

Then your discomfiture will be prominently featured in selectively- edited videos made by antivaxers. I know that Dr. Kroner promised to provide you with unedited video, but ask yourself this: Are you really going to want to have to use that video to show what really happened? Even those of us who are familiar with antivaccine tactics and misinformation can have difficulty defending against a full-on Gish gallop.

If you decide not to back out and decide to go ahead with this, then let me urge you to do a few things to prepare.

Above all, you need to know your opponents:

1. Peruse the websites of your fellow panelists who have websites.
2. Watch at least one video {preferably more than one video) from each of your fellow panelists and have simple refutations ready to the points they make. Del Bigtree, Toni Bark, and Sherri Tenpenny are very prolific video makers and if you Google their names + YouTube you will find a lot.
3. Watch VAXXED. (It's on Amazon Prime unfortunately. If you don't have Amazon Prime, I bet Dr. Kroner would get Del Bigtree to get you access to a screener if you asked.) Then read my review of it: https://sciencebasedmedicine.org/andrew-wakefields-vaxxed-antivaccine-propaganda-at-its-most-pernicious/.
4. Be prepared for the attempt for the argument that "vaccine choice" = "freedom" and school vaccine mandates = tyranny. You will hear it, because Del Bigtree is on the panel and it's one of his favorites.

5.   Keep your messages very simple, very declarative, and avoid our usual scientist's tendency towards nuance. This isn't a scientific conference. It's propaganda battle. Don't let any of them sidetrack you into the weeds.

Thanks for reading, particularly given how long this email is. I hope you will consider what I have said and withdraw from the panel. If you decide to go ahead, I will help in any way I can; that is, if you decide that you want my help. I can also put you in contact with others who have a lot of experience combatting antivaccine misinformation who could help prepare you, like Prof. Dorit Reiss, whom I've taken the liberty of cc:'ing. Best of luck.

David
Cc: Dorit Rubenstein Reiss, PhD
David H. Gorski, MD, PhD, FACS
Professor and Chief, Breast Surgery Section Michael and Marian Ilitch Department of Surgery Wayne State University School of Medicine Medical Director, Alexander J. Walt Comprehensive Breast Center

Catherine Roe replied:

On Sep 13, 2018, at 5:16 PM, Catherine Roe replied:

Dear Dr. Gorski,

OK, I really feel stupid. I thank you for taking the time to inform me. I was all excited because someone was inviting me to a conference to talk about AD and would pay for it! I certainly don't want to be associated with anti-vaccine people. I emailed them to decline, but haven't heard anything back yet. One of the readers of your blog also altered me to your blog of yesterday. You are right, those people would eat me for lunch. Again, thanks for your warning and for saving me from a really unenjoyable few days.

Best, Cathy Roe
Catherine M. Roe, PhD
Associate Professor of Neurology Washington University School of Medicine

Gorski replied:

From: David Gorski
Subject: Re: One Conversation
Date: September 13, 2018 at 6:49:39 PM EDT
To: "Roe, Catherine"

Dr. Roe,

I almost feel kind of bad now, because I hate to make you disappointed. However, I really did think that you should have "informed consent" about what you were getting into.

Don't feel bad. There's nothing to be embarrassed or ashamed of, and there's no reason to feel stupid. Shannon Kroner and Britney Valas oozed sincerity, and I do believe they are sincere, just incredibly misguided. They really do think that putting several antivaccine cranks on stage with real scientists is "balance." (Of course, they don't view the antivaxers as cranks.) Let's put it this way. I'm naturally suspicious of this sort of thing, and they almost had me convinced that attending might not have been as horrible an idea as I usually think it is.

The problem, of course, is that discussing vaccines with people like Del Bigtree, Sherri Tenpenny, and Toni Bark is a highly specialized skill; even people who make their living researching vaccines aren't necessarily good at it. The reason, of course, is that they really believe their misinformation and they are convinced vaccines are harmful. To convince the audience of that they'll distort, cherry pick, and misrepresent without even realizing they're doing it. If you're not intimately familiar with the sorts of bogus arguments they make, you don't stand a chance. Even I hesitate to go into a situation like that. Of course, the kicker is that Bigtree and I are pretty much enemies, and there was no way they could get me on a stage with him no matter who else was there.

I'm curious, though. How did they pitch this to you? How did they get you interested? Alzheimer's disease, while fascinating, isn't a natural area to be discussing on a panel about vaccines, although there is an antivaccine myth out there that the flu vaccine and "heavy metals" from vaccines cause Alzheimer's.

David
David H. Gorski, MD, PhD, FACS
Professor and Chief, Breast Surgery Section Michael and Marian Ilitch
Department of Surgery Wayne State University School of Medicine
Medical Director, Alexander J. Walt Comprehensive Breast Center Barbara
Ann Karmanos Cancer Institute

Gorski then forwarded the email to Dorit Reiss:

On Sep 13, 2018, at 3:57 PM, David Gorski wrote:

FYI.

David
David H. Gorski, MD, PhD, FACS
Professor and Chief, Breast Surgery Section Michael and Marian Ilitch
Department of Surgery Wayne State University School of Medicine
Medical Director, Alexander J. Walt Comprehensive Breast Center Barbara
Ann Karmanos Cancer Institute

Reiss replied:

On Sep 13, 2018, at 7:00 PM, Reiss, Dorit R. wrote:
I'm sorry she feels embarrassed, but better for her and all around. I think your
answer should help there. Thank you.

Gorski replied:

From: David Gorski
Sent: Friday, September 14, 2018 12:58 AM
To: Reiss, Dorit R.
Subject: Re: One Conversation

She seems OK now.
    I think she feels deceived and used. I can't disagree. She's a neuroscientist
studying Alzheimer's disease. She has no background in vaccines. I bet Chris
Shaw was going to pounce about aluminum as a cause of Alzheimer's.

David
David H. Gorski, MD, PhD, FACS
Professor and Chief, Breast Surgery Section Michael and Marian Ilitch
Department of Surgery Wayne State University School of Medicine
Medical Director, Alexander J. Walt Comprehensive Breast Center

And Reiss concluded:

From: Reiss, Dorit R.
Sent: Thursday, September 13, 2018 11:41 PM
To: David Gorski
Subject: Re: One Conversation

I suspect you're right, and I agree she has cause to feels [sic] deceived and use
[sic].

Dorit Rubinstein Reiss Professor of Law
UC Hastings College of the Law

The letter is quite revealing in a number of ways. First, Gorski assumes that the people he is writing to are like innocent babes, hardly able to protect themselves from the anti-vaxxer orc-like beings that they would confront at a meeting like *One Conversation*. According to Gorski's letter, these are creatures who would seek to confuse the faithful and lead to their destruction, or at the very least the loss of their allopathic medical souls. Only someone well trained in fighting anti-vaxxers would have a chance, someone like Gorski himself. (Of course, Gorski also declined to attend, but that was ostensibly for a different reason, namely, not giving the anti-vaxxers the legitimacy that his presence might have implied. Right.)

And of course, had he been there, we'd all have slunk off like the fakes that we are. Right. It's hard not to imagine that if Gorksi had been at the Black Gate, he would have been running the other way as fast as he could . . .

Gorski even mentions me in this letter, as it appears from the email and from some of his other blogs that I and my former research associate, Dr. Lucija Tomljenovic, have mightily irritated him. Given that others of my colleagues have had the same sort of "insolence," however, it is hard to feel particularly special.

As an aside, I really was looking forward to meeting the various panelists on the other side and had no plans whatsoever to discuss aluminum in Alzheimer's disease, and certainly not in regard to vaccines. So it goes.

Taking the time to write to all of the supposedly pro-people on the proposed panel, and at great length, really shows just how obsessed Gorski is in preserving the true faith and protecting the innocent from Sauron's anti-vaccine embrace. With his blogs and his intervention to keep any broader discussion of vaccine safety from happening, one can't say he is not dedicated, the twenty-first century Aragorn-wannabe, albeit less brave it seems at the Black Gate. There will be some who will think this is a cheap shot, particularly among Gorski's devoted followers. No doubt anyone could be psychoanalyzed at a distance and all of us would find that we fit into one syndrome or other. For example, I'm sure Gorski or one of his followers could look up what I have written and get some clinical psychologist to analyze it. Fair enough. And to anyone who wants to do this, I say, in the words of Epictetus, "If that person really knew me and my flaws, they'd have said something much, much worse."

The difference between a cheap shot and a legitimate exposure is that Gorski, like a politician, has made himself into a public figure. And public figures, particularly those who pontificate and mock others, maybe deserve to get some measure of insolence back.

While writing this book, I decided to follow up on the *One Conversation* pro-vaccine dropout of Dr. Roe. I used the FOIA material as the basis for writing to Dr. Roe in order to give her a fair opportunity to explain her reasons and about the correspondence with Gorski. (The full back-and-forth between the two is in Appendix 14.)

My intent was the same as that of a journalist: I had obtained information that helped me understand what had happened prior to *One Conversation* and wanted to be fair by offering Roe a chance to respond.

I hardly think it was an offensive or threatening letter, and indeed it is pretty standard journalistic practice to give someone in a story a chance to tell their side of it. Of note, the correspondence with Gorski does not match what Roe later told Britney Valas, one of the organizers, when Roe said that she was too busy to attend.

Roe never replied to me, as I suspected she wouldn't. Instead, she rushed to Gorski for comfort. Now, I don't know what transpired in this later correspondence, but I do know that the outcome was that Gorski blogged about me trying to intimidate Roe[20] and proceeded to put the entire email exchange between the two of them online. In his blog,[21] Gorski made the entire story about me trying to get *him* and in the process portrayed Roe as an innocent, helpless babe in the woods, someone totally incapable of taking care of herself in an academic setting. In other words, for his own ego, he tossed her under the proverbial bus.

Before this, I had planned to use the letter in this book, but I was seriously contemplating blacking out Roe's name, as it was not my intention to embarrass her. Rather, I wanted to demonstrate the backroom maneuvers that accompany the suppression of vaccine controversies. Gorski's blog solved this issue for me by revealing all.

I sent the blog to the same clinical psychologist I had used before, and his comments are shown in Appendix 13 along with the first evaluation. If any readers had any doubts about Gorski's mental state before, this last evaluation should put these to rest.

One last note: one would imagine that one of Gorski's heroes is Edward Jenner, the father of vaccinology. And with good cause. Jenner, as described in Chapter 2, was actually quite a caring, decent person, and, for his day, a good physician and scientist. But here is the difference: Jenner, in his writings to colleagues and critics, was ever the gentleman, finding it apparently easy to disagree on scientific matters without insolence to others.[22]

This makes Jenner even more admirable. Gorski should have learned more from Jenner's example.

And Now, the Orac "Wannabes"

There are a number of these, and I will only mention a few of the best-known ones just to round out this part of the chapter.

One of these wannabes calls him/herself "Skeptical Raptor" (SR)[23] and writes often about vaccine-related issues. SR delights in attacking any who dare to question vaccine safety. He/she has been writing since 2012.

Who is he/she really? I don't know, nor do I care. However, SR acknowledges that he/she is neither an MD nor a working scientist, but that his/her background is instead in the pharmaceutical industry in research and development. At least that statement is honest. Skeptical Raptor claims expertise in "immunology, microbiology, cell biology, biochemistry, and evolutionary biology" based on an undergraduate degree in biology from "a US research university" and a biochemistry/endocrinology graduate-level degree from a "major US research university," not specified. Beyond that, how seriously should one trust the judgment of someone out of the very industry that the vaccine skeptics critique? I'll let the reader decide.

Dorit Reiss is another person who comes up in attacks on vaccine critics. I will discuss her role further below in the section on the corruption of the peer-review process.

In brief, Reiss is a professor at the University of California Hastings College of the Law in San Francisco,[24] an institution that appears to carry considerably more weight than, say, Wayne State University, where Gorski works. By a curious coincidence, she graduated from Hebrew University of Jerusalem, my alma mater, where I did my graduate degrees.

Reiss holds the James Edgar Hervy Chair of Litigation and writes quite a bit about legislation and mandates concerning vaccination. She is a frequent commentator to Gorski's Respectful Insolence and somehow seems to pride herself on knowledge of things vaccine-related. Is this accurate? Based on some of her commentary on aluminum in vaccines, I'd have to say no and suspect that Dunning-Kruger is alive and well in her soul.

Then there is David Hawkes, who works for the VCS Foundation, Ltd, in Victoria, Australia, as the director of Molecular Microbiology.[25] He claims to be a molecular virologist with a PhD in human immunodeficiency viruses. His main interest seems to be HPV cervical cancer and screening.

Hawkes prides himself as being an actual academic and has cowritten a few papers critiquing the notion that the ASIA syndrome is a true medical condition.[26] In his spare time, he seems to delight in writing to editors of academic journals to complain about articles he doesn't like. The old army expression of "picking fly shit out of pepper" seems best to summarize his efforts. His various attempts to "debunk" the work of real academics like Drs. Gherardi and Shoenfeld have, in turn, been pretty thoroughly debunked themselves. Dunning-Kruger yet again?

There are other lesser personalities, as well, although it might be hard to be lesser than those cited who claim expertise on vaccine-related subjects.

In Canada, the current media darling, at least for the Canadian Broadcasting Corporation, is the University of Alberta law professor Timothy Caulfield. I will mention him again in a later chapter when I consider the CBC itself on the subject of vaccines.

Caulfield writes about various issues, but a lot of his focus is on vaccines. He is the author of *The Vaccination Picture*,[27] a book that got decidedly mixed reviews: of ten reviews on the Amazon listing (at the time of this writing), five gave him "5" stars; two gave him neutral reviews; and three rated him 0.1 to 1 star.

Of those who loved it, one wrote that

> As a family doctor, I welcome all resources available to help dispel the myths associated around vaccinations. This is a unique book with great pictures, images, and writing that address all of the major so–called controversies around immunization. It's a great book to have in the waiting room for patients to read and consider. Tim Caulfield continues to be a great advocate for science and this book is another successful venture to promote critical thinking and reasoning.

One person who did not like it wrote:

> Total waste of time. I accidentally bought this book as it happened to be in my cart when I went to purchase something else. I wouldn't mind learning more about vaccinations – this is not the type of crap I would expect from a university professor. A bunch of art, stupid photos, all thrown together in dis-organized fashion. He basically just says science is on the side of vaccinations and then ridicules anyone who would dare disagree with science. No proof, no evidence, no explanation of the issue. Complete waste of time and money. I don't even think it is worth giving away.

Two quite different opinions. To see who might be right, I took a look. What I found was that the second reviewer was bang on. In essence, *The Vaccination Picture* is basically a comic book, and not a very good one at that. In its one hundred-plus pages, it features cartoons, drawings, and pictures that look like they were put together by a ten-year-old who learned graphic design on the web. As for the science, it's not in evidence. Rather, Caulfield offers us all a series of platitudes. Indeed, it does not look like the product of a university scholar. Yet once again, with Caulfield, apparently Dunning-Kruger rears its head.

As noted, I will come back to Caulfield in later chapters.

Weaponizing the Peer-Review System

The following section is a close variant of an article I recently published in a new journal that deals with vaccine issues.[28]

A topic often misunderstood by the lay public and, sadly, even those in academia as well as publishing concerns the process by which research findings make their way into publication. It is important to understand this process for two reasons. First, it will make clearer to readers how articles, or even books, get vetted for publication.

Second, in light of various retractions that have tended to hit vaccine-skeptical scientists in recent years, understanding how the process is supposed to work will help understand what happens when the process goes astray.

In science, it is common that studies submitted to a professional journal for publication are first reviewed and critiqued by fellow scientists from the same field, very often even the same subfield. In this process, the editors of journals will, upon receipt of a submitted article, send it for "peer review." Peer review is designed to solicit the frank opinions of the authors' scientific colleagues—peers—about the quality of the proposed article. The peer review considers the manuscript's strengths and flaws across a range of criteria. These include the core concepts, the literature cited, the hypothesis to be tested, the study design, methods to be used, the actual data and analysis of these data, and, not least, the interpretations of these data. Also considered are language use and how the paper itself is organized and presented. Each of these items is critiqued, and good reviewers suggest ways that flaws in any of these aspects can be improved.

The philosophical basis of peer review is thought to go back to the seventeenth century and is credited to Henry Oldenbourg (The Royal Society, 1672).[29]

Typically, at least in the biomedical sciences, editors will seek out multiple reviewers, with two usually being a bare minimum.

In most cases, peer review is "anonymous" in that the reviewers' identities are not revealed to the manuscript's authors. In some cases, the identity of the authors is also not revealed. More recently, some journals have gone to a system in which anonymity is not observed for the reviewers at all.

Overall, the process of peer review is intended to be a collegial exercise designed to improve the quality and presentation of scientific data. Largely, at least for work submitted for publication in professional journals, the role of the peer reviewer is supposed to be voluntary, unpaid, and intended to be fair and unbiased.

Journal editors will typically try to choose reviewers who have demonstrated expertise in a given field and who have no conflicts of interest with the author, either positive or negative. This buffering distance is termed "arm's length" and typically is based on a honor system. Positive conflicts of interest might arise if the reviewer and author are related, or if they have published together in the recent past. Negative conflicts of interest might involve existing animosities or might include extreme differences of viewpoint on a particular subject between reviewer and author.

Editors of journals typically have enormous leeway when considering how to deal with manuscript submissions. First, they can decide whether the subject or presentation of the article fits within the domain of the journal. If they choose to do so, they can return the manuscript unreviewed as being "unsuited to our journal." This is not per se a comment on the quality of the manuscript, merely whether it fits current or ongoing journal needs. For example, high impact journals such as *Nature* or *Science* decline to review over 90 percent of submissions. In such instances, it is usually the case where the editor believes that he/she has recently published articles on the same theme and at that point doesn't need more. Or, a specialist journal may not find the subject matter to be of particular interest to their main readership. As an example, a journal focusing on Alzheimer's disease may not view an article on Lou Gehrig's disease as one fitting within the journal's mandate.

If any of these aspects are answered in the negative, editors can decline to send the article out for peer review.

Typically, it is only after one or more editors have reached the conclusion that an article *might be* acceptable that they do send it off to reviewers. However, even at this stage the editor retains considerable control of the process. Part of this control lies in deciding which potential reviewers will be invited to review the article and what the editor may decide to do with reviews received.

An article that gets only positive reviews will normally be published after minor revisions, if any are requested, and only after these have been addressed by the authors in a modified submission.

An article that gets mixed reviews, however, or one that a particular editor chooses to review in person, allows the editor wide latitude and discretion. If the editor wants to counter reviews already at hand if these have judged the submission to be less than satisfactory for any reason, the editor is free to seek still further opinions from additional reviewers. Or, the editor can reject the submission outright.

By selecting reviewers whose work is known to the editor, it is possible at any stage to turn the process either in a positive or negative direction based on the preference of the editor. Additional reviews are commonly requested when one or more of the initial reviewers declare that they are not able to judge a particular portion of the article, for example with respect to statistical or computational complexities, other methods, or some particularly esoteric theory or algorithm. The editor, however, after some or all of the reviews requested are received, is free to accept the recommendations of reviewers, or to effectively weigh in as the final arbiter for, or against, publication.

Once the foregoing steps are completed, the editor will typically communicate with the authors to notify them of the outcome of the reviews and the recommendations of the reviewers. Common editorial responses are: The manuscript is acceptable as is, acceptable with minor revisions, acceptable with major revisions, or rejected.

If the article falls into any but the last category, the authors are given a set period of time in which to provide the necessary revisions. Revised manuscripts are returned to the editor with notes describing the authors' responses to the reviewers' critiques. In most cases, if these are deemed sufficient, the article is considered "accepted" for publication by the editor.

Often, however, the back-and-forth between reviewers and authors may have multiple phases, all adjudicated by the editor.

In either case, in due course, an accepted article leads to a letter from the editor and a contract that usually transfers all or part of the copyright of the article from the author to the journal. At this point, the article is considered to be "in press," as in essence a contract has been signed between the authors and the journal. Of note, most journals charge a publishing fee that can range from trivial amounts to thousands of dollars.

In due course, galley proofs are sent to the author for a final check of contents and to correct any typographical errors. Once completed, the article is printed in the journal, posted to the journal website, or both. Some journals place accepted articles on their webpage prior to the galley stage and then replace them with the proofed versions when available.

The peer review system is not without flaws, as has been long recognized. For example, reviewers may fail to disclose conflicts of interest or may lack the required expertise to effectively judge a submission, particularly in cases where the submission presents new techniques, novel theoretical arguments, or controversial subject matter, findings, or conclusions.

Sometimes, reviewers commit significant errors in their reviews either by carelessness or ignorance of particular subjects. They may sometimes overlook major flaws in manuscripts they review and accept for publication. This is one reason why there are typically multiple reviewers so that any errors by one are ideally negated by better reviews from the others.

Journals usually have procedures in place to deal with real or potential problems that come to light after an article has already been published. If the errors are correctible, the journal may publish a list of errata (sometimes called a corrigenda), that is, a correction of something not correct in the article. In open-source electronic publications, the journal may publish an entirely new version.

In cases of reasonable controversies, contrasting perspectives, and alternative conclusions to be drawn from published research, a common editorial

Population Vaccination Rate for Measles

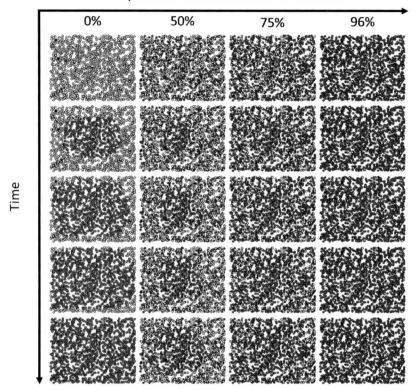

Figure 4.3A. Modeling herd immunity for measles. In each panel of the schematic, the model assumes that all persons are unvaccinated or do not have natural immunity at the beginning of the disease by an infected person (vector) being placed into the population. The model specifications are shown below the panels. Colors indicate vaccine status for the various participants: green, vaccinated; yellow, unvaccinated; purple, the roughly 0.3 percent (whole population) who cannot be vaccinated for whatever reason; red indicates those who have contracted the disease. The initial red point in the first panel on the left indicates the introduction of the initial vector. Moving from left to right, the number of those vaccinated increases from zero to close to the estimated number needed to achieve herd immunity. For simplicity, the model assumes that all induced immunity is vaccine-derived and does not take into account natural disease-induced immunity. Nor does it attempt to compensate for primary vaccine failure.

Population Vaccination Rate for Measles, with increasing secondary vaccine failure

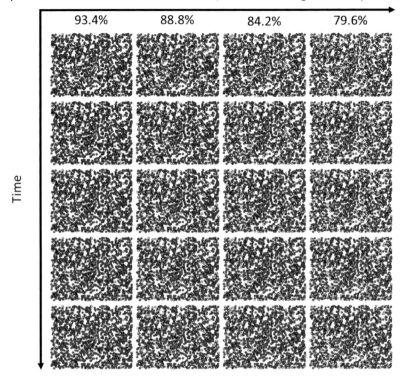

Figure 4.3B. Based on the results of schematic 4.3(a), I now consider the impact of increasing secondary vaccine failure. To do so, various sources have been used to consider the following populations: zero to nineteen years old (approx. 23 percent), and greater than fifty-five years old (21 percent). It is assumed that all of those under nineteen are fully vaccinated; it is also assumed that those over fifty-five have contracted the natural disease and now have natural immunity. Assuming that secondary vaccine failure has not yet occurred in the younger population and that the older population is all naturally immune, we are left with some 46 percent of the overall population in whom secondary vaccine failure can be assumed to exist. These are likely highly conservative values, and any diminution in these values bolsters the case for overall secondary failure. Running left to right, the increase in secondary vaccine failure in the 46 percent of previously vaccinated shows that with between 30 and 40 percent secondary failure in this group, herd immunity largely fails. The extent to which the assumptions stated above are not correct could mean that secondary vaccine failure for measles will occur at even lower values.

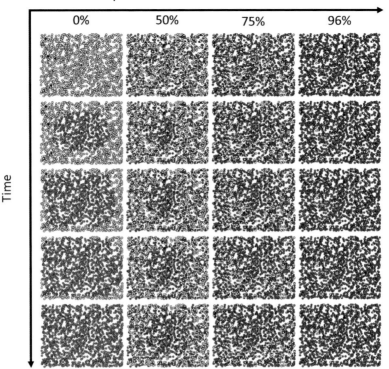

Population Vaccination Rate for Pertussis

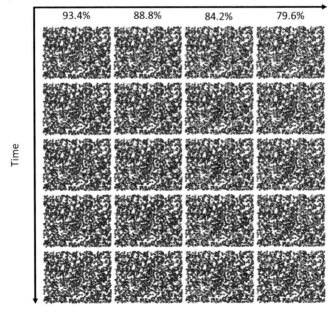

Population Vaccination Rate for Pertussis, with increasing secondary vaccine failure

Figures 4.3C and D. Modeling herd immunity for pertussis. Details as in the first graphs, but adjusted for pertussis.

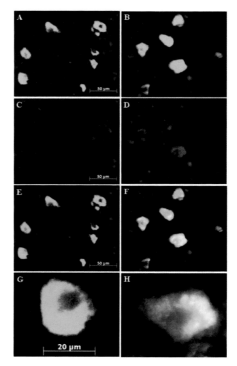

Figure 5.2. Figure and caption from Petrik et al., 2007. NeuN and activated caspase-3 fluorescent labeling in ventral horn of lumbar spinal cord. Green = NeuN; red activated caspase-3; yellow = colocalization of NeuN and activated caspase-3; blue = nuclear DAPI. (A,B) NeuN labeling in control and aluminum hydroxide injected mouse lumbar spinal cord sections, respectively. (C,D) Control and aluminum hydroxide mouse lumbar spinal cord sections labeled with cas-pase-3. (E,F) Merge of NeuN and caspase. Magnification ×40 A–F. White arrow indicates neuron enlarged in (G,H). Enlargement of neurons E,F at ×100 magnification. (I,J) Enlargement of another activated caspase-3 positive motor neuron at ×100 magnification. J, Merged image of activated caspase-3 and NeuN. A–F; Scale bar = 50 μm. G,H; Scale bar = 20 μm. I,J, Scale bar = 10 μm.[1]

1 Petrik, M. S., Wong, M. C., Tabata, R. C., Garry, R. F., and Shaw, C. A., "Aluminum adjuvant linked to Gulf War illness induces motor neuron death in mice." *Neuromolecular Med*, vol. 9, no. 1 (2007): 83–100. PMID: 17114826.

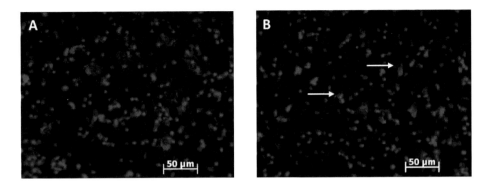

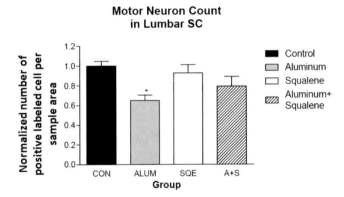

Motor Neuron Count in Lumbar SC

Figures 5.3A-C. Figure and caption from Petrik et al., 2007. Choline acetyltransferase (ChAT) fluorescent labeling in ventral horn of lumbar spinal cord. (A)Con\trol section shows ChAT labeling of motor neurons (×20 magnification). (B)An aluminum-injected animal shows decreased ChAT labeling and abnormal morphology of motor neurons (white arrows) compared with the controls (×20 magnification). Scale bar = 50 μm. (C) Only cells positively labeled with ChAT were counted as motor neurons (n = 32, eight per group). Mice injected with aluminum hydroxide showed a statistically significant decrease in motor neuron number (35 percent) compared with the controls. There was no significant difference in motorneuron counts between all other groups compared with the controls. Data are means ±S.E.M ***p < 0.05 vs control mice using one-way ANOVA.[2]

2 Ibid.

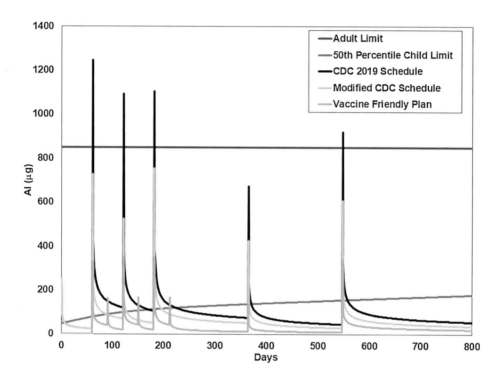

Figure 5.5. Aluminum hydroxide levels from children's' vaccines based on the CDC's recommended immunization schedule (2019). Each spike represents the amount injected at the various time points (x axis) compared to the minimal safe level over time (y axis). Adapted from McFarland et al. (2019)[3]

3 McFarland, G., La Joie, E., Thomas, P., and Lyons-Weiler, J., "Acute exposure and chronic retention of aluminum in three vaccine schedules and effects of genetic and environmental variation." *J Trace Elem Med Biol*, vol. 58, no. (2020): 126444. doi:10.1016/j.jtemb.2019.126444. PMID: 31846784. https://www.sciencedirect.com/science/article/pii/S0946672X19305784.

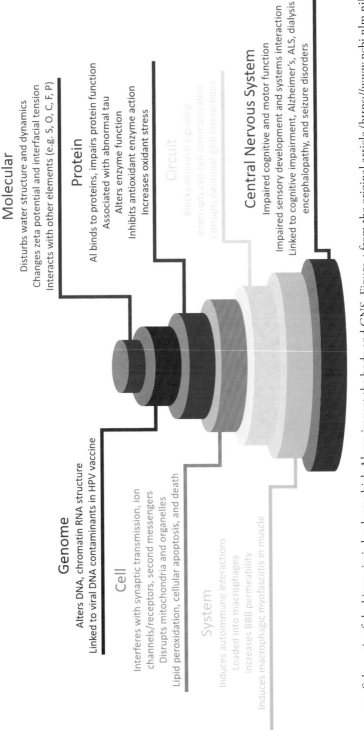

Molecular

Disturbs water structure and dynamics
Changes zeta potential and interfacial tension
Interacts with other elements (e.g. S, O, C, F, P)

Protein

Al binds to proteins, impairs protein function
Associated with abnormal tau
Alters enzyme function
Inhibits antioxidant enzyme action
Increases oxidant stress

Circuit

Blocks neuronal signaling
Interrupts cell-cell communication
Corrupts neuronal-glial interactions

Central Nervous System

Impaired cognitive and motor function
Impaired sensory development and systems interaction
Linked to cognitive impairment, Alzheimer's, ALS, dialysis
encephalopathy, and seizure disorders

Genome

Alters DNA, chromatin RNA structure
Linked to viral DNA contaminants in HPV vaccine

Cell

Interferes with synaptic transmission, ion
channels/receptors, second messengers
Disrupts mitochondria and organelles
Lipid peroxidation, cellular apoptosis, and death

System

Induces autoimmune interactions
Loaded into macrophages
Increases BBB permeability
Induces macrophagic myofasciitis in muscle

Figure 5.7. Schematic of the biosemiotic levels at which Al can impact the body and CNS. Figure 4 from the original article (https://www.ncbi.nlm.nih.gov/pmc/articles/PMC4202242/) has been redrawn.

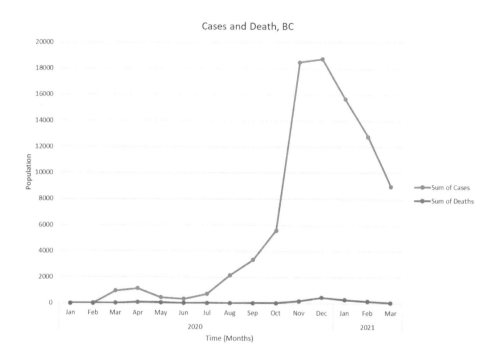

Figure 12.4. Incidence of COVID-19 cases and deaths in British Columbia.[4]

4 https://health-infobase.canada.ca/src/data/covidLive/covid19-download.csv.

approach is to publish "letters to the editor," or one or more "replies" to the published article. Some of these may be solicited by the editor of the journal or may be voluntarily submitted by readers of the journal to be considered for publication, or not, at the pleasure of the editor. In such responses, collegial public debate about methods or interpretations is commonplace.

Traditionally, the article's authors are invited to reply to any critiques. Similarly, it is typical for both the letters to the editor and any authors' replies also to be subject to peer review. In some journals, this back-and-forth can have various stages, enabling the journal's readership to benefit from the academic discussion between scientists.

In cases where reviewers have missed crucial flaws in an article, editors typically retain the right to ask authors to comment on or fix any problems. If these can't be done, the editor may seek a joint "retraction" with the consent of the authors of the article.

In contrast, if the authors don't agree to a retraction, listing an article as retracted basically means that the article is so flawed that it must be removed from the journal. This last is always accompanied by a notice or highly visible stamp saying that it has been retracted and why such an action has been taken.

Conventionally, at least until recently, adequate grounds for retraction included demonstrable plagiarism, duplicate publication of material without attribution or notice to the journal editor, intentional data distortion and/or falsification, or some combination of such infringements.

It should be noted that a published notice of a retraction carries a much heavier pejorative weight than almost any number of published errata or corrigenda. In regard to the latter, authors who find or are advised about errors may correct one or more unintentional errors in any publication with a minimum of public embarrassment.

Although no serious scholar wants to make errors, much less to publish them, such unintentional mistakes can never reach the level of condemnation implied by a forced retraction. The latter is an entirely different matter suggesting deliberate malfeasance of some sort infringing on commonly accepted ethical standards of science. In contrast, the reasons for rejection of manuscripts prior to publication are usually visible only to the journal's editor, possibly a handful of reviewers, and the authors themselves. The public is not privy to the reviews and correspondence leading to the rejection.

A retraction, however, is a public event that inevitably shames all of those involved, especially the author(s), and is likely to have far-reaching negative effects that may be difficult to mitigate and that are virtually impossible to undo. One or more forced retractions can end a promising career.

Further, other scientists will note the published notice of retraction, and the implication that it was grounded in malfeasance follows authors for years and can negatively impact their ability to obtain grants or publish other professional articles. In addition, a forced retraction usually leads to fairly intensive scrutiny by both the authors' home institution and whatever granting agency or agencies may have funded the work.

Some journals are now moving to alternatives to retraction. As cited in an article by Enserink in 2017,[30] there are now options besides outright retraction. These include "retraction and replacement" and "retraction and republication" for articles that may have serious errors, but whose core concepts are still considered worthy of publication. The Stanford University's Meta-Research Innovation Center (METRICS)[31] has devised a more nuanced system to deal with various circumstances and suggests fourteen options. Some of these include "withdrawal," "retired," "canceled," "self-retraction," and "removal," indicating increasing levels of severity based on the perceived or proven errors in an article, with three amendment categories: insubstantial, substantial, and wholesale.

All of the foregoing discussion shows that editors, reviewers, and authors are actively seeking means to devise a system acknowledging that errors can occur and that they range in severity from innocent mistakes to deliberate fraud, with only the latter warranting a forced retraction and the stigma associated with it.

Given reasonable estimates that as much as two-thirds of the published biomedical literature is factually incorrect or misleading,[32] if retraction were indeed necessary for most errors in the published literature, especially of the biomedical sciences, a large portion of that published literature would need to be forcibly removed. However, the bulk of these papers owe their flaws to honest errors in study design, statistics, interpretation, etc. Only a relatively small fraction arises due to actual fraud.

Increasingly, in recent years, however, there has been a trend for the standards of the peer-review process to be violated, not by authors, but by those who have a grudge against the authors, or more often the topic of an article. In some cases, inexperienced journal editors may fail to understand the overall process. In other cases, journal editors may be subject to extreme external pressures to change the basic peer-review process by rereviewing or retracting articles that are already published, and doing so for reasons that would not objectively merit such action.

The following examples discuss some recent failures of the peer-review process and the implications of these failures for fair, unbiased peer review, as well as the broader implications for the independence of scientific inquiry.

Evidence of the misuse of the peer-review process has been growing for years and more recently seems to have accelerated. Often, the retracted articles involved highly contentious subjects such as the safety of genetically modified organisms (GMOs) in food production.

One clear example on this latter topic was the postpublication retraction of an article by Dr. Gilles-Éric Séralini's group on the effects of GMO corn in laboratory rats in the Elsevier journal, *Food and Chemical Toxicity*.[33] The article went on to describe the negative health effects of this diet on the experimental animals. Two months after publication, the article was retracted. *Food and Chemical Toxicity* has ties to the agricultural industry, including Monsanto, then a key developer of GMO crops, and Roundup, a widely used herbicide.

The alleged reasons for the retraction were numerous, including inadequate statistical analyses, small numbers of animals per group, and others. According to Séralini, none of the reasons given for retraction were valid.

While the article may well have had serious flaws, it was the business of the journal's editors and their chosen peer reviewers to determine this *before publication*.

The authors in this case later republished the same article in another journal and successfully sued a European media outlet for suggesting that the data in the article were fraudulent.[34]

My first experience with the same misuse of the process followed the acceptance of an article titled "Behavioral abnormalities in young female mice following administration of aluminum adjuvants and the human papillomavirus (HPV) vaccine Gardasil," by Inbar et al., published in the journal *Vaccine*.[35]

The article detailed results in colony mice receiving the human papillomavirus (HPV) vaccine Gardasil, a vaccine made by Merck.

In due course, the reviews came back. These were largely positive but did require some revisions, which the lead and senior authors provided. On receipt of the revisions, the associate editor accepted the article, which was then posted to *Vaccine*'s website, where it remained for some days before being suspended by the chief editor, Dr. Gregory Poland.

Poland sent the article to three new reviewers for a rereview. These came back within days and were uniformly negative, and on this basis, Poland retracted the article. The authors of the article were not allowed a chance to defend the article from the reviews, and the retraction status remained. Subsequent investigation by the authors revealed that Poland's institution and laboratory had in the past accepted funding from pharmaceutical giant Merck, the company that made the Gardasil vaccine written about in the article.

As in the Séralini case, the article was later republished.[36] However, the damage to the reputations of the various authors had been accomplished.

A more recent example concerned an article that I and some of my colleagues had submitted to the *Indiana University Law Review*. The article dealt with the variation in the assignment of severity of adverse reactions following administration of the human papilloma vaccine (HPV) Gardasil. The comparison was between physicians assigned by the CDC versus independent physicians chosen by the authors. The article simply compared the responses to test cases of the two sets of evaluators and provided a statistical analysis of the difference.

The article was duly accepted for publication to the journal, copyright transfer agreements were signed, and the article was posted to the website of the journal. Several days after this, however, the article was removed from the website, and the authors were informed that the article would be sent for rereview. Inquiries to the journal's editor about the reasons for this decision were not answered. For this reason, we withdrew the article rather than risk the article being retracted.

One of the authors, Mary Holland, filed an FOIA request to the University of Indiana and received the email correspondence between the journal editor, Nicolas Terry, and Dorit Reiss. The email exchange, included in Appendix 11, clearly shows how the normal peer-review process can be corrupted by those seeking to suppress articles that threaten some agenda.

Reiss, as mentioned previously, has a history as a crusader against what she considers vaccine misinformation. However, the problem lies not with her advocacy, but more with the machinations and utter contempt for the peer-review process demonstrated by Terry.

While these examples concern controversial research on GMOs and vaccines, a similar trend leading to attacks on scientific articles from other disciplines has been well documented in the 2002 book *Trust Us, We're Experts*, by Rampton and Stauber.[37] In the cases documented in the book, scientific critiques of various products can lead to attempts to discredit the scientists involved, attempts to cause retractions of their work, and media assaults on their character. Rampton and Stauber highlight examples from critiques of the health consequences of tobacco use, pesticides, the use of lead, biosludge, and climate change. The newer controversies involving GMOs and vaccines merely show that the program described in the book continues into other areas that threaten corporations and that the response to any such threat is now standard and predictable.

The key difference between publishing in the past and in the present is this: in the past, retractions of articles viewed as hostile to industry were infrequent; it now seems that the attempts to force retraction have become the weapon of choice to silence independent scientists. Trolls and bloggers, some of whom may be employed by the relevant industry for this purpose, now actively seek out articles on these topics and almost immediately mobilize attacks through complaints to the publishing journals and to the authors' universities.

The increasing frequency of attempts to force retraction constitutes an abuse of the peer-review process which seems designed solely to eliminate inconvenient articles and authors. In this view, such attacks are less about the accuracy of the literature that the peer-review process was designed to provide than about protecting products and the corporations that make them.

As a collateral side effect, such attacks also threaten the very basis of the scientific peer-review system, which, though sometimes flawed, is still the best system available to evaluate scientific studies.

There are ways to prevent this sort of abuse in the peer-review system from happening. Essentially, these boil down to making sure that editors clearly understand that once an article has been accepted, a contract has essentially been signed. Further, that violations of the contract can have civil consequences, as do any other breaches of contract law.

It is an unfortunate new reality that the solution to ending the weaponization of the peer-review process may lie in the hands of the authors themselves: if authors were to demand a clear statement from editors about retractions, then all parties would be protected. Authors need to be very clear from the beginning of the process that the grounds for retraction are those that have traditionally been observed with deliberate infractions. Second, authors should demand that once accepted, less the traditional infractions, it cannot be sent for rereview. If both were adopted, it seems likely that those seeking to turn the peer-review process into a tool for academic "cleansing" would be forced to find a different avenue for attack.

Considering Sagan's *Demon-Haunted World*

The late astrophysicist Dr. Carl Sagan published a book with coauthor Ann Druyan in 1995 called *The Demon-Haunted World: Science as a Candle in the Dark*,[38] the premise of which was that science is the only force that stands between us and pseudoscience. The goal of the book was to teach lay people the scientific method and to help them learn to think critically and skeptically about science. These are all outstanding goals, just not in the way that Gorski and some of his allies think it should be done.

Sagan and Druyan even provided what they called a "baloney detection kit," designed to help the reader detect real science from fake science-pseudoscience.

The "baloney" detector has twenty items for consideration. As these are lengthy, I put them into Appendix 15, available online, for the reader to peruse.

Also, I will apply this baloney detection kit to some pro-vaccine and supposed anti-vaccine individuals to see where they might actually fit in if one were to score their pronouncements and writings. To do so, I will start with the working hypothesis that this exercise will be a slam dunk for Gorski and his fellow bloggers. Let's see.

1. Wherever possible, there must be independent confirmation of the "facts."

2. Encourage substantive debate on the evidence by knowledgeable proponents of all points of view.

3. Arguments from authority carry little weight—"authorities" have made mistakes in the past. They will do so again in the future. Perhaps a better way to say it is that in science there are no authorities; at most, there are experts.

4. Spin more than one hypothesis. If there's something to be explained, think of all the different ways in which it could be explained. Then think of tests by which you might systematically disprove each of the alternatives. What survives, the hypothesis that resists disproof in this Darwinian selection among "multiple working hypotheses," has a much better chance of being the right answer than if you had simply run with the first idea that caught your fancy.

5. Try not to get overly attached to a hypothesis just because it's yours. It's only a way station in the pursuit of knowledge. Ask yourself why you like the idea. Compare it fairly with the alternatives. See if you can find reasons for rejecting it. If you don't, others will.

6. Quantify. If whatever it is you're explaining has some measure, some numerical quantity attached to it, you'll be much better able to discriminate among competing hypotheses. What is vague and qualitative is open to many explanations. Of course, there are truths to be sought in the many qualitative issues we are obliged to confront, but finding *them* is more challenging.

7. If there's a chain of argument, *every* link in the chain must work (including the premise)—not just most of them.

8. Occam's Razor. This convenient rule-of-thumb urges us when faced with two hypotheses that explain the data *equally well* to choose the simpler.

    Always ask whether the hypothesis can be, at least in principle, falsified. Propositions that are untestable or unfalsifiable are not worth much. Consider the grand idea that our Universe and everything in it is just an elementary particle—an electron, say—in a much bigger Cosmos. But if we can never acquire information from outside our Universe, is not the idea incapable of disproof? You must be able to check assertions out. Inveterate skeptics must be given the chance to follow your reasoning, to duplicate your experiments and see if they get the same result.[39]

Of these criteria, how many would Gorksi violate? Remember, Gorski prides himself on always using the scientific method in his endless fight against the evils of pseudoscience. One would therefore expect that he would have the anti-vaxxers running for cover, much like they would have had he only showed up at *One Conversation*.

Of the twenty Sagan baloney fallacies, I rated Gorski based on his blogs. Doing so, I come in with a score of a minimum of fourteen slices of baloney. This score, however, is generous, as he may also be in violation of items 4, 5, 12, and 15, as well. Only on number 8, the statistics of small numbers, would he possibly not be in error. For this reason alone, his overall violation of Sagan's baloney detector is somewhere between fourteen and nineteen.

How do most so-called anti-vaxxers compare in the same evaluation? My score for them is seven yes, five maybe, for a total of twelve violations.

Nineteen versus twelve leaves little doubt about who is being unscientific. Further, this outcome supports the view held by so many on the vaccine-skeptcial side of the house that as far as actual science goes, they are far ahead, at least of Gorski. Admittedly, that is a low bar to jump.

Sagan ends the chapter with a necessary disclaimer:

> Like all tools, the baloney detection kit can be misused, applied out of context, or even employed as a rote alternative to thinking. But applied judiciously, it can make all the difference in the world—not least in evaluating our own arguments before we present them to others.

It would be difficult to disagree. Indeed, if the pro side would just stow their attitude and accompanying arrogance and seek collaborations with the hesitant side, we might just be able to figure out the vaccine injury issues and make some real progress.

CHAPTER 10

The "Trifecta" of Fear: The Media, the Medical Establishment, and the World Health Organization

We can easily forgive a child who is afraid of the dark; the real tragedy of life is when men are afraid of the light.

—Plato[1]

No power so effectually robs the mind of all its powers of acting and reasoning as fear.

—Edmund Burke[2]

Fear

As I began writing this chapter in the summer of 2020, I was sitting on a BC Ferry Corporation vessel sailing from Victoria to Vancouver. The Ferry Corporation, a private entity that used to be public, had demanded that passengers have face masks and comply with "social distancing." You didn't then actually have to wear your mask unless you couldn't rely on the latter, although as this book goes to press, you do.

The mask wearers on this particular occasion were not confined to any age demographic that I could see: there were the elderly, singly and in pairs, middle-aged folks, and even twentysomethings, teens, and kids, all with masks. Clearly, they seemed to believe something very different about this pandemic than I did. Why was this?

Outside the window where I was sitting was an open deck where people could walk in the sunshine. On that day, it was pretty breezy, which made the fact that most people outside were masked even stranger. I couldn't help but wonder what they thought they were doing, or how they thought the virus could get at them twenty feet from the closest person while standing in a thirty-knot wind.

Figure 10.1. Mask wearing during COVID-19.[3]

The masks were the usual ones that are seen today in our "new normal": some surgical, most cloth, a few fancier ones of various materials. It would probably be the world's safest bet that those wearing masks on the ferry, and elsewhere, had no clear idea about how the masks were supposed to work, what their actual materials' characteristics were, or the possible side effects of wearing them.

Maybe what I saw then, and have seen many times since, was what social media has called "responsible behavior" designed to protect oneself and others. Maybe all of the mask wearers then and now get this and I don't, perhaps making me selfish and irresponsible.

The "selfish and irresponsible" critique is one that I have seen time and time again on social media platforms such as Facebook, almost always from "friends" who are supposed to be left of center. I will come back to that last observation later in the book.

But maybe what I had been seeing—and still am seeing—is not social responsibility based on an actual understanding of COVID-19 transmission, but rather a reflection of some primal fear, a fear approaching hysteria that has little or nothing to do with actually slowing the pandemic.

Did the people on that ferry ride read the scientific literature on COVID-19 and understand viral transmission in general? Did they know which masks work or don't? Do they know the actual, versus possible, health risk they faced if they were to become infected? I think it safe to conclude that few of the masked people on the boat could have answered "yes" to any of the above questions. Instead, they "know" what they know, and what they know comes from what they have heard from the media.

Fear is a natural reaction to threatening situations that sets up a series of biological responses designed to enable us to engage in "fight or flight." In real situations, that is, this is true. But we still don't have such a situation on the ferries. No cougar was prowling the ferry decks ready to pounce, nor was a terrorist squad about to start spraying everyone with bullets.

Here we had, and still have, an invisible entity that may, or may not, be as deadly as the mask wearers seemed to believe.

But fear nevertheless is what we had on that boat and in much of the world. Where does it come from? The short answer is that this sort of fear is learned, not from the actual science on viruses or pandemics, but from three primary sources: the media, the medical establishment, and the big daddy of health organizations that the first two religiously follow: the World Health Organization.

In the pages to follow, I will examine each of these in detail.

The Role of the Mainstream Media in Inducing Fear of the Vaccine-Hesitant and COVID-19

I don't think that many of us would argue that journalism in print or on television today is not of the same caliber as that which our parents or grandparents read or watched. It's not that careless and money-driven journalism didn't exist fifty years ago, rather that at least the lower-grade journalism was more or less compensated for, at least in the United States, by the likes of Walter Cronkite on the nightly news and those reporters who dug into the war in Vietnam such as I. F. Stone and Seymour Hirsh. Or later, in the 1970s, those like Bob Woodward and Carl Bernstein, who took down a corrupt president.[4] Not only was journalism better then in general, but investigative journalism was a real and vibrant part of news reportage.

Agenda-driven news was, however, beginning to come to the fore. I remember being outside of Esteli in northern Nicaragua in 1986 with a group of people who belonged to a mostly American antiwar group, Witness for Peace.[5] I was there as an assistant to a freelance journalist who reported for a radio station in Ithaca, New York. We had wanted to better understand the "Contra" war and had come north from the capital, Managua, to report on the group.

The group had mustered at a small village crossroads just outside of Esteli and were planning to march a few kilometers down a fairly dangerous road to a village that had been recently attacked by the Contras.[6]

As we prepared to march, a squad or so of Sandinista soldiers in jeeps showed up. Their commanding officer offered to accompany the group, noting the danger from Contras still in the area. The Witness for Peace's leaders refused the escort, clearly telling the officer that having government soldiers with them would make it appear that their goal was propaganda for the Sandinista government, rather than trying to protect people in border villages. They argued about it for a few minutes. The officer finally shrugged, told the Witness for Peace people that his soldiers would be fairly close by if needed, wished everyone good luck, and then departed.

As the marchers started their walk, I found myself standing a few meters away from an ABC reporter and cameraman and thus heard what the reporter was saying for his story. Basically, the story that was going to be broadcast back in the United States was that Witness for Peace had demanded that they be protected by the soldiers and that the latter were only too happy to comply for propaganda purposes. In other words, the story being filmed was 180 degrees out of phase from what had actually happened.

My Spanish was not that good and maybe the reporter's Spanish was worse, but to state that the army was going to accompany the group when in fact the soldiers had driven off the other way was a shock and my first real encounter with the corporate media twisting facts to suit some purpose.[7] Not my last such encounter, however.

As the eighties and nineties wore on, investigative journalism in the United States (and Canada) got worse, and truth in the American media went further downhill as described in the book *Into the Buzzsaw* by Kristina Börjesson.[8]

Börjesson included chapters by journalists who had tried to honestly report on a range of issues from Operation Tailwind in Vietnam[9] to bovine growth hormone in milk.[10] In these stories and the others in the book, the clear shift of the media to a more corporately controlled press was obvious, relentless, and sadly prophetic for what the lack of true journalism would look like years later in 2021.

Back in the day, most Americans and Canadians thought they had the freest press in the world. Would many still think so today? It would be pretty hard for any but the most oblivious to deny that currently, true investigative journalism is a rare entity, on life support at best, perhaps already tipped over the brink into extinction at worst.

There are still some holdouts, still trying to be true to their profession. Ben Swann and Sharyl Attkisson, for example, are two such journalists[11] And yet when these journalists try to report on their investigations, too many on the political left, in my experience, denigrate Swann and Attkisson as somehow right-wing ideologues and thus not credible. This dismissal of reporters such as these is based on some perceived place in a narrow one-dimensional political left-right spectrum, not on the evidence of what these reporters are actually saying and verifying to be correct.

The problem does not just exist on the left. Far too many on the right trust news sources such as Fox and *Breitbart,* whose stories are too often slanted heavily in favor of the corporations who pay for the advertisements.[12] And those who pay for the ads are, increasingly, the pharmaceutical industry, as any viewer of American television can readily verify.

The term often used for such media manipulations is "astroturfing," in which some corporate or political player(s) attempt to mold public opinion by pretending to represent grass-roots organizations.

It's Fun for Canadians to Beat Up on the American Media, but Before We Do . . .

The frank corruption of the media is widespread and even encroaches into media outlets that are officially public rather than private. This fact is glaringly obvious in America, leading many people in Canada to smugly laugh at the spectacle south of the border, while believing our own national broadcaster, the Canadian Broadcasting Corporation (CBC), is somehow immune to slanted news and without an agenda. Most Canadians believe this, and I once did, too. But is it true?

Sadly, I've learned over the years that it is not true at all, and although CBC remains a "Crown Corporation"[13] and thus is not subject to the same high-intensity market forces for survival, they often operate as if they were. Increasingly, however, the CBC is being sucked into being a satellite of the corporate world with the creation of new programming they call *Tandem*.[14] Here is what the CBC says about this new programming:

> A true engine of collaboration, Tandem is CBC / Radio-Canada Media Solution's Branded Content service. We combine your brand's subject expertise with our credibility and experience in digital, audio and TV production to create intelligently designed multi-platform Branded Content campaigns, anchored in experiences that leave a measurable impression on our audience.

And just in case you didn't really get what they mean, they describe very clearly what *Tandem* is designed to do:

> Our unique all-Canadian platform of CBC & Radio-Canada properties gives the stories that Tandem makes with you an unparalleled reach.
>
> When developing a campaign, our experts apply research and analysis to find relevant and engaged audiences, and discover where their interests meet your messaging goals. From there, we design content that makes a meaningful connection between your brand and the hearts and minds of Canadians. By working together, we'll propel your message to new heights.

Allow me to translate if this is still not clear enough: We, the CBC, will use our trusted corporation with its reporters and anchor persons to sell you stuff on behalf of corporations, which will pay to use our image as an impartial news source.

So when Canadians say that the US media have surrendered their independence to a corporate agenda, they are correct. But they have no right to feel smug, because we in Canada just did the very same thing.

All of this goes a long way to understanding what has clearly been emerging as a corporate agenda, one all paid for with tax dollars to a national Crown corporation. And it also goes a long way toward explaining the otherwise grossly lazy approach to news that individual CBC reporters have taken in recent years. Now with *Tandem* and, I suspect, years of higher-level CBC planning to become a tool of corporate interests, we can look at the stories that CBC covers and see the fat man behind the curtain.

I'll mention two fairly recent and egregious examples of CBC's emerging corporate friendly agenda as a way to introduce the third. All of these were chosen because they are in relation to topics that I actually know something about.

In each case, the CBC was partially, more often completely, wrong in their fact-checking and interpretations, leaving me with the queasy feeling that if they are so wrong about stuff I do know about, how often are they just as wrong about the rest, where I don't have first-hand knowledge? My gut feeling is that it is quite a bit. *Tandem* is not going to make any of that better.

Example 1: In 2002, Vancouver made a bid for the 2010 Winter Olympic Games. Up to that point, I had no feeling one way or the other about the Olympics or the impact of the Olympics on cities and the environment. A little cursory reading, however, rapidly made me skeptical of the glowing claims that the Vancouver Olympics organizing committee started putting out, claims that were repeated with little curiosity and less evaluation by Vancouver's and Canada's media, including the CBC. In fact, the CBC was often the worst of the lot and pretty much took up the role of an active booster of the Games, both in the bid period and then during the years leading up to the actual Games.

The CBC's outright boosterism came as a surprise to me, as did their inability to read between the lines to see the background or to connect the dots to what was really a giant real estate scam.[15]

CBC's blatant support for Vancouver's Olympics was partially offset by their reporters talking to me or other Olympics opponents, but it was a decidedly lopsided ratio: ten pro Olympic messages would bracket each and every critical statement.[16] Of note, CBC then held, and continues to hold, the Canadian television broadcasting rights for the Olympics, so in this case, their decision can be understood in terms of revenue for the corporation. Much like their American media cousins, the CBC was hardly going to bite the very lucrative hand of the International Olympic Committee.

Example 2: It would be safe to say that CBC's reporters never really understood the Syrian Civil War, nor seemingly did they try to. They continued to miss really fundamental aspects of the story, such as the struggle of the Kurds in the northeastern part of Syria known in Kurdish as Rojava. In Rojava, the Kurds and their Yazidi, Christian, and Arab allies fought a desperate war against the Islamic State and eventually won, at least for a time.

The CBC's inability to tell the story competently became crystal clear to me early in 2018 when the Turkish army invaded the historically Kurdish canton of Afrin. In the course of the invasion, thousands of people from Kurdish and other minority groups such as the Yazidis, soldiers and civilians alike, were wounded or killed, with hundreds of thousands more displaced from their homes.

In May 2018, the Turkish government struck a deal with the Syrian regime of Bashar Assad in which the Syrians would allow "rebel" (read: jihadi) fighters and their families to leave the city of Homs for Turkish-controlled areas of Syria. It was a win-win for the Turks and Syrian regime: The Syrian army was spared having to clear Homs by street-to-street fighting in which they would likely have lost thousands of their own soldiers. The Turks got to resettle the jihadis in Afrin, thus completing the ethnic cleansing of the Kurds.

CBC's reporters clearly understood the first part. They just as clearly did not understand the second part, not that it was hard to figure out. Once again, the possible reasons came down to some preconceived notions or perhaps just to laziness and simply failing to seek the underlying reasons for the transfer of the jihadis to Afrin. The motivation for the Turks was pretty clear to everyone in Rojava, as the CBC would have discovered had they talked to any Kurds, or Yazidis, or Syrian Christians, but they did not appear to have done so. Yes, getting into Rojava was then, as now, difficult and slow, but it was doable, as I well knew because I was there at the time.

Example 3: CBC's position on vaccine safety has never been critical in the least, and their reporters routinely parrot the CDC's or Health Canada's talking points, almost verbatim: Vaccines are safe and effective; the science is settled; we don't know what causes autism, but we can be sure it's not vaccines, etc.

Unlike American TV and radio media, which derive a huge fraction of their revenues from the pharmaceutical industry for the constant advertising of different products, Canada does not routinely allow the pharma to advertise on air, so the profit motive at first glance seems unlikely. Maybe it's simply that the CBC leadership and/or the reporters simply believe the talking points mentioned above, but that should not serve to stifle any critical inquiry. Such inquiry that does exist in CBC's vaccine coverage seems to be restricted to going after vaccine critics, as typified by a story prepared by *Marketplace*.[17] More *Tandem*-inspired "news"? Let's see.

In this segment, reporters for the show went "undercover" to "expose" the anti-vaxxer movement, which they presented under the heavily biased title of "Hidden cameras capture misinformation, fundraising tactics used by anti-vaxx movement." The journalists and their hidden cameras went to an event called Vaccine Injury Epidemic, or VIE, held in Washington, DC, in mid-November

2019. The print version of this segment ran for some seventeen pages, in which the terms "anti-vaxx" or "anti-vaxxer" are used sixteen times.

In journalism, using a term repeatedly to characterize people or events as a certain thing can be an overall sign of what is termed "framing," that is, the intent to present a story in a particular way[18] by using certain words or phrases to create a particular impression. In this case, the repeated use of the term *anti-vaxxer*, as elsewhere in the media and in the mainstream medical community, is intended to convey the message to the audience that anyone so characterized is not to be trusted.

As discussed in earlier chapters, "anti-vaxx" is a pejorative suitcase word intended to convey the notion that the thing or person so described is scientifically illiterate, or worse, willfully distorting the "facts" of vaccines being safe and effective. Ironically, while being undercover and not revealing their identity as reporters to those they interviewed, the same reporters decry the "deceptive tactics" the so-called anti-vaxx movement uses to raise money and spread "misinformation."

The VIE event was described as "ritzy," with the attendees hoping that their pricey tickets would snag them selfies with the big names in the anti-vaxxer hall of fame. The latter, the reporters blithely hint, is all about making the big bucks off the gullible and naive, while driving vaccine hesitancy. And it is dangerous with potentially "deadly consequences," the latter view delivered by a doctor working for Toronto Public Health, Dr. Vinita Dubrey.

The reporters quoted an author, Seth Mnookin, who wrote a book denying any connection between vaccines and autism.[19] Mnookin was quoted as saying this about the anti-vaccine "stars" at VIE: "I think it shows there is an awareness if they try to have this debate in a real way, and try to present the facts and have an honest discussion with parents and with legislators, that their side would lose." And, "They don't have facts on their side, and they don't have science."

As detailed in the last chapter, this would be funny if it were not so absurdly wrong. Mnookin obviously doesn't know, or care, that it is a simple fact that the mainstream medical community continually refuses to have precisely this discussion, taking the view that to do so would somehow "legitimize" the skeptics as having any valid points whatsoever. The cowardly failure of the pro-vaxx team, who had promised to show up at *One Conversation*, as already discussed, illustrates this all too well.

One is left to wonder whether Mnookin would be up for a debate with any of the "anti-vaxx" leaders described in the article. After all, Mnookin did write a book on the topic, so presumably he would be more than able to handle Del Bigtree, or Robert F. Kennedy Jr., also nonscientists. Maybe he could even crush James Lyons-Weiler, PhD, or the notorious Dr. Andrew Wakefield, MD. Easy peasy, right?

So why won't he offer to do so? Maybe for the same reason that the actual pro-side scientists/doctors who were supposed to show up at *One Conversation* didn't: just maybe their "science" is not as settled as they wish it were? And if none of these folks could take on Bigtree and the others, just imagine the utter shellacking they would take if they tried to confront Drs. Romain Gherardi or Yehuda Shoenfeld. I have to admit I'd pay good money to see this play out: Gherardi, ever the Gallic gentleman, would quietly demolish them in any sort of debate; Shoenfeld, a former Israeli paratrooper and a scholar with probably close to two thousand peer-reviewed papers, would have them for breakfast.

One other odd thing about the segment is that it didn't actually feature any scientists able to address the alleged anti-vaxx misinformation.

Instead, in addition to Mnookin, *Marketplace* got quotes from Prof. Timothy Caulfield, a professor from the University of Alberta's Faculty of Law cited in Chapter 8. Caulfield's book *The Vaccine Picture* has made him the Canadian media darling when he tries to dismiss "anti-vaccine misinformation."

Here is what Caulfield's university posts about him:

> Timothy Caulfield is a Canada Research Chair in Health Law and Policy, a Professor in the Faculty of Law and the School of Public Health, and Research Director of the Health Law Institute at the University of Alberta. His inter-disciplinary research on topics like stem cells, genetics, research ethics, the public representations of science and health policy issues has allowed him to publish over 350 academic articles. He has won numerous academic and writing awards and is a Fellow of the Royal Society of Canada and the Canadian Academy of Health Sciences.[20]

This sounds pretty impressive for sure. It would seem a slam dunk that Caulfield could debate the Bigtrees and Kennedys and mop the floor with them. So why doesn't he do so? Unknown, but rather than go mano a mano with them, he provides pithy quotes for *Marketplace*.

Speaking of the attendees at VIE, Caulfield said, "Their goal is to create noise, to create uncertainty. . . . Spreading misinformation will create chaos—and that's exactly what they've done."

What can we derive about Caulfield from this? That he is given to hyperbole? And that he too may not have a clue what he is talking about when it comes to the actual, versus assumed, science?

Since Dr. Andrew Wakefield was at the meeting, *Marketplace* also felt the need to pillory Wakefield further by revisiting the standard tropes.

Wakefield, as detailed in Chapter 4, is the delicensed British doctor whose retracted 1998 *Lancet* case series supposedly spawned the anti-vaxxer movement. Wakefield has long served the media as the ready go-to whipping boy anytime any whiff of vaccine safety concerns arises. This is not new turf for CBC, which drags out the Wakefield bogeyman whenever needed to score cheap points, without ever, as far as I know, actually spending any journalistic effort to determine if the story about Wakefield's alleged scientific fraud is even true. Rather than defame Wakefield directly, *Marketplace* apparently decided to let Caulfield do it for them, a legally safer option for sure.

Speaking about what Wakefield may or may not know about his original claims about autism and the MMR vaccine, as well as the money he has raised to promote this view, Caulfield is quoted as saying:

> If he does know that it is a lie, this is a pretty evil pursuit. . . . If he doesn't know it's a lie, that he's somehow tricked himself into believing this nonsense, it shows an incredible ability to ignore the truth . . . it shows an incredible ability to allow harm to continue despite what the science says.

Brave words, indeed.

So, Prof. Caulfield, how about inviting Wakefield or the others to a debate? A professor of law with (allegedly) 350 published papers should chop up all those anti-vaxx demons into mincemeat. Or maybe not. Maybe what the "science" says is not what Caulfield thinks it says, but then he seems to be basing his views on the comic book he published.

And one last note: As Caulfield is the law professor, not I, were I a lawyer, I might just find in comments such as those above something that might wind up in court. And a word to the wise: Dr. Wakefield now seems to have access to considerable money. For this reason, if I were Caulfield, I'd moderate some of my comments a bit more stringently in the future.

Finally, linked to the broadcast is a segment in which Asha Tomlinson, the lead reporter on the segment; Katie Pedersen, one of *Marketplace's* producers; and Dr. Dubrey of Toronto Public Health all have a nice chitchat about the segment and the safety and effectiveness of vaccines in general.

In order to show that they are fully on top of the issues, they bring up concerns about some vaccine ingredients: mercury in the form of ethyl mercury (Thimerisol, as discussed in Chapter 5), aluminum adjuvants (also from Chapter 5), and formaldehyde, the latter used in the preparation of some vaccines to kill the viruses that some vaccines are made from.

Tomlinson asks about these three ingredients, and Pedersen blithely writes them all off as no reason for concern, at the same time mouthing rather basic errors about the actual science. In particular, aluminum is described as not an issue, since, according to her, it is flushed out of the body within hours. The very fact that this is not true (see Chapter 5) and easily demonstrable that it is not true points to either supreme ignorance of the scientific literature, or to a CBC agenda, or both. I strongly suspect it is both.

So, on three issues—the Vancouver Olympics as a moneymaker for local developers and the International Olympic Committee, Afrin in Syria and what really happened there, and vaccine safety—CBC fails across the board, a series of failures so pronounced that one really does have to wonder how mistaken, misinformed, or agenda-driven they are about *any* issue.

So much for honest journalism in Canada. Fox News move over, there is a new kid in town and one just as likely to sling canned talking points to promote the frame du jour.

In regard to COVID-19: CBC clearly relished the emerging COVID-19 story, shifting their coverage to all things COVID-related. A joke going around in my social circles was that CBC is now an acronym for "COVID Broadcasting Corporation," with extensive and endless coverage on the pandemic and virtually nothing on the wars in Syria or Yemen, wars that mysteriously seem to have vanished.

A more recent CBC trick is to ignore any news that doesn't fit their new shiny corporate-friendly agenda: in August 2020, a huge antilockdown rally was held in Berlin. According to the organizers and verified by drone images, the crowd was estimated at five hundred thousand. Robert Kennedy, Jr. gave a rousing speak in which he evoked his uncle's famous "Ich bin ein Berliner" statement.[21] How did the CBC report this? They didn't. It never happened, as far as Canadians who rely on the CBC for their news would discover.

For the CBC, as well as for many in the mainstream medical community, COVID-19 and the resulting pandemic could only end with a return to "normal" when a vaccine was found. The utter lack of critical thinking involved in simply parroting such a statement highlights the depths to which journalism, at least in Canada, has fallen. In essence, it evokes the magical unicorn notion of vaccine-induced herd immunity along with the prevalent, but incorrect, notion that a vaccine is the solution to any and all health issues related to infectious disease. And in this regard, it would be absolutely essential for all people to get the vaccine(s) when one came about. To hammer this point home, the CBC made certain to castigate a Manitoba education minister who had dared to suggest that taking a vaccine or not should be an individual's choice.[22]

I learned in my advanced-level army courses that one had to ask the "so what?" questions.[23] For example, if you receive orders to capture a bridge, you have to ask yourself the "so what" questions that define how you will do it: Capture a bridge? So what? So I probably need infantry and combat engineers. So what? How am I going to gather these assets and get them there? So what? I need transportation. Is this going to be our own vehicles, or do I need to borrow them from somewhere? Etc., and on and on it goes until you have answered all the questions that allow you to carry out your mission. Thus, when the CBC or any news organization simply repeats the official mantra without asking the "so what?" questions, they have not done their homework and have not considered what the impact of some future vaccine might be, whether it will be truly safe and, if not, how might the adverse effects compare to the damage the disease is doing?

This approach is unprofessional and, sadly, the state of much of media, at least here in Canada. Welcome to *Tandem*, indeed.

I'll revisit the CBC and their lack of journalistic competence and integrity later in the book when I discuss COVID-19 in more detail.

Let me finish this section with another quote from Dr. Mateja Cernic on the role of the media in reinforcing the religious doctrine of vaccinology and the deliberate attempts to paint those who dissent as enemies of the State:

> Roughly, the media representations of vaccination's critics could be divided into themed groups (irrationality, irresponsibility and danger), which interrelate and this interrelation reveals the most obviously the ideological struggles that are waged in the field of vaccination.[24]

The Mainstream Medical Establishment

The mainstream medical establishment has largely fallen into line on COVID-19, just as it has on most things vaccine-related. This is not all that surprising when we consider the essentially religious nature of vaccine belief systems, much of which has spilled over into considerations about COVID-19, including the subjects of masks, vaccines such as the experimental mRNA and viral vector vaccines, and the pros and cons of using antiviral medicines such as hydroxychloroquine.[25]

Much of the medical establishment pirouettes in tune to the pronouncements of the CDC and WHO, turning almost effortlessly to embrace recommendations by these überagencies that seem to change from day to day. Were the consequences not so far-reaching and even dire for people around the world, it would have been as entertaining as watching a politician or medical bureaucrat say one thing one day and deny saying it the next morning. And, of course, as per the discussion in Chapter 8, medical doctors who do not toe the line are almost instantly punished by their peers.

The problem here is really twofold. First, a lot of physicians really don't remember all that well the minuscule amount of vaccine knowledge they learned in medical school, as discussed in Chapter 1. And because they don't, or don't seem willing to revisit or challenge received wisdom, they go along with "groupthink" on the issues of vaccines and COVID-19. Again, this is purely religious behavior whether motivated by fear of censure, pride in their typically unchallenged status as health experts, or maybe just intellectual laziness.

In Canada, the medical establishment nationally and in my province of British Columbia fell into the predictable clichés about vaccines and COVID-19. The media made absolutely no effort to add any critical voices to either subject. Although I don't really follow what happens in other provinces all that closely, it seems likely that it was the same across Canada.

Together, the medical talking heads combined with a media focused on getting the stories out their way simply reinforced each other in a kind of positive feedback loop, making doubly sure that no critical voices were heard.

When the resulting message was for people to panic, panic most did, aided in many cases by fellow citizens who were only too happy to reinforce a lockdown of dissenting voices. That such censorship often came from those self-describing as left of center is something I will address in later chapters.

Here in British Columbia, we were subjected to daily media briefings by our provincial health officer, Dr. Bonnie Henry, whom we will meet in Chapter 13, and Health Minister Adrian Dix. Dix might be better described as a New Democratic Party[26] "apparatchik," not known for particular abilities on any file he has held. At least he seemed to understand his shortcomings as health minister and wisely let Henry run the briefings.

In this, Henry did a pretty good job insofar as delivery went. A lot of people praised her for this and felt reassured by her apparent calm in the face of the pandemic. No one seemed to critique her occasional lack of information or the accuracy of this information, certainly not the media.[27]

Henry has been known to advocate for decriminalizing drugs in general, a wise position, and being concerned for health care among the homeless, also a decent position. Given this, it is hard to understand how she was seemingly blindsided by the nearly 300 percent increase in drug overdoses in the months of the pandemic, the skyrocketing increase in spousal and child abuse, and other entirely predictable outcomes to the ensuing restrictions that the provincial government put in place as COVID-19 panic set in.[28]

During the summer of 2020, it seemed that British Columbia, per capita, was getting off very gently from the COVID-19 pandemic. Then as fall became winter, we were immersed in a "second wave," at least in part driven by PCR testing that seems likely to have been done incorrectly (see Chapter 13).

Henry, widely praised before, began to take some flack for her endlessly wavering statements and for some frankly outrageous comments on so-called "glory holes" for sex and the advice not to leave cookies and milk out for Santa Claus.

All ridicule aside, much as it was deserved, there is still a major question not being asked by the media or the public, namely, what is actually the role of a provincial health officer? There are really two choices. First, to react to emerging situations, manage the panic your own lack of planning generated, and keep playing catch-up during whatever medical crisis may arise.

Or, do what senior staff in most armies and corporations do: war-gaming in the first place, and SWOT (strengths, weaknesses, opportunities, threat) analysis in the second. These are essentially the same processes.

War-gaming is the process by which a staff, usually headed in this exercise by the intelligence officer, considers the operation they have been tasked with, enumerates the resources and history of the enemy, and looks into the likely moves and countermoves that can be expected to occur. For example, if I move my forces here, what is the enemy likely to do? In particular, what have they usually done in the same circumstances in the past, and which of their countermoves would be the most harmful to me now? Then the intelligence officer offers different response options to his commander to counter all the things the enemy *might* do.

Where did that happen before COVID-19 came barreling out of China? Nowhere in British Columbia or Canada, the United States, Europe, or pretty much anywhere else apparently? In other words, most health authorities worldwide did what no intelligence officer would dare fail to do. In short, they didn't prepare for all possible scenarios and were thus left with the sole option of reacting blindly without all of the needed information.

This outcome is well known, and this is precisely why armies, at least in the West, teach war-gaming. Indeed, war-gaming in both conflict and other disaster planning has been commented on since antiquity. Seneca wrote that "Nothing ought to be unexpected by us. Our minds should be sent forward in advance to meet all problems, and we should consider, not what is wont to happen, but what can happen."[29]

The question posed remains. Where were all of these health officials' contingency planning for accurate tests for infection; essential equipment, including ventilators (not actually such a good idea, as we've since discovered); other medications, etc.; planning for the societal and longer-term health effects of lockdowns; and so on? The answer is nowhere. No health agency to my knowledge actually did the war-gaming in advance and, from this, developed functional contingency plans.

Well, maybe there was one organization that did the prior planning and even held a conference on a scenario involving a coronavirus pandemic. This was in mid-October, 2019, in New York City. The Johns Hopkins Center for Health Security was the host of the event called "Event 201." The major donor was the Bill and Melinda Gates Foundation.[30]

It would be hard to find a more prophetic document, one that reads almost like a movie screenplay for a disaster movie about a pandemic.

The bottom line: planning *was* done, just not by the people who would have to respond to COVID-19.

As for the people who planned the scenario and maybe more, we will meet them again in Chapters 13 and 14.

Who Is the WHO, and What Do They Do?

The World Health Organization, WHO, was founded in 1948, as a specialized agency of the United Nations. It is responsible for international public health. Its mandate states that

> The WHO's broad mandate includes advocating for universal healthcare, monitoring public health risks, coordinating responses to health emergencies, and promoting human health and well being.[31]

Further,

> It provides technical assistance to countries, sets international health standards and guidelines, and collects data on global health issues through the World Health Survey. Its flagship publication, the World Health Report, provides expert assessments of global health topics and health statistics on all nations.[32]

The WHO is led by a director-general (DGWHO), who is appointed by and responsible to the World Health Assembly (WHA). The current director-general is Tedros Adhanom Ghebreyesus, who was appointed on July 1, 2017. Previously, Ghebreyesus was the health minister and later the foreign minister of Ethiopia, described by the *Globe and Mail* in a recent article as both an authoritarian and a surveillance state.[33]

Figure 10.2. Tedros Adhanom Ghebreyesus, director-general of the WHO.[34]

Ghebreyesus is neither a physician nor an infectious disease expert. To make up for these gaps in his training, he seems to have acquired the backing of Bill Gates before his election to the WHO directorship.

There are two sources for this allegation, the first from an article in the magazine *Panam Post*[35] by Emmanuel Rincón. In the article, Rincón writes that

> A 2017 article in *Politico*, published before the election of the new director of the WHO, indicates that a French diplomat suggested that Gates supported Tedros and that Gates had also financed health programs in Ethiopia when Tedros was the Health Minister. This raises some suspicions since Tedros was accused of covering up cholera outbreaks in his country during his time as Health Minister. Ultimately, despite all this, the preferred candidate of Gates, the African Union, and also China, was the winner of the race.

The original 2017 article by Natalie Huet and Carmen Paun making that allegation, also in *Politico*, said this: "Last year, a French diplomat suggested that Gates also supports Tedros, having funded health programs in his country when he was health minister."[36]

Huet and Paun go on to write that

> Some billionaires are satisfied with buying themselves an island. Bill Gates got a United Nations health agency in Geneva.
>
> Over the past decade, the world's richest man has become the World Health Organization's second biggest donor, second only to the United States and just above the United Kingdom. This largesse gives him outsized influence over its agenda . . .

And,

> The result, say his critics, is that Gates' priorities have become the WHO's.

Ghebreyesus beat out two competitors for the directorship of WHO. How much had Gates given to help move his chosen candidate along? Heut and Paun cited a figure of $2.4 billion from 2000 to 2017, or about $133 million/year. For an organization with an annual operating budget in 2017 of $4.5 billion, this represents about 3 percent over these years.

Not a vast percentage, but one that almost certainly buys a lot of leverage for an individual like Gates. And the payments and leverage keep growing. Gates Foundation donations in 2018 were $228,970,196; in 2019, $226,337,804, or about 5 percent; in the first two quarters of 2020, $613,779,000, or almost 14 percent if the WHO budget is similar to that in 2017.[37] As the numbers grow, so too does Gates's influence over the organization, almost to the point where his agenda becomes the WHO's agenda, as we will see later.

True to his authoritarian roots in the Ethiopian government, it was Ghebreyesus who accepted as credible the initial reports about the emerging COVID-19 pandemic put out by the Chinese authorities. It was also Ghebreyesus who played a major role in halting worldwide clinical studies on the drug hydroxychloroquine to treat COVID-19 based on two papers that claimed the drug was dangerous. The two papers on which this determination rested were soon retracted, as the data sources could not be confirmed.[38]

The WHO has long had a habit of declaring pandemics during infectious disease outbreaks. They did so with the severe acute respiratory syndrome (SARS) event in 2003 when they suggested restricting travel. SARs died out in 2004 and never really spread beyond Asia. Next up was the 2009 H1N1 swine flu pandemic, for which the WHO received mixed reviews, some noting mixed messaging about the severity of the disease. The WHO's response to the Ebola outbreak of 2014 in West Africa was deemed slow and ineffective. These earlier criticisms have mirrored some current ones about COVID-19.[39] In addition, critics have noted financial ties to the pharmaceutical industry and extensive links to the Bill and Melinda Gates Foundation.

One thing the WHO seems to have been involved in over the years has to do with testing a DpT vaccine aimed at preventing neonatal tetanus that seems in some cases to contain human chorionic gonadotropin (hCG).

In 2017, I was a coauthor on an article titled "HCG Found in WHO Tetanus Vaccine in Kenya Raises Concern in the Developing World."[40] The lead author was a long-time colleague, Prof. John Oller, a distinguished scholar specializing in signaling theory as applied to biological systems (see Chapter 5). Oller had originally been contacted by several physicians from Kenya who were concerned that a push for neonatal tetanus vaccination in Kenya was being used as a smoke screen for birth control. Oller investigated these concerns and found that the parent body for this vaccine push was the WHO. The more he and the rest of the coauthors examined the WHO and their previous campaigns for tetanus, we all came to believe that there was another agenda at play apart from preventing tetanus in newborns.

Oller's research, documented in the article, showed that there had been various studies using the tetanus vaccine as a carrier for hCG. In humans, hCG naturally diminishes during pregnancy, but if it is given to humans or animals during the earliest stages, miscarriages result. If given prior to pregnancy, conception may not occur. The scientific studies published by Talwar and colleagues (cited in Oller et al.) had suggested the conjugate vaccine of tetanus and hCG as one way to both prevent tetanus, but also to diminish fertility in the poorer countries of the world.

The Kenyan doctors, who had brought the issue to our attention, were part of a Catholic doctors group and saw in the Kenya campaign an attempt to decrease fertility in the country. While the goal of preventing neonatal tetanus is certainly laudable, any attempt to decrease fertility would not be as laudable unless the women vaccinated had given consent to receiving the vaccine. This was clearly not the case.

Some of the circumstances of delivering the vaccines to rural clinics were also quite suspect, as detailed in the article. During the vaccination campaign, the doctors had obtained some of the vials of the vaccine batches used and had them analyzed in various independent laboratories for hCG. The results from a number of these analyses came back positive for the presence of hCG, where none should have been.

Oller et al. concluded that either one of four possibilities existed: 1. The chemical analyses were in error, but this possibility was made unlikely by similar outcomes from several laboratories; 2. The doctors who had raised the alarm, or others who might have had access to the vials, had tampered with them by adding hCG. While possible, two factors mitigated against this interpretation: First, the tetanus toxin was conjugated to the hCG in a manner that was inconsistent with someone simply injecting hCG into the vials. Second, the doctors swore in an affidavit that they had not tampered with the vials in any way; 3. The hCG was indeed there but arose from contamination at the manufacturing site; 4. The hCG in the vials was deliberate and arose from trying to use a tetanus campaign as a sterilization campaign in disguise.

In the end, we settled on number four as the most likely scenario based on all of the circumstances and evidence.

The article, in a rather minor journal, attracted much attention and was rapidly attacked by trolls and bloggers. We have recently published a detailed rebuttal to the various critiques in a new journal.[41]

The above suggests that all is not straightforward with the WHO, which seems to have a number of underlying agendas in play. In this regard, it is worthwhile noting that it was not Ghebreyesus who set the WHO on a trajectory for greater collaboration and reliance on the pharmaceutical industry. That distinction likely belongs to Dr. Gro Harlem Brundtland, who was WHO director-general from 1998–2003, a period in which the pharmaceutical industry began to make serious inroads into the organization;[42] critiques of Brundtland's cozy relationship with the pharmaceutical industry can be found in an article by C. Raghavan in 2001.[43]

I will go into the ties between the WHO and the pharmaceutical industry, along with the same ties to the Bill and Melinda Gates Foundation, in Chapter 12.

But for the rest of this section, I want to focus on several documents put out by the WHO under the directorship of Ghebreyesus that may indicate the scope of their plans for the future of world health with a clearly stated agenda that seems to be almost entirely focused on vaccine development and delivery to the entire population of the world. That this goal matches almost perfectly that of Bill Gates is not likely to be a simple coincidence.

The first item is from the WHO's *Weekly Epidemiological Record*, a WHO in-house journal, which is not peer-reviewed. In this article, the authors, all of whom work for the WHO or the CDC, provide some curious calculations for measles and measles fatalities worldwide.[44]

The authors' claim is that in the period of 2000–2018, measles vaccine coverage went globally from 72 percent to 86 percent and that in the same time period measles decreased by 66 percent from 145 million to 49 million. Based on these numbers and their algorithm, which takes reported measles cases and adjusts for what they think the real number should be, the authors estimated the number of saved people who would otherwise have died without the vaccine at 23.2 million (see their Table 2 from the cited document).

The authors make no corrections for anything else that might have changed in the world that might impact measles fatalities, for example water supplies, overall hygiene, nutrition, etc. So, to accept the 23.2 million lives saved by the vaccine, one has to also accept the notion that no other factors were in play during this period, a rather untenable assumption that clearly will not stand up to a detailed analysis.

The authors do acknowledge that there just might be some caveats to their conclusions:

> The findings reported here are subject to at least 2 limitations. First, large differences between estimated and reported incidence indicate overall low surveillance sensitivity, making comparisons among regions difficult to interpret. In addition, the estimates from the measles mortality model might be biased by model inputs, including vaccination coverage and surveillance data.[45]

Translated from "WHO-ese," what they are saying seems to be that they don't have very good data and that they have chosen to model measles coverage at the low end and fatalities at the high end. The outcome of so doing is that the estimate is purely model-dependent and that the authors have made it come out the way they wanted it to by how they chose to put in their parameters.

This is revealing in showing how they do their estimates, estimates that are then taken up by the mainstream medical establishment as demonstrated facts.

The next two documents, also internal and not peer-reviewed, tie together a theme that the WHO has been working on for some years, namely, global vaccine coverage, described below as Immunization Agenda (IA) 2030.

The first of these is titled *"Immunization Today and in the Next Decade, Developing together the vision and strategy for immunization 2021–2030,* draft zero for co-creation by 14 June 2019."[46]

The new immunization strategy as described was to be ready to be endorsed by the World Health Assembly[47] in May 2020, after being vetted by the Strategic Advisory Group of Experts on Immunization (SAGE) and the WHO's Executive Board in October 2019.

It is not clear if these time lines were met due to the pandemic.

The document also states that

> Vaccination is core to preventive medicine, emphasizing the importance of strengthening links between immunization and other key areas of public health, such as water, sanitation and hygiene, behaviour change, and vector control, in coordinated approaches to prevention of infection.

It is clear that as of the publication of this document, Gates was getting his way with the WHO's future planning.

The document goes on to list as one of the priority goals under the rubric of "Systems & Integration":

> Bridging to stronger and more integrated immunization systems: Ensuring that hard fought immunization gains are sustained and that future systems deliver high-quality vaccination services integrated into wider national health systems throughout people's lives, as services are increasingly delivered along the life-course and as part of primary healthcare.

The full set of goals is illustrated from one of the figures in their document.

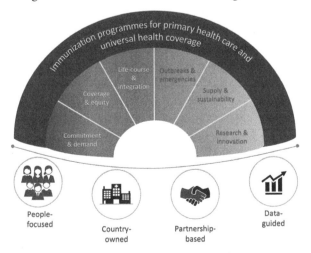

Figure 10.3. Redrawn from the WHO's IA2030.[48]

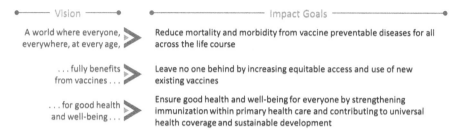

Figure 10.4. Redrawn schematic from WHO's IA2030.[49]

Listing the "top immunization challenges," the document focuses on number ten (of twelve): "Addressing vaccine hesitancy and anti-vaccination activism."

And the response: "Mobilizing public support through CSO, political leaders and champions to raise the appreciation of vaccination and to seed the value, confidence, and demand for vaccination at the community level."[50]

Wouldn't it be more effective to build vaccine confidence by providing the actual research? It appears not.

Instead:

> *Vaccine hesitancy.* Vaccine hesitancy is fast becoming recognized as a global problem. If ensuring that all populations can access reliable and quality vaccination services isn't challenging enough, the reluctance of care-givers to use vaccination services when they are available is a growing concern. The reasons for vaccine hesitancy are complex and context specific but can contribute to large pockets of susceptible individuals, compromising herd immunity and increasing the risk of transmission of diseases.[51]

Or, again translated from WHO bureaucratese: the problem is not the lack of safety studies because "the science is settled," rather how to manage public perceptions. Of interest, this is very much the approach taken by the Vaccine Confidence Project, as noted in Chapter 8.

As weird as this all is, it gets weirder with the WHO's frankly Orwellian document, *"Immunization Agenda for 2030, A global strategy to leave no one behind*, edited draft for WHA (World Health Assembly)."[52]

In IA2030's introduction, the authors cite the document as promoting a "global vision and strategy for vaccine and immunization 2021–2030." The "vision" is summarized as "A world where everyone, everywhere, at every age fully benefits from vaccines for good health and well being."[53]

As for their "Impact goals," these include:

- reduce morbidity and mortality from vaccine-preventable diseases for all across the life course;

- leave no one behind by increasing equitable access and use of new and existing vaccines;
- ensure good health and well-being for everyone by strengthening immunization within primary health care contributing to universal health coverage and sustainable development.[54]

In the draft four version of the document released on April 2, 2020, the authors state that

> Vaccines are critical to the prevention and control of many communicable diseases and therefore underpin global health security. Moreover, they are widely seen as critical for addressing emerging infectious diseases, for example by containing or limiting outbreaks of infectious diseases or combatting the spread of antimicrobial resistance. Regional outbreaks (e.g., of Ebola virus disease), *the COVID-19 pandemic* and the threat of future pandemics (such as with a novel flu strain) have and will continue to strain even the most resilient health systems.[55] [Italics are mine.]

IA2020 goes on to state the following, which is worth quoting in detail:

> The Immunization Agenda 2030 (IA2030) sets an ambitious, overarching global vision and strategy for vaccines and immunization for the decade 2021–2030. It draws on lessons learnt, acknowledges continuing and new challenges posed by infectious diseases and capitalizes on new opportunities to meet those challenges. *IA2030 positions immunization as a key contributor to people's fundamental right to the enjoyment of the highest attainable physical and mental health and also as an investment in the future, creating a healthier, safer, more prosperous world for all.* IA2030 aims to ensure that we maintain the hard-won gains and also that we achieve more - leaving no one behind, in any situation or at any stage of life.[56] [Italics are mine.]

And,

> IA2030 provides a long-term strategic framework to guide a dynamic operational phase, responding to changes in country needs and the global context over the next decade. *This document is therefore just the beginning.* [Italics are mine.]

I suppose this last sentence about IA2030 only being the beginning is correct, and something that we should all worry about, or at least those concerned with personal freedom.

IA2030 does take time to critique the Global Vaccine Action Plan (GVAP):

> GVAP was to be implemented through national immunization programmes, with the support of partners; however, GVAP was only partially successful in influencing national actions, and partner activities were not always fully coordinated globally or nationally. To enhance country ownership, which is critical to the success of the IA2030, tailored strategies will be necessary to respond to the significant differences among countries in size, resources and the conditions in which people live, with consideration of subnational differences. IA2030 will strengthen existing partnerships and build new relations, especially at the country level, such as with a wider range of civil society organizations and the private sector, under the leadership of national programmes.[57]

IA2030 is the successor document to "GVAP, Decade of vaccines, 2011–2020." In GVAP, the diseases targeted were polio, neonatal tetanus (see the previous section), measles and rubella, cholera, viral hepatitis, vector-borne diseases, yellow fever, meningitis, influenza, and rabies.

It would be safe to say that this plan was not a resounding success, particularly in regard to the polio vaccine campaign in India funded in large measure by the Bill and Melinda Gates Foundation.[58]

Nothing in the GAVI document does anything besides praise vaccines across the board in every possible sphere of human life. Nothing in the document suggests the need for any caution or concern, and nothing whatsoever is said about informed consent.

In IA2030, the document zooms in on where GVAP came up short in their "decade of vaccines" and the failure to contain the "anti-vaccine" movement.

In IA 2030, the authors note regarding "sustaining trust" that

> Uptake of vaccination depends on many factors, from the convenience and quality of facilities and services to the spread of misinformation about the safety and effectiveness of vaccines. These considerations must be understood and addressed to enhance and sustain trust in vaccines and immunization services in communities, to increase health literacy with a focus on vaccination at all levels, and to build resilience against misinformation. *The harm being caused by anti-vaccine messaging, especially on social media, should be addressed by understanding the context and reasons for lack of trust and by building and keeping trust, especially in the face of fear and distrust in traditional establishments. Strategic investments to increase trust and confidence in vaccines, in particular through strong community engagement, would increase community support for vaccines and ensure that vaccination is viewed as a social norm.*[59] [Italics are mine.]

Finally, IA20 lists their seven strategic priorities:

> Immunization programmes for primary health care and universal health
> coverage; commitment and demand; coverage and equality; life course and
> integration; outbreaks and emergencies; supply and sustainability; research
> and innovation.[60]

Governed by four guiding principles:

> People focused;
> Country owned;
> Partnership based;
> Data guided.[61]

Of note, the "partnership-based" component is heavily focused on partnerships with the pharmaceutical industry. The "data guided" is only appropriate, apparently, if it gives them the outcome they want, since the document cites no references (of thirteen) that raise any concerns at all about safety.

Nowhere in IA2030 does one find any ethical concerns or the slightest consideration of informed consent. Rather, it is all about manufacturing compliance with the agenda.

Among the truly jaw-dropping claims made for vaccines are the following. According to IA2030, vaccines are responsible for "Economic benefits," presumably because fewer sick people make more money and help the economy and incur fewer treatments needed for disease and help reduce concerns about antibiotic resistance. In addition, the IA2030 agenda will promote the creation of a vaccine industry that will help low- and middle-income countries with jobs. Further, vaccine agendas will increase educational "attainment," promote gender equality, and "mitigate the risk of disease resulting from climate change and new disease outbreaks."[62]

Even more, IA2030's agenda will "Increase social institutions"; help with cleaner technology; and build "Partnerships with civil society, communities and the *private sector*."[63] [Italics are mine.]

One might be tempted to suggest that in the partnerships with the "private sector," the pharmaceutical industry is the true beneficiary.

IA2030 and the other documents that preceded it are very revealing in that they clearly lay out the goal for a fully vaccinated world. Those who support this agenda will see in it a beneficent attempt to bring the wholly believed benefits of vaccines to the entire world and unify the world in equal health across the life span. Those who tend to be more skeptical of the WHO and the agenda as laid out in IA2030 will see in this document something more

346 Dispatches from the Vaccine Wars

ominous: a large-scale effort aimed at bureaucratic control and reliance on the private sector—read the pharma—as a vision for a brave new world of corporate dominance. The new world sought is very much in keeping with Huxley's dark dystopian view of the future where the State dictates everything.[64]

All of this ties in rather neatly to a much broader agenda, namely, that of the World Economic Forum (WEF), the so-called Great Reset, discussed at greater length in Chapters 13 and 14.

What Is the Impact of This Triangulation of Forces on Dissent?

This is a key question that goes to the heart of the current situation concerning critiques of anything related to vaccines, or now the "official" narratives about COVID-19.

As noted in Chapter 8, vaccines have become the sacred cow of medicine, considered by many to be above and beyond all critique. The CDC and the WHO, as typified by IA2030, hammer this message home with each pronouncement. Apart from the WHO, organizations such as the Wellcome Foundation[65] and the Vaccine Confidence Project[66] list vaccine hesitancy as one of the most pressing health threats with nary a concern expressed for any possible harms that vaccines might cause to some individuals.

What those questioning any aspect of vaccine safety are facing at present is, in many ways, nothing new. What is new, however, is that unlike having to face only an industry desperate to protect their bottom line, what is now happening is a full-court press that combines the weight of the media, the medical establishment, and powerful international bodies such as the WHO. The pharmaceutical industry mostly stays in the shadows, conveniently out of sight, and lets its consiglieres and capos take the point positions.[67]

If the pharmaceutical industry is like a mob family, who then are these consiglieres and capos? There are a number of candidates whom we will meet in the following chapters. But who is the mob boss?

CHAPTER 11

Vaccines and the State of Exception

*This struggle [for human liberty] may be a moral one, or it may be a phys-
ical one, and it may be both moral and physical, but it must be a struggle.
Power concedes nothing without a demand. It never did and it never will.
Find out just what any people will quietly submit to and you have found
out the exact measure of injustice and wrong which will be imposed upon
them, and these will continue till they are resisted with either words or
blows, or with both. The limits of tyrants are prescribed by the endurance
of those whom they oppress.* [Brackets for addition are mine.]
— Frederick Douglass[1]

Vaccine and Other Mandates: Implications for Human Natural and Civil Rights

The first question to ask in regard to vaccination policy, particularly mandates
for school children or members of a society at large, is actually part of a much
bigger question. It is a question that human societies have struggled with likely
for as long as there have been humans. In brief, how do we choose to be governed, if at all, and what relationship does the individual have to the State,
however the latter may be defined?

Perhaps the foremost political philosopher of the American Revolution
wrote this in his seminal pamphlet, *Common Sense*:

> Society in every state is a blessing, but government even in its best state is but
> a necessary evil; in its worst state an intolerable one; for when we suffer, or
> are exposed to the same miseries by a government, which we might expect in
> a country without government, our calamity is heightened by reflecting that
> we furnish the means by which we suffer.[2]

Paine was referring in particular to the British government of George III, but
also more generally to any government whose actions overstep the bounds of
agreed-upon governance to impinge on human freedom. Power, and thus the

power to govern, in Paine's view, was either delegated or assumed. The first was the necessary evil the above quote refers to; the latter, an example of a descent into tyranny.

The same question has been posed by many others at various times in human history. In spite of this, numerous societies by choice, or coercion, end up with a form of government in which the "ownership" of a person that should belong to that person, or to a deity, ends up belonging to the State. The latter characterizes various totalitarian States of the past, such as ancient Sparta, or those of more modern times such as Nazi Germany, Stalinist Russia, or even modern-day China.

If the State has control of your body, or that of your children, it can enact various laws, all formally "legal," to compel you to comply with some regulation that the State has promulgated for the "greater good." And it almost goes without saying that the State gets to define what the greater good actually means. An example of this is when any State goes beyond whatever nominal freedoms citizens possess to restrict these freedoms further under what is termed a "state of exception." I will return to this in the sections below, but for now, suffice it to say that most governments, left to their own devices, will at some point expand their power over their people using such excuses. As the late George Carlin once observed in relation to the United States, "Rights aren't rights if someone can take them away. They're privileges. That's all we've ever had in this country, is a bill of temporary privileges. And if you read the news even badly, you know that every year the list gets shorter and shorter."[3]

Carlin, a very talented comic, used comedy as a vehicle to expose those who would abuse the rights of others, be those rights "natural," such as those discussed in the American Declaration of Independence,[4] or civil rights, those that have been negotiated.[5] In his comedy routine, Carlin did not distinguish between the two, viewing them all as privileges to be removed by the State on any whim. Sadly, as we've seen over the last years, especially in the year of COVID-19, Carlin seems to have been quite prescient.

So in relation to the rights you think you have, if the government can make the claim that it "owns" your body or that of your children, in the process denying the most basic natural right of all, then a vaccine mandate makes sense from their point of view.

If we accept that herd immunity (see Chapter 4) can actually be accomplished through vaccination and then add on the assumption that any adverse reactions are extremely rare, we have in a nutshell the State's logic for mandates. From this "logic" arises the conclusion that, no matter how large adverse vaccine effects may actually be, such adverse effects are simply the price that individuals in a society have to pay for the "greater good." And if that conclusion is accepted, then it follows that children, or adults, can be vaccinated against their will by way of mandates. In the case of children, such mandates can be enforced by the threat of punishment for noncompliance, such as the denial of educational facilities. By the same logic of the State, if herd immunity continues to fail due to secondary vaccine failure or other reasons, then similar mandates with penalties for noncompliance can be imposed on adults, as well.

Indeed, such penalties are now being proposed for any future noncompliance that might follow mandates for a COVID-19 vaccine. A recent *Perspective* article in the *New England Journal of Medicine* by Mello et al. (2020) mulled the need for precisely such a remedy.[6]

The authors consider the circumstances under which a mandate for the vaccine might be justified. Their shopping list includes the following:

> Covid-19 is not adequately contained in the state; The Advisory Committee on Immunization Practices has recommended vaccination for the groups for which a mandate is being considered; The supply of vaccine is sufficient to cover the population groups for which a mandate is being considered; Available evidence about the safety and efficacy of the vaccine has been transparently communicated; The state has created infrastructure to provide access to vaccination without financial or logistic barriers, *compensation to workers who have adverse effects from a required vaccine, and real-time surveillance of vaccine side effects*; In a time-limited evaluation, voluntary uptake of the vaccine among high-priority groups has fallen short of the level required to prevent epidemic spread. [Italics for emphasis are mine.]

Of these, the first five guidelines are just the window dressing, in which some are absurdly unrealistic given the history of vaccine administration and compensation as per the italicized section above. What to do?

The authors write that

> Although state vaccination mandates are usually tied to school and day care entry, that approach is not appropriate for SARS-CoV-2 because children won't be a high-priority group. In addition, state mandates should not be structured as compulsory vaccination (absolute requirements); instead, *noncompliance should incur a penalty. Nevertheless, because of the infectiousness and dangerousness of the virus, relatively substantive penalties could be justified, including employment suspension or stay-at-home orders for persons in designated high-priority groups who refuse vaccination.* [Italics are mine.]

At least the authors acknowledge that making noncompliance punishable might be problematic: "Neither fines nor criminal penalties should be used, however; fines disadvantage the poor, and criminal penalties invite legal challenges on procedural due-process grounds."

Other articles, such as one that recently came out in the *Boston Globe,* are far less concerned about human rights violations[7] as well as in some of the comments attributed to so-called bioethicists, as noted in Chapter 6.

Mello et al. appear to partially recognize that imposing vaccine mandates is a slippery slope from a human rights perspective. What they, but none of the other advocates for mandates, seem to realize is that in order to proceed with a mandate, any government has to either ignore, or simply violate, various national and international guidelines that protect the "security of the person."

"Security of the person" is indeed the speed bump that has to be navigated when imposing mandates for any medical intervention. At least in the United States, this natural right is both implicitly and explicitly enshrined in the foundational documents of the country. In countries like Canada, government mandates would run smack into Section 7 of the Canadian Charter of Rights and Freedoms. Much the same would apply to other countries, as well as the United Nations itself.[8]

Given that vaccination is a medical intervention, the same considerations would apply, considerations which are further bolstered by various legal decisions arising from the Nuremburg, Helsinki, and other documents discussed in Chapter 4.

One way to think of mandates in terms of relative rights is to consider how medics perform their duties in either military or civilian life. For example, I was once an army medic and as such was required to do whatever was necessary to save a soldier's life. The soldier had no say whatsoever because he/she belonged to the army, which exercised power over any free choice by the individual.

As a civilian medic, I can't operate in the same way. Indeed, from the beginning of any intervention, I am legally required to ask the injured person if I can help and, if consent is given, to continually ask the person's permission to proceed. It is only if the person is too young to respond, or not able to respond due to the injury, that I can assume "implied consent" and carry on with procedures within my training. The only caveat to this last point is that if a parent or partner is present, I need to ask for their permission on behalf of the injured person. There is no getting around these restrictions, and any medic who does so could face considerable civil and professional consequences.

However, with vaccine mandates we are not talking about people in an army, or those unable to respond, but rather laws that are directed against those normally presumed to be capable of exercising free choice.

What goes hand in hand with the exercise of free choice is that the person facing some treatment has to be aware of what the treatment is designed to do, what it may actually do, or what negative consequences from it might arise.

Again, using the medic example, if I am starting a secondary survey, I have to tell a conscious patient that I want to check them for other injuries and to give them the broader picture that some of the things I may do in checking for injuries may cause pain. If the conscious patient says "no" at any point, I stop, as required.

Where this differs from a population measure such as mandated vaccines is that my requirement as a medic to provide informed consent for an individual is my responsibility; their choice is solely theirs in a free society. In the case of vaccine mandates, this cannot be considered to be a free choice because a state of exception pronouncement by government allows that authority to supersede individual rights.

Governments, at least in notionally democratic societies, skirt around the violation of individual rights by positing relative risk ratios as part of their justification for a state of exception decision. For example, a common claim for measles is that it will kill one in one thousand persons who catch the disease compared to the chance of serious adverse effects being one in a million. Both numbers are wrong, as we saw in Chapter 4, but are used continually by health authorities and the media as if they were correct.

The attempt to nullify rights is done without asking, or answering, various key questions. For example, for disease-related deaths and serious complications, were those so affected susceptible to particular conditions, such as being malnourished or having comorbid conditions such as other diseases? For the one in a million negatively impacted by the vaccine (pretending for a moment that this number is correct), does that person carry a particular gene or DNA enhancer region that makes them more susceptible to vaccine injury?

Without knowing the answers to these questions, the ratio of disease injuries and deaths versus those that might follow a vaccine can't be evaluated properly, nor can it be the basis for truly informed consent.

In other words, with vaccine mandates any government is compelling individuals to comply without full information and leaving them the equivalent of involuntarily betting on the lottery: things will probably be fine, but then, maybe they won't. And if they are not, there is no recourse for the individual in any court but the so-called Vaccine Court (see Chapter 3) in the United States, a highly difficult and time-consuming path even in the best of circumstances. Essentially, all of the risk and consequences if things go wrong lie solely with the person injured.

All of this forms part of the basis of one of the demands of the vaccine resistance movement, which states, "Where there is risk, there must be choice."[9]

Vaccine proponents of mandatory vaccination tend to dismiss these concerns, but it really does come down to a basic philosophical view about human freedom: Who owns your body and that of your children, you or the State?

This is something you either *get* or you don't, and there is no real likelihood of reconciling such divergent philosophies about the nature of the relationship between citizens and the State.

One last point raised by some who oppose mandatory vaccination ties back into the nature of the state of exception and the propensity of governments of all stripes to apply it at will. If mandates are applied and enforced, why should any government stop there? Why not mandate a range of other measures apart from vaccination? Microchipping children for safety? Sure, why not? How about mandatory organ donation to fill the gap from voluntary donation? If there aren't enough voluntary donations, then some of the population will have to get by with just one kidney, etc.

Pro-vaccine advocates will claim that these are ridiculous examples that would never happen in real life and that raising this specter only serves to frighten those sitting on the fence about vaccine mandates. What such proponents can't provide, however, are examples where governments have refrained from continuing their assault on freedoms when and where they can.

The international conventions I cited in Chapter 8 make it abundantly clear that human experimentation without truly informed consent is not permitted, at least not legally. And yet, mandates being fundamentally experiments violate these prescriptions time and time again. Whether the issue is school mandates designed to achieve the illusory vaccine-induced herd immunity, or with the COVID-19 vaccines, which have had no timeline studies of potential adverse events, all are experiments nevertheless, and thus not legal according to the international documents discussed previously.

Just as governments often ignore or find a way around the documents on human experimentation that they have previously ratified, those concerned with mandate and the loss of informed consent—freedom, in other words— can find a way to fight back against such mandates whether they be for measles or COVID-19 vaccines. This last point will be explored in greater detail in Chapter 14.

The quote by Frederick Douglass, which opens this chapter, could hardly be more relevant. As Douglass correctly noted,

> Find out just what any people will quietly submit to and you have found out the exact measure of injustice and wrong which will be imposed upon them, and these will continue till they are resisted with either words or blows, or with both. The limits of tyrants are prescribed by the endurance of those whom they oppress.

And make no mistake, mandates applied to human beings are indeed the stuff of tyranny now and in the future. Worst of all, this seems to be a future that far too many are only too willing to embrace in this age of COVID-19 hysteria. The oft-quoted statement attributed to Benjamin Franklin applies: "Those who would give up essential Liberty, to purchase a little temporary Safety, deserve neither Liberty nor Safety."[10]

The Essential Nature of Governments of All Stripes

It would be safe to say, at least in some circles, that any government, by its very nature, will always seek greater control over the governed. This might not always be true, but then this challenges the skeptic to find examples where it is not. In my view, there will be few such examples. The reality is that any government, intrinsically, seeks greater control, an observation not lost on a host of current and past philosophers.

I will discuss this point in much greater detail when I consider Giorgio Agamben's thesis on the "state of exception,"[11] that is, that the rights of citizens of even the most notionally democratic country can be taken away by the State pretty much anytime the State chooses to do so. Or, as I noted in regard to George Carlin, rights that can be taken away are not rights, but privileges. Wikipedia defines privileges and rights as

> A privilege is a certain entitlement to immunity granted by the state or another authority to a restricted group, either by birth or on a conditional basis . . . By contrast, a right is an inherent, irrevocable entitlement held by all citizens or all human beings from the moment of birth.[12]

In other words, as discussed above, rights are innate. Privileges are conferred for good behavior and can just as easily vanish for bad behavior. This is a lacuna into which the rights of individuals disappear, more easily in authoritarian/ fascist States, but in notionally democratic States, as well.

These distinctions, and the inevitable slide toward authoritarian solutions, will be explored below.

State of Exception

Giorgio Agamben[13] is a professor of aesthetics at the University of Verona and a prolific author of eleven books. His *State of Exception* does not make for easy reading, as it involves a detailed analysis of various philosophers and political theorists in relation to the notion of how States address the issue of human natural and civil rights. Translating it from Italian has made it more difficult still. However, to understand where our societies find themselves in the "new normal" of COVID-19, and the antecedents that brought us to this place, it is an essential resource.

There can be little doubt that we are living through a period of history in which the trend toward more authoritarian control is accelerating. This is true in countries that never had any pretense that they were going to become democratic or ensure the rights of their citizens. Obvious examples are China and Saudi Arabia. There are many more. Perhaps even more appalling is that in virtually all of the countries that have traditionally prided themselves on both being democratic in structure and for respecting the rights of their citizens, the official "rights" of citizens have been badly eroded or even ignored entirely. Australia with their current and planned policies on COVID-19 may be the worst of these at present, although the policies enacted in Israel would make that country a close runner-up.

State of Exception provides the background by which to consider these human rights violations. First, the term *state of exception* has historically gone by other names, for example, "state of emergency" or "state of war." But whatever one chooses to call it, the outcome is the same: the suspension or outright abrogation of civil and human rights by a ruler and/or government when they feel the need to do so.

A state of exception is one that, in Agamben's view, ". . . implies uncertainty as it is a law that negates laws," such as a law that might establish a constitution.

Agamben also noted that a state of exception is itself initially an exception, but to normal laws and constitutions. Over the years, however, states of exception have ceased to be an exception, but rather have become a major force in political systems. Agamben terms it the "liquidation of democracy" and observes that the state of exception seems to have become a new operating paradigm of government.

A key early proponent of state of exception powers of government was provided by Carl Schmitt, a Nazi theoretician and jurist.

Agamben spends much of his book addressing Schmitt's advocacy of states of exception as part of the overall rationale justifying the Nazi regime.

There is a reason why Agamben does so: a state of exception may be instituted by a government or leader in response to some "emergency," real or manufactured. Concerning the latter, sometimes the same ruler or government has created the need for such provisions through fear that follows a "false flag" incident.[14] A prime example of a false flag event was the burning of the Reichstag that allowed Adolf Hitler and the Nazi party in Germany to assume dictatorial powers, allegedly to protect the Reich against communists.[15] Many people believe that the equivalent happened in the United States in the days after September 11, 2001, with the passing of the Patriot Act.[16]

Much of Agamben's focus in his book deals with the aftermath of 9/11 and the sometimes indefinite detention of people identified, correctly or not, as "terrorists" using the deliberately vague term of "enemy noncombatants" by the

American government. The attacks in New York and Washington became the rationale for the prison camp at Guantanamo, the US prisoner-of-war camps in Iraq such as at Abu Ghraib, and the various "black sites" run by the CIA.[17]

In regard to Americans within the United States, the functional suspension of various aspects of the US Constitution was allowed to happen, combined with increased surveillance of Americans at home and abroad.[18]

The bottom line is that once a state of exception has been evoked, the normal rule of law is suspended for citizens and noncitizens alike. In addition, such rights rarely return to where they were before the state of exception was declared. Again, as an example, the various Patriot Acts[19] remain in force almost twenty years after the 9/11 attacks.

Agamben, addressing the aftermath of 9/11, wrote that

> . . . law decrees now constitute the normal form of legislation to a degree that they have been described as 'bills strengthened by guaranteed emergency.' This means that the democratic principle of the separation of powers has today collapsed and that the executive power has in fact, at least partially, absorbed the legislative power. Parliament is no longer the sovereign legislative body that holds the exclusive power to bind the citizens by means of the law; it is limited to ratifying the decrees issued by the executive power . . . And it is significant that though this transformation of the constitutional order (which is today underway in to varying degrees in all the Western democracies) is perfectly well known to jurists and politicians, it has remained entirely unnoticed by the citizens.[20]

And,

> . . . in conformity with a continuing tendency in all of the Western democracies, the declaration of the state of exception has gradually been replaced by an unprecedented generalization of the paradigm of security as the normal technique of government.[21]

Another perfect example of governments employing state of exception diktat comes from Canada in the spring of 2020. In late April, the Liberal Party minority government of Prime Minister Justin Trudeau[22] decided to impose a ban on a series of types of rifles that his government termed "military style" without even bothering to go through the parliamentary process of passing legislation. The term "military style" was an absolute fiction, since no true military-grade automatic weapons have been legal in Canada for decades.

What was the official reason for this decision? The edict followed a shooting in the province of Nova Scotia in which a man using illegally obtained firearms killed twenty-two people.[23] Some reporters actually did their jobs, and their investigations raised the possibility that the man was not just a random crazy person, but rather had been an informer for the national police, the Royal Canadian Mounted Police (RCMP).[24]

As noted, the edict came down from the ruling government without using the legislative machinery of Parliament to pass a law banning the listed weapons. This was odd, since in Canada such a measure would likely have passed with the support of the left-of-center New Democratic Party (NDP) and the Green Party. In this regard, it is important to note that Canada does not share the American fascination with firearms due to a very different national evolution. Many people, particularly in urban areas, are quite opposed to firearm ownership at all.

In spite of the fact that Canadian firearm regulations are already among the strictest in the world, far beyond any US measures, and that the weapons used in the shootings had been obtained illegally, the government simply decided to bypass Parliament and bring down their edict. Part of the official reason was that Parliament had been shuttered due to COVID-19. But rather than an edict, legislation could easily have waited until Parliament resumed sitting, as it did a few months later. The reality was that the man who allegedly did the shooting had been killed and was no longer a menace. Nor was there any concern that some other person might carry out another attack, since the edict applied to legal rather than illegal weapons. It was, in fact, a pure example of the government using a state of exception decree to bolster their future election chances in urban areas without going through the legislative process, a process that would at least have exposed the government's duplicity in pushing forward such legislation.

The bottom line was that the government knew they could get away with it, and therefore did. And they did something similar in August 2020 when they "prorogued" (suspended) Parliament again to diminish another of the endless scandals that have battered the Liberal Party and its hapless prime minister.[25]

With the onset of the COVID-19 pandemic, the notion of state of exception is highly relevant, particularly the speed with which most governments around the world at national, state/provincial levels, and even down to municipalities, have rushed to curtail citizens' rights. These infringements include banning a certain number of people from inhabiting the same space (now euphemistically termed "social distancing"), the curtailment of many business activities, and many more such directives, usually from nonelected authorities, and often bypassing any conventional legislative process.

Many of these measures would not routinely stand up to judicial challenge, and indeed in the section below I will discuss one such civil challenge now ongoing in Canada.

In the United States, many of the measures proposed by the various levels of government are clearly in violation of both the US Constitution and the various state constitutions.

The very fact that such edicts have been allowed to exist during the COVID-19 pandemic clearly shows that the simple evocation of state of exception highlights that constitutions and bills of rights, or in Canada the Charter of Rights and Freedoms, are in such circumstances merely words on paper that have no actual force when rulers decide otherwise.

We don't yet know for certain if we have witnessed a massive overreaction to COVID-19 on the part of governments, or if they have indeed acted in their view for our best interests. Certainly those of a more skeptical frame of mind will suggest that the actual level of fatalities in relation to those actually infected, the paucity of negative outcomes in most people infected, and the measures being taken to chip away at individual or communal rights are far too convenient for this purpose. For instance, the COVID-19 state of exception in Canada has made it absurdly easy in principle to prevent pipeline protests[26] or climate demonstrations, or anything of the like.

The various proposed vaccine mandates were themselves examples on a smaller scale of states of exception, triggered in many cases by the 2018/19 measles outbreaks. These mandates were often successful, for example in New York, California, and Washington, and less so in other states or provinces such as Connecticut and New Brunswick.[27]

COVID-19 promises to further drive mandates forward, particularly for recent COVID-19 vaccines, not to mention the inevitable boosters.

Statements by Bill Gates and a variety of others pretty clearly show the short- and long-range intent: We won't go back to "normal" until everyone is vaccinated against COVID-19.

That such proposals violate various constitutional guarantees is obvious, the key one being that of "security of the person," as discussed previously.

Another violated right is that of freedom of religion, which, in the United States, is also an officially protected item in the First Amendment to the Constitution, easily violated by the State of New York, which proceeded to remove any religious exemptions to mandatory vaccination.

This mandate in particular struck the Haredi Orthodox Jewish community in the state particularly hard. In this case, the religious freedom that the Haredi thought they enjoyed in the United States turned out to be largely illusory.

It is interesting to note that the legislators who have promoted vaccine mandates often cite the concept of herd immunity as the justification (see Chapter 4), believing, erroneously, that such can be achieved by vaccination. It is even more interesting that various governmental agencies are now stating that natural herd immunity is not a good thing for COVID-19 but can only be achieved by vaccination, a view that mirrors the WHO's changed definition.

This notion that herd immunity is only useful when it is induced by vaccines not only betrays the lack of a clear scientific understanding of just what herd immunity actually is, but moreover reveals an underlying agenda at play. With COVID-19, the targets are not merely school children, but entire populations, predictably the eventual outcome of mandates.

As others have noted, enforced vaccination was not necessarily going to be accomplished by police vaccination squads holding one down to administer the shots, but rather by a series of regulations. As with the school mandates, the right to a public (sometimes even private) education is made to depend on vaccine compliance and status. In other countries, to be discussed later, the loss of social benefits and other financial penalties are inflicted on parents who do not comply. And in places like Argentina, one's ability to get or renew a driver's license, get on a plane or train, get a passport, etc., are curtailed in adults without the proper vaccine credentials. This is precisely the sort of world that Bill Gates and the WHO wish upon us all.

Another aspect of this level of social control is evident in the types of censorship employed by Facebook, Pinterest, Amazon, and others about what they deem "misinformation."[28] The fact that some of the actions by various social media companies came at the bequest of US Congressman Adam Schiff points to the growing merger of the state and corporate interests.[29]

No other level of collusion of the State and the corporate world is better typified than that of the pharmaceutical industry and branches of government, notably the FDA and the CDC. It is in this merger that one can clearly discern the specter of fascism, a specter that is increasingly coming into focus with COVID-19 and the governmental regulations that have followed. With these regulations, state of exception goals by those in power have ceased to be a special instrument for use when needed, but rather the modus operandi that is likely to be with us for some time to come.

The State of Exception and the COVID-19 Pandemic

Early in the COVID-19 pandemic, Agamben wrote a brief commentary about the pandemic and noted in particular the impact of fear that led to various emergency regulations by the government.[30] The comments are worth quoting at length:

> *Fear is a bad counsellor, but it makes us see many things we pretended not to see. The first thing the wave of panic that's paralysed the country has clearly shown is that our society no longer believes in anything but naked life. It is evident that Italians are prepared to sacrifice practically everything-normal living*

conditions, social relations, work, even friendships and religious or political beliefs-to avoid the danger of falling ill. The naked life, and the fear of losing it, is not something that brings men and women together, but something that blinds and separates them. Other human beings, like those in the plague described by Manzoni, are now seen only as potential contaminators to be avoided at all costs or at least to keep at a distance of at least one metre. The dead-our dead-have no right to a funeral and it's not clear what happens to the corpses of our loved ones. Our fellow humans have been erased and it's odd that the Churches remain silent on this point. What will human relations become in a country that will be accustomed to living in this way for who knows how long? And what is a society with no other value other than survival?

The other thing, no less disturbing than the first, is that the epidemic is clearly showing that the state of exception, which governments began to accustom us to years ago, has become an authentically normal condition. There have been more serious epidemics in the past, but no one ever thought of declaring a state of emergency like today, one that forbids us even to move. Men have become so used to living in conditions of permanent crisis and emergency that they don't seem to notice that their lives have been reduced to a purely biological condition, one that has lost not only any social and political dimension, but even any compassionate and emotional one. *A society that lives in a permanent state of emergency cannot be a free one. We effectively live in a society that has sacrificed freedom to so-called "security reasons" and as a consequence has condemned itself to living in a permanent state of fear and insecurity.*

It's not surprising that we talk about the virus in terms of a war. The emergency provisions effectively force us to live under a curfew. But a war against an invisible enemy that can nestle in any other human being is the most absurd of wars. It is, to be truthful, a civil war. The enemy isn't somewhere outside, it's inside us." [Italics are mine.]

Agamben went on to state that

The following reflections are not about the epidemic, but what we can understand from people's reactions to it. It is a question, that is, of reflecting on the ease with which an entire society has acquiesced to feeling itself plaguestricken, to isolating itself at home, and to suspending its normal conditions of life, its relationships of work, friendship, love, and even its religious and political convictions.

As we view the real-world consequences of the pandemic, particularly governmental responses to it, some issues come more starkly into focus. These will be explored in greater detail in later chapters, but for now let's enumerate just a few of the most obvious: The impact on the economy with many millions of people reduced to poverty, the consequences for schools and education, and, not least, that of religious freedom.

In addition, it is essential to ask the question of where did all of the money that was lost from the middle class and lower class go? Cui bono, who benefited? What was the impact on overall health, not just the COVID-19 illnesses and deaths, but also the cumulative impact of state of exception measures by the various governments on drug addiction and alcohol deaths, the impact on mental health and perhaps suicide, on spousal and child abuse, and the overwhelming impact of poverty on health overall?

Many journalists and others have compared the COVID-19 outcomes on the economy to the Great Depression of 1929, a depression that largely only abated in the United States once it went into World War II. The media and the various talking heads like Bill Gates and the WHO have been quick to note that the world will only get back to normal once there is a vaccine, while also claiming that antibody levels do not equate to immunity.[31] Is that what it will take, or will it take something far more like an actual war in which those pushing the scantily hidden agenda of COVID-19 clash with those opposing this agenda? I will explore this question in Chapter 14.

But first, I want to consider in brief the nature of the now-growing resistance to the Gateses, Schwabs, Faucis, Ghebreyesuses, and the pharma masters. That is, the people fighting back by various means.

Fighting Back: The Resistance to State of Exception Ramps Up

It seems obvious to me, and to others, that the skirmishes in the vaccine wars that have been going on for a number of years have entered a critical phase with COVID-19. Is it reasonable to state that what is on the horizon is not merely a clash of opposing lay and scientific views on vaccine safety, but part of a brewing war for liberation from the overreach of those driving the "Great Reset"? Hyperbole? Perhaps, but the events since March 2020 suggest that it is not.

I will discuss this viewpoint in much more detail in the final chapter of the book but want to note in advance some key points in this current section. The first of these is that to win a war of any form, it is essential to control information and thus the competing narratives that people are allowed to see. It is also crucial to prevent the "enemy" from having effective means of communications. Controlling information flow, much like the capture of an enemy's capital or a strategic feature, is what in modern warfare is termed the "vital ground."[32]

In today's world, while much information is still in print, a great deal more is digital. Much of the latter, the accurate and inaccurate, is on the various social media platforms.

Watching "news" organizations like Canada's CBC, cited previously, twist reporting in print media, radio, and television to some favored narrative is clearly one way to try to win the battle for information. Watching social media platforms actively surveil and censor information on the web either on their own initiative or at the request of government is another.[33]

Such information manipulation by the usual players may not be novel but is increasing and is unlikely to abate given the stakes in this new war.

As an example, an article in 2019 by Jarrett and Sublett in the *Infragard Journal*[34] deliberately asserts that vaccine hesitancy comprises a national security risk, more akin to conventional forms of terrorism:

> The modern anti-vaxxer movement, composed of people who falsely believe that vaccines are dangerous, started with the publication 20 years ago of a now-retracted study by David [sic] Wakefield that erroneously linked the measles, mumps and rubella vaccine (MMR) to autism . . . And while the Centers for Disease Control (CDC) has released studies that show no link between autism and vaccines or that an aggressive vaccination schedule for children causes autism, many people still believe that there is a connection and refuse to vaccinate their children.[35]

Here is their conclusion:

> What can we do to prevent this scenario? We need to have bipartisan leadership support the scientific evidence. In addition, clinicians, health educators, community and religious leaders, and physicians must be part of a campaign to refute the anti-vaxxers and need to specifically reach out to communities with a high prevalence of vaccine hesitancy. *We also need social media companies to continue to refine the algorithms that power their services to better distinguish quality information from deceptions or otherwise misleading information.* Unfortunately, there is no guarantee that these approaches will be successful. Therefore, public health and emergency planners must now prepare for possible scenarios where herd immunity will not be a tool to control a pandemic.[36] [Italics are mine.]

There are, however, increasing responses that can be used to fight back: alternative news media sources, journals, and broadcasts can be used, many of them digital. And as attempts increase to push these alternative voices off of digital platforms, many of those seeking such alternative narration simple go to the "dark web,"[37] which is vastly more difficult to control.

In blatantly totalitarian states, the control of the media is absolute, and this includes what is presented on the Internet. Democratic parliamentary states, such as Canada, are also prone to frankly oppressive means of information control where the current government has mulled the notion of criminalizing "misinformation" concerning the COVID-19 pandemic.[38]

In response to this, some organizations have taken an activist legal approach as part of their fight back against what they see as governmental overreach—state of exception—by any other name, to the pandemic.

Vaccine Choice Canada (VCC), an Ontario-based vaccine rights group, described as an anti-vaxxer organization by much of the media, filed a 191-page lawsuit in the Superior Court of Justice in Ontario on July 6, 2020.[39] The lawsuit names as defendants Canadian Prime Minister Justin Trudeau and a range of other federal officers including the federal health officer Dr. Teresa Tam. At a provincial level, the lawsuit includes Premier of Ontario Doug Ford and a number of his provincial officials, and the mayor of Toronto, John Tory. Also listed as a defendant was the CBC, discussed in the previous chapters.

The suit alleges that

> As against the Crown and Multiple Defendants, the Plaintiffs claim: 'A Declaration that the "COVID Measures" undertaken and orchestrated by Prime Minister Trudeau ("Trudeau") and the Federal Crown, constitute a constitutional violation of "dispensing with Parliament, under the pretense of 'Royal Prerogative' contrary to the English Bill of Rights (1689) as read into our unwritten constitutional rights though the Preamble of the Constitution Act, 1867, emanating from the unwritten constitutional principles of Rule of Law, Constitutionalism and Democracy, as enunciated by the Supreme Court of Canada, inter alia, Quebec Secession Reference . . .[40]

Cutting through all of the legalese, the lawsuit is basically saying that three levels of government—federal, provincial, and municipal—have violated the civil and natural rights of the citizens of each jurisdiction by compelling, without passing legislation, the following: "self isolation; social distancing; the compulsory wearing of face masks; arbitrary and unjustified closures of businesses; the closure of schools, daycares, park amenities, and playgrounds."[41]

The list of claims in the lawsuit goes on to specify the lack of ability for citizens to access: ". . . education, medical, dental, chiropractic, naturopathic, hearing, dietary therapeutic and other support, for the physically and mentally disabled, particularly special need children with neurological disorders; the closing down of religious places of worship."[42]

The lawsuit alleges that the defendants, in laying down these regulations, violated the Canadian Charter of Rights and Freedoms, particularly the right of association (right 2), life, liberty and security of person (7), unlawful search and seizure (9), arbitrary detention by enforcement officers, and 15 (e.g., "Equality before and under the law").[43]

The VCC legal action goes on to enumerate in much more detail the sorts of pronouncements from every level of government that are in violation.

In regard to listing the CBC as a codefendant, the plaintiffs argue that the CDC has a demonstrated history of rather grotesquely misrepresenting the news. The first charge against the CBC is that

> . . . the CBC, as the publicly-funded broadcaster under the Broadcast Act, "owes a fiduciary duty to be fair, independent, impartial, objective, and responsible, in its news coverage and investigation of the 'pandemic,' and COVID-Measures, which fiduciary duty it has flagrantly and knowingly breached."[44]

The above completely mirrors my ongoing concerns about the honesty and agenda of the CBC.

Finally, the lawsuit also alleges that Trudeau, in his pronouncements on COVID-19, is simply repeating nearly verbatim the same statements already uttered by Bill Gates and the WHO. Scripted and nearly identical bullet points are the hallmark of an orchestrated campaign, as well documented by journalists Rampton and Stauber in their book *Trust Us, We're Experts*, cited previously.

Other legal avenues are being pursued in Canada and other countries.

I recently asked Sonal Jain, a lawyer then at Children's Health Defense, a question concerning vaccine mandates, either for current vaccines, as well as mandates for any future COVID-19 vaccine. In her prétitled *International solutions to unethical medical or scientific experimentation on human subjects* (Appendix 16), Jain phrased the question in this manner:

> Do mandatory vaccination programs constitute unethical human experimentation as vaccines are not duly tested before being injected and there is a complete absence of informed consent? What are the various modes of redress against this policy of mandatory vaccination in international law?

Jain's reply took into consideration the various ethical guidelines cited in chapter 8, including the Nuremberg and Helsinki documents. Jain wrote that

> In the international arena, most documents relating to ethics in human experimentation are merely declaratory. They do not create any binding legal rights or obligations that may be enforced by individuals affected by violations of the rules contained therein. Since they do not confer rights, they also do not provide for any means of redressal for egregious violations of the

principles. Typically, these documents are very highly regarded and are of significant importance as they have helped in shaping policies on medical human research on a national level.

However, the International Covenant on Civil and Political Rights creates legally binding obligations on the member states, and the Optional Protocol to the Covenant also provides individuals an effective means of redressal.

At first glance, this determination appears to suggest that these foundational documents are unlikely vehicles to fight mandates. However, Jain also noted that:

The United Nations Human Rights Committee appears to be the best forum to bring an action for two reasons-first, it is the only international organization that recognizes the ability of individuals to bring human rights violation claims in the international arena and second, it is one of the strongest mechanisms to mobilize public opinion. Even though, its recommendations do not bind the state against which the recommendations are made but it certainly is a very strong forum to get public attention to this issue.

Such cases, either at a state or an international level, then put any government contemplating mandates in the same sort of bind that Mahatma Gandhi's "salt march" put the British Raj.[45] That is, it forces the oppressor to choose, for them, one of two equally unpalatable options.

In the case of mandates, any government would either have to cease and desist *or* acknowledge that they had no intention of honoring the fundamental rights of citizens in regard to informed consent to medical experimentation.

The choice for such governments is thus a stark one: turn their backs on their friends in the pharmaceutical industry and the WHO, or admit that their citizens actually have no rights, merely temporary privileges. Either outcome becomes fraught with danger for any government, particularly the latter choice. Such an outcome would be for many citizens the equivalent of taking the "red pill" scene in the movie *The Matrix*. Governments, of course, always have the option to "up the ante" and move toward greater repression, but that too has severe potential consequences, as we will see in later chapters.

These are only several examples of using the courts to fight government and corporate overreach, both now so currently intertwined that a lawsuit against one, in principle, is basically a lawsuit against the other. Other such actions will be presented in the final chapter of the book.

The Future of Human Freedom in the Age of COVID-19 and the "New Normal"

Those born in contemporary North America *might* be forgiven for assuming that what we consider our natural and civil rights are guaranteed. Such is not the case, however. Nor is it the case that either form of rights is equally distributed among all members of society. As the old, and often misused, adage has it, "freedom isn't free," and whatever we in the United States or Canada have now had to be fought for, often with violence in the past or even today.

Americans celebrate their "freedoms" often without acknowledging the price that was paid in the Revolutionary or Civil Wars and later in the struggles for workers' rights, nor do most Americans acknowledge that some of their historical heroes, such as the brilliant political philosopher Thomas Paine or the remarkable soldier John Laurens, fully understood the fact that the rights they were fighting for did not apply to all Americans of the time,[46] nor do they even now. Canadians tend to be pretty smug about the whole thing and tend to denigrate the United States without turning much attention to their own suppression of freedoms for Native peoples and others.

The bottom line is that most people in both countries have little experience fighting for their own freedom, let alone that of others. And because they don't, they are not well positioned to understand the processes by which rights—or privileges—are taken away.

As I have noted earlier, implicit in the foundational documents of the United States and explicitly stated in the Canadian Charter of Rights and Freedoms is the notion of the "security of the person," perhaps the most fundamental natural right humans have. And yet in both countries, various levels of government contemplate mandates for vaccines generally, or COVID-19 specifically. This situation describes an *aphoria*, defined generally as an irreconcilable internal contradiction or logical disjunction. And yet this disjunction seems not to trouble at all the various politicians who voice demands for mandates. The reason for this may arise from the nature of the state of exception, as described by Agamben.

In his book, Agamben introduced the Roman notion of *iustitium*, that is, the "suspension of law" or a standstill in which one tries to define a law that is not a law. Agamben goes on to note that a state of exception is not per se a dictatorship, rather a "space devoid of law," yet having the "force of law."

Agamben wrote that

> The state of exception is not a dictatorship (whether constitutional or unconstitutional, commissarial or sovereign) but a space devoid of law, a zone of anomie in which all legal determinations—and above all the very distinction between public and private—are deactivated.[47] [Anomie: a lack of the usual social or ethical standards of an individual or group]

This is then where we are with vaccine mandates of any sort: we contemplate laws that provide for a state of exception, a space devoid of law, whose only functional response by those who don't want to surrender their presumed rights is outright resistance of whatever form becomes necessary. As Frederick Douglass, whose quote opens this chapter, clearly understood, taking one's rights in the first place, or taking them back from those who would usurp them, may ultimately demand the same level of response.

The idea that members of society can use all measures up to and including violence to safeguard rights is not actually particularly novel or revolutionary because, as Agamben observed, once a state of exception has become the norm, then the necessary revolutionary response follows. In regard to the edicts issued by the German Weimar Republic, Agamben wrote that

> . . . the end of the Weimar Republic clearly demonstrates that, on the contrary [to protecting the constitution] a "protected democracy" is no democracy at all, and that the paradigm of constitutional dictatorship functions instead as a transitional phase that leads inevitably to the establishment of a totalitarian regime. [Brackets are mine.]

In his book, Agamben quoted from both the Italian and German constitutions on the enshrined resistance to a state of exception:

From the Italian Constitution: "When the public powers violate the rights and fundamental liberties guaranteed by the Constitution, resistance to oppression is a *right and a duty* of the citizen." [Italics are mine.]

The Constitution of German Federal Republic, Article 20, states that ". . . against anyone who attempts to abolish that order [the democratic constitution], all Germans have a *right of resistance*, if no other remedies are possible." [Italics and brackets are mine.]

The US Declaration of Independence basically states the same principle:

> We hold these truths to be self-evident, that all men are created equal, that they are endowed by their Creator with certain unalienable Rights, that among these are Life, Liberty and the pursuit of Happiness. - That to secure these rights, Governments are instituted among Men, deriving their just powers from the consent of the governed, - *That whenever any Form of Government becomes destructive of these ends, it is the Right of the People to alter or to abolish it, and to institute new Government, laying its foundation on such principles and organizing its powers in such form, as to them shall seem most likely to effect their Safety and Happiness.* [Italics are mine.]

It would be hard to be clearer than these examples to show that we have an inherent right to resist tyranny, even a duty to do so.

How then should we resist the attempts to subtract rights from citizens by corrupted, or ignorant, politicians acting at the bequest of some corporate entity?

This last brings us back to vaccine mandates and the rather casual notion put forward in some circles, even by those who claim to be "ethicists," that those who refuse mandates can be compelled to obey by the progressive whittling away of their rights/privileges until they comply. As noted recent articles, the loss of such privileges might include the loss of a job, a passport, or a driver's license, the possibility of getting on an airplane, or the ability to attend public events, etc.

Initially, many citizens may go along with such restrictions out of fear. That same fear may lead them to attack those who won't comply and support state of exception regulations intended to take away the fundamental rights of their peers. Hannah Arendt coined a term to describe the actions of ordinary "good" citizens and how they helped the Nazi regime suppress and slaughter Jews, minorities, and dissidents: Arendt called this the "banality of evil."[48]

The battle lines are now becoming clearer. Those who want to retain even the slightest whiff of freedom will have to choose to resist "by words or blows, or both" any entity seeking to oppress them, whether this be by governments, corporations, or a merger of the two.

We should never forget that fear is a powerful motivator for good or ill. Right now here in British Columbia we are witnessing the spectacle of otherwise very good people, school teachers, for example, asking the provincial government to delay school openings. Similarly, ferry workers have demanded greater mask wearing by people on their boats. Why? Because both groups are afraid. And this fear arises because government spin and media connivance have made them constantly afraid of their fellow citizens.

So what comes next? In my view, the pushback will not only be local or even national, but a global revolt to rid us of the growing suppression of human freedom by parts of the corporate world. Like it or not, this is likely inevitable.

A revolution is not yet at hand, but it may be coming sooner than most of us would like to think because the very notion that some sort of revolution is in our future is discomforting to those of us used to having "privileges." Perhaps mostly for this reason, the revolution will likely be led by those with little, and therefore with little to lose.

The oligarchs of the world, and the lapdogs they have bought, who think they can have their way with freedom might want to reconsider the ultimate cost of seeking to impose an authoritarian vision on other humans.

Before I move on to the next section to consider the nature of the mandates and frank oppression in various other countries, let me cite another comment from Tom Paine: "Tyranny, like hell, is not easily conquered; yet we have this consolation with us, that the harder the conflict, the more glorious the triumph."[49]

Vaccine Mandates around the World

As noted elsewhere, most American states have some level of vaccine mandates for school attendance, typically the main form of control. Historically, exemptions to such mandates have been allowed for medical reasons or philosophical and religious reasons. Following the 2019 measles outbreaks, some states reined in all but the medical exemptions, California and New York, for example, being the most stringent.

In New York, Governor Andrew Cuomo, supported by a Democratic majority legislature that seems to have drunk more than a normal portion of the vaccine Kool-Aid, has proposed a rash of legislative measures. A recent one is a bill to mandate COVID-19 vaccines. This may not succeed, but the future success may hinge on just how scared the public and legislators get.

Other bills in the New York legislature that have been proposed but not passed include a mandate for influenza vaccines,[50] one to mandate HPV vaccines,[51] and two bills to allow minors to get vaccines without parental knowledge or consent (one for vaccines and drugs against STDs;[52] another to allow children fourteen and older to get any vaccine without parental consent or knowledge, which was passed by the Senate Health Committee and then stalled[53]).[54]

It's not that Cuomo and the Democrat majority aren't trying to do the pharma's bidding; they are. But it seems others are pushing back.

In other cases, such as in Connecticut, attempts to remove religious exemptions failed.

In Canada, to date no province mandates vaccines for children or for school attendance, although some have tried. In the province of New Brunswick, a mandatory vaccine schedule for school attendance failed, as previously discussed.

Around the world, most countries have some level of mandates for children, in some cases with financial penalties for parents who do not comply.

Here are a few more examples:

Italy lists ten diseases for which vaccines are mandatory for children zero to six years old to enter preschool or kindergarten based on a 2017 law, #119. Some of these are multivalent vaccines such as the hexavalent vaccine that includes polio, diphtheria, tetanus, pertussis, Hep B, and Haemophilus influenza B, with the vaccine given at two months and repeated twice more in the first year. The same law added the MMRV vaccine to the MMR vaccine that was required to be administered between thirteen to fifteen months. The fines for non-compliance are set at five hundred euros.

Children older than six can enter school without such vaccines, but there may be an "administrative fine" of one hundred to five hundred euros for the parents. Exemptions can occur if the child has had the disease for which the vaccines were developed or for other medical contraindications.[55]

Slovenia has a list from birth to age eighteen of both mandatory (nine) and recommended (four) vaccines. Of these, only four are mandatory for children under five years old.[56]

What is very clear is that in these two examples the number of vaccines in the schedules is significantly lower compared to the CDC's recommended schedule for the US.

CHAPTER 12

Tangled Web: The WHO, Bill Gates, and the Pharmaceutical Cartel

*Humankind has never had a more urgent task than creating broad immunity for coronavirus. **Realistically, if we're going to return to normal, we need to develop a safe, effective vaccine**. We need to make billions of doses, we need to get them out to every part of the world, and we need all of this to happen as quickly as possible.*

*That sounds daunting, because it is. **Our foundation is the biggest funder of vaccines in the world**, and this effort dwarfs anything we've ever worked on before. It's going to require a global cooperative effort like the world has never seen. But I know it'll get done. There's simply no alternative.*" [Bolding for emphasis is mine.]

—Bill Gates[1]

Sorting Out Who's Who in the Herd in the Room

The COVID-19 pandemic[2] has served to focus worldwide attention of a lot of otherwise disinterested people on health issues. Specifically, the media hype and measures by most world and local governments have caused an abundance of panic about this virus, making many people not only afraid for themselves, but for their relatives, especially those who are elderly.

Somewhat more generally, the pandemic has led to greater inquiry by many people about who is really calling the shots, literally and figuratively, as governments around the globe scramble to figure out a proper response to the disease. Much of the global response to COVID-19 will be discussed in the next chapter, and I will leave much of the considerations about how effectively the various countries' responses have been until then.

In the following, I want to consider the first part in isolation. First, Cui bono?, a question posed before. That is, who is getting something out of this situation in terms of greater wealth, power, or both? Maybe no one is, at least not in the sense of fostering the pandemic for such purposes.

In this latter view, some astute individuals and entities have simply had the foresight and luck to succeed financially in these chaotic times. It is, as one of my former political allies used to call it, the "cock up" theory of history: crummy stuff happens, it's not anyone's fault really apart from incompetence/indifference, and it's certainly not a conspiracy. Along the lines of this narrative, COVID-19 came from bats to pangolins in China, then jumped to humans, and then it spread around the world, and it's really no one's fault.

The view that COVID-19 is a "natural" outbreak is certainly the interpretation of much of the mainstream media and governments of all stripes: if there is any blame to be assigned, it may arise from some governments not locking down hard enough immediately. Or we need to blame some people for being too selfish to help out by not following official health guidelines for mask wearing and social distancing. For many people, this is the easiest hypothesis to accept, as it does not require any independent analysis, rather simple obedience to governments that many believe, or want to believe, simply want what's best for you and your family.

It might be someone's fault, however, if the first part of the story is wrong. As I will show in Chapter 13, within parts of the science community a very different view of the origin of COVID-19 is being discussed.

In turn, a more skeptical view of the origin of COVID-19 leads to a counter-hypothesis that if you follow the money carefully enough, you will find the origin, maybe even find the source(s), of the event hiding in plain sight. It's sort of like the joke about obvious things such as the proverbial "elephant in the room" that seem invisible to most people. Or, in this case, maybe it's more like a herd of elephants that some people can't bring themselves to believe are there and thus simply refuse to see even as they are getting trampled.

Readers will probably have noticed by now that I am often given to skepticism in the real sense, rather than acquiescence, and my skepticism tends to look for patterns. One such striking pattern is when media and governments start using the very same words and phrases to describe some event and its outcomes in a way that appears to be scripted.

The very astute and often political Brasscheck TV recently ran a segment in which they examined two examples of such mimicking: the similar pronouncements about the 2003 Gulf War and Saddam Hussein's alleged weapons of mass destruction, and the need for a COVID-19 vaccine to allow the world to go back to "normal," whatever people think that normal actually is.[3]

Such word patterns are important and form part of the screening process that a lot of scientific journals now use to detect plagiarism or even autoplagiarism. In other words, if greater than a certain percentage of words and phrases in two articles is the same, the probability grows that one was copied from the other, or that both are copying from some common source. Listening to the Brasscheck TV segment, it becomes crystal clear that those talking about

Iraq were simply mimicking the narrative of the Bush Administration and the clique of neocons within it. For COVID-19, it becomes equally clear that politicians and the media were simply copying Bill Gates's comments or those of Klaus Schwab.

For this reason, Bill Gates is one of those figures I would put into the room as one of the almost invisible elephants. Gates, as we will see soon, has one of the best reputations that money can literally buy and for this reason largely hides in plain sight. His camouflage is promoted by most of the mainstream media, usually portraying him as a dedicated, sometimes misunderstood, philanthropist only seeking to save us all; if only we'd let him, we'd all be just fine and back to "normal."

Who else would go into the same room? How about the WHO, whose overall perspective on health, in particular on vaccines, was presented in Chapter 10? That chapter detailed the WHO's track record and vision of a frankly Orwellian future, the latter spelled out in great detail in their Immunization Agenda 2030.

As noted previously, the WHO and Bill Gates are seemingly jointed at the hip with the Bill and Melinda Gates Foundation's financial contributions to the organization growing year after year in lockstep with Gates's influence on the organization. It almost goes without saying that the level of financial largesse that Gates provides to the WHO makes his agenda largely theirs, as well. For this reason, attempts to examine Gates or the WHO mean that one has to examine them both at the same time.

Who else? Who will make money on the eventual vaccines or other drugs designed against COVID-19? None other than the various pharmaceutical companies, especially the mega ones, whose functional ownership of the US Congress has expanded manyfold since the 1980s, as will be described soon.

I will examine each in turn in the following sections and show that the entities listed interconnect in far too many ways for their association to be considered accidental; and indeed, they are not accidental. Some may prefer the cock up theory of events in this regard and refuse to see any conspiracy here. In the following sections, we will see.

One thing to note before I begin to look at members of the herd: what follows can only be a summary of the roles and interactions that Bill Gates and his foundation, the WHO, WEF, and the pharma play both generally and in this age of COVID-19. Needless to say, each alone would warrant its own critical book, books that no doubt are in progress.

The Official and Unofficial History of Bill Gates
I confess that I hadn't paid much attention to Bill Gates or the Bill and Melinda Gates Foundation prior to starting this book in October 2019. Even then, all I

knew about Gates was that he was the cofounder of Microsoft, where he had a reputation as a predatory CEO and later had started the Bill and Melinda Gates Foundation when he stepped down as CEO of Microsoft in 2008. I knew that the Foundation gave a lot of money to develop and promote vaccines, mostly in the Third World, and that it had worked with the WHO in several capacities in this regard. I knew that the Foundation also supported GMO agriculture, often again in collaboration with the WHO.

That was about it.

Since COVID-19 first burst into world consciousness in early 2020, we have all seen a lot more of Bill Gates, courtesy of a normally fawning media that typically feeds him the softest of soft pitch questions that allow him to pontificate on a range of subjects that he clearly knows little about. For example, what he claims to know about vaccines is that they are essential, invariably safe, and, in the age of COVID-19, the only thing that will get the world back to "normal." Whether he recognizes, or not, that a great number of people from various countries do not share his enthusiasm for either vaccines for everything or GMO crops as a boon to agriculture is not clear.

Further, what is it that he actually knows about pedagogy that recently led New York Governor Andrew Cuomo to seek his advice on the future of public education in New York? This is less than obvious, and, typically, the media interviewers never ask. When asked about vaccine-tracking technologies or devices for human vital statistics biomonitoring, he tends to deny any interest in developing such technologies. I will show later that in this area, as in so many more, Gates is not being candid.

It is the current pandemic that has really brought Gates into widespread prominence. While many continue to see him as a philanthropist hero with the financial resources and passion to save humanity, a growing number of people across the world are now coming to view him quite differently.

I admit to being one of this latter grouping, and my initial impression on seeing his performances on television is that there is something more than slightly creepy about the man. Maybe it is less his words than how he delivers them with the strange half smile at the oddest of times, such as when talking about the pandemic, or the constant hand moving as he emphasizes some point.

His hand movements alone show something very peculiar. It is not the case that he was raised in a Mediterranean culture where hand gesturing is common, but more like he is weaving a web of some sort. A random search of Google sites on interpreting body language mentions these curious performances. The net conclusion seems to be that some of the things his mouth says really don't seem to match what other parts of his body seems to be indicating. In other words, there is something else going on in his head. Maybe it's only that he is a pretty shy, geeky guy and may get tense during interviews. Or then again, maybe he is being less than truthful about something. Or, just maybe, he is a geeky guy who is also lying.

What's clear, however, is that Gates had a major makeover after he left Microsoft to start his Foundation: gone is the snarly arrogance of his Senate Judiciary Committee hearings into Microsoft's monopoly practices in 1998;[4] instead, we now get the avuncular Bill Gates, casually dressed in a light sweater, answering media questions in a soft voice, accompanied by the endlessly moving hands. This is good old Uncle Bill, the philanthropist who has chosen to use his vast wealth for human betterment, no longer the predatory capitalist of yesteryear.

How was this transition accomplished? I have no doubt that after the Senate hearings and the initial loss in the subsequent lawsuits against the government,[5] Gates found himself a good public relations (PR) firm that began to give him much-needed pointers on how to clean up his public image.

The steps seemingly were these: to begin with, create a charity to dole out money to causes that might evoke the maximum positive publicity, a goal accomplished two years later with the establishment of the Bill and Melinda Gates Foundation. In this way, Gates mimicked Andrew Carnegie, a steel and railroad baron who actively suppressed unions and engaged in monopoly practices a hundred plus years earlier.[6] To solve the image problems he had developed, Carnegie tried, and largely succeeded, in reimagining his reputation by giving away some of his considerable wealth to build libraries, schools, and other public buildings across North America. Where I live, for example, there are Carnegie libraries in both Vancouver and Victoria.

This part accomplished, Gates's PR firm no doubt suggested that he appear to lose the cocky attitude that had come to the fore in his Senate hearings. Next, they tried to teach him to speak and dress more like a regular sort of guy. A lumber jacket would have been taking the advice too far, but a cashmere sweater was likely just about right.

Those of us in Canada have seen a similar kind of makeover with a former prime minister, Steven Harper, trying to soften his image to become just a regular guy in the lead-up to the 2015 federal election.[7]

With Harper, it was a facade much as one might suspect it is with Gates. Whether the changes to the Gates public persona were real or not, nevertheless the new Bill was now fixed in place in the public mind as ready and able to donate his fortune for the betterment of humankind. Mission accomplished.

Now all that was needed was to find a crisis to make Gates a trusted household name.

In the following I will delve into Gates's history, the official version, as well as a more critical version, the latter really only now starting to emerge at the same time as his prominence grows. I will also take a look at the Bill and

Melinda Gates Foundation and their ties to the WHO, the pharmaceutical industry, and a number of other groups that all seem interconnected to vaccines in general and COVID-19 in particular.

Bill Gates as the Media Normally Portrays Him

Gates's biography can be found in various places (e.g., on Wikipedia), most of which more or less agree on the general story: William Henry Gates III was born in 1955 in Seattle, Washington, the son of lawyer William H. Gates II and Mary Maxwell. The Maxwells were a banking family.

As the standard biography has it, Gates dropped out of Harvard in 1975 to start Microsoft with Paul Allen, a childhood friend and fellow computer enthusiast. Gates became the CEO.

In 1998, Gates testified before the Senate Judiciary Committee, which was looking into monopoly practices by Microsoft. The entire timeline of events can be found in a document from the *Washington Post*.[8] In the end, Microsoft lost the litigation battle with the government, and the same year Gates left his role as Microsoft's CEO and began the Bill and Melinda Gates Foundation. Twenty years later, in a curiously timed move, Gates also stepped down from his role as chairman of the board for Microsoft on March 13, 2020, literally days before the COVID-19 pandemic began to seriously impact life in North America. The official reason for doing so was that Gates wanted to devote more time to his Foundation.[9] This timing coming just prior to COVID-19's major impact on Europe and North America might well just be fortuitous, but his resignation from Microsoft's board certainly allowed Gates to become the face of global vaccination efforts, efforts that segued neatly into a similar effort to promote mass vaccination to halt COVID-19 through his multiple investments into companies working on various COVID-19 vaccine candidates. In fact, Gates was the main backer of several world vaccine efforts described in the previous chapter about the WHO.

The Gates Foundation has, in fact, quite a network of connections to vaccine makers and promoters, as illustrated in the schematic of Table 12.1, which was taken from an article by journalist Tim Schwab, to be discussed below.

The schematic shows the overlap of the Foundation's interests in vaccines with the US Department of Health and Human Services (HHS), three of whose component parts, the CDC, FDA, and NIH, all collaborate with the Foundation on a range of vaccine projects. In regard to the NIH, Gates has worked closely with Dr. Anthony Fauci, whose own branch of the NIH, the National Institute of Allergy and Infectious Diseases (NIAID), has partnered with the Foundation to fund a range of vaccine-related projects both in the United States and abroad. Of the latter, the key one in regard to COVID-19 was at Wuhan University's Institute of Virology level 4 containment facility in the city of the same name. This last link will be explored in greater detail in Chapter 13.

Top international pharmaceutical companies

Rank	Company	Headquarters	2019 Total revenue ~USD billions	2019 Sales from Vaccine	Reference
1	Johnson & Johnson	New Brunswick, New Jersey	82.1	N/A	https://www.investor.jnj.com/annual-meeting-materials/2019-annual-report
2	Roche	Basel, Switzerland	63.85	N/A	https://www.roche.com/dam/jcr:a3545548-a7f9-40f4-a70e-7266a363f856/en/ar19e.pdf
3	Sinopharm	Beijing, China	60.18	N/A	http://ir.sinopharmgroup.com.cn/attachment/20200424081601427924 8723_en.pdf
4	Pfizer	New York, New York	51.75	$5.85 billion	https://www.annualreports.com/HostedData/AnnualReports/PDF/NYSE_PFE_2019.pdf
5	Bayer	Leverkusen, Germany	48.02	N/A	https://www.annualreports.com/HostedData/AnnualReports/PDF/OTC_BAYZF_2019.pdf
6	Novartis	Basel, Switzerland	47.45	N/A	https://www.novartis.com/sites/www.novartis.com/files/novartis-annual-report-2019.pdf
7	Merck & Co.	Kenilworth, New Jersey	46.84	$7.39 billion	https://www.annualreports.com/HostedData/AnnualReports/PDF/NYSE_MRK_2019.pdf
8	GlaxoSmithKline	Brentford, United Kingdom	43.92	£7.2 billion	https://www.gsk.com/media/5820/fy-2019-results-announcement.pdf
9	Sanofi	Paris, France	39.28	€5.73 billion	https://www.sanofi.com/en/investors/financial-results-and-events/financial-results/Q4-results-2019
10	AbbVie	North Chicago, Illinois	33.27	N/A	https://investors.abbvie.com/static-files/719318f-9a32-42ee-92ee-a34975edcd19

Top USA pharmaceutical companies

Rank	Company	Headquarters	2019 Total revenue ~USD billions	2019 Sales from Vaccine ~USD billions	Reference
1	Johnson & Johnson	New Brunswick, New Jersey	82.06	N/A	https://www.investor.jnj.com/annual-meeting-materials/2019-annual-report
4	Pfizer	New York, New York	51.75	$5.85 billion	https://www.annualreports.com/HostedData/AnnualReports/PDF/NYSE_PFE_2019.pdf
7	Merck & Co.	Kenilworth, New Jersey	46.84	$7.39 billion	https://www.annualreports.com/HostedData/AnnualReports/PDF/NYSE_MRK_2019.pdf
10	AbbVie	North Chicago, Illinois	33.27	N/A	https://investors.abbvie.com/static-files/719318f-9a32-42ee-92ee-a34975edcd19
11	Abbott Laboratories	Chicago, Illinois	31.9	N/A	https://www.annualreports.com/HostedData/AnnualReports/PDF/NYSE_ABT_2019.pdf
13	Bristol-Myers Squibb	New York, New York	26.15	N/A	https://s21.q4cdn.com/104148044/files/doc_financials/2019/2019-BMS-Annual-Report.pdf
14	Thermo Fisher Scientific	Waltham, Massachusetts	25.54	N/A	https://s1.q4cdn.com/008680097/files/doc_financials/2019/ar/final/2019-Annual-Report.pdf
16	Amgen	Thousand Oaks, California	23.4	N/A	https://www.annualreports.com/HostedData/AnnualReports/PDF/NASDAQ_AMGN_2019.pdf
17	Gilead Sciences	Foster City, California	22.45	N/A	https://www.gilead.com/-/media/files/pdfs/yir-2019-pdfs/2019-gilead-yir_desktop.pdf?la=en&hash=E6571C6C78464A40834E85CA069E9F91
18	Eli Lilly & Co	Indianapolis, Indiana	22.32	N/A	https://investorlilly.com/static-files/34d71960-241f-4160-bd20-86f0a85df4def

Table 12.1. Estimated profits of the top ten pharmaceutical companies overall for the latest year for which data are available.

Two lesser-known organizations that are highly dependent on the Gates Foundation for funding are the Global Alliance for Vaccines and Immunization (GAVI, the Vaccine Alliance, previously the GAVI Alliance) and the Coalition for Epidemic Preparedness Innovations (CEPI). GAVI was founded in 2000 with Bill Gates promising to fund them as early as 1999. CEPI was founded in 2016 and funded by the Gates Foundation in 2017.

What do these two organizations do? Here is what GAVI has on its website as part of their reason for being:

> By the late 1990s, the progress of international immunisation programmes was stalling. Nearly 30 million children in developing countries were not fully immunised against deadly diseases, and many others went without any immunisation at all. At the heart of the challenge was an acute market failure; powerful new vaccines were becoming available, but developing countries simply could not afford most vaccines. *In response, the Bill and Melinda Gates Foundation and a group of founding partners* brought to life an elegant solution to encourage manufacturers to lower vaccine prices for the poorest countries in return for long-term, high-volume and predictable demand from those countries. In 2000, that breakthrough idea became the Global Alliance for Vaccines and Immunisation - today Gavi, the Vaccine Alliance. [Italics are mine.]

GAVI's website goes on to note that it "brings together public and private sectors," a theme that repeats endlessly in Gates's views on health solutions, in particular vaccines, that need the embrace of capitalism in order to succeed.[10]

GAVI further stated that they

> . . . [help to] vaccinate almost half the world's children against deadly and debilitating infectious diseases. As part of its mission to save lives, reduce poverty and protect the world against the threat of epidemics, Gavi has helped vaccinate more than 760 million children in the world's poorest countries, preventing more than 13 million deaths. Gavi has already protected an entire generation of children, and is now working to protect the next generation. [Brackets for addition are mine.]

Crucially, GAVI seems to mirror the very same views promoted by the WHO in their document, IA2030, cited previously:

> By improving access to new and under-used vaccines for millions of the most vulnerable children, the Vaccine Alliance is transforming the lives of individuals, helping to boost the economies of low-income countries and making the world safer for everyone. Gavi's impact draws on the strengths of its core

partners, the World Health Organization, UNICEF, the World Bank and the Bill & Melinda Gates Foundation, and plays a critical role in strengthening primary health care (PHC), bringing us closer to the Sustainable Development Goal (SDG) of Universal Health Coverage (UHC), *ensuring that no one is left behind.* Gavi also works with donors, including sovereign governments, *private sector foundations and corporate partners*; NGOs, advocacy groups, professional and community associations, faith-based organisations and academia; *vaccine manufacturers*, including those in emerging markets; research and technical health institutes; and developing country governments. [Italics are mine.]

Yet again, the GAVI vision is that of the WHO and, in turn, that of the Gates Foundation.

What about CEPI? CEPI is also a major recipient of Gates Foundation largesse and has described themselves and their mission thus:

> CEPI is an innovative global partnership between public, *private*, philanthropic, and civil society organisations launched in Davos in 2017 *to develop vaccines to stop future epidemics.* Our mission is to accelerate the development of vaccines against emerging infectious diseases and enable equitable access to these vaccines for people during outbreaks.[11] [Italics are mine.]

What distinguishes GAVI from CEPI? Not much. What about their plans? GAVI wants to vaccinate the world's children, much like the WHO; CEPI has a more specific goal to prepare for a pandemic:

> . . . CEPI will advance vaccines against known threats through proof-of-concept and safety testing in humans and will establish investigational vaccine stockpiles before epidemics begin-'just in case.' Second, we will fund new and innovative platform technologies with the potential to accelerate the development and manufacture of vaccines against previously unknown pathogens (eg: within 16 weeks from identification of antigen to product release for clinical trials)-'just in time.' Third, CEPI will support and coordinate activities to improve our collective response to epidemics, strengthen capacity in countries at risk, and advance the regulatory science that governs product development.

So CEPI covers off "just in case and just in time." Who pays for this?

> CEPI was founded in Davos by the governments of Norway and India, the *Bill & Melinda Gates Foundation, the Wellcome Trust, and the World Economic Forum. To date, CEPI has secured financial support from the Bill & Melinda Gates*

Foundation, Wellcome Trust, the European Commission, and the governments of Australia, Belgium, Canada, Denmark, Ethiopia, Germany, Japan, Mexico, Norway and the United Kingdom. Additional investment from sovereign governments, the *private sector* and philanthropic foundations has also been provided to support our COVID-19 vaccine programmes. In response to call the Governments of Austria, Australia, Belgium, Canada, European Commission, Finland, France, Greece, Germany, Iceland, Italy, Japan, Luxembourg, Kingdom of Saudi Arabia, Norway, the Netherlands, New Zealand, Serbia, Spain, Switzerland, and the United Kingdom, alongside private sector companies and donations through the UN Foundation COVID-19 Solidarity Response Fund, have pledged $1.4 billion in financial contributions. Close collaboration with global partners is also crucial to the success of our work to develop vaccines against emerging infectious diseases. That's why work with *industry,* regulators, and other bodies to ensure that any vaccines we develop get licensed and can reach the people who need them. [Italics are mine.]

Regarding the donations from the Wellcome Trust to CEPI: it seems that the Wellcome Trust got funds from the Gates Foundation, as well: US $450,000 in 2016.[12]

Note that Davos, mentioned above, is where the WEF meets.

There are other players in the overall constellation of vaccine-delivery boosters, one of which is the Vaccine Confidence Project (VCP), part of the London School of Hygiene and Tropical Medicine, as discussed in detail in Chapter 11.

The VCP has the mission of convincing the world of the benefits of vaccination and warning of the dangers posed by anti-vaccine sentiments. In Chapter 11, I described how the VCP sees anti-vaccine sentiment as one of the major threats to public health in the world. Four online webinars held in the summer to winter of 2020 discussed the problems posed by vaccine "misinformation," particularly in the regard to a future COVID-19 vaccine. No concerns about possible adverse reactions were presented.

The VCP's director is Prof. Heidi Larson, an anthropologist who has worked with both the WHO and GAVI.[13]

Funding for VCP comes from various sources including European and international pharmaceutical companies, prominently GSK and Merck. They also get funds from an entity called the Innovative Medicines Initiative, which in turn gets some of its funding from the Gates Foundation.[14] The amount received was almost eight million dollars from the Gates Foundation in 2016.

So, in partial summary, the Gates Foundation funds CEPI directly and also moves more funding through the Wellcome Trust; the Gates Foundation also pushes money out to the VCP through the Innovative Medicines Initiative. If this sounds suspiciously like one aspect of money laundering, it's because it seems to be true. None of this means the funding is strictly illegal, rather just concealed from public scrutiny by the multiple financial transactions. I would imagine that there are many more such examples that any competent forensic accountant could uncover.

It is not surprising that all of the above entities who work on projects designed to provide mass vaccination for the world's population know one another and may work together to further their largely overlapping goals. This is the same in any field. What is surprising, however, is that the source of funding for some of these groups seems to largely arise from a limited number of sources, namely, the pharmaceutical industry and the Bill and Melinda Gates Foundation for a significant fraction of their budgets. Neither of the latter could be considered disinterested parties, as I will show below.

None of this is to suggest that all of the players above have evil intent in promoting vaccines. Some of them, like the Vaccine Confidence Project, maybe even GAVI and CEPI, seem more the true believers in the doctrine of vaccination, as discussed in Chapter 8. Like the faithful in almost any religion, these groups only see the aspects of the vaccine theology that they think are true and thus meaningful. At the same time, they tend to ignore, or minimize, those aspects that have been shown to be harmful.

Are these entities engaged in deliberate malfeasance? Maybe not: they believe what they believe, perhaps uncritically, and will take money from fellow believers if it advances their respective missions. I get how this works: funding for projects or research is hard to come by, so why not take it from organizations that seem to share your vision? The apparent money laundering aspect, noted previously, makes this interpretation less certain.

As for the governmental agencies like the CDC, FDA, and NIH in the United States and Health Canada in my own country, they also believe what they believe and likely don't see the ways that their original mission has been compromised by their links to the pharmaceutical industry and the Gates Foundation. That alone does not make them evil, just doctrinaire and inertia-bound and perhaps victims of the "sunk cost" fallacy (see Chapter 13).

What about the WHO? In light of all the evidence, including the past nonpandemics that they claimed were pandemics, documents like IA2030, and their joint collaborations with the Gates Foundation and others on vaccines and agricultural products and policies, they definitely belong in the room with the other invisible elephants.

In the following pages, I will focus on those who I do think really do have suspicious intentions. Namely, the Bill and Melissa Gates Foundation, the WHO, the WEF, and the big pharmaceutical industry.

Bill Gates: Captured Media, Captured Audience

Even before COVID-19 burst onto the world stage, Gates was accustomed to receiving the sort of fawning media coverage that few mortals could even dream about. The fawning became more pronounced post emergence of COVID-19.

That this sort of media coverage has largely evaporated since Gates's pending divorce from Melinda is quite revealing. Indeed, since the announcement of the divorce on May 3, 2021, the mostly fawning media coverage of all things Gates has shifted by 180 degrees to an endless stream of articles not only questioning Bill's marital fidelity, but even linking him to the late convicted pedophile Jeffery Epstein. It is well known that powerful men can survive episodes of cheating on their spouses; it is harder by far to survive allegations of sexual assaults on teenage girls. This rapid fall from grace has to beg the question of who is behind the very orchestrated takedown of one of the world's most powerful men.[15]

Prior to this latest meltdown, an example of the typical soft toss genre was published in *National Geographic* and titled "Saving the world with money: the agenda of Bill Gates,"[16] in which the interviewer, Susan Goldberg, tosses Gates the tamest questions designed to allow him to pontificate on the pandemic.

A lot of what Gates has said seemed to be focused on solving inequality and poverty in the world, laudable goals for anyone, even for a staunch capitalist captain of industry. In this he mimics the same sentiments of the WHO, GAVI, and CEPI, with whom he is financially and doctrinally linked.

In this interview, Gates's passion for the role of corporations, mainly the pharmaceutical companies taking the lead on vaccines, became crystal clear.

Speaking about COVID-19, Gates says:

> . . . But yes, the ingenuity of in the pharmaceutical companies, that's why we have six vaccine candidates-several of which are extremely likely to prove safe and efficacious by early next year. If this pandemic had come 10 years ago, our Internet bandwidth wouldn't have let us do our office jobs. The vaccine platforms wouldn't be as far along. So, it's phenomenal we can say that within a few years, with a little bit of luck on the vaccines, some generosity, and a real effort to get the word out that it's safe, this pandemic will come to a close. [It would be] a lot more negative to say, 'Oh my God this is going to continue indefinitely.' Fortunately, because of science and the pharma companies jumping in, that's not the case. [Brackets are mine.]

Bingo. Here is where Gates comes out of his philanthropist persona to reveal his underlying faith in the corporate world, that is, capitalism with a happy face, to solve the problems afflicting humanity. This may not be particularly surprising given his background. It is, however, very much what his stance was when fighting the government over the monopoly practices of Microsoft: governments can't solve these things, only industry can—in this case, the pharmaceutical industry. Nor can philanthropy do it either, at least not alone, perhaps one reason that the Gates Foundation continues to invest in the pharmaceutical sector to deliver vaccines to the world's population and the giant agricultural businesses in order to provide GMO crops to the world's hungry.

Why does *National Geographic* have anything in particular to say about COVID-19 and Gates that would warrant a major interview? Maybe it's because they also take money from the Gates Foundation? In fact, they do, as journalist Tim Schwab has shown.

These sorts of media reports have, until very recently, been typical of most of what Gates receives, even from those outlets that are more likely to be left of center in overall outlook. Some of these, however, may have begun to look a bit more closely at Gates's philanthropy from his eponymous foundation, and the latter's driving purpose, noting that while Gates may not be the major contributor to the WHO, his current 14 percent contribution to the organization's overall budget surely buys him considerable sway as the largest non-State actor. All of this fits neatly into the growing role Gates has carved out for himself over the last months as a COVID-19 guru and savior.

Perhaps the lengthy honeymoon in which any serious questions of the Gates Foundation's long-range intentions are avoided might be coming to an end: even though media such as National Public Radio (NPR) in the United States and *The Guardian* newspaper in the United Kingdom (both recipients of Gates Foundation money) carry on the glowing Gates narrative, lately that narrative has come under serious review.

In March 2020, just as COVID-19 began to ramp up in the West, *The Nation* featured an article, "Bill Gates's Charity Paradox," by journalist Tim Schwab.[17] The article ran under the headline "A *Nation investigation illustrates the moral hazards surrounding the Gates Foundation's $50 billion charitable enterprise.*"

In the article, Schwab took a look at a 2019 Netflix three-part documentary called *Inside Bill's Brain*. Schwab wrote that

> . . . the film reinforces the image many of us already had of the ambitious technologist, insatiable brainiac, and heroic philanthropist. *Inside Bill's Brain* falls into a common trap: attempting to understand the world's second-richest human by interviewing people in his sphere of financial influence.

Schwab went on to write that "Gates has proved there is a far easier path to political power, one that allows unelected billionaires to shape public policy in ways that almost always generate favorable headlines: charity."

Just like Carnegie a century before, Gates and his foundation seemed to have rediscovered the magic formula.

Schwab continued:

> Through an investigation of more than 19,000 charitable grants the Gates Foundation has made over the last two decades, *The Nation* has uncovered close to $2 billion in tax-deductible charitable donations to private companies-including some of the largest businesses in the world, such as GlaxoSmithKline, Unilever, IBM, and NBC Universal Media-which are tasked with developing new drugs, improving sanitation in the developing world, developing financial products for Muslim consumers, and spreading the good news about this work.

And,

> *The Nation* found close to $250 million in charitable grants from the Gates Foundation to companies in which the foundation holds corporate stocks and bonds: Merck, Novartis, GlaxoSmithKline, Vodafone, Sanofi, Ericsson, LG, Medtronic, Teva, and numerous start-ups - with the grants directed at projects like developing new drugs and health monitoring systems and creating mobile banking services.

Much of Schwab's analysis is backed up by Martin Levin writing in *Nonprofit Quarterly*, particularly in regard to the power that the Gates Foundation wields with politicians around the world as the Foundation's money continues to press for corporate solutions to a world crisis.[18]

And there you have it, the list of the key things that the Gates Foundation invests in: pharmaceutical companies making vaccines, indeed almost a who's who of major vaccine companies, ways to monitor human behaviors, and potential means to take control of money. Individually, bad enough, at least the vaccine component. Linked together, maybe a foundation and an individual creating a web of control for all humans everywhere?

It is hard, at least for me, to look at this list and not also see Gates's various interviews with his ever-moving hands spinning webs. Or to be more charitable, maybe the Foundation is simply able to perceive their future opportunities for financial growth tied to these companies? Maybe investments by the Foundation are merely a win-win situation from a corporate perspective: follow an agenda *and* make more money for the Foundation at the same time.

Schwab noted that

By Bill and Melinda Gates's estimations, they have seen an 11 percent tax savings on their $36 billion in charitable donations through 2018, resulting in around $4 billion in avoided taxes. The foundation would not provide any documentation related to this number, and independent estimates from tax scholars like Ray Madoff, a law professor at Boston College, indicate that multibillionaires see tax savings of at least 40 percent—which, for Bill Gates, would amount to $14 billion—when you factor in the tax benefits that charity offers to the superrich: avoidance of capital gains taxes (normally 15 percent) and estate taxes (40 percent on everything over $11.58 million, which in Gates's case is a lot).

For an alleged charity, and for a man whose wealth seems primarily to be used to amass vast power, the payoff for his donations may have found the ultimate sweet spot.

Schwab quoted a Gates critic, James Love, director of Knowledge Ecology International, to emphasize this point:

He uses his philanthropy to advance a pro-patent agenda on pharmaceutical drugs, even in countries that are really poor . . . 'Gates is sort of the right wing of the public-health movement. He's always trying to push things in a pro-corporate direction. He's a big defender of the big drug companies. He's undermining a lot of things that are really necessary to make drugs affordable to people that are really poor. It's weird because he gives so much money to [fight] poverty, and yet he's the biggest obstacle on a lot of reforms.' [Brackets are mine.]

As the figure from *The Nation*, reproduced in Fig. 12.2, demonstrates, the Gates Foundation's fifty-billion-dollar endowment has generated $28.5 billion dollars of income against an outlay of $23.5 billion in "charitable" grants over the last five years. From these numbers, it is clear that one can push forward an agenda, donate to charities, and still reap a significant profit.

Soon after the *Nation* article appeared, Schwab followed up with another equally damning article in the *Columbia Journalism Review* titled "Journalism's Gates keepers."[20] In this article, Schwab took a close look at how the Gates Foundation uses its wealth to manipulate media depictions of both Gates and the Foundation.

One of the news outlets highlighted in the Schwab article is NPR. Schwab notes that in relation to media taking philanthropy donations, it can lead to some often overt manipulation of how reporting is done: "Nowhere does this concern loom larger than with the Gates Foundation, a leading donor to newsrooms and a frequent subject of favorable news coverage."

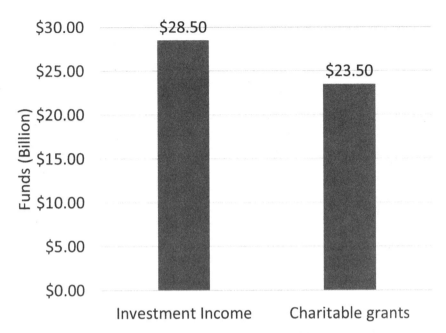

Company	Received donations
Merck	$9.4 million
LG	$53 million
Sanofi	$3.5 million
Eli Lilly	$3 million
Ericsson	$3 million
Takeda	$23 million
Unilever	$2.7 million
Pfizer	$16.5 million
Novartis	$11.5 million
Teva	$11.4 million
Philips	$1.7 million
Lixil	$1.5 million
Medtronic	$100,000

Figures 12.1 and 12.2. Redrawn graphs of changes in the wealth of the Bill and Melissa Gates Foundation.[19]

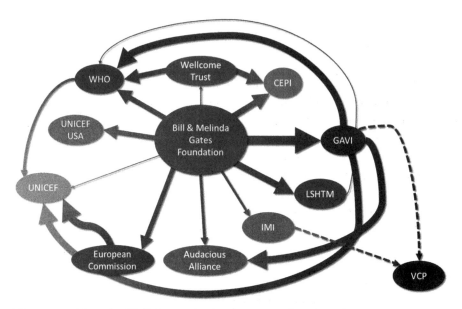

Figure 12.3. The web of Bill Gates: what his fortune touches through the donations of the Bill and Melinda Gates Foundation.

Building on this, Schwab went on to show precisely how the Gates Foundation's donations have changed reporting about Gates overall:

> Gates's generosity appears to have helped foster an increasingly friendly media environment for the world's most visible charity. Twenty years ago, journalists scrutinized Bill Gates's initial foray into philanthropy as a vehicle to enrich his software company, or a PR exercise to salvage his battered reputation following Microsoft's bruising antitrust battle with the Department of Justice. Today, the foundation is most often the subject of soft profiles and glowing editorials describing its good works.

And in our new age of COVID-19, nowhere is this sort of bought pro-Gates media coverage more prominent than with stories about the pandemic and Gates's role in fighting it:

> . . . news outlets have widely looked to Bill Gates as a public health expert on Covid-even though Gates has no medical training and is not a public official. *PolitiFact* and *USA Today* (run by the Poynter Institute and Gannett, respectively-both of which have received funds from the Gates Foundation) have even used their fact-checking platforms to defend Gates from 'false conspiracy theories' and 'misinformation,' like the idea that the foundation has financial investments in companies developing Covid vaccines and therapies.

In fact, the foundation's website and most recent tax forms clearly show investments in such companies, including Gilead and CureVac.

Put another way, not only do media that takes Gates's money report favorably on the Foundation's efforts, but also actively protect him and his Foundation from those asking uncomfortable questions.

Of course, the same media sources made sure to steer clear of linking the eventual COVID-19 pandemic to the October 18, 2019, Event 201 held at the Johns Hopkins Center for Health Security,[21] an event delivered in New York City "in partnership with the World Economic Forum and the Bill and Melinda Gates Foundation."

The World Economic Forum described their mission on their website thus:

> *The World Economic Forum is the International Organization for Public-Private Cooperation.* The Forum engages the foremost political, business, cultural and other leaders of society to **shape global, regional and industry agendas**. It was established in 1971 as a not-for-profit foundation and is headquartered in Geneva, Switzerland. **It is independent, impartial, and not tied to any special interests**. The Forum strives in all its efforts to demonstrate entrepreneurship in the global public interest while upholding the highest standards of governance. Moral and intellectual integrity is at the heart of everything it does.[21] [Bolding is mine.]

Is it just me, or is the need to "shape global, regional and industry agendas" contradictory to "It is independent, impartial, and not tied to any special interests"?

Clearly, the WEF shares the same vision as the Gates Foundation, that is, a world in which the corporate sector plays a dominant role in all aspects of human affairs.

Event 201 featured an "imaginary" scenario about a high-level pandemic with the attendant economic and social disruptions. The similarities between this scenario and the later actual events were remarkably prescient. Or just maybe Event 201 was field testing a script of the future real-world events? I'll revisit this thought in the last chapter.

Two articles critical of Gates and the Gates Foundation in clearly left-of-center journals in the space of several months? Maybe Tim Schwab just has particular hate for Bill Gates? Or maybe he is seeing the Gates machine for what it clearly is? Are any others on the left also beginning to ask questions?

To find out, let's look at a recent article by Vandana Shiva, a legendary antiglobalization activist.

In a recent article published in *Independent Science News,* she writes that, "Gates' 'funding' results in an erasure of democracy and biodiversity, of nature and culture. His 'philanthropy' is not just philanthrocapitalism. It is philanthroimperialism."[22]

Based on Shiva's history and the biography cited in the article, she is not your conventional conspiracy theorist and hardly alt-right. Rather, as described in the article, she is "a philosopher, environmental activist and eco feminist" and "founder/director of Navdanya Research Foundation for Science, Technology, and Ecology." Shiva has also long been a virtual icon of the left in the fight against economic globalization.

Shiva laid out her argument that Gates and his foundation are not the humanitarian entity that most media portray them to be, rather something far more sinister. In the article, Shiva wrote in a caption below a picture of Gates in one of his now customary Uncle Bill sweaters, "As I look to the future in a world of Gates and Tech Barons, I see a humanity that is further polarized into large numbers of 'throw away' people who have no place in the new Empire."

Commenting on Gates's use of a war analogy to describe a future pandemic given in a 2015 TED talk[23] and now concerning the current situation, Shiva observed that

> In fact, the pandemic is not a war. The pandemic is a consequence of war. A war against life. The health emergency of the coronavirus is inseparable from the health emergency of extinction, the health emergency of biodiversity loss, and the health emergency of the climate crisis. All of these emergencies are rooted in a mechanistic, militaristic, anthropocentric worldview that considers humans separate from-and superior to-other beings . . . All of these emergencies are rooted in an economic model based on the illusion of limitless growth and limitless greed, which violate planetary boundaries, and destroy the integrity of ecosystems and individual species.

The "limitless growth and boundless greed" that Shiva speaks about is none other than the expression of another example of disaster capitalism itself, hyperpredatory capitalism by any other name and the very system that made Gates a multibillionaire and whose influence continues to flow through everything his Foundation says and touches.[24] One might argue that this is the latest iteration of capitalism as a system of economics that has reached a particularly dangerous phase for humanity and the planet that in the distant past was sometimes under some rational control. Or, going all the way back to Marxist assessments of capitalism, is the current stage the inevitable consequence of a system designed to maximize profit no matter what the cost to the world and its people?[25] The culmination of all of this is the vision of the WEF for the "Great Reset" or "4th Industrial Revolution." I will come back to all of this in Chapter 14.

Shiva went on to describe how the poor around the world have had their livelihoods all but destroyed by the lockdowns, pushing additional hundreds of thousands into even greater poverty and hunger. Quoting the World Food Program, Shiva stated that one consequence could be three hundred thousand fatalities every day among this already marginalized population.

Expanding on Gates's role, Shiva wrote that

> Health is about life and living systems. There is no 'life' in the paradigm of health that Bill Gates and his ilk are promoting and imposing on the entire world. Gates has created global alliances to impose top-down analysis and prescriptions for health problems. He gives money to define the problems, and then he uses his influence and money to impose the solutions. And in the process, he gets richer. His 'funding' results in an erasure of democracy and biodiversity, of nature and culture. His 'philanthropy' is not just philan-throcapitalism. It is philanthro-imperialism . . . The extended coronavirus lockdown has been a lab experiment for a future without humanity.

Shiva's concerns about the scope of the Gates Foundation's intervention in human affairs is supported by a recent article in *Jacoben* by Jan Urhahn that looks at the Gates Foundation partnership with the WHO to deliver a "Green Revolution to the farmers in Africa."[26]

Shiva noted also a clearly demonstrated fact, sometimes dismissed as yet another conspiracy theory of the alt-right: on March 26, 2020, Microsoft applied for a patent, WO 060606, dealing with the monitoring of human bio-logical activity, such as brain activity, fluid and blood flow, organ function, and eye, facial, and skeletal muscle movement. These parameters were to be potentially linked to a cryptocurrency system. Seeing where all of this might be going, Shiva wrote that

> The patent is an intellectual property claim over our bodies and minds. In colonialism, colonisers assign themselves the right to take the land and resources of indigenous people, extinguish their cultures and sovereignty, and in extreme cases exterminate them. Patent WO 060606 is a declaration by Microsoft that our bodies and minds are its new colonies. We are mines of 'raw material'—the data extracted from our bodies. Rather than sovereign, spiritual, conscious, intelligent beings making decisions and choices with wis-dom and ethical values about the impacts of our actions on the natural and social world of which we are a part, and to which we are inextricably related, we are 'users.' A 'user' is a consumer without choice in the digital empire . . . But that's not the totality of Gates' vision. In fact, it is even more sinister - to colonise the minds, bodies, and spirits of our children before they even have

the opportunity to understand what freedom and sovereignty look and feel like, beginning with the most vulnerable.

Shiva went on to write that in May 2020, New York's Governor Cuomo announced that the state had partnered with the Gates Foundation to "reinvent education," as noted previously. Who knew that Gates was an expert on pedagogy? Surely we must now add this skill set to Gates's skills in medicine.

All of this was widely ignored by much of the left, apparently absorbed in identity politics and hell-bent on seeing any critique of the underpinnings of the pandemic as merely more selfish, irresponsible alt-right conspiracy mongering. I will return to the pathetic responses to the pandemic by much of the left in the last chapter.

There is, however, hope, as people like Tim Schwab, Vandana Shiva, and others begin the process of pulling apart, brick by brick, the Potemkin village edifice of philanthropy that Gates has created with his vast wealth.[27]

The real picture of Gates may be starting to come into focus. What Uncle Bill in his cashmere sweater is weaving with his constantly moving hands is nothing less than a web of control made up of strands from industry and government, all designed to capture humanity by controlling health, food, communications, resources, and indeed our very individuality.

In the next chapters, I will explore the agenda that Gates and his Foundation have served and also try to get a glimpse of the forces who share this agenda but have decided that Bill himself is expendable.

Shiva completed her article with a dire warning and a choice for us all:

> We stand at a precipice of extinction. Will we allow our humanity as living, conscious, intelligent, autonomous beings to be extinguished by a greed machine that does not know limits and is unable to put a break on its colonisation and destruction? Or will we stop the machine and defend our humanity, freedom, and autonomy to protect life on earth?

This is indeed the very question I will probe in the last chapter. But before I do, we need to take a good hard look at the last piece of the puzzle that will serve to answer the question of who benefits. The answer, apart from Gates in his headlong rush for more money and especially more power, is none other than the penultimate elephant in the room: the pharmaceutical industry.

The Role of the Big Pharmaceutical Industry in Human Health and Disease

One of the first serious exposés of the actual designs, goals, and practices of the pharmaceutical industry was provided in Dr. Marcia Angell's revealing 2004 book, *The Truth About the Drug Companies, How They Deceive US and What We Can Do About It.*[28] Dr. Angell back then was certainly no conspiracy theorist and no opponent of vaccination. Rather, she was the former editor of the prestigious, at least back then, *New England Journal of Medicine*, a position she held for twenty years. She was, however, a staunch critic of the pharmaceutical industry as it appeared to her when the book came out sixteen years ago. One wonders how she would find the industry now in the age of COVID-19.

Angell began her book by noting how much money people in the United States spend on drugs. Then, and this is based on data gathered in 2002, the number was two hundred billion dollars per year in the United States with four hundred billion dollars raked in worldwide. Angell went on to note that for the top fifty drugs used in the United States, for the average cost for senior citizens, who often are on many drugs simultaneously for a variety of ailments, an individual's cost could come out to $1,500 a year per drug. For people on fixed incomes, at least in the United States, such expenses sometimes lead to a hard decision about whether to buy pharmaceutical drugs or put food on the table or heat their houses in the winter.

The decision on which to do was often influenced by slick industry advertisements that played constantly on television when the book came out. One can still see such ads on television in the United States. These ads typically feature attractive people doing some sort of fun outdoor activity with the note that a certain medical condition threatens their health, or even their lives. The advertisements then go on to promote a drug that the advertiser (some pharmaceutical company) says may solve the problem. This pronouncement is always followed by a list of potential adverse effects that are shown in text and announced, in brief, against the same background of people having fun. The commercial then fully segues back to the people enjoying life with the final tagline of "Ask your doctor if 'drug X' is right for you." The pharma's marketing research tells them that a significant number of people go on to decide that the drug is indeed right for them and thus demand that their doctors prescribe it.

The pharmaceutical industry had already become the largest lobby group in Washington, DC, by 2004, eclipsing even the defense industry, whose lobbying led to their massive profits from the 2003 Iraq War. Sixteen years later, the pharmaceutical industry's status as the largest lobbying group is certainly still true. The industry then, as now, heavily contributes to members of Congress on both sides of the aisle. In 2002, the pharma employed 675 lobbyists on Capitol Hill. How many pharma lobbyists are there in 2020? Vastly more, as cited by Dr. Toby Rogers, quoted later.

Although the pharma used to devote the bulk of their campaign contributions to Republican members of Congress, Robert F. Kennedy Jr. found that the Democrats had caught up and took equivalent donations from the pharma lobby once their own party restricted the funds they could take from other sources.[29] The industry's shift from focusing on the historically pro-business Republicans to the (then-) more socially-progressive Democrats came about in the 1990s and has continued to the present.

In many ways, at both the federal and state levels, Democrats have been become the staunchest supporters of big pharma. Given this, it is difficult to imagine that the campaign funding they receive does not play a significant role in their support for anything the pharma desires, not least when it comes to vaccine mandates.

Angell traced the dramatic ascendency of the pharma from the early 1980s under Ronald Reagan's presidency with the onset of "technology transfer" legislated in the Bayh-Dole Act. This legislation basically enabled universities to patent discoveries made by their researchers that had, in most cases, been paid for with taxpayer dollars. Most research funding in the biomedical sciences in the United States, then and now, arises from the NIH, described earlier, and to a lesser extent from the National Science Foundation and Veteran's Administration. All of these are taxpayer-funded entities of the federal government. In regard to the NIH, it takes consulting fees from the industry, and even though by doing so they may recoup some tax payer dollars, it also ties NIH financially to the industry.

When universities became able to sell off the discoveries of their scientists to companies, it essentially acted as a huge financial benefit to the pharma, which could buy these discoveries for pennies on the dollar in cash or royalties, or both. Yes, they did have to pay something, but they no longer needed to maintain the same level of research infrastructure in people and material.

As Angell described, when the pharma claims that the exorbitant prices they charge the public for their drugs are due to research and development costs, what they are really talking about is the cost of buying up publicly funded discoveries and then advertising the drugs that arise from these same discoveries.

Universities in Canada do much the same to promote corporate aspirations. When I first joined the Faculty of Medicine at my university in the late 1980s, there was already a highly active University-Industry Liaison Office that not only facilitated the process of transferring intellectual property to drug companies, but actively sought out novel discoveries in order to do so. A recent example of how this corporatization of my university is still alive and well is shown in a webinar announcement that suggests ways for investigators to market their work (see Appendix 9).

The same sort of university collusion with industry was seen in the case of Dr. Andrew Wakefield (Chapter 4), where Brian Deer and others accused Wakefield of being primarily motivated by commercial dreams.

The press conference that Wakefield attended with his dean after his 1998 article came out was clearly part of the Royal Free Hospital's attempt at commercialization.

My university and the Royal Free are hardly alone in this sort of behavior; indeed, it is now virtually universal in academia. The monetization of basic science demonstrated over the years could not but have a profound effect on the ethics and behaviors of universities and their medical schools wherever and whenever such technology transfer arrangements are permitted.

Angell also noted that the changes in the 1980s led to drug sales going up thirtyfold between then and when her book was written. Certainly it has gone up many times since in the ensuing years.

The next act passed by Congress was the Hatch-Waxman Act that served to extend monopoly rights for patented drugs, the patents often issued for the so-called "me too" drugs that were not novel, but rather variations on existing drugs that may have lost patent protection. As a perfect example of bait and switch, Angell described how in 2001 AstraZeneca, one of the major pharmaceutical companies, faced an expiring patent on their bestselling heartburn drug, Prilosec. The solution was to simply replace it with Nexium, another proton pump inhibitor. The essential difference was not the active molecule used, but that with Nexium the company had many future years of patent protection. As a side note, AstraZeneca is the manufacturer of one of the main COVID-19 vaccines to be described in Chapter 13.

As Angell wrote,

> This is an industry that in some ways is like the Wizard of Oz - still full of bluster but now being exposed as something far different from its image. Instead of being an engine of innovation, it is a vast marketing machine. Instead of being a free market success story, it lives off government-funded research and monopoly rights.[30]

And,

> Big pharma likes to refer to itself as a "research-based industry," but it is hardly that. It could be best described as an idea-licensing, pharmaceutical formulating and manufacturing, clinical testing, patenting, and marketing-industry.

Angell also noted that the industry tends to promote diseases that can be made to match the drugs they already have, a point that will be reinforced below in the evaluations of the industry by Cassels and Moynihan.

The drug trials that the pharma does perform do not always serve the purpose that the FDA intended. For example, Phase 4 trials, the so-called "commitment studies" that the pharma is supposed to do as part of longer-term safety surveillance, are really not what they are cracked up to be. First, the companies really have no clear financial incentive in doing these at all, since they cost money. Worse, what if the company doing a Phase 4 study found that there were safety problems associated with the drug that had not been seen prelicensing? Angell stated that this being so, the best the industry could do, from their perspective, would be to use Phase 4 trials to identify other market demographics that could be targeted to buy the drug in the future, perhaps for newly designed diseases.

The FDA, which demands these studies, lacks the capacity to effectively regulate whether these studies get done at all. As a result, the companies simply rarely do them. A case in point concerns HPV vaccine Phase 4 trials of Merck's Gardasil, which have never been done.

One key bit of legislation that made vaccines incredibly risk-free and thus more lucrative for the pharma was the National Vaccine Compensation Act of 1986, described in detail in previous chapters. This Act effectively made the pharma's vaccines bulletproof in that they then became liability-free and hence not a possible target of any legal action by families of the vaccine injured.

In 1992, Congress passed the Prescription Drug Use Fee Act that made drug companies pay user fees to the FDA, functionally putting the FDA in a conflict-of-interest position with the very companies they were supposed to be regulating. As Angell described, user fees paid to the FDA were initially intended solely to expedite the approval of new drugs. However, at $310,000 per new drug (in 2002), such fees rapidly became a major source of FDA funding. Drugs were often rapidly approved by advisory committees in which conflicts of interest abounded.

Angell estimated the value of the fees paid back in 2002 to be about half of the agency's budget. How much of the FDA's budget these fees now produce is anyone's guess, but with more drugs coming up for approval, this amount can only have grown.

Angell completed her description of the pharmaceutical industry's practices and their deep penetration into Congress by quoting Senator Bernie Sanders:

> Even the New York Yankees sometimes lose, and it has been known that, on occasion, the Los Angeles Lakers lose a ballgame. But one organization never loses, and that organization has hundreds of victories to its credit and zero defeats in the United States Congress. And that is the pharmaceutical industry.[31]

Angell's critiques of the pharmaceutical industry caused a small ripple when her book came out in 2004 and then faded from view. The last is not surprising in the least: if you control the media and the media decides what is, or isn't, news, then stories critical of the pharma, or Bill Gates, fade from view amazingly quickly. That is assuming they get reported at all. Sixteen years later, it would be a safe bet that critiques of either would vanish even sooner, less the increasing attacks on Gates after May 2021.

An equally damning portrayal of the industry came out a year after Angell's book in which Alan Cassels, a Victoria-based journalist, and Ray Moynihan, a scientist at my own university, looked at how the industry was not content merely to acquire the work of others for very little capital outlay and manipulate prices on the resulting drugs. Rather, the industry started to take it to an entirely different level by actively *inventing* diseases for drugs they already owned. In many cases, these were drugs that were nearing the end of their patent protection.[32]

The key theme of *Selling Sickness* is that the pharmaceutical industry decided long ago that there were just too many healthy people to allow the industry to reap the profits they hoped to accrue. The answer was obvious, at least to them: invent diseases for which they already had drugs that could be repurposed to these newly invented conditions.

In this scheme, it didn't ever really matter if the drugs worked for the illness or not, only that enough people could be convinced that they had these novel diseases and that the solution lay with the products the industry was peddling.

As the authors stated,

> The idea that drug companies help to create new illnesses may sound strange to many of us, but it is all too familiar to industry insiders. A recent *Reuters Business Insight* report designed for drug company executives argued that the ability to 'create new disease markets' is bringing untold billions in soaring drug sales. One of the chief selling strategies, said the report, is to change the way people think about their common ailments, to make 'natural processes' into medical conditions. People must be able to be 'convinced' that 'problems they may previously have accepted as, perhaps, merely an inconvenience' - like baldness, wrinkles and sexual difficulties - are now seen as 'worthy of medical intervention.' Celebrating the development of profitable new disease markets like 'Female Sexual Dysfunction,' the report was upbeat about the financial future for the drug industry. 'The coming years will bear greater witness to the corporate sponsored creation of disease.'

"If you build it, they will come," to quote from the 1989 movie *Field of Dreams*. And indeed, customers did come to buy drugs for a range of new, manufactured "diseases" including: high cholesterol, depression, menopause, high blood pressure, premenstrual dysmorphic disorder, social anxiety disorder, osteoporosis, irritable bowel disease, and female sexual dysfunction.

Remember, this list was current in 2005. How many more new diseases have been added since then? No doubt many, but the real market with explosive growth since 2005 has been in vaccines.

As Cassels wrote in a recent article quoting the CBC on COVID-19 and the emerging prospect of COVID-19 vaccines:

> Pharma has the best of many worlds when it comes to vaccines: limited, and often biased research to support the approval, limited risks in terms of getting sued for safety reasons and then societies like paediatricians and public health people doing all their promotion for them. All the while a highly fearful, easily manipulated public will be eager customers to demand the vaccines, told from the beginning by the most trusted Public Health doctor, Dr. Henry [we have met Dr. Henry before several times in earlier chapters and will again in the next chapter], that only a vaccine is going to save us. What a license to print money.[33] [Brackets are mine.]

And also from Cassels: "The big difference between drugs and vaccines is that most of the marketing of vaccines is done by public health officials, as seen in this piece on CBC today."[34]

And indeed it is, and indeed the pharmaceutical industry's plans to massively profit from the pandemic, aided and abetted by Bill Gates and the WHO. Cassels went on to consider the pharmaceutical industry's hoped for profits from COVID-19 vaccines:

> . . . I predict Covid will die off and disappear long before any vaccine is developed, but since so much money has already been poured into the research, and so many public dollars banking on it, plus so many people believing it's the holy grail, even a crappy vaccine is likely to be extremely profitable.[35]

Cassels may have been wrong about the first part because of the rushed EUA that Moderna and Pfizer have just received, but he is likely correct about the final financial windfalls from the vaccines.

Cassels went on to note the power of the frame when it comes to vaccines, even on researchers whose own work should make them know better: "If Skowronski, with a straight face, can sell repeated flu shots every year, despite her own research that shows some vaccines have actually increased one's susceptibility to the flu in subsequent years, then she can sell anything . . ."[36]

Skowronski, whom we met in Chapter 8, is the BC CDC researcher who had provided both an epidemiological and an in vivo animal study showing that the H1N1 influenza vaccine of 2009 rendered later influenza vaccines less effective. As Cassels observed, Skowronski now seems to have joined the chorus of health experts recommending getting the newest influenza vaccine to help lessen the impact of COVID-19.

The evidence for massive malfeasance by the pharmaceutical industry continues to grow, book by book, article by article. At least, that is, when a journal or media outlet is not controlled by being recipients of Gates Foundation money. But, as Tim Schwab has indicated, it is getting harder and harder to find media entities not locked into Gates's financial embrace, at least until recently

There are now converging lines of evidence from all over the political spectrum that the pharma is running rampant over our health in the name of profit. From Kennedy, cited in Chapter 7, to Angell and Moynihan and Casals in this chapter, a picture emerges that is not very laudatory.

The latest revelations come from two very divergent directions. The first is from Dr. Judy Mikovits, as revealed in her recent book with Kent Heckenlively, *Plague of Corruption*,[37] which follows a previous book about her, *Plague*, with Heckenlively as the lead author.[38]

Mikovits is not so much going after the pharma in these books, although the pharma's hold on research would have to be considered a central theme. Rather, she is after modern biological science itself for having sold its collective soul to the pharma. As we have seen, at least in medicine, the two are largely entwined.

Before becoming a book author, Mikovits was a highly regarded scientist, the head of a major cancer lab at the NIAID, Anthony Fauci's branch, with multiple grants. She then had the scientific "misfortune" to discover viral contamination in some vaccines as she sought the origin of chronic fatigue syndrome (CFS).[39]

The problem for Mikovits was not the discovery per se, but rather that she decided to be honest and report it. The discovery of the contamination, in this particular case of a virus called XMRV linked to CFS, led, in short order, to her being stripped of her position and even briefly jailed.

Plague of Corruption reads like a spy novel, so much so in fact that it is tempting to view it as fiction or the work of an unbalanced mind. The evidence Mikovits presents, however, as well as the growing realization that vaccine product safety is sometimes careless at best and appallingly criminal at worst as detailed in Chapter 3, makes it impossible to dismiss her evidence.

Backstopping Mikovits's claims are the data on the cancer-causing virus Simian Virus 40 (SV40) found in vaccines, as documented by journalist Debbie Bookchin and Jim Schumacher.[40] Mikovits also dived into the contention that "acquired (human) immunodeficiency" (AIDS) arose from the production of a polio vaccine used in the former Belgian Congo in the 1950s, although this view is disputed.[41]

More recent reports support Mikovits's contentions about contamination: Various vaccines sold in Italy contain a range of contaminants[42] and may not even contain the antigens that they are expected to have in order to trigger an immune response.[43] If correct, such levels of contamination continue to occur without any apparent regard for the consequence to human health.

These results from the two Italian groups have not been replicated to date nor has the second one been peer-reviewed. If replicated by other laboratories, however, both studies would point to significant problems in vaccine production and a cover-up by various authorities. As the cumulative evidence indicates so far, given the demonstrated behavior of the pharmaceutical industry over the last forty years, none of the above should any longer seem improbable.

It would be difficult for any normal person to not recoil from the suggestion of corporate malfeasance at such an audacious level, supported, and even promoted, by the Gates Foundation and the WHO.

One of the clearest examples of WHO mendacity can be found in an article I helped write with a colleague, Prof. John Oller. The article addressed the apparently deliberate attempts at population control using human chorionic gonadotropin (hCG) conjugated to parts of a DPT vaccine given to young women of reproductive age in Kenya.[44]

The official reason for the vaccine campaign, sponsored by the WHO, was to control neonatal tetanus. A number of independent analytical laboratories in Kenya were later able to confirm the presence of hCG where none should have been present.

Could any of the above even be partially true? The answer is that all of it could be. A definitive demonstration that it is was recently provided in a remarkable PhD thesis from Australia whose author, Dr. Toby Rogers, has connected the various strings linking the pharma, the Gates Foundation, and the WHO in trying to hide the origins of modern-day ASD.

Rogers's thesis, titled *The Political Economy of Autism*,[45] follows all of these threads and shows how they interconnect the key players.

A schematic illustrating some of these relationships was shown in Fig. 12.3 (on page 386). One can, of course, deny that all of these entities are tied together to deliver some agenda and that believing otherwise is just utter nonsense "conspiracy theory" material.

But to do so and thus label the evidence as such would simultaneously have to deny any such connections. This would be difficult, if not impossible, to do given how well such connections have been documented in this chapter and elsewhere. Further, to assume that all of these entities are operating independently in pursuit of the same goals stretches Occam's Razor to the breaking point.

Recall that Occam's Razor states that the simpler of the explanations that equally explain the observations is more likely to be correct. To accept that there is no collusion between all the players discussed so far in this chapter would be to massively ignore this principle.

Rogers intended his thesis to follow the autism epidemic, demonstrating, first, that it actually exists, as described in Chapter 4. Rogers then went on to examine possible factors responsible for autism and showed how unlikely it was that autism is a genetic disease. Looking at an environmental level, Rogers examined a range of toxicants that *could* contribute to the rising incidence of ASD and finally examined vaccines and the ever-increasing pediatric vaccine schedules.

Rogers's thesis provides a clear exposition of the role of a largely unregulated pharmaceutical industry that has accomplished what he termed "regulatory capture" of all the official regulatory agencies: the CDC, FDA, and NIH—and even significant parts of the US Congress.

Where Rogers charts new territory is when he explores "political epidemiology," defined by other authors he cites as ". . . the study of the impact of welfare regimes, political institutions, and specific policies on health and health equity."[46]

Rogers's methodology then looked at what he termed "three descriptive epidemiological factors (time, place, and person), as well as the sorts of factors discussed in political epidemiology (toxicants, law, and regulatory institutions)."

Much of Rogers's thesis was devoted to understanding "theories of power" because, as he noted, this is:

> . . . important for the study of any disease or complex health condition because power shapes what is studied, what is not studied, and how matters are studied. Science and society are intertwined in epidemics and the response to an epidemic is a reflection of power operating on many dimensions.[47]

And,

> Lukes' (1974) three dimensional view of power has enormous implications for the study of autism. Autism at first appears to fit the pluralist view of power - a battle between competing groups over resources. But autism also fits the second dimension of power as autism prevention has, thus far, been kept completely off the political agenda. And power also shapes the way that families and autism groups conceive of what is possible politically in the midst of this crisis - focusing for example on insurance reform rather than banning of certain toxic chemicals.[48]

Further, Rogers mirrored Angell in noting that this shift occurred with the passage of the Bayh-Dole act in 1980, an act that was mimicked in other countries as the thrust of neoliberalism swept much of the globe.

Based on his analysis, Rogers concluded that

> Corporate interests, not human health, continue to be the overriding force determining regulatory outcomes. Scientists have known for at least 100 years that lead is toxic; lead poisoning is a preventable disease; and yet both policy-makers and the courts refuse to hold industry responsible . . .[49]

And citing yet another study, Rogers quoted one Bruce Lanphear in stating that ". . . unless it serves the needs of private enterprise, public health is incapable of controlling the causes of chronic disease and disability."[50]

The above considerations are not specific to vaccines and autism, but rather to the nature of capitalist exploitation of the environment and human health, whether from asbestos, tobacco, pesticides . . . or vaccines. In other words, any product that will make the corporations money is going to be promoted regardless of the consequences to human health.

In each case, the companies are aided and abetted by US agencies mandated to protect the health of Americans. These include those we have discussed before: the CDC, the FDA, and the NIH. The process plays out in much the same way in other countries with their own regulatory agencies.

Updating the level of lobbying that makes failure of effective regulation possible, from the Angell book in 2004 to today, Rogers wrote that

> The capture of regulatory agencies is made possible by the capture of elected officials. The pharmaceutical/health products industry spent over $4 billion between 1998 and 2018 to lobby elected officials and candidates, more than any other industry . . . The number of pharmaceutical company lobbyists increased from 729 in 1998 to 1,803 in 2008, before declining somewhat to 1,440 in 2018 . . .

In 2018, 919 (63.8%) of the lobbyists employed by the pharmaceutical industry had previously "worked in Congress or another branch of the federal government . . ."[51]

Converging Lines of Evidence about the Pharma and the Endgame for COVID-19

In science, experimental verification of some hypothesis provides concrete support that the hypothesis actually describes how some aspect of nature works. As discussed earlier, this is not proof, but rather a matter of probability. The level of probability increases with each verification of the hypothesis in similar experiments by replication of the original finding as well as by corollary studies that add details to the emerging understanding of the phenomenon in question. Additionally, different methods provide converging lines of evidence that lend even greater support. For example, if we are trying to demonstrate that aluminum adjuvants have adverse health impacts on the CNS in humans, an in vivo experiment repeated several times in experiments with mice provides this first level of support. If in the same sorts of studies we can show adverse aluminum outcomes at various levels such as behavioral, cellular, and biochemical, it is stronger still. The evidence that the impacts of aluminum in the CNS are detrimental is made stronger if the same sorts of behavioral and neuropathological outcomes can be demonstrated in other species, such as in sheep, rats, or other animals, including humans. The evidence for this hypothesis is further strengthened by cell culture experiments.

In the same way, the converging lines of evidence for the existence of a complex megacartel of interacting players attempting to seize control of human health lends greater support to the idea that such a cartel in fact exists. For this reason, the actions of the WHO as described earlier and in Chapter 10,[52] the revelations about Bill Gates and the Bill and Melinda Gates Foundation provided by various researchers, and the overlapping descriptions by Angell, Moynihan and Cassels, Mikowits and Heckenlively, and Rogers of the manipulations of the pharmaceutical industry all lead to a similar conclusion.

Crucially, the Rogers thesis provides a starting point for understanding the clear, albeit complex, web of interconnections at play in the development, and potential for remediation, of autism. As Rogers pointed out, a key player here is the pharmaceutical industry, a multibillion-dollar entity that seeks to make humans (and animals) endless consumers of their products. It does this by making drugs for real diseases and selling them at exorbitant prices, as discussed by Angell, or by manufacturing diseases for which they already have a product. The latter, as described by Cassels and Moynihan, works as well as, or better than, prescribing for real illnesses.

In essence, the goal of the pharmaceutical industry is to make every human a lifelong customer, from womb to cradle to grave. They have not always gotten away with this audacious scheme, but, as Rogers described, the coming of Ronald Reagan and the neoliberal agenda in the United States and around the world set the stage for the capture of "regulatory control" by the pharma.

The pharmaceutical companies' control of the various levers of political power at all levels was the next milestone in their long-term goals. It took them forty years to accomplish, but to a large extent this goal has been achieved. The same companies next turned their eyes on medical organizations and medical schools and successfully, for the most part, were able to extend their control in the medical sphere as well, as fully detailed by both Rogers and Angell.

While these regulatory control measures apply to the overall business plan, nowhere is this scheme more apparent than when dealing with the ever-growing pharma cash cow: vaccines.

It is here in the vaccine wars that we see the full agenda playing out. It did not begin with vaccines, rather with AIDS in the 1980s and the hoax that perpetrated the use of AZT, largely led by Anthony Fauci of NIAID, as the most effective treatment.[53] Later came a series of drug scandals such as Vioxx[54] along with a host of others in which the pharmaceutical industry was seen to have deliberately covered up fatal drug side effects, effects that they knew were likely to occur and did nothing to prevent. In regard to such malfeasance, Robert F. Kennedy Jr. has described the big pharmaceutical companies as "serial felons," a description that seems more than apt in light of the extensive fines they have been made to pay for their harmful products.[55]

The Endgame Trajectory

As the pharma capitalized on human ego and frailty with new diseases invented, they discovered vast new markets in erectile dysfunction, depression, shyness, and a host of other ailments that previously had simply been part of the human condition.

The problem was that the pharma had no way to force people to take these drugs and had to rely on advertising and paid-off physicians to convince their patients. How much better would it be to make their products liability free for any adverse outcomes? How much more lucrative might mandates be as a means to make everyone become a patient?

The industry solved the first problem by engineering the National Childhood Vaccine Injury Act, described in Chapter 3, which was designed to remove pharma liability to vaccine lawsuits. With the burden of adverse vaccine effects transferred to the taxpayer, the pharma then launched a wave of newer vaccines without fear of any civil consequences. To ensure that all vaccines were sold as widely as possible, the pharma lobby then went to work on vaccine mandates across the United States and in many other countries, as well. To a large extent, that push succeeded with almost all American states, and many countries, now having some form of school vaccination mandates.

Further, in the last few years, various American states have significantly diminished the grounds for vaccine mandate exemptions by first removing personal and then religious exemptions to mandated vaccination.

Overall, these measures of control have advanced to the point where the pharma now feels that it can do whatever it wants with little fear of exposure, let alone having to face any sort of justice.

Even for an industry so emboldened, these gains were still not enough. With COVID-19, it must have seemed to them that the heavens had opened up to herald not only a continuation of their already vast regulatory control, but even more goodies: the Holy Grail itself in the form of the complete control of human health. Vaccines may not be the only aspect of future control that the pharma dreams of, but surely it is the pointy end of the spear, which, if successful, would enable all the other health mandates that are intended to follow.

How much money was the pharmaceutical industry poised to make, even before the COVID-19 pandemic and world hysteria? Here are some numbers from the past: vaccine profits in 2014 were $32.2 billion according to *Statista*, with an anticipated $54.2 billion in 2019 and $59.2 billion in 2020, the latter number before COVID-19. With this, any claims that vaccines are not enormously profitable can be discarded.[56]

The main thing standing in the way of vaccine profits now and in the future is vaccine uptake. It's a simple equation: the more people who take whatever vaccine is on offer, the more money the pharma makes. Perhaps in this view, voluntary vaccine uptake is vaguely fine, but compulsory vaccine mandates are enormously better.

There is, however, one small speed bump, namely, the idea that pushing vaccines, COVID-19 and any others, on all people might infringe basic rights, particular "security of the person." This concept, considered quaint by some parties, is mostly ignored by mainstream media and health agencies. Increasingly it is ignored by governments, many of whom seem to be gripped by the same COVID-19 hysteria.

One of the "inconvenient" speed bump documents here in Canada is the Canadian Charter of Rights and Freedoms, particularly Section 7, which deals with security of the person in detail. Section 7 says that

> Security of the person includes a person's right to control his/her own bodily integrity. It will be engaged where the state interferes with personal autonomy and a person's ability to control his or her own physical or psychological integrity . . . or imposing unwanted medical treatment.[57]

For Americans, the notion of security of the person is explicit in their foundational documents: the Declaration of Independence and the Bill of Rights of the Constitution.

What to do about speed bumps? Well, that is what bulldozers are for; in political terms, that is what state of exception decrees are for (Chapter 11).

It is in this sense that the battle about vaccines is key to understanding the desired end state of the pharmaceutical industry and their main partners, the Gates Foundation, the WHO, and the WEF.

This fight is thus not inconsequential, but rather instead the "vital ground" that cannot be surrendered if human beings are to have any control over their bodies, and thus their lives.[58]

The players in this scheme start with the pharmaceutical industry itself, which dares not be seen to be the "man behind the curtain" pulling the strings. This is where the Gates Foundation and the WHO come in: the first is viewed by many as purely a humanitarian organization, the second as a beneficent world body only concerned with global health for the benefit of all humankind. While both depictions are beginning to fray around the edges, it would be safe to say that the majority opinion about both is still positive, at least in North America.

Tie the above into a bought media that promotes the cartel's message of fear, which leads, in turn, to terrified populations demanding action from their governments. The final part of the puzzle is that of the politicians who will, first, mimic Bill Gates's statement that there is no going back to "normal" until the whole world is vaccinated as they deliver through state of exception legislation the ultimate prize: worldwide mandates for vaccines to prevent COVID-19 and, in due course, any vaccine for any COVID-19 variants or other pandemics that may come down the line in the future.

At least in the planning of the cartel's desired endgame, this is how the future is supposed to unfold.

This last brings us back to where I started the chapter: is all that we've seen since COVID-19 burst forth from Wuhan simply a colossal cock up or an event of deliberate design? I've tried to make the case that it is the latter with Bill Gates and his foundation playing a prominent role, a notion supported by recent critical investigations. Martin Levine (cited earlier) recognizes the power that the Gates Foundation wields but posits that the Gates Foundation has enjoyed their vastly expanded influence by dint of years of meticulous planning, the building of connections, and, yes, their financial clout. In this view, the Gates Foundation and Bill Gates himself have made themselves the best-prepared people at the right time and place to help the world navigate through this crisis when most governments seemed indecisive.

So with two competing hypothesis, which is correct? Once again, Occam's Razor: in this case, the first hypothesis may be simpler. But does it equally address all aspects that have been discussed so far in this chapter? Does it, for example, explain Event 201, whose timing is curious to say the least? Does it explain the vast transfer of wealth that has now occurred to those already multi-billionaires? Does it explain how most of the media simply parrot the words of the Gates Foundation? If all of these aspects can be easily explained, then hypothesis one wins. If not, then two wins, or at least some version of it.

In the final chapter, I will explore in more detail these two possibilities and consider in greater detail why I still think COVID-19 is part of a greater scheme and how various people and groups around the world are now starting to organize to thwart the dystopian future that this scheme was created to provide.

Before I do, however, I want to take a more detailed look at the COVID-19 pandemic and how, in relation to the vaccine wars, all that occurred pre-COVID-19 pandemic was merely the early skirmishing. COVID-19 is not just a skirmish: we have here the actual invasion and the attempt to deliver human health to the corporate world. Human health here has multiple dimensions: what you put into your body, as in vaccines and other drugs; what you feed your body, as in the food you are able or allowed to grow; and how you control which parts of your body are under the surveillance of others. Should it be a surprise that all of these aspects of human life are marked for control by the same megacartel: the pharmaceutical industry, the Gates Foundation, the WHO, and the WEF?

As described by Vandana Shiva, what we do about this invasion will, in large measure, describe our immediate, and maybe long-term, future as free human beings with agency over our own bodies and those of our children.

CHAPTER 13

The Age of COVID-19: Fear, Loathing, and the "New Normal"

Who defined the crisis and its orthodox meaning?
Those who fashion the narrative.
Why did they choose that meaning?
So the government could claim the right to specific new powers.
Where will this lead?
To the exercise of those new powers.

—R. K. Moore[1]

Introduction to COVID-19

Much what follows is a work in progress. In large part, this is due to the pace of events that may make much of this chapter dated by the time this book comes out. It is partially for this reason that I have excerpted parts of this chapter for publication in the *International Journal of Vaccine Theory, Practise, and Research* before the book publication date.[2] This has been done with the permission of both the journal and Skyhorse Publishing.

COVID-19: The Early Days of a Pandemic

By now, we've all heard words and phrases about COVID-19 repeated endlessly, words and phrases most of us never want to hear again. Some of these include *unprecedented*, used to describe the endless pandemic and the ensuing panic, confusion, and government ineptitude; *social distancing*, the notion that you can protect yourself and others by keeping a two-meter distance from other humans; *social shaming* and *virtue signaling*, indicating the opprobrium that those who don't obey guidelines for social distancing and masks often face from those who do; and *flatten the curve*, a concept originally intended to convey the danger of overstressing a hospital's personnel and resources with Covid-19 patients. There are others.

To understand these additions to the world's lexicon, let's go back to the beginning.

The WHO announced "COVID-19" as the name of a new virus from the corona family of viruses on February 11, 2020, using guidelines previously developed by the World Organisation for Animal Health (OIE) and the Food and Agriculture Organization, both parts of the United Nations.[3] It was formally named "severe acute respiratory syndrome coronavirus 2," or SARS-CoV-2, but informally the name most people use is coronavirus disease 19, or COVID-19. This virus is closely related to the coronavirus that gave rise to SARS in 2003. COVID-19 is thus the newest member of the coronavirus family (see Chapter 2 with Table 2.2), the virus family responsible for a large fraction of what we consider to be the common cold.[4] Other coronaviruses of this family can be fatal to humans, e.g., MERS and SARS.[5]

The COVID-19 pandemic,[6] so declared by the WHO on March 11, 2020, came about as this book was entering a rewrite phase.

COVID-19 was not in the book's original proposal, but events overtook it, and us all.

What we thought we knew about the disease in those early days consisted of scraps of information filtered through the Chinese government and various international media, and, notably, backed up by the WHO.[7]

The official story then, and now, from all of these sources was that COVID-19 is a zoonotic coronavirus that jumped to humans from either bats or pangolins, or both, probably from a "wet market" selling live animals or their carcasses in Wuhan, China. It was merely coincidence that the nearby Wuhan Institute of Virology had a Level 4 containment facility where researchers worked on coronaviruses. It was merely another curious coincidence that these scientists were involved in what is called "gain of function" (GOF) research designed to alter the gene sequences of viruses to see how they might change in nature, or to protect against those bad actors who would try to alter the viruses to make them more deadly as biological weapons. Of course, the flip side of such research is to do exactly that.[8] It is simply another coincidence that many such studies have been paid for by the NIAID headed by Dr. Anthony Fauci.

Even more curious coincidences seemed to be lurking in the background, as well: on July 5, 2019, Dr. Xianggue Quo and Keding Cheng,[9] a husband-and-wife team at Canada's Level 4 facility in Winnipeg, were escorted out of the building by the Canadian federal police, the Royal Canadian Mounted Police,[10] for reasons that were never fully disclosed. Both Quo and Cheng appeared to have long-established ties to the Level 4 facility in Wuhan.[11]

All of this background makes the official narrative of the entire range of events in 2020 seem increasingly dodgier.

It gets better, or worse, depending on one's perspective: on October 18/19, 2019, Johns Hopkins University's Bloomberg School of Public Health, Center for Health Security, hosted a two-day tabletop simulation exercise about a pandemic that closes down most of the world. This exercise, called *Event 210*, was in large part paid for by the World Economic Forum and the Bill and Melinda Gates Foundation.[12]

There were fifteen main participants drawn from business, medical, pharmaceutical, and governmental organizations. Of the fifteen, four stand out: Dr. Chris Elias from the Bill and Melinda Gates Foundation; Prof. George F. Gao, the director general for the Chinese Centers for Disease Control; Stephen C. Redd from the US CDC; and Adriane Thomas, vice president of Global Health from Johnson and Johnson, one of the world's giant pharmaceutical companies. These four, based on their credentials, seem to be the main players. As for the other eleven, it is not clear why they were there at all, but it may be surmised that they were not simply chosen at random to fill empty chairs.

In the face of growing numbers of victims in Wuhan and elsewhere in Hubei province, the Chinese government did a hard lockdown of the city and province. The official death tally for China was reported as 4,636 with 89,272 cases as of January 13, 2021,[13] in a population of 11.08 million in Wuhan,[14] 58.5 million in Hubei province,[15] and 1.398 billion in China.[16] By April 26, 2020, the pandemic in China was officially over[17] but had spread around the world, very much as in the *Event 201* scenario.

And with that, the pandemic in China seemed to have ended, just as the rest of the world went into initial waves of infection with higher case levels of COVID-19 infection and death rates, various levels of lockdown, and state of exception decrees. In most countries, a second wave and even a third wave soon followed. Whether these additional waves are real or not will be addressed later.

At this time, there are a number of things we don't fully know, although more information appears daily, albeit much of it preliminary, or even just wrong. Some of these questions have, as yet, no clear answers: what were the real illness rates and fatalities in China, and why do these seem to be so much lower per capita than in parts of Europe and the United States? Next, if we had been led to believe that COVID-19 would be far deadlier for those with comorbid conditions, why were the refugee camps of the world not decimated by the disease?

We don't know the answers to these latter questions.

In the following pages, I will try to fill in some of the answers to questions such as: Where and how did COVID-19 really originate, particularly was it a natural virus or one modified by researchers as a GOF manipulation?[18]

What is the actual pathophysiology of the disease, including impacts on the CNS? What are the age, sex, national, and ethnic demographics of

COVID-19 infection? What are the real numbers, a question that ties directly to the question of methods to test for the virus and/or the surrogate markers of prior infection? How effective have the measures taken by various medical authorities been, e.g., masks, social distancing, lockdowns?

In the following, I will take these topics on one at a time.

The Still-Unanswered Questions about COVID-19: A Preliminary Overview

The Origins of COVID-19

A number of studies have suggested that COVID-19 arose as a zoonotic virus that jumped from other species to humans first in the animal "wet market" in Wuhan.[19] This remains the dominant narrative. A contrary study, however, has cast some doubts on this view. This is the work of Zhan et al. featured in an article by Rowan Jacobson in early September 2020 titled "Could COVID-19 have escaped from a lab?"[20] The Zhan paper, "SARS-CoV-2 is well adapted for humans. What does this mean for re-emergence?," noted some oddities about what the authors term "evolutionary dynamics."[21]

The article has not apparently been peer-reviewed, so this is an obvious concern in terms of credibility. However, the authors certainly have the credentials that make their work likely to be taken seriously, at least once some of the panic about COVID-19 wears off. First author S. H. Zhan was at the Department of Zoology and Biodiversity Research Centre at my own university (UBC) before moving over to a company called Fusion Genomics Corp.; second author B. E. Deverman is a professor at the Stanley Center for Psychiatric Research, Broad Institute of MIT and Harvard; the senior author is Y. A. Chan, formerly a PhD student also at UBC and now a postdoctoral fellow at the Broad Institute with Deverman.

In the article's abstract, the authors write that

> In a side-by-side comparison of evolutionary dynamics between the 2019/2020 SARS-CoV-2 and the 2003 SARS-CoV, we were surprised to find that SARS-CoV-2 resembles SARS-CoV in the late phase of the 2003 epidemic after SARS-CoV had developed several advantageous adaptations for human transmission. Our observations suggest that by the time SARS-CoV-2 was first detected in late 2019, it was already pre-adapted to human transmission to an extent similar to late epidemic SARS-CoV. However, no precursors or branches of evolution stemming from a less human-adapted SARS-CoV-2-like virus have been detected.

In other words, COVID-19 showed up in the human population far too well adapted to humans to be a virus that jumped from other species. Further, there seemed to be no earlier variants that would be typical of a new virus.

The authors suggest several ways this might have occurred. While they don't dwell on it and indeed are very cautious in how they interpret their data, one of these ways might be if COVID-19 was deliberately engineered as part of some GOF research project, perhaps at the Level 4 facility at the Wuhan Institute of Virology.

Jacobsen in his article explores all of this and opens with the following: "The world's preeminent scientists say a theory from the Broad Institute's Alina Chan is too wild to be believed. But when the theory is about the possibility of COVID being man-made, is this science or censorship?"

It's a good question. Jacobsen explores the oftentimes hostile responses that the Zhan paper received and how Chan as senior author handled it. Once again, as discussed earlier in Chapter 8 when looking at orthodoxy in vaccine research, the possibility that COVID-19 arose in a laboratory clashes with the story that many scientists and much of the media want to tell.

The story has morphed once again with Denmark reporting that COVID-19 had jumped from humans to farmed mink and the fears that it may jump back again to humans, perhaps a not trivial concern if it were to become more pathogenic. What is sure, however, is that the pangolin origin speculation is simply not correct, as no COVID-19 precursors have been found in that species.[22]

The results of Zhan et al. are fully backed up by a more recent exhaustive study by Dr. Steven Quay in January 2021.[23]

It is worth considering that GOF research on various pathogens is nothing new and has gone on for the better part of the last hundred years in the United States and increasingly in other countries that compete with the United States for world dominance. China is one such country. The rationales for GOF research and its history are presented in detail by Oller.[24] Oller also cites the "law of maximum entropy," in this case suggesting that the more GOF research any State does, the greater the likelihood of a release of some engineered pathogen[25]

The stunning illogic of GOF research to create more pathogenic viruses in order to protect against them has also been explored by Oller, who notes that such research is the equivalent to the fire department of a major city setting large parts of the city alight in order to prepare for a major conflagration. Rather than actually being protective, such a research program virtually assures that a deliberately constructed GOF virus will eventually find its way into the human population either by accidental release or by the deliberate action of "bad actors."

What Are the Pathological Impacts of COVID-19?
Respiratory

When the victims of COVID-19 first started to fall ill, first in Wuhan and then in the West, the initial medical response was aimed at the most obvious symptoms, which are primarily those of the respiratory system. For many people the general symptoms were at about the level of a common cold or even yearly influenza: cough, sore throat, fever, and others. Some of those affected, primarily at that time the elderly and those with significant comorbid respiratory and cardiovascular conditions, developed a condition known as acute respiratory distress syndrome, or ARDS. This was similar to other acute deadly coronavirus infections of the past: Middle East Respiratory Syndrome (MERS) and Severe Acute Respiratory Syndrome (SARS). In each case, the virus' "spike protein" seemed to primarily attach to epithelial cells in various parts of the respiratory system as well as epithelial cells on veins and arteries of the respiratory system at a receptor binding site termed ACE2 (angiotensin converting enzyme 2). In turn, the infection triggered severe inflammatory responses and cell death leading to fluid buildup in the lungs. Moreover, the infection often provoked what was termed a "cytokine storm" in which the immune system overreacts and releases various inflammatory cytokines such as IL-6 and TNF-α and others that actually damaged lung alveolar cells. The result was often lung failure leading to multisystem organ failure followed by death in very severe cases.

The initial response was to place severely ill patients on ventilators, a treatment some physicians now realize often did more harm than good.

This simplistic version recently has been challenged by Dr. Russell Blaylock, who postulates that the cytokine story is not simply an immune response gone awry, but one that also involves damage to cells by glutamate release and the attendant toxicity.[26]

Cardiovascular

Research coming out of Italy made it clear that the infection was not confined to the respiratory system, but also involved other systems such as the cardiovascular. Once again, the ACE2 receptor was the molecular target for the virus inducing direct cell death as well as indirect cell death by a cytokine storm as in the lungs. A key vascular feature in some of the affected was a coagulopathy, defined as an abnormal blood flow with either excessive bleeding *or* clotting.[27]

Renal and Hepatic Systems

Significant negative impacts of COVID-19 viruses have also been reported in both the kidneys and liver, once again through ACE2 receptors. However, it should also be clear that even without direct infection, serious impacts on respiratory and cardiovascular function will in turn damage other organs, as well.[28]

Nervous System

A surprising feature that began to emerge in some patients seemed to involve the nervous system, in which initial phases of the disease seemed to be associated with the loss of smell or taste, headache, dizziness, and what was described as a "brain fog." More severe nervous system outcomes included stroke, multiple sclerosis, Guillain-Barré, and encephalitis.[29]

COVID-19 can impact different parts of the nervous system, including the CNS, the peripheral nervous system, and the musculoskeletal system. Overall, 3.46 percent of patients with COVID-19 infection were found in one Chinese study to have CNS outcomes (24.8 percent of those with overall nervous system involvement).[30] Sometimes such deficits seemed to remain after other symptoms had abated, including smell and taste loss as well as cognitive changes.

The nervous system outcomes have not to date been as extensively investigated as those for other organ systems, but it seems clear that there are multiple pathways by which the virus can gain access, including through the olfactory epithelium[31] and by transport along the vagus nerve from the GI system.[32] This latter is perhaps particularly concerning given that emerging speculation about the origin of Parkinson's disease involves the same route from the GI system, up the vagus nerve and into the brain of a misfolded protein called alpha-synuclein.[33]

Overall, such outcomes in the CNS may be taken to indicate the serious potential for triggering longer-term progressive neurological disorders through a range of mechanisms, including inflammatory processes triggered by binding to ACE2 receptors, direct impacts on astrocytes and microglia, and a cytokine storm, as already cited.[34]

How Severe is COVID-19 as an Infectious Disease in Humans?

The COVID-19 infection can be extremely severe in some individuals and can lead to death, and may have longer term impacts insofar as the nervous system may be permanently damaged. It should not therefore be assumed that the disease is trivial for all people in its manifestations.

However, there are still fundamental questions that remain to which there are, at this time, no conclusive answers. These include the following: What is the percentage of the population affected by the virus; how does age, sex, ethnicity/race impact infection by the virus?; and how severe is it overall for most people?

Percentage of the Population Affected

A now-dated early summary of the number of infected was provided by reporter Ben Swann.[35] Basically, Swann takes the WHO's own COVID-19 case numbers and shows how that organization was manipulating data, leading to a level of panic in society about a pandemic that had rarely been seen in peacetime. As Swann noted at the time, the implications for the economy of the world alone were daunting and continue to be so today. And that was only the first "wave" of the disease.

Part of the panic that the WHO, various governments, and the mainstream media spawned arose from two main things: the first was how the number of deaths per case infected was calculated. Problem one within this arose from trying to determine how many people had actually died from COVID-19 as a causal factor, not merely a comorbid condition.

In the early days of the pandemic, there was clearly a tendency to attribute any death for someone who showed any of the respiratory symptoms of the disease to be a COVID-19 case.[36] This procedure, often accompanied by various dubious laboratory analyses (see section later), was all it took for such fatalities to be assigned to a COVID-19 mortality category. The distinction between "dying from COVID-19 versus dying with COVID-19" remains highly relevant. For example, if one has COVID-19 and it impacts your organ systems as previously described to the point where death occurs, this is clearly a true COVID-19 fatality. More indirectly, if one has a significant and documented respiratory or cardiovascular disorder and the disease triggers the cascade of events that further damages an already-stressed system leading to death, then it is also legitimate to call that a COVID-19 death. So what are the real fatality numbers? It depends, and therefore it must be assumed that we don't really know.

Problem two arose from trying to determine how many people had actually contracted the disease regardless of any overt symptoms linked to COVID-19, if any at all (the so-called "asymptomatic" cases). And the only way to assess the real numbers has been to do the mass screening by either polymerase chain reaction (PCR) for active virus or serology looking for antibodies to the virus in blood or saliva. In the early phases of COVID-19, such tests were not reliably done and, as discussed later, for PCR still are not likely to be accurate for total case determinations.

For these interrelated reasons, if you don't know the numerator (the number of fatalities) and don't really know the denominator (the total number of people infected), you can't calculate the rate of fatalities in the population. Prof. John Ionnidis, mentioned several times in the book, ran some antibody tests back in the spring of 2020 and came up with COVID-19 infection rates that were much higher than the medical authorities in California were claiming at the time.[37] What the latter were then doing was to take the presumed COVID-19 fatalities and divide that number by the estimated number of those infected. Together, the product gave an absurdly high fatality rate per case that was nowhere near accurate, with some such estimates being something in the range of 3 percent to 10 percent fatalities per total infections.[38]

The media, of course, grabbed such numbers and ran with them, thus vastly inflating public panic and turning COVID-19 into the modern equivalent of the Black Death. In some sense, the erroneous modeling data that came out of Imperial College in London, in part funded by none other than the Gates Foundation, vastly overestimated the expected number of deaths.[39]

Similarly, not knowing the second value is not likely to help you make sensible public policy about things like mask wearing, economic closures, and so on.

It is worth recalling that our own British Columbia provincial health officer, Dr. Bonnie Henry, who brought in the various "essential" measures to control COVID-19, was previously caught on camera at a June 2, 2020, news conference telling journalists that the disease was equivalent to a bad case of influenza (the flu).[40]

To look a bit more into Henry's more recent "fun with numbers" campaign, consider the first graph in a January 18, 2021, press conference[41] and compare what the graph shows for actual deaths versus "cases." (see Fig. 13.1 in the photo insert). Scrolling in the various categories along the time line from the earliest cases in March 2020 to the time of writing, one discovers a clear trend: the ratio of deaths early in the pandemic to the present has increased about sevenfold as the overall number of cases has gone up. At the same time, the number of alleged cases has gone up about twenty-five times. Several possibilities come to mind to account for this discrepancy, accepting for the moment that Henry has either the numerator or denominator of the disease correct: first, over time, treatment for COVID-19 has become a lot more effective, thus reducing deaths. Or, as discussed below, testing has produced a lot of false positives. Clearly, something is wrong here, and this is a perfect example of using data, however flawed the sources, to energize a desired narrative.[42]

So, let's look at the actual numbers by age, ethnicity, and sex, where possible broken down by demographics for both the United States and Canada as redrawn from the official sources and summarized in Appendix 17.[43]

Looking at the numbers in Appendix 17 gives us a picture of the numbers of infected and the deaths attributed to COVID-19. None of this, however, evaluates the accuracy of the data, particularly given some of the issues for diagnosis and testing cited in the section below. Indeed, it should be obvious by now that these numbers based on questionable testing data are highly uncertain. On top of this, there are serious concerns that the CDC, Health Canada, and other agencies have been merging actual deaths due to COVID-19 with deaths from other diseases such as influenza, and/or attributing to COVID-19 deaths from completely different disease conditions if COVID-19 samples test positive.

Assuming we want to accept anything attributed to the CDC as accurate, a fast glance at these numbers for the United States shows evidence for the observations that have been made from the beginning of the pandemic, namely, that most of those diagnosed with COVID-19 (CDC, as of Dec. 4, 2020, showed 10,483,169 cases) concentrated in the eighteen-to-sixty-four age range (75.6 percent) with actually very few cases under age four, 1.8 percent, and only 8.2 percent in the five-to-seventeen age bracket, confirming that those under eighteen are not the main disease vectors. Somewhat surprisingly, the elderly above sixty-five, considered in some reports to be the most at risk of catching the disease, come in at 14.5 percent. However, those above fifty-six are most likely to die, with the peak found in those above eighty-five at 32.4 percent. Total deaths in various age groups attributed to COVID-19 stood at 198,788, oddly a number about 70,000 less than usually reported at that time by the media.

These COVID-19 death numbers reflected what has been seen in various Western countries and may indicate either greater mortality due to age, due to living conditions as in care homes, or both.

The male/female ratio was pretty close at 47.9 versus 52.1, respectively.

In terms of ethnicity, white non-Hispanics made up the bulk of cases at 52.5 percent, followed by Hispanics at 24.2 percent and with 14 percent for African Americans. These latter numbers, however, cannot be taken to be definitive, because they are not adjusted for proportion of the overall population, nor does the CDC consider that reporting is consistent between states and communities.

Canada during the same period reported 372,409 cases (same problems as previously mentioned) and approximately 12,285 deaths. There was a relatively even case spread until age sixty, after which the infection rate declined. Male cases predominated over female from ages above thirty-nine. Health Canada did not report on ethnicity.

More Numerical Comparisons

As of December 2020, the death toll in the United States is somewhere between 198,000 to 269,000 people in a population of 331 million,[44] or 0.0006 to 0.00081 percent of the population; in Canada, it comes out to be 12,470 deaths in a current population of 37,742,154, or 0.00033.[45] Are these numbers significant? Maybe not in a statistical sense, but surely they are for the families of the dead. Of course, the latter presumes that the tests by which the physicians determined that the cause of death was COVID-19 were accurate. As I will show next, this may not be correct.

Comparisons are also subjective. How significant are the cases of Alzheimer's disease in the United States, for example? It turns out there are over five million Americans of all ages with Alzheimer's disease, the most common age-related neurological disease, at 0.015 percent, thus vastly higher than for COVID-19.[46] Alzheimer's disease, by the way, makes daily life a struggle for both victims and their families, and while it may not kill you, per se, it vastly diminishes quality of life in your final years. Apart from the families of those impacted and physicians and researchers working on the disease, does anyone really care all that much? Not really. Why not? Maybe because the media has not seen any advantage in hyping the Alzheimer's disease numbers.

Juggling the Numbers of the Dead

Is there any evidence to suggest that there is some funny number handling going on? One thing that supports the notion is from a relatively recent report by Johns Hopkins University that came out in an article on November 22, 2020, in the student paper, *The Johns Hopkins News-Letter* (Appendix 17).[47]

The article was retracted by the university on November 27 for potentially providing "misinformation."

In the original article, writer Yanni Gu interviewed Genevieve Briand, director for the Master's in Applied Economics. Briand had presented her data in a PowerPoint webinar. In compiling the data, Briand had used CDC data from mid-March 2020 to mid-September. Her conclusion was that the impact of COVID-19 on mortality across all age groups had not changed from previous years. Oddly, what had changed was the cardiovascular death rate along with other diseases. Briand had wondered why this might be so and came to the conclusion that somehow some disease numbers put out by the CDC and others might have been "misleading."

Could Briand have been wrong in her analysis and conclusions? Yes, for sure. But how is it misinformation when a scientific presentation deviates from an official narrative? We've seen this same response to apostasy in previous chapters, so it should hardly have come as a surprise that the Briand material and the newspaper report got retracted by those in control.

The November 27 retraction notice was quick to point out that "Briand is neither a medical professional nor a disease researcher."

Assays to Evaluate COVID-19: PCR versus Serology

A key concept to keep in mind in the following is that viral infection in an uninfected person largely depends on both the amount of virus they are exposed to and the time period over which they are exposed. There are other factors such as the virulence of the virus, as well.[48]

Testing Protocols
PCR

Most jurisdictions around the world have been using variants of the PCR to determine if someone is infected with COVID-19. The basic technique is that a researcher can take a very small amount of genetic material, for example from a virus located in a human patient, usually from the mouth or throat, and run this through a series of chemical steps in a device termed a thermal cycler. Each cycle is one amplification doubling of the original signal. In British Columbia, as elsewhere, the preferred PCR method of many is called real-time reverse transcriptase PCR (rRT-PCR).[49]

There are numerous ways this protocol can go wrong, but in the hands of qualified researchers, such problems should be minimal.

In order to detect an RNA nucleotide sequence, for example that which gives the particular regions of the spike protein of COVID-19, a number of amplification cycles are typically undertaken.

The threshold for detecting viral RNA that might be from infectious viruses is called the cycle threshold or "Ct." Typically, researchers use twenty-five to thirty-five cycles, but not typically for detecting *active* virus. Going toward the higher end and beyond increases the risk of rejecting a true null hypothesis, that is, of getting a false positive answer (Type I error); going too low-end risks failing to reject a false null hypothesis (Type II error). The relative dangers applied to COVID-19 testing are that a false positive might characterize people as being infectious when they are not, while a false negative risks missing someone who is.

Alan Cassels, a Victoria-based journalist we've met before, took a look at potential PCR testing problems here in British Columbia.

Writing in *Focus on Victoria*,[50] Cassels took a look at disease numbers and deaths since March 2020. He wrote that

> What stands out from these numbers [of number of "active cases," hospitalizations, deaths, etc.]?
>
> An extremely low likelihood of death by COVID-19 in BC. Certainly lower than any annual toll of the flu. Certainly lower than the numbers of people who have died from cancers, heart attacks, overdoses, suicides and the myriad of other things that take life every single day. If you take 2019 as an average, 132 people per day die in BC, from all causes. That was the last full year without a pandemic virus. [Brackets for addition are mine.]

And,

> With less than one person per day dying of COVID in BC, one is tempted to ask if we're making a mountain out of a molehill. I'm increasingly surprised by the general subservience of the populace and the absence of thoughtful dissent against emergency measures that are undoubtedly causing all kinds of other suffering, wreaking long-term havoc on our society, our livelihoods and our economy.

In his article, Cassels took the view that serious COVID-19 illness and death, while tragic, tends to focus health professionals on one disease and thus one health outcome. In the process, however, both they and the province's politicians may neglect others suffering from a range of health-related problems. For example, neglected health concerns may include the increasing impacts of drug use with the attendant mortalities, spousal and child abuse, and increasing poverty due to COVID-19 lockdowns' economic impacts. In regard to this latter situation, false positives tend to make politicians go to extreme measures to control disease spread, as they are now doing in various countries, while exacerbating the other issues that have long ranging implications for overall health.

Cassels went on to elaborate about the crucial issue of false positives:

> I consulted a molecular biologist (who asked me to withhold her name as she works as a provincial government biologist) who said that we have to be very cautious in interpreting these tests because the reverse transcriptase enzyme has poor efficiency in converting RNA to DNA. She told me that if we do over 30 to 35 cycles 'we can't culture a live virus from the sample.' Basically, she added, 'a high cycle threshold means we're finding meaningless fragments that say nothing about the infectivity of the patient.'
>
> This is an expert who uses the RT-PCR test everyday in her work doing forensic science, so I trust she knows its limitations. She was quite forthright in saying that possibly as many as 90 percent of those testing positive for COVID-19 are probably not infectious. Which is to say they may have had 'fragments' of the virus, but they couldn't possibly spread the virus to anyone else.

These comments are backed up almost exactly by a recent study by Jaafar et al.[51] and even the British government's Public Health England.[52] An even more damning evaluation of PCR methods was recently provided by the New York law firm of Siri and Glimstad in a letter to US Health and Human Services that not only questioned the validity of the methods of PCR being recommended, but also the Ct.[53]

Jaafar et al. compared Ct values for patients versus active virus recovery from samples in cell culture experiments: at twenty-five cycles, the real positive rate was 70 percent; at a Ct of thirty, it was down to 20 percent; at thirty-five cycles, it was less than 3 percent. This means that when running thirty-five cycles, the false positive rate is 97 percent.

To add to this, an article that came out last year shows just how incorrectly British Columbia, and likely a host of other testing states/provinces/countries, is performing its PCR tests to determine case numbers. In British Columbia, not only do they test samples with thirty-five cycles, but they are also apparently not consistent in the number of cycles used.

Ryan et al. (2019) noted the first two major mistakes to not make with PRC:

> Mindful of the recommendations contained in series of existing review papers on eDNA . . . we offer the following suggestions for standardizing eDNA techniques in light of our own findings. *To maximize diversity detected with a given primer set, minimize PCR cycles, preferably fewer than 35; Keep PCR protocols strictly consistent across samples you wish to compare.*[54] [Italics for emphasis are mine.]

In other words, to get accurate measurements, one needs to do the exact opposite of what the health authorities in British Columbia, and almost everywhere else, have done to date.

The British Columbia Centre for Disease Control, which does PCR testing for COVID-19 for the province, runs their PCR tests at thirty-five cycles, as noted above. From solely a perspective of knowing who has had COVID-19 and thus getting a better grip on numbers at any time period, this might be useful in calculating the true overall death rate due to infection.

However, from the perspective of determining how many people might still be infectious and thus require more severe population control measures by the authorities, this is simply inaccurate. As described earlier from Cassels, it's simply a crude shorthand way to identify positive PCR outcomes as COVID-19 "cases" when they are clearly nothing of the sort, at least not individuals who can infect anyone else. What, however, they may be are people who had the virus and now only have residual noninfectious fragments remaining in their bodies. It remains possible, however, that these PCR tests might also identify people who are at the beginning of an infection and hence potentially cases.

To confirm what I had heard from Cassels, I asked the BC CDC directly. Here is what came back:

> PCR testing
>
> The cycle threshold number used to diagnose COVID-19 may vary based on the test used but we typically use a cutoff of 35 cycles and simultaneously detect two targets (the RDRP and E gene) [some of the nucleotide sequences] and certain assays use cutoffs of 40 or even more cycles.
>
> Cycle threshold represents how many rounds of amplification are required to detect COVID-19 RNA in a sample. More cycles mean less copies of virus in the sample hence there are concerns about being overly sensitive.

However, this is a very complex issue. There is good evidence that when more than 24 to 30 cycles are required to detect virus the virus concentration is so low that it becomes difficult to cultivate the virus. However the cells used in the laboratory to cultivate the virus aren't equivalent to the cells in the *nasopharynx or the lungs in people. So just because one can't culture the virus in a laboratory that does not mean that it won't transmit. Many believe that with low copy numbers (high CT) values the virus is not likely to be transmitted.*

But it is also important to understand that it is not that the test sensitivity is being inflated, rather having a very sensitive test helps address missing infected people because of poorly collected samples (collecting adequate samples is difficult, and samples such as saliva typically have less virus especially in outpatients).

In the literature and first hand we have seen a number of cases COVID-19 in British Columbians where the person is early on during their infection course and the initial sample had a very high CT value ~35 (low virus RNA concentration) and the next day the CT was ~14 (high virus RNA concentration). Setting the detection threshold to [sic] low seems appealing until one misses that early case that can transmit infections to multiple people. [Italics and brackets are mine.]

So, in other words, the BC CDC recognized that they are on the very high end of detection but felt that they could balance the resulting high false positive rate by their worries about going too low and by their reliance on a particular cell culture method.

As for the statement that the cell cultures used by some researchers may not reflect the ability of low levels of actual virus to infect cells of the nasopharynx or lungs, it's a simple problem to solve: go to a cell culture repository, such as American Type Culture Collection (ATCC),[55] and get the right kind of cells. It's easy, and I did it in five minutes: open the website, go to *Cell Lines for COVID-19 Research*, find the section on *Primary Cells* that you want, and then take your pick. It is literally about that hard. All of this is again begging the question of why BC CDC chooses not to find the cell types that they claim they need rather than make excuses. Notably, Jaafar et al. in their analysis used cultures of nasopharyngeal cells from COVID-19 patients.

It is worth noting in regard to the overall discussion about PCR that the coinventor of the PCR, Nobel Laureate Dr. Kary Mullis, completely agreed that running too many cycles allows one to find anything that one wants to find, whether it is meaningful or not.[56]

All of this raises the next question: does the provincial health officer, Dr. Henry, mentioned earlier, really not know the broader literature, such as the

Jaafar et al. study, or is there something else here, apart from a rickety bureaucracy, that the rest of us are missing?[57]

Wouldn't a better Ct have been in a sweet spot of detection within acceptable false positive and false negative levels, for example, somewhere between twenty and twenty-five? Maybe not, but back at the press conference discussed previously, they were apparently diagnosing COVID-19 with serology, not PCR.

There is another explanation for the BC CDC/Henry inertia on this file: it's called the "sunk cost fallacy," that is, the continuation of a plan even after it has been shown to be the wrong plan. The reason? Because whoever is doing it is already so invested in doing it one way that they continue to do so regardless.[58]

As a follow-up, I went back to the BC CDC to ask them if they retested the previously positive COVID-19 patients after their quarantines had ended. I also asked them which computer models they were using to predict viral spread. They didn't answer either question. However, the answer to the first question came care of a colleague who does contact tracing for the government. His answer: they don't retest those deemed positive and instead assume that the latter are now immune and no longer capable of spreading the virus. What, however, if these individuals were false positives, as now seems highly likely? It means, in brief, that the government of British Columbia does not correct their potentially inaccurate case numbers that drive the wide-ranging and restrictive health measures.

A recent CBC article tried to defend the use of a high Ct, at the same time slamming critics as misstating the value of PCR tests. As usual, the article demonstrated that CBC is driving an agenda by creating a straw man argument, namely, that PCR was an effective method of measuring viral nucleotide sequences and hence the false positive concerns floating around on social media were simply "conspiracy theories." And yet, the argument put out by most critics was never that PCR could not be used to detect the fragments, but rather that at a high Ct one could not be sure that a positive result for any individual was a real "case."[59]

Serology

Another way to test for COVID-19, or any other infectious disease, is to use antibody methods, in other words, to look at antibodies created by your immune system in response to infection with some pathogen, such as a virus. As cited above, this seems to be what BC used to do. Serology is, at best, a surrogate marker, as described in Chapter 2. Positive antibody levels do not tell you if you are immune, per se, merely that your immune system has responded to some pathogen. Antibody serology alone also does not tell you about the response of immune memory cells.

In the context of COVID-19 case evaluations, serology can, however, tell you if you have encountered the pathogen. Antibody screening is not a method, however, to tell you if you have an active infection, but rather, later, that you have had one.

A key concern with antibody testing is that the testing "kits" must have high sensitivity and selectivity. What this means is that to have any valid predictive value, any such test needs to be able to detect the antibodies to some disease with high levels (*sensitivity*) and not confuse them with another binding site (as in another virus). This latter is selectivity, and it is crucial. Based on these values, the ability of any serology antibody test is highly variable[60] and can easily give false results.

Some antibody detection products are better than others, and finding those that give reliable outcomes has been shown to be generally difficult, particularly in the COVID-19 pandemic. One particular example, as cited earlier, involved the much-criticized antibody tests conducted by Dr. John Ioannidis in his initial screening for antibody levels for COVID-19, in an attempt to derive the correct percentage of the population affected by the virus. The criticisms of this work seemed to arise mostly from a disagreement with the ongoing mainstream narrative about how deadly COVID-19 is, rather than the methods employed.

Halting Disease Spread by Various Means: How Effective Are These Measures?

Different countries took different strategic and tactical approaches to halt the spread of COVID-19. Typically, these involved various levels of lockdowns of movement and restrictions of association, mandates for masks, and other supposed health measures. One fairly fierce debate was whether Sweden, which did a minimalistic protocol for COVID-19 containment, fared better than other Western European countries that opted for more stringent measures.

The initial results seemed to suggest the opposite.[61] At the time of this writing, the outcomes are still not clear.[62]

A similar question can be asked about China, where the pandemic began. How many died, and did they get control of the pandemic with a harsh lockdown regime? The answer is that we don't really know.

How Contagious Is COVID-19 versus Influenza?

The official answer to this question leads, inevitably, to some of the control measures cited below. The mainstream medical version seems to be that the main problem with COVID-19 is its longer incubation period of two to fourteen days compared to one to four days for influenza.[63] These sorts of comparisons are the basis for the two-week quarantines.

Masks

The key question to ask is this: does wearing masks by the population at large help slow COVID-19's spread, or can it hurt the individuals who do so? The first part of the question presumes, as often stated, that COVID-19 is more contagious than a viral disease like influenza. This assumption is based on the common wisdom, which is promoted heavily by the mainstream media and health officials, that mask wearing definitely helps and can't possibly hurt. This leads, in turn, to members of the public "virtue signaling" or even actively shaming those who won't, or can't, wear masks with a comment that many of us have heard over the last year: "Just wear the damned mask already."

But is it even true that mask wearing actually helps slow transmission? The answer is maybe yes, maybe no, and it depends.

This is confusing, to say the least, and much of this confusion arises from assuming that *all* masks are equally effective at blocking viral transmission in either direction. In the following, let's look at the main types of masks in order of effectiveness, from lowest to highest, where the main criterion of effectiveness is viral permeability due to the materials used in mask construction. The second criterion is how the masks are worn.

The most common mask types now seen in public are the *cloth masks* that come in a variety of designs and colors. These are typically made of one or two cloth layers of different types. Depending on the weave of the cloth mask, these might diminish viral transmission in either direction by blocking large droplets or even phlegm. That, of course, would be good. Is that, however, how most COVID-19 is transmitted? No: it's rather mostly transmitted by the smaller, longer-traveling droplets called aerosols. Cloth masks do pretty much nothing to stop these latter droplets and the viruses that might go in either direction, that is, from wearer to someone else, or vice versa. Canada's federal health officer, Dr. Teresa Tam, recently suggested that people who wear cloth masks should sew an additional third layer of some material, such as diaper filler, into their mask.[64]

Next up are *surgical masks*. As with some cloth masks, these are several layers thick, usually of three as described in advertisements on the Internet: a melt-blown polymer, such as polypropylene,[65] between an inner and an outer non-woven fabric. Is this sort of mask able to stop viral transmission? No, the very same conditions apply as for cloth masks: a surgical mask is not designed to stop viral transmission in either direction. As the name implies, these masks are worn during surgical operations to keep material from the wearer out of the sterile field of a patient's open wound; they also serve to keep blood and tissue from the patient away from the nose and mouth of the mask wearer.

KN95 masks resemble surgical masks. These are Chinese knockoffs of the N95 variety, discussed below. As the name implies, they are listed as being

95 percent effective in stopping viral transmission. This may not, however, be completely correct.

N95 masks are, however, 95 percent effective at stopping viral transmission to the wearer *if they are fitted properly*. Since these masks allow the wearer to breathe freely out, they do not stop viral release to the outside. Like the cloth masks and surgical masks, they will stop large droplets/phlegm from coming out.

Finally, there are the more elaborate masks and face shields, and full body coverings sealed at hands and feet, much like Hazmat suits, designed to keep pretty much everything out. These will do just that but are obviously not realistic for COVID-19 control in the population at large, but rather for staff in a Level 4 containment facility.

One thing to remember with masks of the types described above is that covering one's nose and mouth will prevent bacterial and some viral transmission, but since both can also enter through the eyes, this protection is limited far more than most proponents would claim. In addition, if one touches a surface (a fomite) where the virus might be present, and then touches the mask, the entire benefit, if any, of the mask is compromised.

An argument can be made that, just like that for the usually low effectiveness influenza vaccines, anything is better than nothing. In the latter case, this might be true if there were no potential harms associated with Thimerosal, but as discussed in Chapter 5, this is not likely to be correct.

Are there any possible health consequences to mask wearing? For the surgical, KN95, and N95 masks, probably not, apart from whatever psychological issues a person may have now or in the future based on previous trauma to rape and abuse victims.[66] However, with the cloth masks, the harms are actually more physical than purely psychological: as you breathe into a cloth mask over the space of hours, you are depositing your respiratory system's bacteria into the warm, moist inside of the mask. Bacteria love such environments and will happily start to breed. You now inhale these bacteria, some of them pathological and in more abundance than before. What might be a consequence? A greater colonization of your respiratory system. It may be important to observe that the primary cause of death in the "Spanish flu" 1918–19 pandemic involved secondary bacterial pneumonia (see later for discussion and reference).

It gets worse: virus infection may also go up, as described in a 2015 article in the *British Medical Journal* (*BMJ*).[67]

The editor of *BMJ* decided to add an editorial note to the original article, showing that the authors had decided in the face of COVID-19 that any kind of mask against the virus was better than nothing: "The authors of this article, published in 2015, have written a response to their work in light of the COVID-19 pandemic. We urge our readers to consider the response when reading the article."[68]

In other words, forget what the original article said: that was then, this is now, and COVID-19 hysteria trumps previous data if they don't conform with the currently official panic levels.

Is there other evidence from the medical literature that mask wearing actually diminishes viral, or particularly COVID-19 virus, spread? Apart from the previously stated, no, not really. The Mayo clinic released a document called "COVID-19: How much protection do face masks offer?"[69] basically reiterating the above considerations of mask types and efficacy, stressing that masks are *only part* of the overall process, including hand washing, for stopping viral transmission.

In addition, Drs. Tom Jefferson and Carl Heneghan, both members of Oxford University's Centre for Evidence-Based Medicine, recently penned two articles on the subject of mask effectiveness against COVID-19 spread. Jefferson was previously the head of the Cochrane Collaboration's Vaccine Field group that used to take a generally balanced look at vaccine issues.[70]

Jefferson and Heneghan evaluated the evidence for the utility of masks against COVID-19 and concluded that there is a general lack of evidence in their favor. This doesn't mean that masks might not be effective in some measure as discussed, just that the evidence is not yet all that solid.[71]

In a more recent article, the same authors reviewed a current small Danish RCT study called DANMASK-19 that failed to find a significant difference in COVID-19 infection rates between mask and nonmask wearers.[72]

Looking at the DANMASK-19 study in more detail, it is easy to see why Heneghan and Jefferson came to this conclusion.[73] In the DANMASK-19 study, the authors conducted an RCT, the so-called gold standard for clinical evaluations. In brief, the enrolled 3,030 participants were instructed to wear a standard three-ply surgical mask for 4.5 hours per day when out and about. Controls consisted of an additional 2,994 people who did not wear masks. Of the first group, the authors eliminated those who did not consistently wear their masks for the required time or in the right way. At the end of the trial period, all participants were tested for COVID-19 by their symptoms along with PCR and antibody tests. The results: both groups showed about a 2 percent COVID-19 infection rate, but no apparently serious illnesses or deaths. The rate of infection noted here is marginally higher than that observed by Moderna in their second press release on their efficacy trials, discussed later, and likely reflects phases of the pandemic, as well as differences between countries.

The authors acknowledged a number of limitations to their study, one being that there was no control for what the participants were doing the other 19.5 hours per say. The journal also featured seventeen critiques by readers, some of which were quite valid. For example, one commentator wrote that there was no control for interactions with family members during those nonmasked hours.

There are two take-home messages here: One, as noted by the authors, mask wearing alone is not going to change the outcomes, and indeed this is precisely the problem that should be acknowledged by governments mandating mask use. Second, the number of people infected with variable symptoms in both cases, coming in at 2 percent, hardly describes the media's tendency to portray COVID-19 as a massive modern scourge.

As Jefferson and Heneghan stressed in their first article, overall the way in which medical and lay people evaluate the efficacy of masks is very much in keeping with politics rather than pure science. As we will see throughout this chapter on COVID-19 and even more generally about vaccines, this sort of "trust the science" perspective seems to only apply when "the science" fits with preconceived beliefs.

Another issue, at least here in British Columbia, is that the masking policies seem capricious: masks are now required for adults inside any public building, but not for anyone under twelve years old. Are the medical authorities stating that children can't get CVOID-19 and thus transmit it? If so, this is a contradiction to what they have said throughout the pandemic and indeed to what the CDC's own COVID-19 infection tables shown (see earlier).

Social Distancing

Social distancing is the notion that if people stay a minimum of two meters, roughly six feet, apart from each other, then a person infected with a virus can't spread the virus to another. As described in a recent publication,[74] this two-meter number is a vast oversimplification of the actual physics of virus dispersion that arose from a much older experiment using more primitive measuring techniques. The methods in the older study could not distinguish particle size and distribution very accurately.[75] More realistically, the actual distance measurements depended on the size of the particles released, that is, larger ones as well as aerosols, and the ambient conditions such as wind, temperature, humidity, etc. Making a two-meter rule may seem like a safe middle position to take, but in reality, it is simply arbitrary. Nor does it take into account virus particles on fomites, or what happens within homes where presumably a family is not expected to observe any such distancing. So on the one hand it seems to be something we can all do to limit the spread of COVID-19, or any other virus, and on the other simply a way for the authorities to signal to all of us that their regulations show that they are on top of the problem.

Of course, it should also be noted that the demands for two-meter spacing tend to be ignored if the same authorities have sympathy with events involving masses of people, such as with the various Black Lives Matter (BLM) demonstrations in the United States and Canada, or if they don't, for example with groups protesting the lockdowns. Accepting that BLM supporters are protesting en masse for very valid reasons, the fact that our own health authorities allow them to do so while condemning other groups simply reveals that the two-meter policy is quite capricious, with the key variables being politics and the likelihood that enforcing restrictions would lead to protests getting out of control. In other words, this is not about medical practice; it is about population control. The fact that the media generally go along without question merely emphasizes the extent to which the media have become spokespeople for agencies of the State rather than actually doing real journalism to examine the regulations with a critical eye.

Lockdowns

This brings us to lockdowns. Do they work? The answer, as with masks, is yes, no, and maybe. How can this be? The yes part is simple: if you keep everyone away from everyone else, any infected people will not be able to infect others. Eventually, those who are ill and recover will gain at least temporary immunity and no longer be infectious.

There are various caveats to this, however. The first, obvious one is that the longer you do it, the greater the likelihood of economic disruptions that will tend to cascade into myriad economic and social problems, as we've seen around the world. If you want to crash an economy, then this might be the way to do it.

A key question then becomes how long do you wait before removing the lockdown? Is it the standard quarantine period of fourteen days? If so, how does one explain the second COVID-19 wave in countries such as Italy that were completely locked down with closed borders during the early phase of COVID-19? You can't, unless you also postulate secret reservoirs of virus that are somehow evading the expected time line for the infectious phase. The standard answer often given for the difference when comparing Italy and China is that the first was slow to lock down, the latter did it sooner in Hubei province. But Hubei is part of China, and once the lockdown there ended, the rest of the population of China could go to Wuhan and those in Wuhan could go out. To accept this explanation, one would also have to accept that no one at all outside of Hubei province was infected with COVID-19. This last is patently absurd given that the virus was supposed to have infected the world, starting in Wuhan.

What about Sweden? They didn't lock down but tried to find a middle way. Did they succeed in controlling COVID-19 with much more modest measures, or not? Did they get a second wave, or did they basically get it done all at once? The emerging picture may suggest that they did not,[76] although more than anything it calls into question the understanding of how herd immunity actually works, either for disease epidemics in general or following mass vaccination.

The next thing to consider with lockdowns is what else happens while people are locked down. For example, the economic costs and the inevitable increase in poverty,[77] not to mention the failure to treat other medical conditions (ironically, including kids not getting other vaccines on the CDC schedule), the psychological impacts on everyone, maybe particularly children and the elderly, the increasing levels of depression and anxiety, child and spousal abuse, drug overdoses, and more. If we were dealing with a truly horrible pathogen such as Ebola or Marburg virus, then full lockdown measures might indeed be the better option. But as we were not, as shown in the upcoming section on real cases versus fatalities, then a full lockdown is not in anyone's best interests, apart maybe from those who sponsored a lockdown and stand to profit from the upward flow of wealth in the "Great Reset." I will discuss the latter in more detail in Chapter 14.

Monitoring

The monitoring of COVID-19 cases, including by "contact tracing," has been a train wreck from a scientific perspective. Contact tracing involves determining if someone has COVID-19 and then locating and testing everyone with whom the "infected" person has been in contact, including their families. At a superficial level, this seems like a straightforward way to control the pandemic before it spreads further. But, as cited previously, if your confirmation is a RT-PCR test in which you have set the cycle number too high, all you generate are further false positives. If your goal is something besides actual disease control, this is fine. If, however, it is about disease control, it is not.

Reporting and Snitching

If the authorities were hoping to make ostensibly free societies act like the former German Democratic Republic, this is one way to do it. If that is not what you want, then such recommendations are not for the benefit of anyone but the State.

Vaccine Passports

This idea is increasingly being floated by various entities, including governments, airlines, and others. The idea here is that you will have some sort of electronic device that has recorded your COVID-19 health status: either you have had a vaccine for the disease, or you have had a negative PCR test.[78] If you are listed as "green," you are good to participate in "normal" life. If you aren't, or will not comply with the vaccine or the test, then you will have a "yellow" or "red" status and you won't be allowed participate in many "normal" activities.

The obvious flaws in such a health passport scheme are the following: since a number of vaccines tend to have limited effectiveness over time (secondary vaccine failure), how often do you need to get tested? What about whether the vaccine didn't work in the first place (primary vaccine failure)? Will you now have to have an antibody test to see? Will you have to be revaccinated if not? How often will your "health passport" need to updated? These are the sorts of things that various corporations will love, as well as will State bureaucrats whose lives revolve around making their fellow citizens functional prisoners. Since the flaws in testing will be huge, it clearly is not a health measure and cannot accurately tell us who is infectious or not. Rather, this is clearly a control measure that creates two classes of citizenship.

Legislation

British Columbia passed Bill 19 in the provincial legislature early in the pandemic, mirroring similar legislation in other jurisdictions. Here we have the purest expression of a state of exception, as outlined in Chapter 11.

COVID-19 Vaccine Mandates

As discussed elsewhere in the book, the call for vaccine mandates appears frequently in various legislatures and in the media. Looking back at Chapter 11 on the ideology of vaccination, recall that one way we know that we are dealing with a cult religion is that *everyone* is required to join the cult, freely ideally, or by compulsion if not. (As an aside, I have never once heard those opposed to any aspect of vaccines say the opposite: that everyone has to refrain from being injected. This is a fundamental difference between cult behavior and free choice.)

The 2015 measles outbreak in Disneyland and the later wave of measles outbreaks in 2019 spurred calls for mandatory vaccination, often for infectious diseases other than measles. The panic associated with COVID-19 has made these calls all the louder, proportional to the intense fear that COVID-19 has generated.

As the pandemic continues, we can expect demands for mandates to accelerate due to two factors. The first is fear. The second will come when authorities realize that they are not achieving anything close to herd immunity for curtailing the disease with voluntary vaccine uptake.

From the perspective of true herd immunity, this was never going to be possible with vaccines at all where secondary vaccine failure exists, let alone with some fraction of the population yet to be infected and develop natural immunity. A COVID-19 vaccine, however, was never intended to be only for kids, but for the whole population. Once the authorities inevitably see that they can't get there with the carrot, out will come the stick.

As I've discussed previously, this will not likely be forced vaccination, but rather the withdrawal of your privileges to have your routine daily life of restaurants, bars, movie theaters, and malls, getting on an airplane, freedom of travel within your own country, etc. The hope by those in authority is that the withdrawal of your privileges will force those on the fence to comply with vaccine mandates.

Perhaps such restrictions will work, at least for a time. However, while the same authorities don't realize it, most of the main current vaccine candidates require at least two shots initially and then likely boosters with some frequency for life, maybe several times per year. The companies making the vaccines will love this, of course.

What will be less to love is the almost certain increase in adverse reactions that can be fairly confidently predicted. What then? It will be Gandhi's march to the sea paradox, described in the next chapter, in that any action taken by the authorities using their meager understanding of the disease will be the wrong one: the disease will spread, or autoimmune reactions will spread. Pick one; or better yet, don't go down this path at all, but telling that to a politician or medical bureaucrat is about as useful as talking to a wall, maybe less so.

Social and Medical Consequences of COVID-19 Control Measures

In addition to the above control measures, to a large extent medical and political authorities have relied on fear to drive many otherwise decent people to act as de facto medical rule enforcers for everyone else. Aspects of this were seen in what was termed "social shaming" to blame people not fully complying with the regulations du jour, all the while virtue signaling by mask wearing, regardless of whether the mask worn did anything positive at all. This last was common.

What were the impacts of such actions? As fully predictable, these included making children more fearful of others and their world, a feature that will surely come back to haunt us in the future. These actions include making addicts so fearful of other people that they neglected to use so-called "safe injection" monitored sites, at least here in British Columbia, leading to a significant increase in overdose deaths;[79] increases in spousal and child abuse;[80] increased morbidity and mortality due to people not feeling safe or being willing to access medical care;[81] and the multiple impacts of social isolation on the elderly. In addition, the increased levels of financial distress much of the population suffered, including job loss with loss of income, exacerbated much of this.

Much of this analysis is supported and amplified in a new article by Lyon-Weiler.[82]

Gaslighting the "Proles" for Fun and, Especially, for Profit

A widely stated view by the medical and political establishments is that "normal" only returns once a vaccine is available and taken by most people. Note that the companies making the COVID-19 vaccines have all received liability protection from various governments as part of the price for making their future vaccine available to those countries.[83]

At this time, the following are considered to be the main vaccine contenders (out of dozens or more in development) with their respective vaccines: Moderna (mRNA-1273), Pfizer-BioNTech (BN5162), AstraZeneca (AZD1222), and Johnson and Johnson (Ad26, Cov2.S).

The first two companies use messenger RNA (mRNA) platforms for their vaccines. Basically, mRNA platforms are a novel, largely experimental, means of immune stimulation in which a laboratory-made version of the mRNA for the crucial COVID-19 spike protein (the part that allows the virus to attach to ACE2 receptors) is encapsulated in artificial lipid nanoparticles. The vaccine containing this construct is injected into the muscle and is meant to move into muscle cells, come out of the lipid covering, bind to organelles called ribosomes, and thus cause the mRNA/ribosome complex to generate the spike protein. From there, the protein is expected to migrate to the surface of the cell, where it will trigger an immune response, ideally by both B and T cells, as well as eventually creating neutralizing antibodies by the former.

Animal and Human Studies Pre-COVID-19

Earlier mRNA vaccine studies have been carried out in humans and animals. In earlier human trials, Alberer et al. (2017) of Curvac reported on an mRNA vaccine against rabies and conducted a Phase 1 trial of 101 volunteers for both safety and tolerability that ran from 2013 to 2016. The study reported only mild to moderate adverse effects and good immunogenicity, but not if the vaccine was *injected* intramuscularly. The researchers did not examine the volunteers over an extended time period for adverse effects.[84]

Bahl et al. (2017) from Valera looked at an mRNA vaccine for two avian influenza strains thought to be particularly lethal for humans. The researchers performed their study across a range of species including humans, cynomolgus monkeys, ferrets, and mice, the latter of the Balb/c strain. In this study, the investigators used an mRNA construct against one of two influenza strains, H10N8 and H7N9, identifying coat glycoproteins of the viron that enables the virus to enter the host cell.

In the mice, they used either *i.m.* or *i.d.* injections. Of particular interest from the standpoint of future adverse effects with *i.m.* injections, some of the construct mRNA migrated away from the injection site (see their Table 1 in the study), first to the proximal lymph nodes, and then to the spleen and liver, suggesting to the authors that the lymphatic system was the primary conduit. The mRNA could also be found in various levels in other organs, including the brain. The numbers were not high in the latter. Nevertheless, should we have any concern at all that mRNA was in the brain given the potential for autoimmune reactions? Supposedly, mRNA is degraded very rapidly, so what happened in this experiment?

One reason that this might be a problem comes from how the researchers gathered their distribution data across a range from one hour after injection up to eleven days. At that point, the researchers seemed to have simply combined the data from each time point for each organ examined. Given this, we have no way of knowing how much mRNA might have been in the brain at any time point before eleven days, or after.

In all cases with the four species, the researchers found a strong immune reaction measured by antibody levels. In mice, these responses peaked at twenty-one days and remained high for at least eighty-four days.

The ferrets were challenged with live H7 virus, and the vaccinated animals were shown to have smaller lung viral loads than the placebo animals.

In none of the animals used was there any discussion of possible adverse effects on any organ system, but then again the researchers didn't seem to look. In the small human trials of thirty-one people (twenty-three vaccinated, eight controls), adverse effects were noted in the vaccinated, but the authors claim that none of them were serious during the time frame of the study.

Valera Therapeutics was at the time a branch of Moderna.[85]

Similar results were found by Edwards et al., 2017, from Sanofi-Pasteur, but again no safety data were reported.

Moderna's more recent rhesus macaque monkey study for the mRNA vaccine used twenty-four male and female animals divided into mixed groups of three. In the vaccinated groups, they used either a 10µg or 100µg dosage. Animals were injected twice, first at time zero and again four weeks later, then challenged by being exposed intranasally to live COVID-19 virus. The study claimed that the vaccinated groups showed effective prevention of infection of the lungs in both groups, but only in the nose with the 100µg dose, an observation that makes little biological sense. In both cases, the claim was that effective production of neutralizing antibodies occurred along with increases in T1 helper cells. However, the study also noted that monkeys do not get severe COVID-19 disease.[86]

In addition, the study did not examine any possible adverse effects beyond presumably simple observations of the animal's behaviors, assuming they did that at all prior to sacrifice.[87] Another paper from Moderna, posted to a prerelease journal website (i.e., not peer-reviewed), reported the same high antibody response, but again no long range safety data were obtained.[88]

The Other COVID-19 Vaccines

The vaccines being developed by the other companies mentioned, as well as the Chinese company Sinovax and the Russian vaccine Sputnik V, are so-called viral vectors vaccines. Viral vectors vaccines were discussed earlier in Chapter 2. In these, researchers take a weakened virus for another disease and incorporate into it the DNA for the spike protein. These vaccines are supposedly more resistant to temperature degradation than the mRNA vaccines. Like the mRNA vaccines, however, they are also experimental.

Human Trials: Efficacy Data for Moderna

The first mRNA vaccine efficacy data coming out of Moderna's Phase 1–3 studies and similar claims by Pfizer have been widely touted in the media and by people like Anthony Fauci of NIAID. For this reason, it is important to realize just how such numbers can be manipulated to make outcomes seem more favorable than they may actually be, and thus palatable to the public in an attempt to boost future vaccine uptake.

To help with this, here are some definitions from the *British Medical Journal:*

> Risk terms: AR (absolute risk) = the number of events (good or bad) in treated or control groups, divided by the number of people in that group. ARC = the AR of events in the control group. ART = the AR of events in the treatment group. ARR (absolute risk reduction) = ARC - ART. RR (relative risk) = ART / ARC. RRR (relative risk reduction) = (ARC - ART) / ARC. RRR = 1 - RR NNT (number needed to treat) = 1 / ARR.[89]

In the initial data cited in a press release on November 16, 2020, Moderna claimed an efficacy of 94.5 percent for the vaccine in preventing COVID-19 infection. The Phase 3 trails had enrolled thirty thousand people in a one-to-one ratio of treated versus controls in what Moderna claimed was a double-blinded study. A double-blinded study is one in which neither the participants nor the experimenters know who gets the treatment or the placebo.

Let's assume for a moment that there were fifteen thousand in both the control arm (the claim is that the controls actually received a saline placebo, rare as that is in vaccine Phase trials) and the treatment arm of the vaccine mRNA-1273 given at the middle dose regime of 100µg based on their Phase 1 trials. Two weeks after the second injection, Moderna monitored the participants for about two months out in the real world and recorded those who became infected with COVID-19.[90] The primary endpoint of this monitoring gave ninety-five total infections, with ninety in the control group and five in the vaccine group. How the infection was determined was not specified in their press release but would have been useful to know given the issues discussed in the sections above. One would also like to assume that diagnosis was not simply by a physician observing any symptoms.

If these numbers are correct, and remember this is science by press release with no raw data included, then the Relative Risk indeed comes out at 94.5 percent efficacy. Put into percentages, the controls had a rate of infection of 0.006 percent and the vaccinated group 0.00033 percent. (These data illustrate quite clearly the difference between relative risk and absolute risk.) How about severe cases of COVID-19 found in the control arm? In the same approximately fifteen thousand control subjects, eleven were diagnosed as having a severe case of COVID-19 or 0.0007 percent, or about 12 percent of the total group of the control infected.

Two weeks later in their updated November 30 press release, Moderna updated the previous report for a new efficacy calculation of 94.1 percent with 196 cases of COVID-19. In the control group, the number was 185 cases (0.012 percent) and eleven cases in the vaccinated group (0.00073 percent). Severe cases in the control group were thirty (0.002 percent) and included one death (0.000067 percent), compared to the number of US deaths (taking the upper value of 269,000 as of early December 2020) in a population of 331 million: 0.00081 percent, the latter number suggesting once again that the CDC's death rates from COVID-19 may be inflated, in this case by at least a factor of twelve.

In the space of two weeks, the rate of infection in the control group had gone from 0.006 to 0.01; in the vaccinated group the number of cases had doubled, thus accounting for the decreasing efficacy reported. The severe cases in the control group had gone from 0.00073 percent in the first report to 0.002 percent in the second.

It all sounds very dramatic at first glance, but it may also be smoke and mirrors given the comparison of the percentages at the two time points in the two groups: the infection rate in the controls hovered around the level of influenza of a mild to moderate year, in this case the year 2018 to 2019. Using this year, so as not to confuse any influenza cases with COVID-19, gives a rate of influenza infection in the US population at the time of 328.4 million of 0.01 percent.[91]

In the control group, the severe cases had almost tripled but were still a tiny part of the total. The only real bright spot actually here for Moderna was that their vaccinated group had no severe cases, at least at that point.

It is also important to keep in mind that these data do not reflect certain populations who will certainly be included in a general vaccine program: pregnant women, infants, children, and adolescents.

One thing that may be emerging from these numbers is that they may actually show us that the real percentage of the population at risk overall from COVID-19 is about 0.01 percent, as well as those who will become severely ill and/or die at 0.002 percent. But in order to save that 0.002 percent of the population, how many people would we have to vaccinate, and what would be the trade-off against possible adverse effects?

Looked at from a more skeptical perspective, both groups had a rough doubling of COVID-19 infection, which may lead to speculation that the vaccine does not actually prevent the disease, but merely slows down its expression.

Dr. Peter Doshi, writing in a medical blog about the Phase 1 data,[92] stated the obvious:

> Let's put this in perspective. First, a relative risk reduction is being reported, not absolute risk reduction, which appears to be less than 1%. Second, these results refer to the trials' primary endpoint of COVID-19 of essentially any severity, and importantly not the vaccine's ability to save lives, nor the ability to prevent infection, nor the efficacy in important subgroups (e.g., frail elderly). Those still remain unknown. Third, these results reflect a time point relatively soon after vaccination, and we know nothing about vaccine performance at 3, 6, or 12 months, so cannot compare these efficacy numbers against other vaccines like influenza vaccines (which are judged over a season). Fourth, children, adolescents, and immunocompromised individuals were largely excluded from the trials, so we still lack any data on these important populations.

Doshi is right, of course, and his critiques equally apply to the efficacy studies of Pfizer, which I cover in greater detail next. Doshi also raises the concern about the blinding in the study given that the adverse effects reported by actual vaccine recipients could have clued them to which group they were in and thus modified their behaviors.

The truth is that we won't be able to independently review the data until they are provided in peer-review publications, ideally with the raw data accessible on demand.

None of the apparent efficacy data should really be all that surprising. Readers will recall that in Chapter 2, I showed that various vaccines that could be evaluated had significant levels of effectiveness against a declining slope of infectious disease illnesses.

Human Trials: Efficacy Data for Pfizer

Pfizer's combined Phase 1/2 data and claimed high antibody titers after both doses with antibody levels rising higher after dose two.[93] In the later Phase 3 study,[94] their calculations of efficacy of greater than 90+ percent were based on approximately eighteen thousand people in each of the vaccinated and control arms (and the nature of the placebo injection is still not clear). In this report, Pfizer claimed that after following the subjects for two months, eight of the vaccine-treated group were diagnosed with COVID-19 compared to 162 in the control group. These ratios resemble those of Moderna and form the basis of their efficacy report.

As it currently stands, at least from a regulatory viewpoint, such data allow Moderna and Pfizer, and no doubt eventually the other companies, to hype the efficacy of their vaccines while smoothly sidestepping the actual dangers of the disease, let alone potential short- and long-term adverse effects for the vaccinees. In other words, those who have been saying that COVID-19 infection is at about the same level as influenza seem to be correct.

Johnson and Johnson and AstraZeneca Efficacy Data with Their Viral Vector COVID-19 Vaccines
Johnson and Johnson
This is just more "science" by press release, claiming 98 percent seroconversion but no efficacy calculations as of October 4, 2020.[95]

AstraZeneca
In a press release titled "AstraZeneca: AZD1222 vaccine met primary efficacy endpoint in preventing COVID-19" on November 23, 2020,[96] the company claimed that

> One dosing regimen (n=2,741) showed vaccine efficacy of 90% when AZD1222 was given as a half dose, followed by a full dose at least one month apart, and another dosing regimen (n=8,895) showed 62% efficacy when given as two full doses at least one month apart. The combined analysis from both dosing regimens (n=11,636) resulted in an average efficacy of 70%.

It's looking like the companies making the viral vector COVID-19 vaccines are losing the efficacy battle to the mRNA vaccine makers and thus the propaganda war for marketing, as well.

Summary for the Experimental COVID-19 Vaccines concerning Efficacy
Could all of these calculations be wrong? Sure, they could be, but then we would also have to reject all of Moderna's and Pfizer's claims of efficacy for their vaccine. You can't really claim both, at least not honestly. Could the number of those infected change with longer surveillance? Absolutely, and this seems to be happening, but so too do the efficacy data. We will only know in the future if Moderna and Pfizer keep sending these updates. At least, however, Moderna and Pfizer published their data, unlike the other companies.

All of the above notwithstanding, the efficacy data for COVID-19 vaccine results still says nothing about Moderna or the other companies' safety results, the other real test of the vaccine, which will be addressed below.

mRNA Vaccine Safety Studies: Moderna, Phase 1

Moderna's Phase 1 safety data, produced in a two-volume report of 1,015 pages, were released to Aaron Siri following a FOIA request to the US Department of Health and Human Services[97] and were derived from the postvaccine medical surveillance of eighty-five participants of both sexes divided into seven cohorts. There were no controls, as is typical for many vaccine studies.

All participants received two injections of mRNA-1273, one on the first day, the second twenty-nine days later. All patients were followed up by physician visits on days one, two, and four after each dose, and afterward at three and six months after the second dose. A twelve-month follow-up is still pending. Forty-five of these eighty-five subjects, ages eighteen to fifty-five, were divided into three cohorts with dosages of 25μg, 100μg, and 250μg of the mRNA construct. There was also a fifty-six-to-seventy-year-old age group with twenty people and one at a more-than-seventy-one-year-old age group with the same number. These latter groups received only the 25μg or 100μg doses.

In the first cohort of forty-five participants, thirty-two (71 percent) had possible adverse reaction to the vaccine (see Chapter 2 for grading scales that include "mild," "moderate," or "severe" categories). At 25μg, five of fifteen had adverse events; with the 100μg group ten of fifteen; and in the 250μg group twelve of fifteen.

In the fifty-six-to-seventy age cohort, fourteen of twenty (70 percent) had possible adverse effects; in the seventy-one-plus age cohort, ten of twenty (50 percent) of participants had potential adverse effects.

Moderna's physicians then decided which of the possible adverse effects were really from the vaccine. How they did so was not obvious from the report, and without knowing these details, there is not much point in trying to evaluate nonrelated from vaccine-related adverse events.

In this mass of data compiled from very few people, there are some takeaway messages: first, as the dose of the mRNA increases, the percentage of real adverse vaccine events appears to go up, an outcome that is not really a surprise in any dose-response function. The higher adverse effects in the 250μg group are likely a reason why this dose was dropped in the Phase 3 trials.

The second point is that there were so many moderate and severe reactions overall, an outcome that will be far from trivial if Moderna's vaccine is put out to the general population, including to groups not included in this study, namely, pregnant women, infants, children, and adolescents.

In general, small sample sizes, as in these Phase 1 safety data, are prone to suffer from potentially large errors in interpretation.

mRNA Vaccine Safety Studies: Moderna, Phase 3

Moderna's Phase 3 data were submitted to the FDA on December 17, 2020, for review by the Vaccines and Related Biological Advisory Committee Meeting as an FDA briefing document.[98] This report repeats much of the Phase 3 efficacy data cited above in Moderna's November 30 report. The safety data included the result from 30,350 participants, roughly half of whom were in the vaccinated (all at the 100µg level), versus placebo groups. Ages tested were eighteen to sixty-four, and sixty-five and over. In some cases, these age ranges are merged in the tabulated results.

Participants were monitored after both the first and second injections, in the latter case for up to seven to nine weeks with evaluations of "solicited adverse effects" versus "unsolicited adverse effects," the difference basically being how participants responded to questions posed by Moderna's medical staff compared to complaints that were not on the standard list of questions.

Adverse events typically were greater in number following the second injection.

The report states that for solicited adverse events, "Adverse events were reported in a higher proportion of vaccine recipients than placebo recipients, and this imbalance was driven by reactogenicity (solicited AEs) reported in the 7 days following each dose of vaccine."[99]

Moderna's Table 21 lists adverse effects for dose one and dose two for the eighteen-to-sixty-four age group by the following categories, each of which lists an overall adverse value by number and percentage and a Grade 3 response with the same values. These include the following: any adverse event, pain, erythema (redness of the skin), swelling, and lymphadenopathy (enlargement of lymph nodes). Factor differences for the overall percentages between groups range from about four times to 22.5 times higher in the vaccine group compared to the controls. In terms of Grade 3 reactions, the range is greater, from incalculable due to nil outcomes in the control group to up to forty-two times higher than controls in the vaccinated group.

Other measures such as fatigue, fever, headache, myalgia (muscle pain), and arthralgia (joint pain) showed similarly higher vaccine versus control values.

Similar outcomes were seen in the greater-than-sixty-five age group.

Of particular interest were cases of Bell's palsy (an inflammation of the seventh cranial nerve [facial]) causing part of the face to droop, with a report of four cases overall, three in the vaccinated group and one in the control group.

In the unsolicited adverse events, the numbers were approximately equal between groups, but differed in type: the vaccinated group included myocardial infarction (heart attack), cholecystitis (inflammation of the gall bladder), and nephrolithiasis (formation of kidney stones). In the control group, pneumonia was higher, as was pulmonary embolism (blockage of one of the pulmonary arteries). Moderna simply calculated the unsolicited events in both groups, even though they differed in type, and concluded from this that there was no difference between the two.

Table 19 in their report itself is curious: "Participants Reporting at Least One Adverse Event, Among All Participants and by Baseline SARS-COV2 Status (Safety Set),"[100] which gives numbers of vaccinated and controls who are COVID-19 positive. The numbers presented in this table do not seem to match the efficacy data cited above and strongly suggest a larger faction of the vaccinated participants were infected with the virus.

Overall, once again, without peer review we are left with scientific evaluations of safety and efficacy data based on what the company scientists themselves report. Nevertheless, based on the results of Moderna's Phases 1–3 safety study and the above cited efficacy data, Moderna recently applied for an Emergency Use Application for the mRNA-1293 vaccine from the FDA.[101]

mRNA Vaccine Safety Studies: Pfizer

Two reports have come out on the safety of Pfizer's BNT62b1 mRNA vaccine. The first is by Mulligan et al.,[102] and the second was a Phase 3 safety and efficacy report, first put out to the FDA for EUA approval. This document was evaluated by FDA scientists.[103] These data were also published in the *New England Journal of Medicine (NEJM)*.[104]

Let's look at each in turn.

Mulligan et al. on Phase 1/2: The study looked at forty-five male and female participants, eighteen to fifty-four, in approximately equal numbers. Safety and tolerability were followed for fourteen days after the second of two doses for 10µg and 30µg; at 100µg there was only one dose. The groups are therefore small, particularly the placebo group (N=3 per dose group), and it is not clear what the placebo actually consisted of.

Injection site pain was reported for all vaccine groups, some of it categorized as severe. Some placebo volunteers had minor localized pain at the injection site. The vaccine also induced fever in 75 percent of vaccine-treated patients after the second dose, with two participants experiencing high fever in the 30µg group. Some in these treated groups also experienced sleep disturbances, joint pain, headache, fatigue, and lymphadenopathy, and four cases of Bell's palsy were found in the vaccine groups versus none in the control group.

Pfizer's FDA Briefing Document

Many of the same adverse effects were seen in the Phase 3 data as in the initial safety evaluation by Mulligan et al. The resulting tabulated Phase 3 data are summarized in the report's Tables 17/18 (eighteen-to-fifty-five age group and greater-than-fifty-five group, respectively), listing a range of adverse effects including fever, fatigue, headache, chills, vomiting, diarrhea, new or worsened muscle pain, and new or worsened joint pain. And, as mentioned, there were cases of lymphadenopathy and Bell's palsy.

The latest interpretation of some of the adverse effects seen in the trials from both companies has all the flavor of trying to spin the outcomes to make them somehow seem like a good result. Specifically, statements now coming out in the media quoting physicians was that an adverse effect simply shows that the vaccine is working and your immune system is responding, as it should.[105] This might be true to some extent in that many conventional vaccines with live or attenuated viruses do have the capacity to create a mild version of the disease with the view that your immune system response will prevent a worse real response to a pathogen. This follows from Jenner and makes some sense. However, such cannot be made to easily apply to an mRNA vaccine where no real infectious agent is present to give you the disease.

The alternative explanation is not that the vaccine is effectively activating an immune response to fight a future COVID-19 infection, but rather that your body is responding to something toxic, and/or triggering an abnormal immune response, such as an autoimmune reaction.

A suggestion of this comes from both company's Phase 3 trials with vaccine recipients developing Bell's palsy. Bell's palsy has several triggers, including viral infection, and although it is usually temporary, it is not always so. These results, combined with higher levels of joint pain and headache in the vaccine recipients, suggest that the mRNA vaccine used by companies can have musculoskeletal and neurological consequences in the CNS, both of which should alert us to the prospect that other longer-term consequences may arise with longer surveillance. It is of interest that Pfizer's physicians decided that the Bell's palsy was not related to the vaccine because the numbers seen in these trials were consistent with what might be expected in the general population. In other words, they simply ignored their own controls to use a pseudo control population that they must have known would serve to negate the vaccine-induced Bell's palsy. In other words, they cheated.

The Phase 3 trials were published soon after in the *New England Medical Journal*.[106] These data were a kind of "lite" version of the FDA document cited above. As shown in the article's supplementary Table S3, the vaccinated group received 30μg of the mRNA compared to a vaccinated group in the FDA report. Although the data were viewed as showing safety, it is notable that the table shows an adverse event ratio of 26.1 percent to 12.2 for vaccinated and control, respectively, with 20.1 percent considered to be related versus controls of 5.1 percent; those adverse events reported as severe were 1.1 percent versus 0.6.

The above Pfizer study analysis was supported by a commentary of internal medicine physician Dr. Yves Smith. The doctor and some of his medical colleagues took a look at Pfizer's safety data published in the *NEJM* article and promptly raised a red flag.[107] Dr. Smith had a number of things to say about these phase trials. One issue raised concerned the validity of Pfizer's exclusion criteria; another was about the fact that in the vaccinated group there was at least one case of extreme fever (Grade 4) and a few cases of anaphylaxis, a potentially severe and life-threatening allergic reaction, the latter an outcome not mentioned in the article.

Smith noted that neither of these instances can be taken lightly, particularly in relation to proposals to vaccinate the entire population. Smith commented as well that those vaccinated did not seem to have acquired "sterilizing" immunity, meaning that they could potentially still transmit COVID-19 to others.

Smith also referred to Dr. Marcia Angell, whom we met in Chapter 12, and commented about the utter corruption in the review process by the *New England Journal of Medicine*—which takes paid ads from Pfizer—in reviewing these data. He concluded by stating that

> Please look at Dr. Angell's seminal article from 2009 [and her book of 2004]. She predicted in her works, all of this and more. My profession has been captured by a cabal of corporatist MBA clones, rapacious and unethical pharmaceutical entities, and an academic elite addicted to credentialism and cronyism. They have over the years bought off and infiltrated all of our government health care regulating agencies and our public health system. And they are completely incestuous. I believe where we are now to be worse than Dr. Angell could have ever dreamed. Even more depressing, I see no way out. [Brackets are mine.]

In spite of such concerns, Pfizer and the regulatory bodies seem to have brushed off these concerns and applied for, and received, the equivalent EUA from the British medical authorities based on their initial efficacy and safety studies, both from Phases 1–3. Since then, their EUA equivalent from the United Kingdom and then a US EUA and one from Canada were granted before the New Year.

What Could Go Wrong with mRNA Vaccines?

In principle, not much, if we are going to stay with the mainstream Crick model for DNA triggered protein synthesis.[108] Going back to Chapter 2, where mRNA vaccines were discussed, a laboratory-created mRNA based on the genome of the COVID-19 virus should be simple: the injected lipid-encapsuled mRNA enters the target muscle cell, enters the target cell, binds to ribosomes, the complex makes protein, and the spike protein is extruded to the surface of the cell to trigger an immune response. After this, the mRNA-ribosome complex comes apart, the mRNA is degraded, and that's supposedly the end of it.

All good, and maybe apart from the adverse effects seen in both Moderna and Pfizer phase trials that's all there is to it.

What, however, if there is another, more accurate, model of how DNA transcription to protein works, one in which there are "recursive" feedback loops compared to what the Crick model considers to be a linear process—that is, DNA to RNA to protein? What if instead, as in the Pellionisz model,[109] feedback occurs at all levels: DNA to RNA and back, RNA to protein and back, and then protein back to DNA. Could the artificial mRNA have an impact on DNA? This seems unlikely,[110] but what about the protein made? Is it all exported to the cell surface, or degraded? If not, what then? It must certainly be the hope—maybe prayer would be the better word—of the molecular biologists at Moderna and Pfizer that Crick was right and Pellionisz is wrong. Finally, what are to make of Bahl et al.'s data that the mRNA construct can be found in organs distant from the injection site, including the brain? With mRNA degradation being efficient, this should not be happening.[111] Add to this: what is the possibility that the mRNA found in the brain, if it were to infect neurons, might not create a proteinopathy similar to those that are common in most age-dependent neurological disorders?[112]

A second concern is that of antibody-dependent enhancement (ADE), also termed pathological priming. The main idea here is that an initial exposure to a SARS-like disease, or other viral diseases that have primarily respiratory symptoms, can result in pathological outcome to a second exposure to the same pathogen. Lee et al. (2020) consider ADE in the context of COVID-19[113] in which nonbinding antibodies bind to the virus surface and thus enable the virion to more easily enter and infect macrophages, thus increasing the overall replication of the virus. Another mechanism proposed is that of heightened immune activation, which can result in the cytokine storm, mentioned previously.

This has been seen neither in human COVID-19 victims nor in mRNA vaccinees to date but based on the animal studies is not something to ignore moving forward. For example, an older article by Tseng et al. (2012) looked at a variety of vaccine preparations for SARS in a Balb/c mouse model. An mRNA vaccine was not part of these experiments. After vaccination, the mice were challenged with SARS virus, and all of the vaccine groupings developed pathological changes in the lungs. The authors concluded that "Caution in preceding to application of SARS-CoV vaccines in humans is indicated."[114]

While the above may not apply to mRNA vaccines, rushing any experimental vaccine into widespread use may be foolhardy. What consequences may exist for those vaccinated now in a world still containing COVID-19 virus or the emerging variants? We don't know. The precautionary principle should apply[115] but likely won't, because COVID-19 is considered so deadly that rushing new vaccines to market is viewed as the priority. That the overall lethality is not true for most people has been demonstrated in detail.

Safety Data for the Viral Vector Platforms: Johnson and Johnson and AstraZeneca

Johnson and Johnson

The same Johnson and Johnson press release[116] claims only minor adverse reactions such as mild pain at the injection site. Unlike the Phase 3 reports filed by Moderna and Pfizer, it is difficult to evaluate the rigor of the Johnson and Johnson data from a press release.

AstraZeneca

Here is what the company said in their press release: "No serious safety events related to the vaccine have been confirmed. AZD1222 was well tolerated across both dosing regimens."[117]

Sure, maybe. One guesses we will all have to just wait and see when their phase trial data are finally presented.

Summary of the Safety Data

As above, it very much looks like the mRNA vaccines are winning this horse race, as possibly concerning as their safety data are. All we have are vague statements from the makers of the viral vector COVID-19 vaccines so far and no real safety results to compare to Moderna and Pfizer.

Stability of mRNA Vaccines

Another issue concerns the cold storage requirement of the two vaccines. Moderna's vaccine seems to be transportable at -20°C and storable for a short period at temperatures similar to those in a normal refrigerator (approximately 4–6°C); Pfizer's vaccine needs -80°C, making both transportation and storage much harder logistically to get the vaccines en masse to those who want them or those whom the governments want to have them.

Why the difference? No one really knows, although one guess would be the temperature sensitivity of the lipid capsule encasing the mRNA, or the stability of the various mRNA constructs.

But here is the rub:

> The "cold chain" requirements also call into question how labile the vaccine may actually be, both the mRNA itself and the lipid covering, in the long run inside the human body at a typical temperature of 37C°. This may be one reason why various entities, including the NIAID, have suggested that multiple booster doses may be required.[118]

What make less sense in regard to mRNA stability are the animal data cited for Bahl et al., which claimed to find the construct in various organ systems up to eleven days after injection. In other words, preinjection, the vaccine with its presumably unstable encapsulated mRNA needs very cold storage because it is so labile, but postinjection into a human or animal at normal body temperatures, the mRNA is still detectible. Were I a reviewer in a peer-reviewed journal, this apparent discrepancy would certainly be one I'd want to see addressed.

Insider Trading?

In a move that raised media and public eyebrows and even got the US Security Exchange Commission involved, executives at Moderna and Pfizer began dumping their shares as their press releases were praising the emerging data, and various health officials such Dr. Anthony Fauci, NIAID director, and much of the corporate media piled on.[119]

As Children's Health Defense (CHD) documented in a detailed evaluation of the COVID-19 pandemic (see later below), these stock sales seemed to herald the foreknowledge of some greater plan. CHR wrote,

> . . . a few intrepid journalists have begun calling attention to Big Pharma's pandemic profiteering, even pointing out that 'insiders at companies developing experimental vaccines and treatments . . . aren't waiting until they finish the job to collect their reward' (Wallack, 2020). An October piece in the *Boston Globe* cited the example of Moderna, one of the companies that has rushed a candidate vaccine into clinical trials (Wallack, 2020). It took Moderna a mere three weeks after Bill Gates' initial funding installment to send its first batch of experimental vaccine to research and patent partner, the National Institute of Allergy and Infectious Diseases (NIAID), leading to an immediate surge in share price of 28 percent (Lee, 2020; Loftus, 2020). By early April, Moderna's CEO had become an overnight billionaire, and by October he had sold nearly $58 million in stock (Tognini, 2020; Wallack, 2020). Meanwhile, Moderna's chief medical officer has been 'systematically liquidating all of his company stock'—about $70 million —'in a series of pre-planned trades that have made him roughly $1 million richer each week' (Wallack, 2020). Thus far this year, company insiders have sold $309 million in stock versus under $2 million in 2019, fueling suspicion that they may be 'downplaying possible obstacles to goose stock prices—and increase their personal profits' (Wallack, 2020). Also among those who sold Moderna stock options was Moncef Slaoui, the former Moderna board member and former GlaxoSmithKline executive who now heads up Operation Warp Speed (Rozsa & Spencer, 2020).[120]

In regard to Operation Warp Speed, those whose vaccine hesitancy led them to Trump in the first place reveled in his withdrawal from the WHO, seeing the latter as corrupt. At the same time, many of the same people ignored it when Trump put Slaoui, the ultimate pharma insider, in charge of vaccine delivery in the United States. Similarly, the money Trump took away from the WHO went straight to GAVI, an organization discussed in Chapter 6. This too seemed to have passed under the radar of some.

With all of the above, it is best to remember the quote usually attributed to Mark Twain: "If you tell the truth, you don't have to remember anything." How much more complex it seemingly becomes when one is trying to bolster a story built on a collection of almost absolute misinformation and collusion at a massive level.

The Normal Trajectory of Viral Pandemics and Epidemics: Evidence from the Past

According to various sources including the epidemiologists associated with the Great Barrington Declaration and others,[121] the world is handling the COVID-19 pandemic in a way that is not consistent with the history of such pandemics.[122]

Essentially, viral pandemics have predictable phases, an initial spike of disease cases, followed by a dampening in cases and fatalities, in turn followed by later waves. Eventually, a viral pandemic achieves some sort of natural herd immunity, sometimes bolstered by vaccination, with many of the infected never becoming in the least symptomatic.[123]

The path to true herd immunity can be catastrophic, of course, but the infection rate, severity of the disease, and rate of fatalities of COVID-19 as discussed in previous sections do not suggest that this disease is one of those that are of such a variety, in spite of the media hype. The response of governments around the world, however, may be the actual catastrophe in terms of the myriad other health consequences that may ultimately take a larger human toll than this virus alone.

A Look at Pandemics and Epidemics of the Past

In the following pages, I will look at two events from the last hundred years to see if any of these have aspects that might not fit with most mainstream narratives for COVID-19.

These are the "Spanish flu" epidemic of 1918/19 and the rise of the Guamanian neurological disease spectrum in 1945–54, ALS-PDC, compared to the recent COVID-19 pandemic that is still playing out as this book goes to press.

The "Spanish Flu" Pandemic

First of all, the influenza pandemic did not begin in Spain; rather, the conventional story is that it arose on the Western Front and spread around the world as soldiers deployed home after the November 11, 1918, armistice. Accepting this narrative on the origins of the disease for the moment, why was it so virulent and pathological in its impact, and why did it never come back? The answer to the first question was provided, somewhat surprisingly, in a paper whose senior author is Anthony Fauci.[124]

The article gathered case reports from the years of the pandemic and compared them to actual autopsy reports and bacterial analyses to come up with the conclusion that the vast majority of deaths during the pandemic arose from secondary bacterial pneumonia. In other words, the virus created conditions that led to a severe respiratory bacterial infection and a multisystem organ shutdown. As for why the disease never came back, it may be that true herd immunity prevented it from doing so, or perhaps it simply mutated to a less pathogenic form.

There is an alternative explanation for the origin of this pandemic. This view holds that the epidemic actually originated in an army base in Fort Riley, Kansas, following an experimental meningitis vaccine administered by doctors from the Rockefeller Institute. Some soldiers fell ill and died. Those who appeared to be asymptomatic were deployed to fight in Europe and thus spread the disease.[125] Is this narrative correct or incorrect? The data suggesting that the epicenter of the disease was Fort Riley seem to be correct and are validated by the Merons et al. paper cited previously.

Guam and ALS-PDC

Elsewhere in this book and in a previous one, and in a recent book chapter,[126] I have discussed the nature and most common view of the origin of this neurological disease spectrum. However, here, as with other disease-origin data, there is evidence for a viral trigger, perhaps even evidence that a vaccine was involved.

To examine this more closely, let's look at the time line of the disease on Guam.

Aspects of what later was called ALS-PDC had been known to occur occasionally for at least a century before being "discovered" by Dr. Harry Zimmerman, a doctor with the US Navy posted to Guam in 1945 at the end of World War II.[127] Zimmerman observed and wrote about a form of ALS that seemed abnormally high among the native Chamorro people compared to ALS incidence in the continental United States.

In 1947, an epidemic of Japanese encephalitis broke out on Guam, infecting people as well as various domestic animals.[128]

Some forms of encephalitis are known to produce a kind of Parkinson's disease, in many ways similar to that observed on Guam with PDC.[129]

In 1949, Dr. Albert Sabin, later of oral polio vaccine fame, devised a Japanese encephalitis vaccine that was passaged numerous times though mouse brain cell cultures (see Chapter 2). Mice are known to harbor several viruses (Theiler and Safford) that infect their nervous systems and produce some neurological outcomes that strongly resemble those of ALS-PDC.[130] Viral contamination of vaccines is hardly an unknown phenomenon, for example the simian virus 40 (SV40)[131] contamination of the later Jonas Salk polio vaccine.[132]

Given the above with Zimmerman's observations of high levels of ALS in 1945, these speculations may not fit for the ALS part of ALS-PDC. But it certainly does fit for PDC, which Zimmerman did not observe and which was only later described by Kurland and Mulder in 1954.[133]

In the early 1960s, Sabin's JE vaccine was replaced by other versions. Of note, no one born after 1960 has developed ALS-PDC.[134]

A key question is whether either Theiler or Safford virus can act in zoonotic transmission to humans. No one really knows, although this is hardly impossible, given the known rodent transmission of Hanta virus to humans.[135]

A viral and a vaccine-adverse effects etiology for a part of ALS-PDC is thus feasible and might serve to explain some of the features of the disease, including the cluster of disease cases within families and the spectrum's ultimate disappearance.[136]

Both the 1918/19 influenza pandemic and the Guamanian ALS-PDC disease epidemic deserve further scrutiny, although insofar as medical authorities might discover something they really don't want to see, it is unlikely that any such study would be funded by any government agency.

Finally, as Oller[137] mentions, the realistic possibility that previous coronavirus outbreaks or pandemics, including those for influenza, the 2009 H1N1 outbreak, and the later SARS and MERS coronavirus outbreaks, may have arisen from similar GOF research.[138]

Predictions on the Pandemic: Three Time Points in the First Year of COVID-19

Below are some bullet point notes I made during various phases of the COVID-19 pandemic. Let's see how these stack up to the events that have followed.

March 23, 2020

1. The COVID-19 panic will spur legislation in many countries that will be intrusive in a number of ways, including impacting freedom of movement, association, speech, Internet freedom, and bodily autonomy. I predicted that there will be vaccine mandates such as in various countries or worse that include adults and take away privileges such as having a driver's licence or a passport not permitted for non-compliers. This will extend to other vaccines as well, measles being

a key one, but HPV and others to be included. Influenza may be included, as well.

2. Travel restrictions will come into place and freedom of movement will be curtailed without permission from the State. The Internet will be more regulated to remove and prevent "misinformation." Police and intelligence services will have greater power to monitor all communications, especially for those whom the authorities have IDed as potential troublemakers.

3. Those who publish or speak about things deemed misinformation will be fined or jailed. This will include vaccines, vax schedules, economic recovery plans, critical industries, etc. Herd immunity will be the given reason in regard to vaccines.

4. The economic state for many individuals in the United States and Canada will become more precarious. The rich will have bounced back; the poor not, even including with stimulus packages. Many small and medium businesses will fail for good.

5. Governments in both countries, reeling from costs, will declare certain projects designed to help the banks/oil industry and the pharma, etc., as protected in the "national interest"; protesters will increasingly be characterized as terrorists.

6. There will be general cutbacks in service for all government agencies.

7. The next ripple in health scares will trigger a full collapse.

8. Terrorists and those fighting for social policy changes will realize their potential to paralyze the economy. Terrorists will use this; others will simply think about it.

One thing that is clear is that rights lost/suspended will not come back on their own. Police powers will not revert to pre-2020 levels. Health care in Canada will cost more, including more costs for items that previously were free.

May 2020

The following will be an interesting exercise in early May, 2020, depending on how you count, six to eight weeks into the COVID-19 events. The following is an attempt to put all of this into context, fifteen months before this book goes to press. In this year, a lot may change.

What I want to do now is to put the entire COVID-19 story, as we now understand it, into what have emerged as two competing narratives of the existing, and still changing, facts about the disease and the national and international response to it. So in the following pages, I will first give what might be deemed the "official" narrative in that it is in line with what the WHO says. The second narrative is that of scientists, journalists, and lay people who have very different interpretations. Once I have listed the facts of the case, I will note below the "official" versus the "unofficial" explanations. When this is done, I will return to the consideration offered by Occam's Razor (Chapter 1) and discuss which of the two narratives is the simplest and thus more likely to be correct.

Here are the narratives:

1. COVID-19 originated in Wuhan, China.

   Official: The virus is an example of a virus that can jump between species (zoonotic), and it originated in bats or pangolins that were infected with a bat/pangolin coronavirus and sold in "wet markets" of dead and living animals. From there, the virus spread to humans. The infection spread from Wuhan to other regions of China and then around the world. The fact that it emerged in Wuhan is just a coincidence.

   Unofficial: Yes, Wuhan is the site of origin, but why Wuhan versus numerous other locales in China and Asia with wet markets? What makes Wuhan unique is that it has a Level 4 containment facility at the Wuhan Institute of Virology that was looking at coronaviruses and doing gain of function studies.

   Score so far: The official version needs transmission from animal species to humans; such transmission is known. The unofficial version needs the virus to have somehow come out of the Wuhan Level 4 lab, and various reports suggest that this facility was known to be lax in security. This sort of describes the "accidental release" hypothesis, versus that which says the virus was "engineered." There is evidence for both.

   Outcome: Tied.

2. The death rate after infection is not known. Various government and CDC figures put the rate as up to 10 percent. Other sources put the figure considerably lower.

   Official: We are assuming that the fatality rate is at the high end. We base this calculation on the modeling data out of Imperial College London from the group headed by epidemiologist Neil Ferguson.

   Unofficial: You don't know the real fatality rates until you know how many people were infected. We don't. Hence, the number is likely much smaller, and this interpretation is supported by various other studies (for example, Prof. Ioannidis at Stanford).

   Outcome: Time will tell which numbers are correct as the data around the world come in; that is, assuming that the various health departments can determine how many people were infected overall (the denominator). For the moment, tied.

3. Various countries have instituted more or less complete lockdowns of their populations. Sweden is one of those that used very limited measures.

   Official: People had to be kept apart to prevent the spread of the virus. Governments that acted on the advice of the WHO kept the pandemic from getting worse and "flattened the curve" so that medical services were not overrun. Yes, this had a negative impact on the economy of most countries, but lives are worth more than money.

   Unofficial: Sweden kept most of their economy open and had minimal social distancing guidelines. Their serious COVID-19 illness and deaths were not enormously higher than countries of roughly equivalent populations and GDPs, e.g., Belgium. Thus, the curve flattened on its own regardless of social distancing, much as other viral infections seem to do.

   Outcome: We will have to see what happens with the Swedes, but if the numbers hold up, then their "model" was clearly better than that of the WHO based on the concept of natural herd immunity.

4. The lockdowns need to be in place until we have a treatment for COVID-19, either a vaccine(s) or some antiviral agent.

   Official: One of these two options is going to have to be in place before we get back to "normal."

   Unofficial: Every day you leave the lockdowns in place, the more people globally you push into poverty. Poverty goes with malnutrition and comorbid health conditions. We disagree about the fatality rate, but it is a certainty based on the outcomes in the former Soviet Union and other places that the fatalities from the collapse of the world economy are going to be somewhere between higher to vastly higher than those from COVID-19.

   Outcome: The official position is based on assumptions on whether a vaccine or therapeutic will work and be safe. Given that the companies that are rushing to find a vaccine are (a) not doing serious safety testing and (b) will not have had a significant time line to look for adverse effects before releasing them to the market, all of the faith in a vaccine-based solution arises from assumption and dogma, not from fact. In addition, the development of both RNA and DNA vaccines to fight COVID-19 will be a vast experiment in humans; it is possible, rather probable, that a range of adverse reactions will occur outside the safety testing time frame. The unofficial side can quote the numbers for how human health and mortality changed in Russia after the fall of the Soviet Union.

   Given this, this point goes to the unofficial side.

5. Herd immunity will ultimately control this and other viruses in addition to the therapeutics in (3) above.
   Official: Herd immunity will be achieved through vaccines.
   Unofficial: Natural herd immunity happens all the time and is well demonstrated. Vaccine-induced herd immunity has never been demonstrated either in animals or humans.
   Outcome: Unofficial wins, as the official version demands evidence not yet provided by experiment.

6. A level of antibody production to either the infectious agent or to a vaccine will indicate that the individual has been exposed and is presumably now immune to that infectious agent.
   Official: Antibodies may not indicate immunity (WHO).
   Unofficial: Are you serious? This is the basis of virtually all vaccine efficacy studies that you support, as antibody levels are considered to be surrogate markers of protection. You just threw your own "science" under the bus.
   Outcome: To claim that antibodies only work as markers of disease protection when you want them to sounds more like an agenda-driven statement than one of fact.
   The point goes to the unofficial version.

7. The private sector will make a great deal of money for vaccines and pharmaceuticals to combat COVID-19 and other viruses to come.
   Official: That's irrelevant. The companies developing these products are working to save us all, and we should be thankful.
   Unofficial: Some of the companies developing vaccines and drugs have demonstrated time and again that they put corporate profit above human safety. Why would this time be different?
   Outcome: Given the well-documented history of corporate malfeasance in drug/vaccine development, the point goes to the unofficial version.

8. Governments, both totalitarian and officially democratic, have mostly adopted the social control measures to diminish the spread of COVID-19.
   Official: These measures were necessary, regardless of the temporary impact on civil liberties and the economy.
   Unofficial: Governments around the world adopted state of exception directives without evidence that the disease was as serious as claimed. And the impact on both the rights of people and the economy are going to be severe and long-lasting.
   Outcome: At this point, we don't know. Tie.

9. Immunization Agenda 2030 (IA2030), version 4, of the WHO was due to be voted on by the World Health Assembly in the summer of 2020.

Official: The COVID-19 pandemic shows just how important it is to implement IA2030.

Unofficial: How bloody convenient that the pandemic starts just as this document was about to be given the green light.

Outcome: Tie.

Totals: Tied: four; official version: zero; unofficial: four.

Based on this analysis, it seems that the unofficial version has the fewest assumptions and the most evidence going forward. The official version, unless all of the current ties come out their way, is more complex and demands that a greater number of assumptions be correct.

June to the End of November, 2020

Governments did not do war-gaming but rather panicked. War-gaming that all of their militaries normally do would consider the classical considerations of the enemy: the capabilities and weaponry, what they have done before, what they are likely to do, and which of these actions is most dangerous to us.

Based on these, potential scenarios should have been created that would have taken into account the actual severity of the disease, the numbers of those hospitalized and deaths, the various symptoms by age, etc. They then would have compared these to projected impacts of economic lockdown: poverty, mental issues, addiction and alcohol, diet, stress, impact on children, etc.

If only looking at the disease itself, then it would make more sense to respond to the worst possible outcome, rather than the least, since at least then more lives could in principle be saved. However, not doing the full assessment, an overreaction can be as bad as, or even worse than, the disease itself.

In British Columbia, Dr. Bonnie Henry and Health Minister Adrian Dix will be hoping that they can claim that without extreme measures, matters would have been worse, but since we don't know the overall impact of the disease alone, we can't say. So rather than do the war-gaming and come up with options, they panicked and did what the WHO told them to do. Henry and Dix may have a lot to account for.

At the Canadian federal level, no advance planning seems to have been done, either. Had the "Feds" done their jobs, they should have known a pandemic was inevitable someday and that they needed to have stockpile drugs of various kinds, ventilators (maybe), and trained the necessary personnel. They didn't, hence the panic. An army intelligence officer who did the same and didn't give his commander appropriate options would get fired.

What does all of the current information on COVID-19 really mean? Again, by the time this book is published, we may know, or at least certainly we will know more. For example, is there evidence that laboratory tinkering with a wild type coronavirus has some purpose? Maybe, if Dr. Alina Chan and her colleagues are correct. Was the release an accident? Occam's Razor would tend to support this interpretation given what we know about the multiple accidents that do happen in Level 4 containment facilities.[139]

One clear take-home message is that we all realize just how vulnerable our civilization actually is to pandemics, real or imagined. And, if we realize this, so do the terrorists of the world who will see a clear path to trying to exploit future fears to promote their agendas. In turn, the very specter of this will inevitably lead governments to enact ever more restrictive legislation to further the spying on their own people and to diminish human rights.

In this latter context, we know that governments of all stripes have used the pandemic to give themselves extraordinary powers, claiming it is for the "greater good," outcomes that Nazi jurist Carl Schmitt would certainly have agreed with and completely in line with historical precedent.

In fact, as this book goes to press, the federal foreign minister, Marc Garneau, still seems to be seriously mulling using some form of Emergency Act to limit movement within Canada[140] on top of restrictions already in place at a provincial level.

This last is not all that surprising given that, at least in my opinion, governments by their very nature are always seeking to expand their powers at the expense of the civil and natural rights of their citizens, even in notionally democratic countries, as cited in Chapter 11. And with such state of exception decrees come the likelihood that some of these measures will remain after the current crisis is over. These might include continued restrictions on association, public meetings, vaccine mandates for COVID-19, or anything else the governments of the day may want. And, it is a fairly reliable rule of thumb that rights once removed rarely return, at least without a struggle.

One of the almost inevitable consequences of the COVID-19 crisis is that the push for mandatory vaccination once a vaccine is available will surely occur, mandates driven by the usual players already documented in this book. With politicians certain to be far less concerned about violating human rights, the question is when, not if, mandates succeed, what then will prevent mandates for the whole range of human coronaviruses (seven at last count), coronaviruses that can cross to humans through zoonotic transmission as COVID-19 allegedly did, and the myriad other viruses deemed existential threats to civilization? The short answer is that there will be no limit on such mandates, all for the greater good, at least until such time as the adverse effects become so obvious that only those with vested interests in vaccine propagation remain unconvinced.

Summary

Were these projections wrong? For the most part, no. But how accurate they were in total will only be determined in the near future.

Triangulating the WEF and the "Great Reset"

To "reset" means to fix or adjust something in a new or different way. The World Economic Forum (WEF), introduced in the previous chapter, is engaged in a worldwide reset of their own design.

Understanding who the members of the WEF are is the first step to understanding what it is that the organization hopes to achieve. From this, perhaps we can understand what the various motives are.

So, first, who is who in the WEF? What is the core constituency? It's simple: it consists of a number of major corporations in various fields, including the pharmaceutical industry, fossil fuels, technology (especially Internet and artificial intelligence technology), and others described in an article by Children's Health Defense that came out recently in the *International Journal of Vaccine Theory, Practice and Research*.[141]

The annual meeting of the Forum is also attended by high-ranking politicians from around the world and by various NGOs.

The founder of the WEF, and still the executive chairman, is Klaus Schwab.[142]

Schwab seems to have recruited members of this elite club to join him each year in Davos, Switzerland, to plot a way for capitalism to project a greener, more socially just face in order to stave off what might likely be the future demise of both capitalism and the elite shareholders that have become billionaires by playing the capitalist game so very well. The official version of the WEF is that Schwab recruited members with the lure of philanthropy, something along the lines of "You have benefited so much from the system; It is now time to give back to the less fortunate."

A more likely recruiting pitch might have been: "If we keep killing the Earth and/or keep impoverishing the peasants, then sooner or later you and your family will die from a depleted Earth or angry peasants who will come put their pitchforks into you." I suspect a backup to the second pitch was to point out to the movers and shakers that they could hobnob with their fellow captains of industry and the politicos who run most of the capitalist-oriented world, including China, the latter notionally "communist." That sort of one-on-one, first-name-basis glad-handing is one reason that various organizations have annual conferences, since it builds future relationships and opportunities for oneself or for your organization. That the 0.01 percent do this as well should not be a surprise, and Davos is the apparently the venue to do it in.

Apart from the highly promoted economic and environmental benefits attributed to the WEF, many of their projects and speculations seem to focus on "transhumanism," a philosophical construct that argues for transforming humans using various advanced technologies in order to change and/or enhance human physiological or mental function.[143]

Where does the WEF funding come from? The participants, for one, including various governments. And, in large measure, a lot of their money arises from the Bill and Melinda Gates Foundation, a fact that probably should surprise no one.

Is it bad that Schwab and the WEF proclaim the need to prevent catastrophic climate change, or to ensure social and economic inequality? Not at all, indeed the opposite. But using the very tools that had a major hand in creating these conditions is very odd, to say the least.

The above is but one example of the inherent contradictions that are rampant in Schwab's vision: in a nutshell, in order to save capitalism and the privilege of WEF members, the Reset seems designed to save the world *and* keep owning it at the same time. Mouthing the right phrases about social inequality and a green environment is the way to get people on your side, particularly perhaps what remains of the left and progressive movements. Indeed, this is what the various speakers at the meetings do: rattle off the "woke" talking points, all the while ignoring the key question of who is really benefiting from the proposed plan. And to ensure that the joyful tidings emerge from Davos in the correct format, the captured media is there to help out while largely ignoring the fact that most of the woke members praising the environment flew into Davos on their private jets.

How does COVID-19 figure into this? Schwab and his coauthor, Thierry Malleret, tell us in a recent book put out by the WEF.[144]

The idea that the book proclaims mirrors that of the WEF website,[145] namely, that with COVID-19 as the "accidental" driving force, the pandemic has forced on humanity the need to restructure economies and societies and essentially rearrange the world to build "shareholder capitalism" in a "shareholder economy." Or, in other words, capitalism in a new form for the twenty-first century, using politically correct language to help people understand that while capitalism was the problem back pre-COVID-19, miraculously it is the solution now if we green it up and add some social justice phrases.

It is an interesting book to be sure, as the ideas that Schwab and Malleret want to convey are written in a style designed to be readily understandable to a lay audience. The three key sections deal with Macro Reset, Micro Reset, and Individual Reset, each with subsections. The Macro Reset portion occupies the first 172 pages of a 280-page book.

Schwab and Malleret begin by trying to put the COVID-19 pandemic into the context of past pandemics or financial crises. Several main themes emerge that serve as the foundation for all that follows. One is that the Black Plague, for example, triggered an evolution of the political and economic structures of the time and thus led to much that we appreciate in these structures in the modern world. In doing so, the authors evoke the concepts of complex adaptive systems, a tenet of chaos theory, and the nonlinear nature of such systems. They then show a diagram of the complex net of interactions of all aspects of the modern world from financial institutions to social structures to food supply and the interactions of countries and within countries, all now subject to a "cascading" failure as some part of the web collapses. In their view, the pandemic triggered the various lockdowns of the economy of the world with extra-devastating impacts on resource economies, or "commodity countries" in their terms. These latter impacts will inevitably lead to social strains within these societies, evoking, in short order, violent social unrest.

This sort of systems dynamics is known in principle from studies of other complex forms of interaction, from how one power grid failure can trigger the collapse of other grids, or how in brain function the degeneration of one neural subset triggers the deterioration of others.

None of this is new; the novelty here is the application of such theorizing in a modern, highly interconnected world to the impact of COVID-19. COVID-19 then acts as a seed crystal or, to use a biological example, more as a prion-like spread that ripples throughout the organism spreading disorder in its wake.

The authors at least are honest enough in the first pages of the book to address the actual severity of the pandemic. They write:

> . . . we look for precedents, with questions such as: Is the pandemic like the Spanish flu of 1918 (estimated to have killed more than 50 million people in three successive waves)? Could it look like the Great Depression that started in 1929? Is there any resemblance with the psychological shock inflicted by 9/11? Are there similarities with what happened with SARS in 2003 and H1N1 in 2009 (albeit on a different scale)? Could it be like the great financial crisis of 2008, but much bigger? The correct, albeit unwelcome, answer to all of these is: no! None fits the reach and pattern of human suffering and economic destruction caused by the current pandemic.[146]

It sounds like the pandemic is worse by a long shot, right? In contrast, Schwab and Malleret then try to put it into ever-greater perspective about the number of deaths to be expected: "Even in the worst-case horrendous scenario, Covid-19 will kill far fewer people than the Great Plagues, including the Black Deaths, or World War II did."[147]

What then accounts for the devastation wrought by the pandemic that allows Schwab and the WEF to seize on it as a bridge to the bright new world of the Great Reset? *Fear.* This is the very fear that allows corporations the ability to accumulate even greater wealth; and the same fear that allows politicians around the world to launch state of exception measures for greater control.

The question that then has to be asked is this: If the interconnectedness of the world is so fragile, as indeed it may well be, and if the whole structure can be toppled by fear, who is responsible for that fear? Who knows that the pandemic, compared to past disasters, is far less destructive to human lives than the media and most mainstream medicine proclaim? And who had the plan to launch the Great Reset before COVID-19 and later found in COVID-19 the ideal trigger to do so? None other than Klaus Schwab and those he works with on his Reset plan for humanity.

One might say that the Schwab and Malleret book can be summarized in a phrase attributed to the eighteenth-century English writer Samuel Johnson: "Your manuscript is both good and original; but the part that is good is not original, and the part that is original is not good."[148]

The website for the 2020 meeting proclaimed that:

> The COVID-19 crisis is affecting every facet of people's lives in every corner of the world. But tragedy need not be its only legacy. On the contrary, the pandemic represents a rare but narrow window of opportunity to reflect, reimagine, and reset our world to create a healthier, more equitable, and more prosperous future.

There you go: win-win, we are all happy now.

One year of COVID-19 later, the January 25–29, 2021, virtual WEF meeting being held proclaimed that it is time to "Get ready for the future" and highlighted a range of talks bracketing what they term the "Davos Agenda."[149]

This year in a virtual Davos, the meeting featured a host of world leaders, the notable names being the functional dictators of Russia (Vladimir Putin) and China (Xi Jinping) and the authoritarian prime minister of India (Narendra Modi), all mingling online with the heads of banks and the pharmaceutical industry.

Much like the earlier meeting, the Davos Agenda seeks to transform every aspect of human life on this planet, and, like last year, the COVID-19 pandemic is the ticket through which this change will be accomplished. When COVID-19 (or any of the variants) has been defeated, the world we reenter with the slogan of "Build Back Better," will be fairer, more just, and definitely greener, at least in theory if not in reality.

The reality, of course, is that the pandemic pushed the world in the opposite direction: greater wealth concentrated upward, more authoritarian rule, and the inevitably greater environmental destruction that is certain to accompany any rebooted form of capitalism.[150]

Watching various governments adopt the WEF "Build Back Better" motto and parrot Schwab is not only curious, but betrays the extent to which a variety of politicians and governments have been utterly corrupted, not that it was that hard to corrupt them in the first place. Schwab's statement in his book that we won't go back to normal until a vaccine saves us all is routinely mimicked by politicians who have either read the book or who are simply reading from the same script.

So, apart from the nice words, what we are seeing with the "new normal" of COVID-19 is nothing more or less than a rolling world coup d'etat created by those same special interests: "big data, big telecom, big oil and chemical, big finance and [the] global public health cartel"[151] as graphed by Children's Health Defense (CHD) [Brackets are mine.] The role of the US military-intelligence apparatus in all of this cannot be ignored.

As CHD wrote,

> The Pentagon's involvement in coronavirus-related efforts goes well beyond DARPA-funded research. Four-star General Gustave Perna is serving as chief operating officer of Operation Warp Speed alongside chief advisor Moncef Slaoui (see below). General Perna, in charge of US Army Materiel Command, oversees the global supply chain for over 190,000 US Army employees (HHS, 2020b). For the first time ever, the distribution of the eventual coronavirus vaccines is being planned as a "joint venture" between the CDC and the Pentagon, with the latter overseeing "all the logistics of getting the vaccines to the right place, at the right time, in the right condition . . ."[152]

These connections are illustrated in Table 13.1.

As the article by CHD makes clear:

> In fact, global financial patterns and pronouncements point to a seismic overhaul of governance and financial systems that is playing out beneath the surface of the pandemic, reaching far beyond the health domain. These developments highlight a disturbing push for global technocracy - a form of centralized, expert-led control over resource production and consumption that the *Wall Street Journal* characterizes as "anti-democratic rule by elites who think they know better.[153]

And, citing *The State of Our Currencies* and other pandemic-related writings by Catherine Austin Fitts,[154] CHD quoted Fitts as saying,

Date	Amount	Company	Funding Focus
March 30	$456M	Johnson & Johnson/Janssen	Vaccine
April 16	$483M	Moderna	Vaccine (Phase 1)
May 12	$138M	ApiJect	Syringes
May 21	$1.2B	AstraZeneca/University of Oxford	Vaccine
June 01	$628M	Emergent BioSolutions	Vaccine/drug manufacturing
June 11	$204M	Corning	Glass vials
June 11	$143M	SiO2 Materials Science	Glass-coated plastic vials
July 07	$450M	Regeneron	Antiviral antibody treatment
July 07	$1.6B	Novavax	Vaccine
July 22	$1.95B	Pfizer	Vaccine
July 26	$472M	Moderna	Vaccine (Phase 3)
July 27	$265M	Fujifilm/Texas A&M University	Vaccine manufacturing
July 31	$2B	Sanofi/GlaxoSmithKline	Vaccine
August 04	$160M	Grand River Aseptic Mfg (GRAM)	Vaccine/drug manufacturing
August 05	$1B	Johnson & Johnson/Janssen	Vaccine (manufacturing)
August 11	$1.5B	Moderna	Vaccine (manufacturing)
August 14	n/a	McKesson (existing contract)	Vaccine distribution
October 09	$486M	AstraZeneca	Monoclonal antibodies
October 13	$31M	Cytiva	Vaccine "consumables"

Table 13.1.

. . . [concerning the pandemic] emphasizes the importance of accepting that what is transpiring in the financial, tech, and biopharmaceutical sectors is interconnected. Part of this involves recognizing that the coronavirus vaccines currently dominating the headlines represent something likely to go far beyond the simple health intervention being held out by scientists and officials as a panacea. Instead, the evidence suggests that Covid-19 vaccines are intended to serve as a Trojan horse to transport invasive technologies into people's brains and bodies. These technologies could include brain-machine interface nanotechnology, digital identity tracking devices, technology that can be turned on and off remotely, and cryptocurrency-compatible chips.[155] [Brackets are mine.]

What do the various players want from all of this? Maybe we can figure it out by examining Maslow's "Hierarchy of Needs" concept.[156]

No one who goes to Davos in person or on Zoom is on any of the bottom steps; rather, all are at the level at least of building wealth, and from wealth, power and control. The wealth step is where I would put the companies that show up. That of power, the politicians. Tedros Ghebreyesus of the WHO and others like him are also somewhere at that level. The NGOs are mostly neither rich nor powerful, but their attendance allows the WEF to check off the affirmative action box labeled "societal concerns" and in the process gain public support for the illusion that the WEF really does mean us all well. The NGOs, for their part, are likely hoping for any scraps from the table.

Finally, among the assorted hangers-on are the people and organizations there to ensure that the "trains run on time," those in GVAP, CEPI, GAVI, the Wellcome Foundation, and the Vaccine Confidence Project, all with Gates Foundation money in their pockets.

And, of course, we should not neglect the mainstream media, who have invested a lot of time and energy pushing the COVID-19 narrative endlessly in the form of "psychic driving."[157]

This brings us to the top of the cui bono pyramid to ask what Klaus Schwab and Bill Gates, as the key drivers of the WEF, really want. With Schwab it is hard to tell, although clearly he sees the pandemic as the means to get to his environmental goals and away from the old normal for which a new capitalism is the means and COVID-19 is the excuse: ". . . the COVID-19 crisis will have gone to waste as far as climate policies are concerned."[158]

The idea to start the WEF was clearly Schwab's baby, although the evolution of the WEF was a group effort, and it would be interesting to know who else lurks behind the scenes. Maybe Schwab saw in the beginning a way to preserve capitalism and then just kept improvising until COVID-19 gave him the opportunity to push the Davos Agenda hard and fast. Or, just maybe, Schwab sees his future in more messianic terms.

As for Bill Gates, what is his motivation? It can't be wealth alone, since he has more than most other mortals, even more than most of those at Davos even after his recent filing for divorce announcement. Wealth like Gates's buys power and control. But control of what? The only answer that makes sense can be the control of humanity and its future direction. This is off the Maslow chart into something quite different; perhaps, like Schwab, a messianic vision in which he, William F. Gates, III has come as our savior? Or maybe, as has been suggested with Gates's ties to the late pedophile Jeffrey Epstein, it was a Nobel Prize that Gates was seeking?[159]

What are we really looking at here? A visionary matched up with a puppet master with endless wealth, now joined together to put forth a vision of a kinder, gentler, shareholder "Fourth Reich" in order to manage the world? Time will tell, but it would certainly be worthwhile for some real investigative journalist to begin a detailed probe into the Gates/Schwab connection.[160]

Converging Lines of Evidence on the Great Reset

Triangulation is the process by which one finds a point or source by observing the overlapping vectors, from three separate locations. Where the vectors all converge is the source. We are now close to that. The Children's Health Defense team has produced the remarkable evaluation cited previously. This vector that CHD draws intersects one from the Weston A. Price Foundation,[161] and one from the work of an independent journalist, Tessa Lena, in New Your City.[162] If three vectors weren't enough to definitively pinpoint the problem, there are others such as the webinars from Alison McDowell[163] and social commentator Russell Brand, the latter laying out the overlap of forces in a recent video.[164]

All lines of evidence point to the same source, Davos, and the same key players, Schwab and Gates and those who may be above them in the food chain.

With these sources, and more certain to follow, it is now coming into stark relief that those at the top pushing for the Great Reset are aiming for a far more tyrannical world than one we have known for generations in the West.

Some may ask how such a plan, a plan that some might term a conspiracy, on such a grand and global scale could even be possible. It's actually not, however, all that hard to imagine: just think of the existence of any Mafia grouping or drug cartel. None of this is to say that all members of the cartel always get along or even like one another. But nothing about what is happening is about like or dislike; rather, it is all about power and control, megawealth for those at the top, with less for those below, often vastly less.

Did it take a long time to put together this plan? Undoubtedly. The coordination of all the moving parts across sectors demonstrates with little doubt that the plan has been in the works for years, vastly longer than we have been led to believe. In brief, COVID-19 is not the cause of the Great Reset, merely the excuse for it to launch, a point that Schwab's book clearly acknowledges.

Viewed in this way, we can see that the actual reality of COVID-19 and COVID-19 vaccines, let alone other vaccines, is not the whole beast, merely the "pointy end of the spear" behind which stands an array of special interests pushing the spear forward on behalf of their joint and individual interests.

The Essential Role of the Captured Media in the Great Reset

In regard to the latter, the following transcript pretty much fills out all the details on how the media see their role. We've already explored some of this previously and in earlier chapters, but the following makes the nature of a captured media crystal clear, if it wasn't already.

Anthony Fauci of NIAID appeared on *PBS News Hour* to discuss the rapid spread of the virus as well as to give hope to others about the development and eventual role out of a future (now current) COVID-19 vaccine.[165]

This is how some of the interview went, with Fauci responding to the PBS interviewer's questions. First, speaking about the normal speed for getting a vaccine to market:

> **Fauci:** . . . Generally that would take a couple of years to get to that point. We're already there. We're going into a phase three trial at the end of the month.
>
> When you're dealing with vaccines you can't guarantee things, but you can say based on the science and the way things are going, that I'm cautiously

optimistic that we can meet that projection that I made months ago. And that is, by the end of this calendar year and the beginning of 2021, I feel optimistic that we will have a vaccine, one or more, that we can start distributing to people. Because if you look at the infections going on right now and phase three trials that are now starting at the end of the month, we could get a signal of safety and efficacy by as we get into the late fall and early winter.

If we do, by the beginning of 2021, we could have a vaccine.

PBS: Available to hundreds of millions of Americans?

Fauci: . . . Start making doses before you know that the trial works. Which means that if it works, you've saved months.

We think we can start getting doses in the beginning of 2021, and the companies have said hundreds of millions of doses within that year.

PBS: Do you have a worry though Dr. Fauci that the anti-vaccine movement could interfere with this timetable?

Fauci: Yes, I do because we have to admit and realize that there is an anti-vaxx movement that we've had to struggle with in this county. I believe the solution to that is community engagement and community outreach, to get people that are trusted by the community to go out there and explain to them the importance of not only getting engaged in the vaccine trial, but the importance of when the vaccine is shown to be safe and effective, to actually take the vaccine because it could be lifesaving and it certainly would be the solution to this terrible pandemic.

Speaking on the dangers of the "anti-vaxxers," Fauci carried on with the notion that such people are the true enemy:

We've got to do a considerable amount of community engagement and community outreach because there is this reluctance to get vaccinated.

I think it has to do with a lot of things that we can clarify. We're moving at a very rapid speed because of the urgency of the situation to develop a vaccine. We want to make sure that we're very transparent, that people appreciate that that speed is not compromising safety, nor is it compromising scientific integrity.

In addition, superimposed upon that is something that we have to face the reality of. It's true, it's unfortunate is the general anti-vaccine feeling among certain segments of our society.

Then there's the issue of people not wanting to be told what to do by authority. It's a bit of the anti-authority, anti-science approach in this country.

Those are all obstacles we have to take head on and we've got to make as much open, honest and transparent outreach to the community to convince them that getting vaccinated is for their benefit and the benefit of the community. And everything about the vaccine development and implementation will be transparent.

The mainstream media is not the sole player by any means in the attempts to blame vaccine dissidents for a potential failure of the COVID-19 vaccine, or any vaccine for that matter. One of the main cheerleaders at the Vaccine Project (see Chapters 8 and 12) is Dr. Heidi Larson, who heads up the Vaccine Confidence Project, which appears to be replete with Gates Foundation money. This is also the group that has joined the chorus claiming "anti-vaxxers" are a major hazard to public health.

On Mandates and Lawsuits

The restrictions on daily life imposed by politicians across the world are not abating. Indeed, they seem to be accelerating. What started out in the early COVID-19 days as full or partial lockdowns to "flatten the curve" and intended for several weeks have largely been maintained. Some of these that may have been suspended in part or in whole when COVID-19 caseloads seemed to be diminishing have now come back in force in countries or regions facing a second or third wave of the disease. For example, here in British Columbia and other parts of Canada, we were experiencing gradual easing of social distancing and other restrictions, but when cases started to rise again, the public health officers began to panic yet again. Our own Dr. Bonnie Henry looked at the rising false positive numbers, ignored proper PCR testing methods, and decided, as has become her wont, to treat British Columbians as social lepers. She has been allowed to do so thanks to Bill 19, an Orwellian piece of provincial legislation that allows her to dictate the lives of over five million people. Here in BC, many continue to see her as a heroine; increasing numbers think she is a petty bureaucrat whose power of the executive diktat has gone to her head.

Spain and Italy locked down hard again; Sweden also increased its relatively soft restrictions.

Around the world, politicians left, right, and center clamor for much-anticipated vaccines to arrive to deliver that most magical of imaginary beasts, vaccine-induced herd immunity.

For this reason, it seems virtually inevitable that mandates for COVID-19 vaccines will soon become the norm. These will be enforced not by overt force, but by escalating restrictions on those who won't comply, that is, by taking away the "privileges" of people who are now awakening to the notion that this is all they really have ever had.

There are, however, some reasons for hope: the Great Barrington Declaration has now been signed by tens of thousands of academics and hundreds of thousands of lay people. This declaration made much of the self-proclaimed left apoplectic, denouncing it as an alt-right attempt to punish poor people. This was an odd critique given that it states precisely the opposite. In brief, the key Barrington idea is to allow most economic activity to go on as normal while ensuring protection for those most vulnerable to the disease.[166] Other newer documents are beginning to appear. One of these is the Victoria Declaration started here in British Columbia.[167]

Lawsuits against governments have been filed in Canada and Germany and elsewhere, and surely many more will follow.

And resistance, covert and overt, is growing against the world coup.

Lessons Learned from the COVID-19 Pandemic

Here are some lessons learned during the last year, and learned by most of us the hard way. That they should have been learned this way is a significant issue, but one best dealt with elsewhere.

The first is that health officials globally, some more than others, utterly failed to anticipate and prepare for this pandemic given that we have known for over a hundred years since the "Spanish flu" that a pandemic was inevitable sooner or later.

This lack of preparation led to inadequate supplies and training. Since these aspects are, or should be, part and parcel of the responsibilities of health officials worldwide, the lack thereof is simply inexcusable.

Next, models of disease spread and intensity, such as that of the Imperial College of London, were woefully inaccurate, often driving unprepared health authorities into panic. This panic was then communicated to the public with clear consequences. These consequences included, as cited above, an almost complete neglect of the other health concerns during the pandemic: drug overdoses, anxiety, depression, spousal and child abuse, alcohol and drug abuse, and increasing poverty, to name just the most obvious ones. In this sense, the failure to understand disease spread led to lockdowns and social isolation that contributed to many of these additional health outcomes.

Additionally, in terms of medical interventions, overventilation may have killed many of the ill in the early days of the pandemic. The failure to explore other early treatment options due to an overreliance on questionable data about therapeutics such as the use of hydroxycholoquine and ivermectin was telling. Instead of this, most health authorities continued to await the salvation from the still-experimental COVID-19 vaccines, delivered from the magical forest of the pharmaceutical industry.

To top that off, there seems to little concern or preparation across health ministries of any country about the almost certain wave of adverse effects that will arise due to these vaccines.

The above is but a short list of lessons/failures to be addressed for future pandemics, be they of natural or of human origin.

The Future of COVID-19 and Us

It must always arouse suspicion when the cloistered rich and powerful with an agenda to push conveniently find the means to execute their agenda. September 11 and the neocons come to mind; the misery that their actions unleashed on the Middle East still lingers twenty years later.

Was the onset of COVID-19 just a convenient and coincidental occurrence that allowed the concept of a social and economic reset to be implemented? As with 9/11, the Great Reset arose in the imaginations of the rich and powerful, people like Klaus Schwab and Bill Gates.

Should we trust the WEF version about the origins of the pandemic, or should we, the 99.99 percent who are affected, consider alternative hypotheses? What would we do if these other hypotheses turned out to be true? Would we resist, and how?

As this section is written, and even later when it goes to press, it seems likely that we will still be dealing with COVID-19 or others diseases like it. At present, various entities from the Canadian Army to governments around the world note that there may be future outbreaks of COVID-19 or other viruses and that we need to be prepared to face these outbreaks for months or years to come. We are indeed on the brink of a "brave new world," in which lockdowns, social isolation and distancing, and all of the societal changes now proposed for the current pandemic will be with us for a long time to come, perhaps forever.

We will endlessly be asked to remember that all of these measures are for the greater good and that "we are all in this together," except of course for the billionaire class, whose concept of being in this all together seems remarkably different from people trying hard each month to pay their rent or mortgage and feed their families.

Now we are asked to protect seniors and those with various comorbid conditions. Next time the vulnerable might be children or those in their twenties or thirties.

In this new world, there will always be new threats and new things for us all to fear and more things to sacrifice for the greater good.

The response to COVID-19 will have charted the path forward for governments working with, or more likely on behalf of, big corporations to increasingly dictate what we do, with whom, how, and when.

And if that is not the sort of world we were warned about by George Orwell and Aldous Huxley, I don't know what is: trust Big Brother, don't question, don't listen to dissenting voices, do what we tell you for the greater good (and we, the government, will decide what that is).

We have arrived for the second time in nineteen years at that new world. September 11, 2001, ushered in the attempted dominance of US society toward a state of perpetual war for the benefit of the defense industry. COVID-19 ushers in the attempted dominance of the pharma cartel and the billionaire elite over the entire world. The former could imprison you at will, even kill you. The latter seeks to control your very existence from conception to grave.

The world has indeed been turned upside down by the confluence of COVID-19 and the proposed Great Reset. It is hard not to imagine that the latter inspired in principle, if not in reality, the former.

Regardless of how fatalistic many of us seem to be in the face of the events of 2020 and the continuing saga in 2021, there is always the hope for resistance to the WEF, Bill Gates, and the other corporate henchmen who may have made this all possible. Before continuing, I want to give this megacartel a name: WEF(P)-G-WHO, in other words the WEF and Schwab, inside of which sits the block of pharmaceutical companies, Gates in the center of it all using his vast wealth to pull the strings, and the WHO linked to WEF through Gates, and all of them linked to someone higher up the scale of power.

Some of these options for resistance and renewal are discussed in the next—and last—chapter of the book.

Epilogue

Finally, in this chapter I want to consider a few emerging issues that are coming up as this book goes to press. There will certainly be many more before this book is available.

Some of these include:

- The requirements for vaccine "passports" for shopping or for boarding a bus, a train, or an airplane or for the ability to cross international borders, maybe even in Canada for provincial borders;
- The levels of adverse effects from the various vaccines that will certainly grow in number, both as the vaccines are given and over the long term time frame as autoimmune problems are identified.

  As a sidebar, which adverse effects may come out of Novavax's recent recombinant protein vaccine adjuvanted with a new proprietary saponin compound, saponins being forms of glycosylated sterols? A number of such steryl glycosides have been shown to be neurotoxic. Although the amount in a vaccine is very low, if it can serve to stimulate the immune system, it can just as well impact the nervous system.[168]
- The building hysteria about the numbers of COVID-19 virus variants now rapidly emerging in various countries;

- The implications of this last point for the official version of herd immunity if the current crop of vaccines does not prevent the vaccinated from transmitting the disease;

And some mysteries may be better understood in the near future, including:

- Where did the 2020–21 influenza season disappear to?
- Why did the expected decimation of refugee camp populations or those living in poverty in our inner cities not occur?
- Not least, how is it that in China, where COVID-19 originated, the cases recorded and the death rate are so much lower than elsewhere in the world, particularly the United States?
- Is there any validity to the emerging hypothesis in some circles that those vaccinated with the various EUA vaccines can shed something that impacts those not vaccinated? As yet, this is anecdotal and perhaps just rumor. However, it would be an easy experiment to do in an animal model of either mice or ferrets in order to find out.
- And finally: Why is Bill Gates buying up farmland all over the United States through shell companies[169] and speaking about a future small-pox epidemic? Maybe he is simply prescient; or maybe he knows things that we won't know until they happen. Concerning the farmland, Children's Health Defense has had a thorough look at "Farmer Bill" in an article that should be required reading for anyone concerned about the man and his plans for us all, albeit likely modified by his divorce from Melinda.[170]

CHAPTER 14

Future Tense:
The Lady or the Tiger?

. . . 'UFOs' and 'extraterrestrials,' can elicit feelings of fear and uncertainty, which was perhaps most evident from Orson Welles' notorious 1938 broadcast of H. G. Wells' science fiction novel The War of the Worlds. *The anticipated reaction to his bit of radio drama demonstrated that* **the perceived threat of attack by an alien life form can create a state of anxiety and mass hysteria among the easily agitated.** [Bolding for emphasis is mine.]

—Michael P. Masters[1]

Never has our future been more unpredictable, never have we depended so much on political forces that cannot be trusted to follow the rules of common sense and self-interest-forces that look like sheer insanity, if judged by the standards of other centuries. It is as though mankind had divided itself between those who believe in human omnipotence (who think that everything is possible if one knows how to organize masses for it) and those for whom powerlessness has become the major experience of their lives.

—Hannah Arendt[2]

Introduction to the End State

The quotes from Masters and Arendt, respectively, pretty much summarize where I think we are at this juncture: At a time and place where most of us fear an alien invader, albeit the invader is a virus from our own planet. Further, many of us are seemingly stuck in a space between the ever-changing fatuous words of so-called experts and a rising chorus of dissent, between those who trust political authority and those who don't. Or, as I have begun to think about it in this year of weird political groupings, betwixt the complacent left and the alt-right, that is, somewhere on the path to a third way, the latter to be discussed in a section below.

In this last chapter, I will try to address what I think are the key issues arising from the other chapters. I will also offer my perspective on where I think the future will take us, a future that seems at this moment to have only two outcomes, like the old story of the lady or the tiger.[3]

Let me first take a look at a sentiment that has been emerging on mainstream media and in some science journal commentaries, perhaps as trial balloons. In brief, when a COVID-19 vaccine comes out, even if only partially effective, everyone has to take it to get to herd immunity, and there will be penalties for those who don't. The article is titled "When a COVID-19 vaccine finally arrives, the government must make people take it." And below this is the message: "Polls show enough people are unwilling to take the vaccine to prevent herd immunity. We can't let that happen."[4] This opinion piece mirrors in lay fashion the more scientific one cited in Chapter 11 from the *New England Journal of Medicine*.

The current article's author, Tom Keane, goes through the typical shibboleths about herd immunity, clearly failing to realize that this has never been achieved by vaccination in the past. Trying to be "helpful," Keane raises the question of what happens if a COVID-19 vaccine(s) is not particularly effective. Answer: it's still better than nothing. What would be his answer to get to his projected 75 percent of the population needed for herd immunity: revaccinate them as needed until the required surrogate markers get high enough? What about vaccine injuries such as those already reported in some of the phase trials and in the initial vaccine rollouts? No concerns, doesn't happen, move along, nothing to see here.

Keane views "anti-vaxxers" like flat Earth advocates, ignorant people who don't believe in science. If you don't want to get the shot even if Anthony Fauci and the other talking heads say it's safe, fine, then pay the price with the loss of your freedom. Here is what Keane concludes the price of not accepting a COVID-19 jab might look like:

> The same model [the current mandatory influenza vaccine in Massachusetts to go to school] can be used for the soon-to-come coronavirus vaccine. Want to go to work? Get vaccinated. Want to go shopping, eat inside a restaurant, or get on a plane? The same. Indeed, the simple requirement can be that if you want to leave your house, you must get vaccinated. Maybe that seems intrusive. But it's actually far less so than what we endured at the height of the lockdown from mid-March through mid-May, when the state shut down its entire economy.
>
> Maybe a few diehard anti-vaxxers will resist and simply stay forever in their homes. Fine by me. As long as they can't mix with the general population, they won't pose any threat. While normalcy returns for the rest of us,

and we work and play as we used to, they can stay stuck inside, isolated and alone, glaring out their windows as life passes them by. [Brackets for addition are mine.]

It is difficult to conceive of a purer statement of medical fascism, indeed real true fascism, than this. Sadly, as I've shown throughout the book, such views are widespread.

So, Mr. Keane, what happens if you get the same double-digit percent of serious adverse reactions already seen in the clinical trials? What then? Will the medical system in the United States cope with it or sweep the problem under the rug by saying that the autoimmune or other neurological disorders that appeared weeks or months later were just coincidences? If history is any judge, we know precisely what medical authorities will say, as they already are saying about adverse reactions to the mRNA vaccines.

The mRNA vaccines are now being widely distributed as this book goes to press, so we will see what executive decisions come about over the next months.

This Book and Me

I began this book in October 2019, just a few months before COVID-19 came boiling out—or was pushed out—of Wuhan, China. In short order, the virus evolved from a health issue, as discussed in the previous chapter, into a form of mass hysteria not unlike that which the *War of the Worlds* radio broadcast sparked almost eighty years ago, but this one was worldwide.

The book I had initially envisioned was going to discuss, rationally I hoped, the history and trajectory of vaccination theory and practice. In this process, I planned on delving into a number of issues like vaccine injuries, including autism, and the role of the pharmaceutical industry and their various allies and partners in creating, in my view, this spectrum of disorders. Ultimately, I hoped to cover the issue of what Dr. Toby Rogers, introduced in previous chapters, has called the "political economy of autism," an issue that is really related to everything about vaccines, not just autism.

That was then.

In hindsight, all of these prior conversations and debates about the merits or detriments of vaccines, or nondebates, as they usually happened due to mainstream medicine's failure to engage, seem like ancient history now. Basically, the book was going to be a glimpse into what, at the time, seemed a highly important, albeit contentious, subject.

The subject is still an open question, one that may not be answered anytime soon. However, the emerging battle that is COVID-19 threatens to redraw battle maps and redefine a lot of assumptions most of us took for granted about medical and scientific institutions and practices, the role of government, the guidelines by which society is organized and governed, and the very nature of human freedom.

Much of the original intent of the book can be found in Chapters 1–12, albeit rewritten and adjusted to the new reality of a world dealing with a pandemic. Whether that pandemic is real or to some measure illusory is still an open question, and the reader surely knows my opinion by this point in the book from Chapters 12 and 13, where it began to take on a new focus related to the emergence of COVID-19. This final chapter completes that trajectory.

ASD, The Resolution: Did Vaccines Cause It?

The answer to this question is not the entirety of this book, as it has been for others, e.g., Robert Kennedy's book on Thimerosal.[5] It was, however, the impetus that in turn led to all of the myriad spin-off topics that have been discussed in the various chapters. In other words, ASD may be a pivotal battle in the vaccine wars, but these are themselves only part of a much larger struggle, as will be discussed soon.

In brief, the answer is simultaneously yes, no, and maybe. Going back to review the Simpsonwood conference, the various IOM documents, and the content of Chapter 5, some things become clear. The MMR vaccine probably had no population level impact on ASD rates, which is not to say that it may not have impacted individual children. Thimerosal may well have had some impact, likely in concert to other factors. As for aluminum adjuvants, the evidence cited in Chapter 5 summarizes my views that aluminum is involved, but again with some synergistic interactions with other factors.

That's the simplistic answer.

The more complex answer is the same one I would give in relation to age-related neurological diseases at the other end of life: Alzheimer's disease, Parkinson's disease, and ALS. What causes these? Approximately 10 percent of each has a very clear genetic mutation that alone is sufficient to produce the clinical outcomes; 90 percent of each are classified as "sporadic," that is, of unknown origin.[6] Although these ratios may shift with future discoveries, it seems quite likely that in the end the sporadic fraction will remain the largest. This implies some environmental toxicants as causal, but probably not alone. The missing component ties us back to DNA, not to genes per se, but to the noncoding regions, particularly those described as enhancer regions.[7]

Viewed in this way, this also comes back to ASD. To date, no causal genes have been found, but rather multiple DNA subregions that differ between those who develop ASD and those who don't and further differ between individual ASD cases.

In both ASD and age-related neurological diseases, we therefore have the probability, at least in my opinion, of a synergistic impact of DNA noncoding regions making the impact of various toxicants more likely to harm the nervous system in some people at some stage of life.

Is this all? No, there is also a clear sex difference in all of the above with males more likely to have any of these disorders. (This is not true for multiple sclerosis, nor for MMF.) In addition, there are a range of other factors also likely to be involved, as detailed elsewhere.[8]

Rogers wrote that there are some truly unusual features about ASD from the perspective of a typical epidemic of disease:

> What is striking about the autism epidemic is how universal it is in developed countries and how it does *not* fit traditional patterns of an epidemic. It is increasing rapidly in all developed countries . . . autism is diagnosed at similar rates across all income groups. . . and there does not appear to be a distinction between urban, suburban, and rural areas. This is unusual. . . . And a wide range of variables in different countries including weather, diet, pollution, and healthcare systems usually produces different health outcomes. But not with autism.[9]

Is it important that we find out what is causing the increase? Yes, if ASD rates are indeed rising, then yes it is. How do we find out? Rogers gave us an answer:

> Solving the riddle of autism would seem to require figuring out patterns, understanding correlations, identifying risk, and using massive amounts of data to select likely targets for regulatory action. In an earlier era, that might have been beyond the capacity of government agencies. But in the age of terrorism, governments in the US, the UK, and Australia are actually extremely good at using big data to figure out patterns, correlations, risk, and likely targets. It is just that most of those resources are devoted to (military) defense rather than public health. The point is, solving a problem this complex is no longer beyond the *technical capacity* of government. It appears to be a problem of *political will*, not institutional ability.[10]

In these views, Rogers is likely to be correct. We could solve the problem, in principle, but we choose not to. Part of the reason we choose not to is that other issues seem more pressing. A bigger part of the reason, in my view, is that we don't want an answer that would force us to make economic choices we'd rather not make. Or, to put it more precisely, to make economic choices that would target companies polluting Earth and the bodies of our children.

Rogers compared the rather lackadaisical approach of government to ASD to what a determined effort would actually look like, citing efforts to control cholera epidemics in London in the 1850s. To do so, physicians had to look at every possible factor. Rogers concluded, much like my comment above, that

> . . . a proper understanding of the 1854 and 1866 cholera epidemics *would lead one to be on alert for the possibility of a disease of the remedy - that somehow unwittingly, some step towards scientific progress might have unleashed unintended consequences.* Finally, the 1854 and 1866 cholera epidemics *teach one to examine autism in its political, economic, cultural, and historical context while specifically examining the possible role of corporations and capitalism in the rise of the disorder.*[11] [Italics are mine.]

ASD, like the other neurological diseases mentioned, is thus likely a disease of our own design in that it arises from our economic decisions that impact human health and persists because our political structures have been captured by those who make these economic decisions.

Recommendations for Future Vaccine Safety Studies

There are various ways to proceed from where we are now, at least solely in terms of ASD and vaccine safety, at least pre-2020. COVID-19 makes this conclusion vastly more complex because, as we will see, the conflict is not purely about vaccines and mandates, but rather about a far broader and even darker agenda.

However, let's pretend the simpler version is correct: how do we prevent the continual rise of ASD?

The first part of the answer should be easy: put the full resources of any government on the problem, seek a diversity of opinion, don't take any aspect of the problem off the table just because it hews too closely to some corporate agenda, create a truly independent working group (much like that originally proposed to Trump by Robert Kennedy Jr.), study the problem intensively using *all of the literature*, draft a report (ideally detailing an opinion that most broadly agree with), identify the steps needed to resolve the ASD problem, and then, finally, take whatever decisive action may be called for. Those who do not agree with the majority report would be free to draft their own dissenting report.

It's complex, but it is not that hard. For example, the investigating panel has to be truly neutral. That is, you want scientists, physicians, and lawyers who are truly independent; you don't want anyone connected to the pharmaceutical industry in any way, or anyone with a potential dog in this fight, such as long-term vaccine critics. Imagine it being sort of like the Stratton et al. IOC working group on vaccine safety,[12] just more independent and without the statements of faith that opened their document. If required, and if justified, statements of faith can come at the end of the report, just not at the beginning.

The panel can still talk to and receive written commentary from those pro, con, and skeptical about all aspects of vaccines, just that those representing these groupings can't sit on the panel. In addition, those on the panel should be drawn from a range of disciplines, not just epidemiology and vaccinology, as these fields may be hopelessly biased by ingrained assumptions and dogma. The chair of this panel has to be someone of impeccable credentials, one whom those from various points of view will respect and consider fair.

How many panelists? Use the Stratton report numbers as the guide: ten plus or minus.

What should the panel review? Everything, going back to the initiation of vaccination, i.e., the original works of Jenner, Pasteur, and others, and then they should move forward from there. All hypotheses should be considered; legal aspects need to be in a final Appendix depending on which conclusions the panel reaches.

If this were any other subject, the successors to the IOM document could, and likely would, do this, but in the current vaccine climate they won't. The social, economic, and political consequences are potentially too great.

However, let's pretend such a panel came into being, did the investigations, and issued a report. What would the report probably say? Likely, it would have come to view vaccines as a medical intervention, much like many others, and recognize that they are useful in some preventative medical applications, just not all of them. In other words, it would treat vaccines much as we now treat antibiotics: In a personalized medicine context, used discretely with constant surveillance, and not applied to every sniffle.

Another recommendation from this imaginary panel might be to take the money out of vaccine production. In other words, rather than relying on the very industry that makes vaccines to police themselves, create a Vaccine Institute, the equivalent of a Crown Corporation in Canada at both federal and provincial levels. Crown corporations at both levels are government-owned, but independent entities that deliver a service on a cost recovery basis. That is, they don't have to make money, but any that they do make goes into governmental treasuries. A Vaccine Institute at provincial and federal levels as a Crown Corporation would research, develop, and distribute vaccines at close to cost. In the United States, the idea of a government-run corporation might not work from a philosophical perspective, but there are certainly alternative solutions between the pure "for-profit" companies now controlling vaccines and one of governmental control.

This institute would follow their phase trials with extended and comprehensive time line studies looking for any possible adverse effects. With no profit motive, this institute would have no clear reason to manipulate the data, apart from maybe religious vaccine zeal.

Finally, recognizing that vaccine injuries might still occur, the overall program would provide financial compensation for those injured or the families of those who died. Rather than going through the endless red tape of the Vaccine Court in the United States, all that one would have to prove would be a temporal association of the vaccine to any negative health outcome.

Finally, the institute would champion the notion that it could not mandate its products, but that people could choose to take or reject any vaccine produced. In other words, the panel would respect free choice of the critical right of "security of the person."

It's a sweet idea, and one bound to be instantly rejected by the companies making so much money on vaccines and by the politicians and legislators already bought off by the same industry. The next section shows just why any such recommendations are likely to fail.

Speaking Truth to Power

In relation to the previous recommendations, one concept that tends to come up in various human and civil rights struggles is that of "speaking truth to power." The notion here is that if one reasonably and patiently explains the errors or immorality of some law or procedure to those in power, then the equally reasonable, albeit unknowing, persons who have passed the law or procedure will relent and reverse it.

The two assumptions that undergird the speaking truth to power mantra are two key words: *reasonable* and *unknowing*. The questions to be asked are these: Can those who promulgate such laws be moved by reason? Or are they following some agenda that really doesn't care what level of reason one brings to the discussion? Next, are they truly not knowledgeable about what their law is intended to do? Maybe some politicians are not knowledgeable, but is it likely that Gates and the WHO have no idea what the implications of their actions are likely to be?

The last is possible, of course, if highly unlikely.

Historically speaking, however, it was pretty unlikely that white supremacists had no idea that they were keeping black people down or that the various pharmaceutical companies had no idea that some of their products were harming people.[13] Rather, it seems far more likely that those who abused power then, and do so now, are completely aware of what they are doing. Further, they are doing it for reasons that make perfect sense to them, and these typically involve power or money or both. Because of this, pointing out the "truth" is not alone likely to move the abusers who have long ago realized what the truth is and simply don't care, particularly if changing their behavior interferes with what they had hoped to achieve in the first place. A perfect example of this would be to consider the WEF.

If the powers that be won't change their behaviors, what options does any resistance movement have? If that movement is moral, it will necessarily choose to start with "soft" resistance, using first letters and petitions, then demonstrations and eventually civil disobedience. Gandhi's march to the sea comes to mind as one of those brilliant middle steps that can embarrass governments into relenting,[14] but this sort of outcome is not always the way things evolve.

If none of these steps work, then the only two options remaining are to surrender or actively fight back.

In regard to COVID-19, events to date may have already moved past these early stages of what had already been shaping up to be a fairly conventional civil rights battle in regard to school vaccine mandates for measles and other CDC-scheduled vaccinations.

This is no longer the case. Before COVID-19, vaccine resistance of whatever form was a marginal part of the overall resistance to corporate control. The combination of the various entities discussed in Chapter 12 and their actions during the pandemic have prompted two widely divergent responses that have been amplified by governments to control all aspects of human life: some people default into obedience by rationalizing to themselves that it is for their own good, not to mention society's, to follow governmental prescriptions; others have started to question these prescriptions and, in the process of doing so, have begun to see the larger picture in vaccine mandates, and that these were only the tip of the iceberg.

The entities discussed in Chapter 12, in my view, seem to have launched a full-court press in an attempt at total control over health. In doing so, they risked awakening a lot more people, people who might now become part of a more general resistance.

If anything, the *Event 201* meeting should have taught the participants that this was indeed a possible outcome, and one that was the opposite of the future they had hoped to deliver. The choice in the near future for most of us may be just as stated above and just as bipolar: surrender or fight.

Let me now consider the obstacles in the path of those who choose to fight. The first of these is the Dunning-Kruger syndrome, something I have discussed previously.

Some Thoughts on Dunning-Kruger

Writing this book has been enormously revealing on a number of fronts. Apart from demonstrating just how much the subjects interconnected to vaccination theory and practice actually are, I've also learned a lot about the growing opposition to vaccine coercion. A major revelation for me was just how often people don't actually know what they are talking or writing about whether the subject be vaccines or human rights: The Dunning-Kruger effect[15] seems alive and well with the most ignorant opinions often expressed the most confidently.

Dunning-Kruger is often quoted when attempting to denigrate those who oppose any aspect of vaccination, yet in reality it applies across the board just as often to those doing the denigrating. It includes not only lay people, but also some scientists, administrators, and bureaucrats, as well as a range of politicians from any side of the aisle.

In regard to politicians, I will admit that my sense is that most go into politics to actually help their fellow citizens and mean to do well. Politics is a tough business in that to be good at it, one has to be relatively knowledgeable about a range of subjects. This is difficult, as most of us are not good at many things and indeed are lucky if we are good at one or two. Politicians, as a rule, are no different. The key here is that unlike other mortals whose opinions may not matter much, politicians actually have the power to pass legislation that then impacts all of their constituents. Ignorant opinion leads to lousy, sometimes tyrannical, law.

Another thing I've noted is the degradation of language use, and a lack of language precision. For example, the term *fascist* is tossed around a great deal as an epithet, mostly to describe those with whom one disagrees.

An example that occurred during the COVID-19 events typified the flexible use of the word *fascist* for me. On the one hand we saw demonstrators protesting the COVID-19 lockdowns as examples of governmental overreach, which they termed fascist. On the flip side, a number of people on the left with whom I share a variety of opinions described these same protesters as fascists.

For me, this was odd and suggested one of two things: first, that everyone is a fascist (which I'm sure would make real fascists happy) or second, the term doesn't really mean what people think it means. In other words, fascism is not just a particular economic-political system, but rather any idea or action that you or your particular bubble don't like.[16]

Thus it becomes easy to see how pro- and anti-lockdown people have no problem calling each other fascists. This is not to say that in both solitudes the "other" may not manifest some aspect of what true fascism is. For example, a number of pro-lockdown people seem only too ready to demand and seek the silencing of those who don't agree and further support even more stringent governmental regulations. Further, in my experience, people pretty firmly anti-fascist in the actual sense of the word, think there is nothing wrong with the WHO and Bill Gates dictating health policy.

The end result of this lack of actual knowledge and language precision leads to predictable results: the inability of people to dialogue across the perceived divide by people who, in actuality, may have more in common than not, in wanting to preserve human rights.

Division inevitably leads to defeat.

One part of this division, as sad as it is for me to admit, comes from those on the progressive left side of politics.

The Hibernation of the Left

Before I begin this section, it is worthwhile to consider terms. What, for example, do any of us mean when we speak of being on the left or the right of the political spectrum? Mostly, we speak in vague economic terms on some sort of spectrum where extreme left is communism and extreme right fascism. But such a one-dimensional spectrum is absurd. A better depiction, although not a perfect one, is a two-dimensional spectrum in which the X axis describes where one is economically and the Y axis depicts where one stands on political and social freedoms and individual liberty. Thus, if your beliefs put you in the bottom right quadrant, you are pretty close to real, versus faux, fascism; if you are in the left bottom quadrant, you are probably a supporter of state socialism as practiced, at least in theory, by the former Soviet Union. The upper left quadrant likely takes you closer to anarchism; upper right, libertarian. With this in mind, let's consider what these categories might mean in the Age of COVID-19.[17]

As someone who tends to identify as left-leaning politically and socially (upper left), I have been more than astonished in the last few years, during the year of COVID-19 in particular, by the positions taken by those I would normally consider as allies in the fight for human rights and against the rising power of the corporate world.

Prior to COVID-19, I sometimes clashed with fellow "lefties" about vaccines, particularly regarding mandates, and found it more than passing strange that people completely able to see clearly through corporate or governmental deceptions about war, environment, poverty, and a host of other issues seemed to attach rather firmly to the mainstream line when it came to vaccine subjects.

It was also galling to me that those with whom I did not often share many social beliefs, such as those on the right, typically tended to be vastly better informed about vaccine safety in general, and autism specifically, than those on the left. It was as if, paradoxically to me, those on the right were often more scientifically literate than those on the other side. Somewhere, somehow, the left had become the mouthpiece for mainstream medicine and, by default, the pharmaceutical industry. Challenging them on any of this led to some frankly absurd comments about vaccine issues and some rather glaring ad hominem comments about "anti-vaxxers."

Several examples come to mind. Asking some of those who had taken a hostile position about "anti-vaxxers" to provide references denying vaccine injury produced some fairly stereotypical responses: first, an utter reluctance to engage, with statements like "I'm not going to follow you down that rabbit hole." Second, the "rabbit hole" commentator unfriending me on Facebook.

I'd like to say this was an isolated incident, but sadly it was not.

Charles Eisenstein pretty much nailed this down in the early phases of COVID-19:

> The whole phrase 'right-wing conspiracy theorist' is a bit odd, since traditionally it is the Left that has been most alert to the proclivity of the powerful to abuse their power. Traditionally, it is the Left that is suspicious of corporate interests, that urges us to 'question authority,' and that has in fact been the main victim of government infiltration and surveillance. Fifty years ago, if anyone said, 'There is a secret program called COINTELPRO that is spying on civil rights groups and sowing division within them with poison pen letters and fabricated rumors,' that would have been a conspiracy theory by today's standards. The same, 25 years ago, with, 'There is a secret program in which the CIA facilitates narcotics sales into American inner cities and uses the money to fund right-wing paramilitaries in Central America.' The same with government infiltration of environmental groups and peace activists starting in the 1980s. Or more recently, the infiltration of the Standing Rock movement. Or the real estate industry's decades-long conspiracy to redline neighborhoods to keep black people out. Given this history, why all of a sudden is it the Left urging everyone to trust 'the Man'-to trust the pronouncements of the pharmaceutical companies and pharma-funded organizations like the CDC and WHO? Why is skepticism towards these institutions labeled 'right wing'? It isn't as if only the privileged are 'inconvenienced' by lockdown. It is devastating the lives of tens or hundreds of millions of the global precariat. The UN World Food Program is warning that by the end of the year, 260 million people will face starvation. Most are black and brown people in Africa and South Asia. One might argue that to restrict the debate to epidemiological questions of mortality is itself a privileged stance that erases the suffering of those who are most marginalized to begin with.[18]

The coming of COVID-19 has not removed either the silence or hostility of much of the left, even the radical left, against those who see COVID-19 as yet another assault on freedom of choice and those who oppose shutdown measures. Indeed, the hostility has only grown in spite of the abundantly clear link between current affairs and the corporate—read, pharma—world. Anyone who opposes shutdowns is now routinely described in leftist circles as "alt-right," as Eisenstein has also observed.

It was thus refreshing in the extreme to come upon an article by Darren Allen that stated,

> A deadly pandemic, we are told, swept across the planet, forcing governments to massively enhance state and police power, lock everyone up in their homes and bork the economy. National governments, transnational institutions and all media outlets were of one voice. Panic. We just *had* to put millions and millions of people out of work then shut them up in a heavily policed panic room. Anyone unable to perceive the foundations of the unofficial left might imagine that they would have interrogated this extraordinary situation, that they would have critically appraised official accounts of the severity of the 'pandemic,' that they would have asked themselves what the likely effects might be of putting so many people out of work; or that it would have been the perfect time for 'radicals' to seriously question the functioning of the system, to explore wider questions about its stability and to critically investigate vested interests; perhaps also take a look at the universal denial of death and how easily people can be manipulated by playing on their fears, or even explore the possibilities for genuine revolt as the economy contracted. They would have been disappointed.[19]

This is precisely correct. I have quoted much of the key premise of the article because it illustrates the issues quite clearly. All I can add to this is: Hey, Progressives, thanks for showing up at last. Where have you been?

This article now joins some of the others by those now awakening on the left, including the recent articles by Tim Schwab and Vandana Shiva, as cited in Chapter 12.[20] To these we can now add commentary by various authors on Klaus Schwab and the Great Reset from various places on the political spectrum.[21]

And last, to my remaining friends on the left: if you really want to fight fascism in all of its manifestations, start looking at how the pharmaceutical cartel, Bill Gates, the WHO, and the WEF are carving up the world for their own benefit. If you can't see the writing on the wall, let alone in the very clear declarations of purpose as shown in previous chapters, it might be better if you went back to sleep and abdicated the struggle for human freedom to others.

If, however, one wants to take a closer look, consider the following sections. Before that, however, I want to consider how the WEF(P)-G-WHO pulled the COVID-19 End Game off . . . so far.

Battle Procedure and COVID-19

In NATO military forces, soldiers in noncommissioned officer and officer ranks are taught Battle Procedure. This is a multistep method by which a commander at any level receives his/her orders, analyzes how he/she will accomplish the mission, thinks about what the "end state" when the mission is accomplished looks like, and prepares the soldiers to carry out the mission. Some of the steps include considerations of the timings involved, the materials and personnel needed given what is available, the geography of where the mission is to be carried out, and a range of other factors. One key aspect is to build in time to conduct rehearsals, since no plan ever goes perfectly. And a crucial concern is "war-gaming," in which the commander considers what the enemy will do if he/she does certain actions and what he/she will do in return. A key part of war-gaming is to understand your enemy: their capabilities in resources and people, what they have done in the past, what they are likely to do in the current circumstances, and what action by them would be most detrimental to your own forces and mission. In essence, it is like chess, in which a successful player thinks many moves ahead and plans the future moves appropriately.

From all of this, a commander at any level builds the plan. The level of complexity of all of the above depends on the level of organization, but the general process is the same. Once all of this is done, the commander issues his/her own orders to subordinates and, after giving them time to prepare their own soldiers, launches the mission.

Similar kinds of analyses are common in business and other fields, but replace the word *enemy* with *opponent* or *competitor*.[22]

A consideration of Battle Procedure and war-gaming leads to the almost inevitable conclusion that the COVID-19 pandemic was a planned operation, not necessarily by any military, but rather by people in industry who understand the concepts well. Clues to this were the speed at which the COVID-19 events rolled out and the rehearsal, in this case *Event 201*, in what is called in the military a tabletop or cloth model exercise, in which one sees what works and what doesn't and then adjusts the plan accordingly.

In terms of the rollout, the actions came so fast, one right after the other, that it became clear that the unfolding events were not random, but coordinated to create a maximum level of fear and disruption. The term used in military circles is that of "getting inside your enemy's decision cycle." In other words, make the enemy (most of us) reactive, rather than proactive, forced to deal with events after they have already occurred and rendering them incapable of anticipating the next step of the rollout, and thus unable to respond appropriately. This is how a military force loses battles and ultimately wars.

And at the end of an exercise or war comes a candid evaluation of what worked or didn't work. This can be termed an "after action report" or is sometimes more casually called a "hotwash."

So what worked in the COVID-19 rollout? Panic worked, likely vastly more than the planners anticipated. In the end it was absurdly easy to scare the bulk of the population into complying with a disease that any rational appreciation would have realized was not the death sentence that the WHO made it out to be. Even Klaus Schwab realized this, as cited in the last chapter. On the negative side, the COVID-19 planners failed to gain control of the communication systems of their adversaries. In other words, the failure to control the Internet allowed doubts about the entire COVID-19 narrative to spread and threatened to derail the hoped-for End State. One can be sure that future pandemic plans and resets by the WEF and their allies will proactively attend to such failings. Indeed, maybe this is what an event hosted by none other than the WEF called Cyber Polygon 2021 is all about.[23]

Vaccine Safety Pre- and Post-the "New Normal" in the Age of COVID-19

The past and current state of affairs is this: before COVID-19, the failure of the medical establishment to do the definitive experiments that would have resolved a possible role of vaccines in ASD combined with the arrogant dismissal of parental concerns had led to pushback by those affected. This resistance could seemingly only be overcome by legal force as legislators sought to enforce vaccination agendas through mandates for school-aged children. By legal force, I mean the term in only the narrowest sense. After all, as we've seen in Chapter 11, the State, however defined, can pretty much do whatever it wants to do when it declares a state of emergency. In this regard, I refer again to the late comedian George Carlin's monologue in which he stated that we "don't have rights, we have privileges."[24]

Chapter 11 detailed this in depth, terming the emerging paradigm of government as a perpetual "state of exception," that is, a place that Agamben described as residing between a state of law and nonlaw. This last fits rather well with the use of state of exception regulations, which, as various scholars have pointed out, may be "legal" because some government says so but still do not translate into being moral.

Since mandates are in reality all about compulsion and force, this renders the notion of morality here a moot point: the State will do what the State wants to do, in the process framing a clear admission that in the absence of real evidence for a stated goal, or without the ability to convince citizens by rational persuasion, the State has no other reply. De facto, the powers that be, the legislatures, the CDC/FDA, and the pharmaceutical industry have revealed that force is the only tool they have left to ward off what would otherwise be an outright defeat of a commercial venture cloaked in the robes of a science-turned-cult religion facing off against parents of vaccine-injured children.

History, however, should teach us that when the State resorts to such force, it risks a backlash far greater than whatever it was they thought to suppress: States have power, that of the police and courts and even the military. That is, they flaunt this power up until the minute that they don't . . . and at that point it all unravels.

There are too many examples in history of how power dissolves to list here, but readers will surely know many of them from the fall of the Berlin Wall, to the last days of apartheid in South Africa, or even on a smaller scale to American police forces falling back from the rage of Black Lives Matter demonstrators.

These were the main themes of this book before COVID-19 took center stage and upended my, and the world's, expected plans. It's not that the world was not heading this way anyway, merely that COVID-19, and those who profited or hoped to profit from it, simply accelerated the pace of events already in motion.

Now, with the pandemic in full force and with a state of exception being the "new normal" governmental default response, the world seems poised between possible futures of which I see two main ones: a full-on Brave New World, or a renewal of real democracy and choice.

The words of Charles Eisenstein are again pertinent here:

> A million forking paths lie before us. Universal basic income could mean an end to economic insecurity and the flowering of creativity as millions are freed from the work that Covid has shown us is less necessary than we thought. Or it could mean, with the decimation of small businesses, dependency on the state for a stipend that comes with strict conditions. The crisis could usher in totalitarianism or solidarity; medical martial law or a holistic renaissance; greater fear of the microbial world, or greater resiliency in participation in it; permanent norms of social distancing, or a renewed desire to come together.[25]

Maybe Eisenstein's "million forking paths" really do boil down to my two paths: we don't know at this time which one we will get. It might be that we go back to "normal," whatever that term means, or meant. But will it be a normal world in which ecological destruction is still the norm, in which economic disparities are still accepted as inevitable, in which some of the top billionaires listed in Chapter 12 have more wealth than billions of "lesser" humans combined? Is it the normal that demands endless consumption as pre-COVID, or normal in that those who have not been pushed into poverty post-COVID are now asked to consume more to protect the economy while they hover one paycheck away from that same poverty that has engulfed others?

Might it be that we enter a world far more dictatorial than any in which we now live, at least in the global North, where everything that makes any of us free has gone under the control of the State, in which we control nothing, not even the illusion of having agency over our wealth, our bodies, or our children? Does it become a world where society is dictated from on high without the slightest concern about any form of actual democracy or the natural rights we thought we had from our Creator?

Or perhaps, when the dust of COVID-19 has settled, we find that we don't want normal again, whatever we thought it was before. Maybe we discover that we don't accept an iron fist but choose instead to reinvent ourselves as a civilization and a species and make the true love of Mother Earth and all of her species, and one another, our overriding mission?

With this in mind, I find it quite odd, bizarre even, how the concluding paragraph of Schwab and Malleret's *COVID-19: The Great Reset* is in synchrony with the goals of many of us for the world after COVID-19:

> We are now at a crossroads. One path will take us to a better world: more inclusive, more equitable, and more respectful of Mother Nature. The other will take us to a world that resembles the one we just left behind-but worse and constantly dogged by nasty surprises. We must therefore get it right. The looming challenges could be more consequential than we have until now chosen to imagine, but our capacity to reset could also be greater than we had previously dared to hope.[26]

The problem, Klaus and Thierry, is not the sentiment, but the delivery by those who have hoarded wealth at the expense of billions of people, have been elected by no one to anything, and now propose to use this vast wealth to modify our lives. Perhaps you can see why skepticism to your lovely words is growing?

The last option of a better world is not guaranteed, and, even if most of us were to desire it, even Schwab and Mallaret, it remains far from clear that it could come to pass. Social revolutions in history, in mentality and behavior, are not normally ushered in by peaceful means. It may be that to get to such a future, we will wind up removing from power those who by inaction or design let the crisis come into being.

As the American revolutionary and political philosopher, Thomas Paine, once wrote: "We have it in our power to begin the world over again."[27]

Paine was, of course, writing about the enormous challenges faced by the American colonists in North America when they took on the most powerful empire on Earth: Britain. To those fighting at that time it must have seemed an insane venture to fight that global empire. But they did, and they won, but not without enormous sacrifice.

We now face another empire: that of the intertwined forces described so well by the Children's Health Defense, as cited in Chapter 13. Not least of these is the grouping that makes up the World Economic Forum.

To be sure, the new empire is as daunting as any enemies of the past. And yet, it remains vulnerable to those who have decided, and remain committed, to fight back, no matter what the price, and to pay what used to be termed in war the "butcher's bill."[28]

I will explore this last thought in more detail later in the chapter. Before I do, however, I want to revisit and further explore some of the consequences that faced the vaccine resistance movement before COVID-19 and what that struggle now looks like. It is in this review that we will find the reasons why the battle against the pharma and their allies has been fought, and must continue to be fought.

Social and Medical Consequences of Health/Vaccine Mandates

In chapter 8, I reviewed the work of Dr. Mateja Cernic on the ideology of vaccination and the implications for vaccination to be increasingly seen as a test of the faith, faith that comes with strict consequences: if you believe in vaccines in the required manner, particularly if you obey the diktat to take vaccines as required by higher authority, then you fall into one class of human beings for whom life can go on as "normal" or what those in power want to be normal; if you don't believe, and particularly if you don't obey, you fall into another class entirely where your life will be quite abnormal.

Cernic wrote,

> Vaccination, especially when required by law, is primarily a social and political issue with considerable ideological implications. It seriously interferes with individuals' rights over their own bodies and the bodies they are responsible for . . . since a partial or complete refusal of the state-prescribed vaccination schedule-particularly when vaccination is made compulsory by law-is often sanctioned.[29]

Leaving aside the rather blatant human rights issues, what might be the medical consequence of greater governmental interference in personal choices about health? The mainstream medical organizations, WHO, CDC, NIH, and others, would likely claim that there are no negative consequences at all, merely the chance to escape infectious diseases for the individual and, in turn, protect the larger community. Apart from the totalitarian nature of any State making this a forced choice, what if there are negative consequences to health, at least for some people?

The preceding chapters should have demonstrated that some people have extremely negative reactions to *some* vaccine ingredients, if not the entire collection of vaccines antigens, adjuvants, and excipients in individual vaccines or the cumulative CDC schedule. Increasing the numbers of these vaccines, whether to COVID-19 or any other viruses, is not going to make these consequences go away. Quite the opposite. Increasing the numbers of mandated vaccines, either for school children or the population at large, will almost certainly push more people into adverse reactions, including various autoimmune disorders.

Is this a guarantee? No, certainly not, and indeed it may turn out that parts of the independent skeptical vaccine research are not correct. However, that would be an experiment, and experimenting on people without their truly informed consent, where the risk of adverse reactions is possible (if not probable), is truly immoral. While not addressing vaccination per se, human experimentation without clear informed consent has been banned, at least in theory, by a number of international conventions, as listed in Chapter 8.

It goes without saying that any mandate for vaccines or anything else cannot, by its very nature, be voluntary. It is for this very reason that vaccine mandates along with the removal of most exemptions in parts of the United States, and as increasingly called for in Canada, may be skating perilously close to human rights violations using "state of exception" or "emergency" dictates or laws to supersede individual liberties for "the greater good."

The idea that pushing vaccines on all people might infringe basic rights, particularly "security of the person," is mostly ignored by mainstream media and health agencies.

Aaron Siri, mentioned previously, is a lawyer practicing in New York and Arizona. Here is what he said about vaccine mandates: "The basic facts and government data regarding vaccine efficacy, safety, and trends, do not support eliminating vaccine exemptions."

And,

> . . . it is untenable that anyone would seek to replace the right of informed consent (which arises from the Nuremberg code's obligation to inform and obtain voluntary consent) by eliminating vaccine exemptions and replace it with an obligation to vaccinate under duress of exclusion from school or worse.[30]

The documents on human experimentation considered really leave little room for doubt about whether vaccine mandates are truly legal in the broadest sense of the word. Of course, any State may choose to say that what is legal is for them alone to decide through state of exception. However, in order to do so, a country such as the United States would have to violate its own founding documents such as the Declaration of Independence and the Bill of Rights, both with explicit references to personal security. In Canada, any government would have to violate our Charter of Rights and Freedoms vis-à-vis "security of the person." For example, Section 7 of the Charter states that

> Security of the person includes a person's right to control his/her own bodily integrity. It will be engaged where the state interferes with personal autonomy and a person's ability to control his or her own physical or psychological integrity . . . or imposing unwanted medical treatment.[31]

Whether one is talking about school children or the population at large, the argument really boils down to this: Is your body yours and do you have the fiduciary responsibility for your children, or do both belong to the State, or perhaps even more likely today, the State as an arm of industry?

The COVID-19 pandemic has not replaced the ongoing struggle between the corporate influences attempting to commodify all aspects of modern medicine versus those who choose to exercise the right to security of the person for themselves and for their children. Whether the latter are correct about their concerns about health generally, about vaccines, GMO crops, 5G, or anything else is simply irrelevant to the issue of free choice. Totalitarian states don't care about the latter, of course, but in the notionally democratic countries we do, or at least we pretend that we do.

What COVID-19 has done, however, is to vastly accelerate the process of removing free and informed choice, with the WEF(P)-G-WHO leading the charge. The surprise here is not that these entities have taken on a much more aggressive stance in these regards, more that with so much of the corporate agenda rapidly being realized, the various players felt that the time was ripe to push even harder. Suggesting to people, as Bill Gates, Klaus Schwab, and Tedros Ghebreyesus have done, that "normal" does not return until there is a vaccine for COVID-19 that all seven billion plus people on Earth must take is a mantra happily mimicked by many politicians and much of the media.

This sort of rhetoric has created the fertile ground for those who are skeptical that a larger agenda is being foisted upon them ostensibly in response to the COVID-19 pandemic. In response, a very predictable resistance has emerged in the form of a growing skepticism about the very nature of the COVID-19 emergency, the steps taken by most governments, and the overall price being paid by ordinary people. A growing antilockdown movement, apart from what had previously been the nascent vaccine-hesitant movement, is now morphing into something far more militant. Rallies and marches in France, Germany, and the United Kingdom have put thousands of people into the streets. Similarly, albeit on a smaller scale, protests are growing in Canada and the United States.

The growing skepticism of those in the streets includes the very realistic fear that an essentially experimental COVID-19 vaccine(s) will be mandated as the sole way out of the lockdowns.

What happens next? How does any government actually enforce such mandates and with which level of coercion or actual force? What happens then? Do the people in the streets meekly submit and hope for the best, or do they become even more militant, even revolutionary? After all, if any government can use a state of exception paradigm to take away one's most fundamental rights, then the choice to those in the streets really becomes a choice between submission and resistance. Governments tend to think they have a monopoly on violence to enforce their will. That may often be true, but is not always true, as history clearly teaches. However, a far better State strategy to prevent any revolution from succeeding is to foster infighting within the movement itself.

Social Movements and Infighting: The Vaccine-Hesitant Movement and the Lurch to the Right

In another section of the book (Chapter 7), I considered the apparent belief systems of those in the opposing vaccine camps. One thing that has become quite clear to me in the last few years is that a number of those who would self-identify as anti-vaccine or vaccine-hesitant often show a leaning to a libertarian perspective with all the subthemes that go with such views. A number of those who seem to be the most vocal in the vaccine-hesitant world are white; many are practicing Christians. It is worth noting that in recent years they have been joined in their resistance to vaccination by a number of Orthodox Jews. Overall, what this may mean is that people of both faiths adhere to a higher moral calling where faith in God supersedes the demands and prohibitions of the secular State.

Within these ranks were many who had supported Donald Trump based on earlier hopes that he would seriously address vaccine injuries. From the latter also came entities such as QAnon[32] and the fellow travelers who seem even further to the right; it has not been uncommon to hear from within this grouping a decided antipathy toward those of other races and religions. This last is problematic from the perspective of wider movement building, since racist attitudes or perceptions that are held may drive away others who share the same views on vaccines and medical freedom in general. And, it should be obvious that the media loves picking up on apparent bias as a means to sidestep the real issues that should be in plain view.

Divisions in social movements are not unusual. In most of the political events that I have participated in over the years, divisions that were either organic to the movement or those that were fostered by outside players were common. The first often arise from ego; the second as a result of the State attempting, and often succeeding, in creating chaos within the ranks.

As an example, I should mention that I have been a social activist for most of my adult life, so I know the phenomenon well and have seen it play out from the anti-Vietnam movement of the 1960s and '70s, at Standing Rock in 2016, and in the medical groups working in Mosul in 2017 and Rojava in 2018 and afterward.[33]

With this background, I can't say I was surprised to find infighting in the vaccine-hesitant ranks and see how it emerged to the detriment of the overall movement.

All social movements go through similar evolutions from when they begin until they finally achieve their goals, or until they are destroyed. Some of the divisions that have appeared in the vaccine-hesitant movement were discussed in Chapter 7 and are just as relevant to any movement organized to fight back against the COVID-19 states of exception.

Cui Bono, Redux

Assuming that all the statistics cited by the CDC and various health authorities were correct about COVID-19, about half of people surveyed say that they would likely take a COVID-19 vaccine, even one deemed still experimental, as long as there were some minimal data on safety and efficacy. Apart from a few zealots who think they are part of a program to save the world, most of those willing to take the vaccine would primarily do so to get back to "normal." That is, a normal in which they could walk into a grocery store without a mask, be able to board a flight to visit family, etc.

Being willing to get the shot is obviously going to be predicated on the actual levels of adverse reactions, and of which types, seen in the various phase trials now underway or completed.

Let's imagine that a lot of people, maybe the majority, eventually agree that the risk is worthwhile. Who benefits? One answer is that potentially the vaccinated do because in an ideal case they are no longer susceptible to the disease nor presumably able to pass it to others. Other beneficiaries in this scenario would be those who have a severe immunocompromised status, making them potentially at risk from those who are not immune for whatever reason. Of course, as cited in Chapter 13, it is far from clear that any of the COVID-19 vaccines to date prevent catching the disease again or the infectivity of the recipient.

Let's leave this for a time and consider: who are really the major beneficiaries, that is, who actually profits from making this, or any, vaccine?

The take-home message from the last two chapters is that there is one main winner: the pharmaceutical industry whose vaccine profits are vast.[34]

If many people agree to take the vaccine when it comes out, and if we agree that the pharma is the main beneficiary, at least financially, the main thing standing in the way of huge vaccine profits now and in the future is vaccine uptake. It's really a very simple equation: the more people who take whatever vaccine is on offer, the more money the pharma makes. Perhaps in this view, voluntary vaccine uptake is great, but compulsory vaccine mandates are vastly better, since they capture, in principle, 100 percent of the potential market across all demographics. If people can be made to take such vaccines multiple times, perhaps for the rest of their lives, so much the better.

The very clear profit motive for the pharmaceutical industry now ties us back to the notion of vaccine mandates, a key driver of vaccine resistance, the subject of the next section.

The Pharma and the Expression of Medical Fascism

Mussolini described fascism as, in part, a merger of corporations and the State, whether the corporations that seek control are those of the defense industry, the agribusiness, the oil and liquefied natural gas extraction industry, or any other. Dwarfing them all at this time is the pharmaceutical industry, which has found a way to control the health of many of those on the planet and hopes to expand its scope to capture them all from precradle to grave. Health, above and beyond anything besides food, is perhaps the greatest motivator of all for most people.

Karl Marx correctly identified the end state of capitalism as the emergence of a frankly predatory form, one without any real limits or concerns for human welfare, solely concerned with maximizing profit regardless of the consequences to people or the Earth. This end state has now come to pass, and it is increasingly obvious that a range of corporations are unable, or unwilling, to see or care about the damage they do to the whole world, the "woke" psychobabbling of Klaus Schwab of the WEF notwithstanding.

Some of this has fostered mass demonstrations against the fossil fuel industry in regard to climate change, in many ways approaching the levels of concern and protest about the overall power of corporate control of trade and "globalization" that swept the world in the late 1990s and early 2000s. These protests faded in the wake of 9/11 with governments of the world then focused on the bogeyman of Islamic terrorism.

In much the same way, the large 2019 climate change demonstrations seemingly vanished post-COVID-19 pandemic. Would it be a conspiracy theory to suggest that when a particular line of protest appears on the way to becoming too influential and thus detrimental to the corporate sector, a newer, overwhelming crisis of another sort rears up to silence the critics?

It could have been the defense industry that brewed up resistance when people's sons were sacrificed for profit in the endless wars that the United States and other countries have fought around the globe. Or it could have been the giant agribusinesses that sought, and still seeks, control of the food supply while people became ill from Roundup Ready crops. Neither of these two industries, however, occasioned much widespread resistance in the twenty-first century.

The pharmaceutical industry had faced a small but growing vaccine resistance movement based on parents of vaccine-injured children. The movement may have been small, but it was gaining traction. More and more people were asking pesky questions, some independent scientists were starting to conduct more incisive studies and to speak out, and demonstrations against school mandates were on the rise and sometimes even successful.

Then COVID-19 came riding in and the pharma became the knight in shining armor here to slay the pandemic. It changed the channel from discussions of vaccine and drug harms to one that focused on defeating a dreaded virus, in the process marginalizing most critics. Who in their right mind, after all, could possibly be anti-vaccine when a vaccine was the only way to save humanity from COVID-19?

The pharma's regulatory capture of much of the media and medical establishment, the endless pontificating by Bill Gates, and the ostensible authority of the WHO all made it seem as if the game were over: the industry has won, the naysayers have lost, and those too foolish to see the writing on the wall can just sit in their homes by themselves and rot.

The COVID-19 crisis showed quite clearly the emergence of the pharma's hoped-for version of a "New World Order" built on foundations laid down over decades: endless drugs and vaccines for an obedient world population delivered by the industry, aided by the Gates Foundation and supported and advanced by the WHO, the latter two with their media-created auras of philanthropy. And the overall agenda is ultimately backed by State power wherever necessary for the "greater good." Add to this some of the emerging technology for further control coming out of companies like Microsoft about monitoring human body activity linked to cyber currency rewards (found online in Appendix 17), and only the willfully blind will fail to see the possible future that the masters of the Great Reset have in store for the rest of us.

In other words, the ultimate and predicted segue from capitalism as our parents knew it to outright and practically unabashed fascism had finally arrived: if the pharmaceutical industry was its shepherd, then the shadowy World Economic Forum (WEF) was the master touting the "Great Reset" of "Build Back Better" and the Fourth Industrial Revolution, thanks to the pandemic. What is the Fourth Industrial Revolution? Luckily, we don't have to guess, as the WEF tells us:

> The Fourth Industrial Revolution can be described as the advent of 'cyber-physical systems' involving entirely new capabilities for people and machines. While these capabilities are reliant on the technologies and infrastructure of the Third Industrial Revolution, *the Fourth Industrial Revolution represents entirely new ways in which technology becomes embedded within societies and even our human bodies. Examples include genome editing, new forms of machine intelligence, breakthrough materials and approaches to governance that rely on cryptographic methods such as the blockchain.*[35] [Italics are mine.]

And there you have it, the Brave New World of the cartel hoping to govern Earth.

Is this outcome with the pharmaceutical industry in control of human health and the WEF in control of the economy the true ending? No, it's not, because, as I will show next, the resistance to this dystopian world vision is active and growing.

Is There a Growing Merger between Vaccine Resistance and Resistance to Corporate Control?

I remember a story from when I lived in Jerusalem about the two philosophical directions that some Holocaust survivors took in their religious observance. The first, albeit maybe not the largest grouping, was to become more religious, relying on faith to explain the horrors of being in the camps or losing family members. Those in the other group became far less religious, seeing in the Holocaust clear evidence that there was no God because, if so, how could a just God have allowed it to happen?

This story may be apocryphal, but if it is not, I think the same thing will happen here for those concerned about vaccines, especially in reference to the COVID-19 pandemic. Indeed, I already see it playing out on Facebook and on other social media sites.

While not anything approaching a valid survey, there seem to be two groupings: the first are those refusing to accept restrictions due to COVID-19, at least in principle, and hewing to the belief that the entire pandemic is being used to push additional vaccines, with mandates if necessary, on the population at large. To them, the events of 2020 smack of the pharmaceutical industry in league with Bill Gates, the WHO, and WEF to finally crush the still-nascent vaccine-hesitant movement once and for all as part of a much larger push for world control.

The other grouping might better be described as those who have been fence-sitting, but now the fear of the virus has forced them into the mainstream camp. Such people tend to take the view that their respective governments are doing the best they can in trying circumstances, that the WHO is really a beneficial organization, and that circumstances will improve for us all once a vaccine is available and widely accepted.

So the question really is: how will either group respond to the inevitable pressure to conform to the guidelines imposed by any State, including the necessity, if it comes to that, to vaccinate against COVID-19? And how, if facing any resistance, is the State likely to respond?

The second group will comply for the "greater good" defined by the State. If history is any guide, the first group, at least some of its members, will continue to resist, and the State, in panic at this resistance, will crack down harder. Certainly the latter outcome would seem to fit in well with state of exception/emergency procedures, as cited previously.

As already discussed in Chapter 7, a significant fraction of the vaccine-hesitant groupings in the United States tends to overlap with Second Amendment advocates, an observation that will make any State crackdown more problematic from an enforcement perspective compared to here in Canada, where far fewer in the population are armed.

Further, in Canada, our societal views are far closer to those of Western Europe, with a greater proportion of the population inclined to accept governmental health decisions. Even in Sweden, which by US standards is far to the left of mainstream American politics, people tend to trust their government. As an example, when Sweden charted its own course to respond to the pandemic, it did not lock the country down as hard as the United Kigndom, for instance, and Swedes tended to support that decision.

So while some people might resist vaccine mandates to deal with COVID-19, most people in most countries will likely go along.

That outcome will be good news for various governments that may have drunk the WEF(P)-G-WHO's Kool-Aid, but it still leaves a problem if the number of dissidents is too high to allow the willingly vaccinated to achieve the mystical kind of herd immunity from vaccination.

The bad news for governments then becomes the problem of assessing just how hard they want to be on those who still refuse to comply, specifically whether to use a carrot or a stick, or both.

Inevitably, this problem with crackdowns on dissidents in general, wherever they may occur, is that hard measures tend to breed more resistance in the long run. Evidence for the latter can be seen in the outcomes of most counterinsurgency wars.

This is not to say that crackdowns can't be effective in the short term. And, while the resistance may be largely passive rather than confrontational at this time, it will surely serve the purpose of making more people question the "man behind the curtain," or who is really calling the shots, literally, in the response to COVID-19 or any other pandemic that may come to pass.

Much as in the movie *The Matrix*, State power will almost surely push people toward the blue pill of ignoring "conspiracy theories" and accepting government decisions, or toward the "red pill" of connecting the dots between the various corporate and State players and thus resisting in one form or another.

Does resistance to State/corporate control have a chance? The answer is yes, but it will take a concerted effort by many people and will test the determination of these people to fight arbitrary power, which, as Frederick Douglass wrote in regard to any form of oppression, must be "resisted with either words or blows, or with both."[36]

There will be some, maybe many, who will mock as just another conspiracy theory any idea that suggests that governments in the United States, Canada, or Western Europe might consider taking extreme measures to control the lives of people for the "greater good."

Would the loss of civil rights, even if fairly drastic, in such cases approach the horrors and utter abomination that was slavery in the antebellum United States or the Nazi Holocaust? No, probably not, but this would be a world that many of us would prefer not to occupy, where failure to comply leads less to uniformed thugs kicking down your front door at 2:00 a.m. than to a denial of everyday freedoms to work, travel, and associate with others as one would.

It might also be a world in which children are considered to belong to the State, in which the State alone decides which medical and other treatments they receive, what and how they learn, and in which compliance to government is considered the basic civic duty. If this sounds far-fetched, one only has to read some of the articles cited previously about the necessity of COVID-19 vaccine mandates.

Many Canadians, as citizens in a country without much of a revolutionary history, might comply given that one of the key national mantras is "Peace, Order, and Good Government."[37] Would Americans? Clearly some would not. A recent telecast by Robert F. Kennedy Jr. made his willingness to fight, and maybe die, quite clear.[38] Some would indeed fight back, testing the determination of government to suppress by whatever means necessary those who will not bend.

With this, the possibilities for serious social and political mayhem are perhaps greater than most not conversant on the ongoing vaccine wars realize.

The Third Way

The retired Detroit physician Dr. Gary Kohls writes frequent columns under the rubric of "Duty to Warn" for the *Duluth Reader*.[39] His columns seem to have a wide reach. Several aspects of his columns have caught my attention over the years: Kohls's exposure of pharma industry lies, particularly around vaccines, along with his advocacy for human freedom for all. Kohls calls this a "third way," in other words, one can be for health freedom along with freedom for all peoples in other ways, both economically and socially. In brief, it's not a win-lose game. In fact, to win this battle, it has to be the opposite.

The movie *Pride*, which came out in 2014,[40] made this point quite brilliantly. The story chronicles a Welsh miner's strike in 1984 during Margaret Thatcher's reign in Britain. In the movie, and in actuality, LBGTQ individuals recognized that the miners were on the right side in the strike and came up to Wales to support them and their families. Initial suspicion and bigotry gave way to mutual respect. At the end of the movie, and again in real life, Welsh miners turned up in London to support a pride march, typifying what Kurdish freedom activist Dilar Dirik has said before: "Solidarity is not charity; it is self-defense."[41]

This is precisely what has to happen if those fighting for any form of human freedom hope to succeed: vaccine mandate opponents need to start showing up at other freedom events as well as environmental protests; members of those other organizations and others on the left in the various freedom struggles need to start showing up at mandate and lockdown protests. Freedom is freedom, and defending your freedom defends my own.

As Robert Kennedy Jr. said in a recent broadcast:

> If you're a Republican or a Democrat, stop talking about that. Stop identifying yourself. The enemy is Big Tech, Big Data, Big Oil, Big Pharma, the medical cartel, the government totalitarian elements that are trying to oppress us, to rob us of our liberties, of our democracy, of our freedom of thought, of our freedom of expression, of our freedom of assembly, and all the freedoms that give dignity to humanity.[42]

Kennedy is completely correct. In reality, there is no other way forward against the power of the pharmaceutical and other cartels in the WEF's Great Reset. Failing to unite around common goals of human freedom will doom us all. Acting together gives us a chance to win and realize that together we indeed ". . . have it in our power to begin the world over again."[43]

The Lady or the Tiger?

As Thomas Paine wrote in one of his most powerful essays, *The Crisis,* "Tyranny, like hell, is not easily conquered; yet we have this consolation with us, that the harder the conflict, the more glorious the triumph."[44]

We all like to think that in the various historical conflicts, we would have stood on the side of justice and shown the necessary courage to defeat tyranny. From Thermopylae to the American Revolution, to Vichy France and Nazi Germany or even to Stalinist Russia. But this is not at all true. Most of us would have gone along with whatever the authorities told us to do. The public response to COVID-19 has taught us this sad fact, as if we really needed to learn it again.

Most of this blind obedience arises from fear—fear of the government, fear of our neighbors, and social shaming, to name but a few of the obvious sources. Fear is contagious, far more so than any natural pathogen. It is more contagious even than human-made pathogens, including those where governments tell us to be afraid, then offer the solution to this fear: Obey and do what we tell you, and you will be safe.

However, just as fear is contagious, so too is courage. It has been seen throughout history and shows up now in those questioning lockdowns and the pharma agenda that is coming into clear view. The hundreds of thousands of people who took to the streets of Berlin in August 2020 are a case in point.[45]

We will need such courage, because we are approaching a crucial point in the struggle against a world corporate agenda. It is at this point where people with different perspectives on the current crisis, coming from divergent interests, have to begin to realize that the enemy each thought they faced alone was the same one, albeit in a different guise. This is the point where those who fought for children injured by vaccines and those who opposed school mandates meet those who are struggling against innate racism in the United States and elsewhere, to realize that the same enemy is the one who profits from medical mandates, institutionalized racism, and ecosystem destruction. This is the point where people fighting oppression in all of its forms realize the truth in what Dilar Dirik said.

Where Do I Stand?
Fiction

Early in 2018, I began to write a novel, sort of a murder mystery, roughly based on a character who in some ways resembled me, or at least a hybrid of me and some of my other science colleagues. The scientist, who has had some academic troubles arising from a retracted paper, was secretly contacted by a stranger who tells him that the events leading to the paper retraction were part of a set up to discredit him and to remove his voice from growing skepticism about vaccines and their potential role in neurodevelopmental disorders, including autism. The source tells him this is Plan A, to get rid of critical voices, in advance of a more sinister plan B: the release of a viral pathogen.

In my story, the virus was intended to trigger mass panic in the public and force governments, always ready to go into state of exception behaviors, to launch lockdowns and speed the development of a vaccine for the disease. Civil and human rights violations followed with the intent to make all vaccines, starting with the new untested one, mandatory for all people. The proposed penalties for not taking the vaccine(s) were to be extreme. A condition of corporate-State fascism would then descend on much of the world.

In my fictional version, the pathogen was something like an Ebola or a Marburg virus. Clearly, I wasn't thinking broadly enough and went for the easy and scarier pathogens, not realizing that even a common cold virus can be made terrifying with some selective GOF work combined with a tidal wave of media-induced hysteria. Add to that mix some ignorant and self- serving politicians, toss in a big helping of panicked people ready to sacrifice freedom on the flimsiest of evidence, then shake well and serve. (Note: I am tempted to think that various terrorists of all sorts got this memo for future attacks.)

As the book progressed, I found myself stuck at a plot point where the virus is first released. As I was starting to think how the rest of the story would play out, I decided to take a break, since I had just received the contract for this book (September 2019) and wanted to get some of *Dispatches from the Vaccine Wars* done, at least the structure, before returning to my novel.

As *Dispatches* began taking shape, COVID-19 exploded onto the world. My novel was now moot, as real world events had overtaken fiction and what had been fiction was now on the daily news.

In my book, the hero was going to prevent the pharma from using their disease to exercise control of world health. In the real world, the pharma and their political allies pulled it off so smoothly that it appears to have been planned and rehearsed, as indeed it likely was.

In hindsight, my fiction project now seems like art, perhaps bad art, predicting an equally bad but real future. Or maybe I, like many others, had intuited the way the future might unfold from what even then was an obvious corporate End State.

Nonfiction: Bench Science

During this last year, my laboratory has redoubled its efforts to provide clear evidence for possible vaccine harms. Using the entire CDC vaccine schedule for children from newborn to age eighteen months, but adjusted for weight in newborn mice, we have administered the exact same vaccines in the same way as would be done in human children. The results now published in the *Journal of Inorganic Biochemistry*[46] show a range of neurobehavioral abnormalities in the vaccinated mice that stratify by age and sex. Many of the initial deficits decrease over time; others do not. In a companion study that should also be in press by the time this book comes out, we measured cytokine levels in the blood of the same vaccinated and control mice at two time points. The key finding in this study was that a number of cytokines are clearly changed soon after vaccination, the most striking one being IL5, a microglial activator. Microglia are the resident immune cells of the CNS and play key roles in neurodegeneration.[47]

In addition, my colleague Prof. John Oller and I have started an independent online journal called the *International Journal of Vaccine Theory, Research and Practice*.[48] As this book comes out, the journal should be on Issue 3, and some of the papers in this issue have been quoted in these last chapters.

Finally, we have been seeking funding through the Institute of Pure and Applied Knowledge and Children's Health Defense to fund a massive parametric study of the vaccine adjuvant aluminum hydroxide and its potential impact on the CNS at behavioral, cellular, and biochemical levels for a range of concentrations of aluminum, from below to beyond the amount in the current zero-to-eighteen-months CDC pediatric vaccine schedule. What will we find? I don't know, but if done well, the study will provide definitive measures of aluminum dose in relation to possible CNS changes. We are completely aware of the possible confounds and limitations of an animal study to humans yet feel that this study, if carried out, would offer definitive evidence for, or against, aluminum involvement in CNS disorders of the young.

For now, these are the ways I fight back. It is a small part perhaps of the overall struggle that may help make what comes out of the gate the lady, not the tiger, at least concerning vaccines.

My Views on Vaccines and Autism

Although I have addressed a range of issues in this book, the reader may still have two main questions.

The first of these is this: what do I really feel about vaccines? It's a reasonable question and one that I have not addressed so far for the simple reason that I wanted readers to come to their own conclusions about vaccines without being prejudiced by what they thought my position might be. Of course, astute readers have likely figured out the bulk of it, or at least think they have based on the previous pages. Let's see.

What I think about vaccines is quite nuanced, but ultimately simple: I think vaccines are a useful medical adjunct that can be enormously helpful in the control of infectious diseases. I view them like other useful medical interventions, including antibiotics, salicylic acid, a range of surgeries, and in general any form of prophylactic medical practice. I also think that vaccines are massively overused, particularly in the young; that they have the potential to harm as well as serve prophylaxis; and in this sense that they are overhyped for their benefits and underestimated for their harms.

No medicine or drug is immune to the impact of dose, either acutely or chronically. What is true for antibiotics is true as well for vaccines. It may well not be the antigen alone, rather some of the actual toxicants in vaccines such as aluminum, and formerly Thimerosal. The dose indeed makes the poison in this regard: if you give enough of any toxicant in a short enough time frame to a developing nervous system, how can one not expect problems to emerge in some fraction of the treated population? The surprise here is not that some children succumb to this, but that more do not. This last issue is where individual genetic profiles come into play, along with sex and a host of other potential risk factors.

To put the issue into COVID-19 or any infectious disease terms, what makes a person likely to be infected? The answer is that it is a dose/time consideration. In this view, exposure to a lot of the virus in a short time frame is more likely to induce infection than the same amount of virus spread out over a longer time period.

Another aspect of the dose issue is that of failed biological signaling, biosemiosis, as cited in Chapter 5 and considered in greater detail in Oller and Shaw, 2019.[49] The take-home message here is that there is likely to be an upper limit to how many times one can deliberately fool the immune system with vaccines before pushing that system into autoimmunity.

In my view, these dose/time and signaling problems are not solvable by ignoring them.

I think vaccine benefits in general have been subverted by the heavy hand of market forces, which, like much else in modern medicine, are primarily about profit rather than people. I think that leaving the control of vaccines, and other drugs, in the hands of the pharmaceutical industry is rank madness. Remember, this is an industry that at best is corrupt (see Chapter 10 and Dr. Marcia Angell), at worst outright felonious (Chapter 7 and Robert Kennedy Jr.). In regard to either, my view is that executives of such companies should be put on trial and, if convicted of a range of crimes, spend much of their future lives breaking rocks under the hot sun.

The second question is this: what have I learned in writing this book? I learned a very great deal about vaccines and the machinations behind their manufacture and use. That was the academic part. In the personal part, I learned, as I had over most of my life, that most people are basically good and decent, but that if they can be made afraid, they change to become quite different. COVID-19 reinforced beliefs I had from much earlier in life: Fear is contagious, and if enough of it is activated, it makes otherwise good, decent people submit without question to tyrants big and small; courage is also contagious and makes the most unexpected people stand up against the same tyrants.

I learned that a large fraction of the population really doesn't understand science, even if they appear to worship it. I learned that a distressingly large number of my fellow citizens lack any semblance of *any* ability for critical thinking. And I learned that they don't know they lack this ability *because* they lack this ability.

On Justice

The Old Testament, sacred to three religions, says this: "Justice, justice thou shalt pursue" [Deuteronomy 16:20], a proscription that must lead to the question of what to do with those who have planned and tried to execute the Great Reset. It also applies to those who were complicit in this conspiracy to dominate the human population of the Earth. In some measure, it even applies to those in politics, in the health professions, and in the financial world who went along with it, perhaps without realizing what they were party to.

One suggestion would be to do what South Africa did after the end of the apartheid regime: convene panels of Truth and Reconciliation and ensure that those guilty pay some price. For those who were the generals of the Great Reset, you will go on trial. If found guilty, then you may spend much of the rest of your lives in prison. As I believe that all humans have some possibility of eventual redemption, then these people may have some chance of it, too. In my view, this first group includes Bill Gates (maybe Melinda, too), Tedros

Ghebreyesus, Klaus Schwab, and all the others cited in the Children's Health Defense article cited in Chapter 13. Of a certainty, they should also forfeit all of their assets to help assuage the suffering of so many they have harmed.

Those in the political world and the health world, and others who were active conspirators, should also go on trial. Their penalty, if convicted, might also include prison, but a more likely penalty would be to ban them from any role in any of their fields for life. Those who were merely complicit, perhaps out of ignorance, may be required to make amends in some fashion, perhaps through some term of community service.

Would that be justice? I think it would be, and indeed a far lesser punishment than some might deem necessary. The goal is not to seek vengeance, but justice, and to be an example to any future tyrants that there is a price to pay, right here on Earth, for such crimes.

Epilogue

I finish this book with two quotes from Thomas Paine, one of my political heroes. The first, cited previously, is worth repeating again: "We have it in our power to begin the world over again."

Even more, we should not despair that the struggle against true fascism has arisen in our day, rather we should be glad that the struggle falls upon us, not our children: "If there must be trouble, let it be in my day, that my child may have peace; and this single reflection, well applied, is sufficient to awaken every man to duty."

The same call now goes out to us all while we still have choices that can be made. There is no doubt that there will be a reset. But a

Thomas Paine, American revolutionary and author of *Common Sense*.[1]

reset to what? What comes out of the door, the lady or the tiger? I think we can now clearly see which one it will be if the powers that be have their way with the Great Reset.

The year of COVID-19 was in many ways a terrible year for some people, but in the sense of Stoic philosophy it was an amazing year: it was a year in which people were able to test their commitments to truth and justice and, from this, their choice of submission or resistance.

Many of my generation grew up wondering if they would have joined the various heroes of the past to fight tyranny. In 2020, we all found out what we really would have done. And most would not have done any of the above, whatever they said in the past. Most would have docilely submitted to power, as so many have in our day to the COVID-19 diktats of governments.

The year 2020 was a very good year for self-realization: by your actions, now you know who you really are.

As we go forward into 2021, we still have two choices: Submit and hope you can keep some of your privileges as a reward for your obedience. Or act like those heroes of the past you always claimed to admire: Resist, fight back, and maybe, just maybe, preserve your world, or even better, make it the world you want it to be. If you choose the latter—accept the challenge of our age—you will act as you always dreamt you would if called forth to fight tyranny. You can become the stuff of legend that your descendants will remember with pride.

In your choice, act as if your very world depends on it.

It does.

Act as if the lives and futures of your children, grandchildren, and seven future generations beyond you depend on it.

They do.

Act as if your choice of freedom or submission affects us all. It most certainly does. Choose one path, dear reader.

Choose well.

Endnotes

Preface

1. As quoted in W. W. Norton & Company; Second Edition (May 17, 1973), New York City, pg. v.
2. Petrik, M. S., Wong, M. C., Tabata, R. C., Garry, R. F., & Shaw, C. A. (2007). Aluminum adjuvant linked to Gulf War illness induces motor neuron death in mice. *Journal of Neuromolecular Medicine*, 9(1), 83–100. doi:10.1385/nmm:9:1:83; Shaw, C. A., & Petrik, M. S. (2009). Aluminum hydroxide injections lead to motor deficits and motor neuron degeneration. *Journal of Inorganic Biochemistry*, 103(11), 1555–1562. doi:10.1016/j.jinorgbio.2009.05.019.
3. https://www.who.int/vaccine_safety/committee/reports/Jun_2012/en/; The two studies that GACVS critiqued were these: Tomljenovic L, Shaw CA. Do aluminum vaccine adjuvants contribute to the rising prevalence of autism? *J Inorg Biochem*, 2011; 105: 1489–1499; and Tomljenovic L, Shaw CA. Aluminum vaccine adjuvants: are they safe? *Current Medicinal Chemistry*, 2011; 18(17):2630–2637.

Chapter 1

1. Carl Sagan, "The Fine Art of Baloney Detection." *The Demon-Haunted World: Science as a Candle in the* Dark, (New York: Random House, 1995), 201–19.
2. Murray Bookchin, *The ecology of freedom: the emergence and dissolution of hierarchy*, (Palo Alto, Calif.: Cheshire Books, 1982), 385.
3. M. Knapp, A. Flach, E. Ayboga, D. Graeber, A Abdullah, and J Biehl, *Revolution in Rojava : democratic autonomy and women's liberation in Syrian Kurdistan*, (London: Pluto Press, 2016), https://www.jstor.org/stable/10.2307/j.ctt1gk07zg; https://www.jstor.org /stable/10.2307/j.ctt1gk07zg.
4. "Evidenced-based medicine" encompasses the notion that medical practice should be firmly based on experiment.
5. See, for example, Robert F. Kennedy's *Thimoersal: Let the Science Speak*, cited in Chapter 5.
6. Peter C. Gøtzsche, R Smith, and D Rennie, *Deadly medicines and organised crime: how big pharma has corrupted healthcare* (2013).
7. *Scientific method*. Retrieved from Wikipedia, https://simple.wikipedia.org/wiki/Scientific _method.
8. *Scientific theory*. Retrieved from Wikipedia, https://en.wikipedia.org/wiki/Theory#:~: text=In%20science%2C%20the%20term%20%22theory,to%20make%20falsifiable% 20predictions%20with.
9. *Hypothesis*. Retrieved from Oxford, Lexico, https://www.lexico.com/definition/hypothesis.
10. J. Lyons-Weiler, personal communication.

11. Kuhn, T. S. *The structure of scientific revolutions*.xv, 172 p. Phoenix Books. Chicago: University of Chicago Press, 1962.

12. Science in any field is always evolving. For example, in the neurosciences I learned in graduate school that no neurons were born in the CNS of adults and that the immune and nervous systems were completely separate. Both ideas were wrong, as shown by experiments in the decades that followed. Even earlier battles within the field about the nature of synaptic transmission that were thought to have been resolved were, in fact, not. In much the same vein, medical practices change over time to reject some older treatments and embrace newer understanding and technologies. Two key examples of the latter involve "blood letting" as a way to treat various diseases states, or the use of leaches.

13. Gladwell, M. *Outliers : the story of success*. 309 p. 1st ed. New York: Little, Brown and Co., 2008. Contributor biographical information http://www.loc.gov/catdir/enhancements /fy0837/2008032824-b.html. Publisher description: http://www.loc.gov/catdir/enhancements /fy0837/2008032824-d.html.

14. Epstein, D. J. *The sports gene : inside the science of extraordinary athletic performance*.xiv, 352 pages. Paperback edition. New York, New York: Current, 2014.

15. *How To Be Proven Wrong*. (2020). From "Daily Stoic". Retrieved from https://dailystoic .com/how-to-be-proven-wrong/.

16. The notion from game theory that one person's loss is the other person's equivalent game such that the net change is zero.

17. See, for example Holiday, R.H and Hanselman, S., *Lives of the Stoics: The Art of Living from Zeno to Marcus Aurelius*, New York: Penguin Random House, 2020.

18. From Stanford Encylopedia (https://plato.stanford.edu/entries/ockham/): "William of Ockham (c. 1287–1347) is, along with Thomas Aquinas and John Duns Scotus, among the most prominent figures in the history of philosophy during the High Middle Ages. He is probably best known today for his espousal of metaphysical nominalism; indeed, the methodological principle known as 'Ockham's Razor' is named after him."

19. Bennett, B. *Logically Fallacious: The Ultimate Collection of Over 300 Logical Fallacies (Academic Edition)*. Sudbury MA: Archieboy Holdings, LLC., 2011.

20. *Lancet*. 2014 Mar 15; 383(9921): 999–1008. Published online 2013 Sep 29. doi: 10.1016/S0140-6736(13)61752-3. The Framingham Heart Study and the Epidemiology of Cardiovascular Diseases: A Historical Perspective. Syed S. Mahmood,Daniel Levy, Ramachandran S. Vasan, and Thomas J. Wang.

21. *MMR and MMRV Vaccine Composition and Dosage*. (2019). From "Vaccines and Preventable Diseases". Retrieved from US Centers for Disease Control and Prevention, https://www.cdc.gov/vaccines/vpd/mmr/hcp/about.html.

22. *Complications of Measles*. (2020). From "Measles (Rubeola)." Retrieved from US Centers for Disease Control and Prevention, https://www.cdc.gov/measles/symptoms/complications .html.

23. R. Sears, personal communication.

24. *Electronic Support for Public Health-accine Adverse Event Reporting System (ESP:VAERS) (Massachusetts)*. From "Agency for Healthcare Research and Quality." Retrieved from https://digital.ahrq.gov/ahrq-funded-projects/electronic-support-public-health -vaccine-adverse-event-reporting-system.

25. Here is the calculation for the amount of time a physician takes to get fully qualified versus a research scientist: Physician: In North America, four years undergraduate degree in usually some science; four years further undergrad medical degree; five years on average for residence (family practice is two years in British Columbia); two or more years

doing more specialized training in fellowships. Total: fifteen. Research Scientist: four years undergraduate studies; two or so for M.Sc.; three to four more for PhD; two to three (or more) as a postdoctoral fellow. Total: thirteen or so years. The two streams are pretty comparable in terms of time learning their respective trades.

26. Poland, G. A., and Jacobson, R. M. "The age-old struggle against the antivaccination-ists." *N Engl J Med,* vol. 364, no. 2 (Jan 13 2011): 97–9. doi:10.1056/NEJMp1010594 PMID: 21226573.

27. Eidi, H., Yoo, J., Bairwa, S. C., Kuo, M., Sayre, E. C., Tomljenovic, L., and Shaw, C. A. "Early postnatal injections of whole vaccines compared to placebo controls: Differential behavioural outcomes in mice." *J Inorg Biochem,* vol. 212 (Nov 2020): 111200. doi:10.1016/j.jinorgbio.2020.111200 PMID: 33039918.

28. J. Oller, personal communication.

29. Problem based learning (PBL) is a less didactic form of instruction in which eight students, at my university, work through a weekly "case" from initial presentation to resolution with an assigned tutor. In the course of the week, the students learn the necessary anatomy, physiology, biochemistry of the organ systems involved and consider, as well social issues that may arise. PBL has now been replaced by case based learning (CBL), somewhat the same format, but more focussed on delivering practical medical knowledge and treatment.

30. Masic, I., Miokovic, M., and Muhamedagic, B. "Evidence based medicine—new approaches and challenges." *Acta Inform Med,* vol. 16, no. 4 (2008): 219–25. doi:10.5455/aim.2008.16.219-225 PMID: 24109156.

31. *Debunk.* Retrieved from Merriam-Webster, https://www.merriam-webster.com/dictionary/debunk#:~:text=English%20Language%20Learners%20Definition%20of,idea%2C%20statement%2C%20etc.

32. Wakefield, A. J., Murch, S. H., Anthony, A., Linnell, J., Casson, D. M., Malik, M., Berelowitz, M., Dhillon, A. P., Thomson, M. A., Harvey, P., Valentine, A., Davies, S. E., and Walker-Smith, J. A. "Ileal-lymphoid-nodular hyperplasia, non-specific colitis, and pervasive developmental disorder in children." *Lancet,* vol. 351, no. 9103 (Feb 28 1998): 637–41. doi:10.1016/s0140-6736(97)11096-0 PMID: 9500320.

33. Ioannidis, J. P. "Why most published research findings are false." *PLoS Med,* vol. 2, no. 8 (Aug 2005): e124. doi:10.1371/journal.pmed.0020124 PMID: 16060722.

Chapter 2

1. Jenner, E. "History of the Inoculation of the Cow-Pox: Further Observations on the Variolae Vaccinae, or Cow-Pox." *Med Phys J,* vol. 1, no. 4 (Jun 1799): 313–18. https://www.ncbi.nlm.nih.gov/pubmed/30489938 PMID: 30489938.

2. Smith, K. A. "Edward Jenner and the smallpox vaccine." *Front Immunol,* vol. 2 (2011): 21. doi:10.3389/fimmu.2011.00021 PMID: 22566811; Riedel, S. "Edward Jenner and the history of smallpox and vaccination." *Proc (Bayl Univ Med Cent),* vol. 18, no. 1 (Jan 2005): 21–5. doi:10.1080/08998280.2005.11928028 PMID: 16200144; Stern, A. M., and Markel, H. "The history of vaccines and immunization: familiar patterns, new challenges." *Health Aff (Millwood),* vol. 24, no. 3 (May-Jun 2005): 611–21. doi:10.1377/hlthaff.24.3.611 PMID: 15886151.

3. Gross, C. P., and Sepkowitz, K. A. "The myth of the medical breakthrough: smallpox, vaccination, and Jenner reconsidered." *Int J Infect Dis,* vol. 3, no. 1 (Jul-Sep 1998): 54–60. doi:10.1016/s1201-9712(98)90096–0 PMID: 9831677.

4. Riedel, S. "Edward Jenner and the history of smallpox and vaccination." *Proc (Bayl Univ Med Cent).*

5. Barquet, N., and Domingo, P. "Smallpox: the triumph over the most terrible of the ministers of death." *Ann Intern Med,* vol. 127, no. 8 Pt 1 (Oct 15 1997): 635–42. doi:10.7326/0003-4819-127-8_part_1-199710150-00010 PMID: 9341063.

6. Riedel, S. "Edward Jenner and the history of smallpox and vaccination." *Proc (Bayl Univ Med Cent).*

7. "A Theory of Germs." In *Science, Medicine, and Animals.* The National Academies Collection: Reports funded by National Institutes of Health. Washington (DC), 2004. https://www.ncbi.nlm.nih.gov/pubmed/20669472.

8. Ibid.

9. Artenstein, A. W. "The discovery of viruses: advancing science and medicine by challenging dogma." *Int J Infect Dis,* vol. 16, no. 7 (Jul 2012): e470-3. doi:10.1016/j.ijid.2012.03.005 PMID: 22608031.

10. It is worth considering that some serious critiques about Pasteur have emerged with the discovery of his notebooks by the late history of science Professor Gerald L. Geison in *The Private Science of Louis Pasteur* (1995), Princeton Legacy Library. Geison's contention is that an examination of Pasteur's laboratory notebooks tells a very different story of his work and discoveries from the conventional view. Called into question are both Pasteur's personal and scientific ethics. In turn, Geison has come under critique by Mark Perutz in a review in the *New York Review of Books,* Dec. 21, 1995.

11. Marrack, P., Mckee, A. S., and Munks, M. W. "Towards an understanding of the adjuvant action of aluminium." *Nat Rev Immunol,* vol. 9, no. 4 (Apr 2009): 287–93. doi:10.1038/nri2510 PMID: 19247370.

12. Mahmood, S. S., Levy, D., Vasan, R. S., and Wang, T. J. "The Framingham Heart Study and the epidemiology of cardiovascular disease: a historical perspective." *Lancet,* vol. 383, no. 9921 (Mar 15 2014): 999–1008. doi:10.1016/S0140-6736(13)61752–3 PMID: 24084292.

13. The most common age-related neurological disorders are, in order, Alzheimer's disease, Parkinson's disease, and Lou Gehrig's disease (ALS). See Shaw, C. A. (2017). *Neural Dynamics of Neurological Disease. Boston:* John Wiley and Sons.

14. Mahmood, S. S., Levy, D., Vasan, R. S., and Wang, T. J. "The Framingham Heart Study and the epidemiology of cardiovascular disease: a historical perspective." *Lancet.*

15. Virus: "Virus refers to a small parasite that consists of a nucleic acid molecule covered by a protein coat. It consists of a non-cellular organization and specific modes of reproduction. A virus is essentially a non-living entity as it does not show any metabolic activity. Typically, a virus particle consists of either a single-stranded or double-stranded, DNA or RNA genome, covered by a protein coat known as the capsid." (Lakna. (2018). Difference Between Virus and Virion. Retrieved from Pediaa: https://pediaa.com/difference-between-virus-and-virious/#Virus). A virus is larger than a virion and is found inside the host cell.

16. Virion: "Virion refers to the complete, infective form of a virus outside the host cell. A virion is a complete virus particle, composed of either a DNA or RNA genome, and covered by a protein capsid."

17. Wernher von Braun.

18. Gomez, P. L., and Robinson, J. M. "Vaccine Manufacturing." [In eng]. *Plotkin's Vaccines*, (2018): 51–60.e1. doi:10.1016/B978-0-323-35761-6.00005–5 PMID: PMC7152262.

19. *Vaccine Types*. Retrieved from vaccines.gov, U.S. Department of Health & Human Services, https://www.vaccines.gov/basics/types.; *Types of Vaccine*. From "Vaccine Knowledge Project". Retrieved from Oxford Vaccine Group, https://vk.ovg.ox.ac.uk /vk/types-of-vaccine.; *Vaccine Supply and Innovation*. Washington (DC), 1985. https: //www.ncbi.nlm.nih.gov/pubmed/25032425.; Clem, A. S. "Fundamentals of vaccine immunology." *J Glob Infect Dis*, vol. 3, no. 1 (Jan 2011): 73–8. doi:10.4103/0974-777X.77299 PMID: 21572612.

20. *Vaccines Licensed for Use in the United States* (2020). Retrieved from US Food and Drugs Administration, https://www.fda.gov/vaccines-blood-biologics/vaccines/vaccines -licensed-use-united-states.

21. Wappes, J. *Poor late-season protection limited flu vaccine impact for 2018–19* (2019). From "Influenza Vaccines." Retrieved from CIDRAP News, https://www.cidrap.umn .edu/news-perspective/2019/07/poor-late-season-protection-limited-flu-vaccine -impact-2018-19#:~:text=For%202017%2D18%2C%20when%20the,seasons %2C%20according%20to%20CDC%20data.

22. In the UBC Medical School's Brain and Behaviour block, usually taught in the second year, the tutor handbook used to cite examples of Guillain-Barré, sometimes arising from the 1976 "swine flu, influenza A virus, subtype H1N1."

23. Phillip, A. *Hillary Clinton likens anti-vaxxers to science deniers: 'The earth is round, the sky is blue, and #vaccineswork'*. (2015). From "Morning Mix". Retrieved from the Washington Post, https://www.washingtonpost.com/news/morning-mix/wp/2015/02/03/hillary-clinton -likens-anti-vaxxers-to-science-deniers-the-earth-is-round-the-sky-is-blue-and-vaccines- work/.

24. Ibid.

25. Behar, A. *Anti-Vaccine Activist Says Trump Asked Him to Head Commission on Vaccine Safety*. (2017). From "Politics News". Retrieved from NBC News, https://www.nbcnews .com/politics/politics-news/trump-meets-anti-vaccine-activist-after-raising-fringe-theory -trail-n705296.

26. *Trump retains Collins as NIH director*. (2017). From "Science". Retrieved from American Association for the Advancement of Science, https://www.sciencemag.org /news/2017/06/trump-retains-collins-nih-director.; Yong, E. *Trump's Pick for CDC Director Is Experienced But Controversial*. (2018). From "Science". Retrieved from The Atlantic, https://www.theatlantic.com/science/archive/2018/03/trumps-pick-for-cdc -director-is-experienced-but-controversial/556202/.; Eddy, N. *Senate confirms Trump appointee Stephen Hahn to lead FDA*. (2019). From "Global Edition Government & Policy". Retrieved from Healthcare IT News, https://www.healthcareitnews.com/news /senate-confirms-trump-appointee-stephen-hahn-lead-fda.

27. Choi, Q. C. *Strange but True: Earth Is Not Round* (2007). From "Space". Retrieved from Scientific American, https://www.scientificamerican.com/article/earth-is-not-round/.

28. *Why Is the Sky Blue?* (2020). Edited by NASA Science. From "Space Place: Explore Earth and Space!". Retrieved from NASA, https://spaceplace.nasa.gov/blue-sky/en/.

29. *Causality*. Retrieved from Cambridge Dictionary, https://dictionary.cambridge.org /dictionary/english/causality.

30. *Coincidence*. Retrieved from Cambridge Dictionary, https://dictionary.cambridge.org /dictionary/english/coincidnece.

31. Je, B., and Seneff, S. "The Possible Link between Autism and Glyphosate Acting as Glycine Mimetic - A Review of Evidence from the Literature with Analysis." *Journal of Molecular and Genetic Medicine,* vol. 09, no. 04 (2015). doi:10.4172/1747–0862.1000187.

32. Cox, P. A., and Sacks, O. W. "Cycad neurotoxins, consumption of flying foxes, and ALS-PDC disease in Guam." *Neurology,* vol. 58, no. 6 (Mar 26 2002): 956–9. doi:10.1212 /wnl.58.6.956 PMID: 11914415.

33. Bradley, W. G., and Mash, D. C. "Beyond Guam: the cyanobacteria/BMAA hypothesis of the cause of ALS and other neurodegenerative diseases." *Amyotroph Lateral Scler,* vol. 10 Suppl 2 (2009): 7–20. doi:10.3109/17482960903286009 PMID: 19929726.

34. Shaw, C. A. *Neural Dynamics of Neurological Disease.* Boston: John Wiley & Sons, 2017. doi:https://doi.org/10.1002/9781118634523.ch1.

35. Hill, A. B. "The Environment and Disease: Association or Causation?". *Proc R Soc Med,* vol. 58 (May 1965): 295–300. https://www.ncbi.nlm.nih.gov/pubmed/14283879 PMID: 14283879.

36. Model systems in biology typically include, from most effective to least: in vivo using living animals; in vitro, usually cell culture; and in silico, usually computer models.

37. *John P. A. Ioannidis.* From "Profiles". Retrieved from Stanford, https://profiles.stanford. edu/john-ioannidis.

38. Ioannidis, J. P. "Why most published research findings are false." *PLoS Med,* vol. 2, no. 8 (Aug 2005): e124. doi:10.1371/journal.pmed.0020124 PMID: 16060722.

39. Shaw, C. A. *Neural Dynamics of Neurological Disease;* Seok, J., Warren, H. S., Cuenca, A. G., Mindrinos, M. N., Baker, H. V., Xu, W., Richards, D. R., Mcdonald-Smith, G. P., Gao, H., Hennessy, L., Finnerty, C. C., Lopez, C. M., Honari, S., Moore, E. E., Minei, J. P., Cuschieri, J., Bankey, P. E., Johnson, J. L., Sperry, J., Nathens, A. B., Billiar, T. R., West, M. A., Jeschke, M. G., Klein, M. B., Gamelli, R. L., Gibran, N. S., Brownstein, B. H., Miller-Graziano, C., Calvano, S. E., Mason, P. H., Cobb, J. P., Rahme, L. G., Lowry, S. F., Maier, R. V., Moldawer, L. L., Herndon, D. N., Davis, R. W., Xiao, W., Tompkins, R. G. Inflammation, and Host Response to Injury, L. S. C. R. P. "Genomic responses in mouse models poorly mimic human inflammatory diseases." *Proc Natl Acad Sci U S A,* vol. 110, no. 9 (Feb 26 2013): 3507–12. doi:10.1073/pnas.1222878110 PMID: 23401516.

40. *Amyotrophic lateral sclerosis (ALS).* From "Patient Care & Health Information Diseases & Conditions". Retrieved from Mayo Clinic, https://www.mayoclinic.org/diseases -conditions/amyotrophic-lateral-sclerosis/symptoms-causes/syc-20354022#:~:text =Amyotrophic%20lateral%20sclerosis%20(a%2Dmy,who%20was%20diagnosed%20 -with%20it.

41. Morrice, J. R., Gregory-Evans, C. Y., and Shaw, C. A. "Modeling Environmentally-Induced Motor Neuron Degeneration in Zebrafish." *Scientific Reports,* vol. 8, no. 1 (2018/03/20 2018): 4890. doi:10.1038/s41598-018-23018-w.

42. Kabacoff, R. *Power Analysis.* From "Quick-R Statistics". Retrieved from Datacamp, https://www.statmethods.net/stats/power.html.

43. Mcleod, S. *What a p-value tells you about statistical significance.* (2019). From "Statistics p-value". Retrieved from SimplyPsychology, https://www.simplypsychology.org/p-value .html.

44. Kuo, M. T. H., Beckman, J. S., and Shaw, C. A. "Neuroprotective effect of CuATSM on neurotoxin-induced motor neuron loss in an ALS mouse model." *Neurobiol Dis,* vol. 130 (Oct 2019): 104495. doi:10.1016/j.nbd.2019.104495 PMID: 31181282.

45. Boodman, E. *Researchers rush to test coronavirus vaccine in people without knowing how well it works in animals.* (2020). From "Health". Retrieved from Stat, https://www.statnews.com/2020/03/11/researchers-rush-to-start-moderna-coronavirus-vaccine-trial-without-usual-animal-testing.

46. Haseltine, W. *Beware of covid-19 vaccine trials designed to succeed from the start.* (2020). From "Opinions". Retrieved from the Washington Post, https://www.washingtonpost.com/opinions/2020/09/22/beware-covid-19-vaccine-trials-designed-succeed-start/.

47. Thiese, M. S. "Observational and interventional study design types; an overview." *Biochem Med (Zagreb),* vol. 24, no. 2 (2014): 199–210. doi:10.11613/BM.2014.022 PMID: 24969913.

48. Singal, A. G., Higgins, P. D., and Waljee, A. K. "A primer on effectiveness and efficacy trials." *Clin Transl Gastroenterol,* vol. 5 (Jan 2 2014): e45. doi:10.1038/ctg.2013.13 PMID: 24384867.

49. Ibid.

50. *Fox Trial Finder.* Retrieved from The Michael J. Fox Foundation for Parkinson's Research, https://www.michaeljfox.org/trial-finder.

51. Tomljenovic, L., and Shaw, C. A. "Do aluminum vaccine adjuvants contribute to the rising prevalence of autism?". *J Inorg Biochem,* vol. 105, no. 11 (Nov 2011): 1489–99. doi:10.1016/j.jinorgbio.2011.08.008 PMID: 22099159.

52. Mawson, A., Ray, B., Bhuiyan, A. R., and Jacob, B. J. F. I. P. H. "Vaccination and Health Outcomes: A Survey of 6- to 12-year-old Vaccinated and Unvaccinated Children based on Mothers Reports." vol. 4 (2016).

53. Levin, K. A. "Study design V. Case-control studies." *Evid Based Dent,* vol. 7, no. 3 (2006): 83–4. doi:10.1038/sj.ebd.6400436 PMID: 17003803.

54. 2019. *Human Papillomavirus (HPV) Vaccines.* Edited by National Cancer Institute. From "Infectious Agents". Retrieved from National Institutes of Health, https://www.cancer.gov/about-cancer/causes-prevention/risk/infectious-agents/hpv-vaccine-fact-sheet.

55. Wakefield, A. J., Murch, S. H., Anthony, A., Linnell, J., Casson, D. M., Malik, M., Berelowitz, M., Dhillon, A. P., Thomson, M. A., Harvey, P., Valentine, A., Davies, S. E., and Walker-Smith, J. A. "Ileal-lymphoid-nodular hyperplasia, non-specific colitis, and pervasive developmental disorder in children." *Lancet,* vol. 351, no. 9103 (Feb 28 1998): 637–41. doi:10.1016/s0140-6736(97)11096-0 PMID: 9500320.

56. *What is a Systematic Review?* From "Systematic Reviews & Other Review Types". Retrieved from Temple University Libraries, https://guides.temple.edu/c.php?g=78618&p=4178713.

57. "Clinical equipoise, also known as the principle of equipoise, provides the ethical basis for medical research that involves assigning patients to different treatment arms of a clinical trial. The term was first used by Benjamin Freedman in 1987. In short, clinical equipoise means that there is genuine uncertainty in the expert medical community over whether a treatment will be beneficial. This applies also for off-label treatments performed before or during their required clinical trial"; Clinical equipoise. (2020). Retrieved from Wikipedia: https://en.wikipedia.org/wiki/Clinical_equipoise.

58. 2019. *Human Papillomavirus (HPV) Vaccines.* Edited by National Cancer Institute. From "Infectious Agents".

59. Mawson, A., Ray, B., Bhuiyan, A. R., and Jacob, B. "Vaccination and Health Outcomes: A Survey of 6- to 12-year-old Vaccinated and Unvaccinated Children Based on Mothers Reports." vol. 4 (2016).

60. Natural seroconversion also occurs following infection with a pathological agent. The presence of antibodies, as a surrogate marker, is taken to indicate that the disease occurred in that patient.

61. *Correlates of vaccine-induced protection: methods and implications* (2013). Retrieved from World Health Organization, Initiative for Vaccine Research (IVR) of the Department of Immunization, Vaccines and Biologicals, https://apps.who.int/iris/bitstream/handle/10665/84288/WHO_IVB_13.01_eng.pdf.

62. *Correlates of vaccine-induced protection: methods and implications* (2013). Chapter "1. Introduction". pg. 1. Retrieved from World Health Organization, Innitiative for Vaccine Research (IVR) of the Department of Immunization, Vaccines and Biologicals, https://apps.who.int/iris/bitstream/handle/10665/84288/WHO_IVB_13.01_eng.pdf.

63. Ibid.

64. Ibid, p. 3.

65. Ibid, p. 15.

66. Ibid.

67. Patel, R. *Canada's top doctor warns against relying on herd immunity to reopen economy.* (2020). From "Politics". Retrieved from CBC News, https://www.cbc.ca/news/politics/herd-immunity-should-not-be-supported-tam-says-1.5545332.

68. *Correlates of vaccine-induced protection: methods and implications* (2013). Retrieved from World Health Organization, Innitiative for Vaccine Research (IVR) of the Department of Immunization, Vaccines and Biologicals.

69. Ibid, p. 43.

70. 2019. *Human Papillomavirus (HPV) Vaccines.* Edited by National Cancer Institute. From "Infectious Agents".

71. Djurisic, S., Jakobsen, J. C., Petersen, S. B., Kenfelt, M., and Gluud, C. "Aluminium adjuvants used in vaccines versus placebo or no intervention." [In eng]. *The Cochrane Database of Systematic Reviews,* vol. 2017, no. 9 (2017): CD012805. doi:10.1002/14651858. CD012805 PMID: PMC6483624.

72. Svensson, S., Menkes, D. B., and Lexchin, J. "Surrogate outcomes in clinical trials: a cautionary tale." *JAMA Intern Med,* vol. 173, no. 8 (Apr 22 2013): 611–2. doi:10.1001/jamainternmed.2013.3037 PMID: 23529157; Baker, S. G. "Surrogate Endpoints: Wishful Thinking or Reality?". *JNCI: Journal of the National Cancer Institute,* vol. 98, no. 8 (2006): 502–03. doi:10.1093/jnci/djj153 %J JNCI: Journal of the National Cancer Institute.

73. Ward, B. J., Pillet, S., Charland, N., Trepanier, S., Couillard, J., and Landry, N. "The establishment of surrogates and correlates of protection: Useful tools for the licensure of effective influenza vaccines?". *Hum Vaccin Immunother,* vol. 14, no. 3 (Mar 4 2018): 647–56. doi:10.1080/21645515.2017.1413518 PMID: 29252098.

74. Sabbe, M., and Vandermeulen, C. "The resurgence of mumps and pertussis." *Hum Vaccin Immunother,* vol. 12, no. 4 (Apr 2 2016): 955–9. doi:10.1080/21645515.2015.1113357 PMID: 26751186.

75. Kulkarni, P. S., Hurwitz, J. L., Simoes, E. A. F., and Piedra, P. A. "Establishing Correlates of Protection for Vaccine Development: Considerations for the Respiratory Syncytial Virus Vaccine Field." *Viral Immunol,* vol. 31, no. 2 (Mar 2018): 195–203. doi:10.1089/vim.2017.0147 PMID: 29336703.

76. (2018). From "21st Century Cures: Announcing the Establishment of a Surrogate Endpoint Table; Establishment of a Public Docket; Request for Comments". Retrieved from Federal Register: The Daily Journal of the United States Government,

https://www.federalregister.gov/documents/2018/10/30/2018–23641/21st-century-cures
-announcing-the-establishment-of-a-surrogate-endpoint-table-establishment-of-a#:~
:text=Section%20507(e)(9,be%20used%20to%20support%20traditional.

77. *Table of Surrogate Endpoints That Were the Basis of Drug Approval or Licensure.* (2020). From
"Development & Approval Process | Drugs Development Resources". Retrieved from
US Food & Drug Administration, https://www.fda.gov/drugs/development-resources
/table-surrogate-endpoints-were-basis-drug-approval-or-licensure.

Chapter 3

1. One of the most influential Stoic philosophers (55–135 AD). See: Holiday, R. &
Hanselman, S. (2020). *Epictetus the Free Man. In Lives of the Stoics: The Art of Living from
Zeno to Marcus Aurelius* (pp. 251–268). New York: Penguin Random House.

2. Shaw, C. A. *Neural Dynamics of Neurological Disease.* Boston: John Wiley & Sons, 2017.
doi:https://doi.org/10.1002/9781118634523.ch1.

3. For a good primer on immunology, albeit one a bit skimpy about vaccines, see: Janeway,
C. A., Walport, M. J., Travers, P., & et al. (1999). *Immunobiology: The Immune System
in Health and Disease* (4th ed.). New York: Elsevier Science Ltd/Garland Publishing;
Sompayrac, L. M. (2019). *How the Immune System Works* (6th ed.). Wiley-Blackwell.

4. Janeway, C. *Immunobiology : the immune system in health and disease.*xix, 635 p. 4th ed.
New York, NY, US: Current Biology Publications; Garland Pub., 1999. text.

5. Bairwa, S. C., Shaw, C. A., Kuo, M., Yoo, J., Tomljenovic, L., & Eidi, H. (2021).
"Cytokines profile in neonatal and adult wild-type mice post-injection of US pediatric
vaccination schedule." *Brain, Behavior, & Immunity-Health*, 15, 100267.

6. Walsh, J. G., Muruve, D. A., and Power, C. "Inflammasomes in the CNS." *Nat Rev
Neurosci,* vol. 15, no. 2 (Feb 2014): 84–97. doi:10.1038/nrn3638 PMID: 24399084.

7. Duncan, J. A., Gao, X., Huang, M. T., O'Connor, B. P., Thomas, C. E., Willingham, S.
B., Bergstralh, D. T., Jarvis, G. A., Sparling, P. F., and Ting, J. P. "Neisseria gonorrhoeae
activates the proteinase cathepsin B to mediate the signaling activities of the NLRP3 and
ASC-containing inflammasome." *J Immunol,* vol. 182, no. 10 (May 15 2009): 6460–9.
doi:10.4049/jimmunol.0802696 PMID: 19414800.; Ichinohe, T., Lee, H. K., Ogura,
Y., Flavell, R., and Iwasaki, A. "Inflammasome recognition of influenza virus is essential for adaptive immune responses." *J Exp Med,* vol. 206, no. 1 (Jan 16 2009): 79–87.
doi:10.1084/jem.20081667 PMID: 19139171.

8. Mariathasan, S., Weiss, D. S., Newton, K., McBride, J., O'Rourke, K., Roose-Girma,
M., Lee, W. P., Weinrauch, Y., Monack, D. M., and Dixit, V. M. "Cryopyrin activates
the inflammasome in response to toxins and ATP." *Nature,* vol. 440, no. 7081 (Mar
9 2006): 228–32. doi:10.1038/nature04515 PMID: 16407890.; Martinon, F., Petrilli,
V., Mayor, A., Tardivel, A., and Tschopp, J. "Gout-associated uric acid crystals activate the NALP3 inflammasome." *Nature,* vol. 440, no. 7081 (Mar 9 2006): 237–41.
doi:10.1038/nature04516 PMID: 16407889.; Halle, A., Hornung, V., Petzold, G. C.,
Stewart, C. R., Monks, B. G., Reinheckel, T., Fitzgerald, K. A., Latz, E., Moore, K. J.,
and Golenbock, D. T. "The NALP3 inflammasome is involved in the innate immune
response to amyloid-beta." *Nat Immunol,* vol. 9, no. 8 (Aug 2008): 857–65. doi:10.1038/
ni.1636 PMID: 18604209.

9. Hornung, V., Bauernfeind, F., Halle, A., Samstad, E. O., Kono, H., Rock, K. L.,
Fitzgerald, K. A., and Latz, E. "Silica crystals and aluminum salts activate the NALP3

inflammasome through phagosomal destabilization." *Nat Immunol,* vol. 9, no. 8 (Aug 2008): 847–56. doi:10.1038/ni.1631 PMID: 18604214.

10. Mariathasan, S., Weiss, D. S., Newton, K., McBride, J., O'Rourke, K., Roose-Girma, M., Lee, W. P., Weinrauch, Y., Monack, D. M., and Dixit, V. M. "Cryopyrin activates the inflammasome in response to toxins and ATP." *Nature.;* Halle, A., Hornung, V., Petzold, G. C., Stewart, C. R., Monks, B. G., Reinheckel, T., Fitzgerald, K. A., Latz, E., Moore, K. J., and Golenbock, D. T. "The NALP3 inflammasome is involved in the innate immune response to amyloid-beta." *Nat Immunol.*; Martinon, F., Burns, K., and Tschopp, J. "The inflammasome: a molecular platform triggering activation of inflammatory caspases and processing of proIL-beta." *Mol Cell,* vol. 10, no. 2 (Aug 2002): 417–26. doi:10.1016/s1097-2765(02)00599-3 PMID: 12191486.

11. Petrilli, V., Papin, S., Dostert, C., Mayor, A., Martinon, F., and Tschopp, J. "Activation of the NALP3 inflammasome is triggered by low intracellular potassium concentration." *Cell Death Differ,* vol. 14, no. 9 (Sep 2007): 1583–9. doi:10.1038/sj.cdd.4402195 PMID: 17599094.; Zhou, R., Yazdi, A. S., Menu, P., and Tschopp, J. "A role for mitochondria in NLRP3 inflammasome activation." *Nature,* vol. 469, no. 7329 (Jan 13 2011): 221–5. doi:10.1038/nature09663 PMID: 21124315; Hoegen, T., Tremel, N., Klein, M., Angele, B., Wagner, H., Kirschning, C., Pfister, H. W., Fontana, A., Hammerschmidt, S., and Koedel, U. "The NLRP3 inflammasome contributes to brain injury in pneumococcal meningitis and is activated through ATP-dependent lysosomal cathepsin B release." *J Immunol,* vol. 187, no. 10 (Nov 15 2011): 5440–51. doi:10.4049/jimmunol.1100790 PMID: 22003197.

12. Halle, A., Hornung, V., Petzold, G. C., Stewart, C. R., Monks, B. G., Reinheckel, T., Fitzgerald, K. A., Latz, E., Moore, K. J., and Golenbock, D. T. "The NALP3 inflammasome is involved in the innate immune response to amyloid-beta." *Nat Immunol.*; Codolo, G., Plotegher, N., Pozzobon, T., Brucale, M., Tessari, I., Bubacco, L., and De Bernard, M. "Triggering of inflammasome by aggregated alpha-synuclein, an inflammatory response in synucleinopathies." *PLoS One,* vol. 8, no. 1 (2013): e55375. doi:10.1371/journal.pone.0055375 PMID: 23383169.

13. Shaw, C. A. *Neural Dynamics of Neurological Disease.*

14. Cherry, J. D., Olschowka, J. A., and O'Banion, M. K. "Are 'resting' microglia more 'm2'?". *Front Immunol,* vol. 5 (2014): 594. doi:10.3389/fimmu.2014.00594 PMID: 25477883.

15. Martinez, F. O., and Gordon, S. "The M1 and M2 paradigm of macrophage activation: time for reassessment." *F1000Prime Rep,* vol. 6 (2014): 13. doi:10.12703/P6-13 PMID: 24669294.; Mosser, D. M., and Edwards, J. P. "Exploring the full spectrum of macrophage activation." *Nat Rev Immunol,* vol. 8, no. 12 (Dec 2008): 958–69. doi:10.1038/nri2448 PMID: 19029990.; Colton, C. A., Mott, R. T., Sharpe, H., Xu, Q., Van Nostrand, W. E., and Vitek, M. P. "Expression profiles for macrophage alternative activation genes in AD and in mouse models of AD." *J Neuroinflammation,* vol. 3 (Sep 27 2006): 27. doi:10.1186/1742-2094-3-27 PMID: 17005052.

16. Fenn, A. M., Hall, J. C., Gensel, J. C., Popovich, P. G., and Godbout, J. P. "IL-4 signaling drives a unique arginase+/IL-1beta+ microglia phenotype and recruits macrophages to the inflammatory CNS: consequences of age-related deficits in IL-4Ralpha after traumatic spinal cord injury." *J Neurosci,* vol. 34, no. 26 (Jun 25 2014): 8904–17. doi:10.1523/JNEUROSCI.1146-14.2014 PMID: 24966389.

17. Martinez, F. O., and Gordon, S. "The M1 and M2 paradigm of macrophage activation: time for reassessment." *F1000Prime Rep*, vol. 6 (2014): 13. doi:10.12703/P6-13 PMID: 24669294.

18. Mosser, D. M., and Edwards, J. P. "Exploring the full spectrum of macrophage activation." *Nat Rev Immunol.*; David, S., and Kroner, A. "Repertoire of microglial and macrophage responses after spinal cord injury." *Nat Rev Neurosci*, vol. 12, no. 7 (Jun 15 2011): 388–99. doi:10.1038/nrn3053 PMID: 21673720.

19. The reasons that replication is not more routinely done basically boils down to time and money versus personal benefit. As important as replication is in principle, the bottom line is that there tends to be little in the way of career advancement for those doing the replication experiments. First, it pretty much does not help with getting an article in a good journal. Next, it does not aid one's chance for promotion by showing that a previous study is correct. In addition, rarely is there money to do so from the various granting agencies, although there have been multiple calls for such a fund to be created. This leaves scientists with the option of using the funds they have for some other studies. Naturally, to do so drains funds from ongoing projects, and there can be serious negative consequences for this. The only way it becomes feasible from both career and financial perspectives to replicate the work of another group is if your results end up calling into question some critical, ideally famous, study. But guessing which such studies are going to fall to your better experimental techniques is a risky business.

20. "Regressive autism occurs when a child appears to develop typically but then starts to lose speech and social skills, typically between the ages of 15 and 30 months, and is subsequently diagnosed with autism."—Regressive Autism. (2020). Retrieved from Wikipedia: https://en.wikipedia.org/wiki/Regressive_autism#:~:text=Regressive%20 autism%20occurs%20when%20a,is%20subsequently%20diagnosed%20with%20 autism.

21. A good overview of the use of placebos in clinical trials can be found here: Gupta, U. & Verma, M. (2013). Placebo in clinical trials. *Perspect Clin Res*, 4(1), 49–52.

22. Neurodyn's testing of Alzheimer's disease drugs uses true placebos in their phase 1 and 2 trials (Neurodyn, CSO Dr. Denis Kay, personal communication).

23. Wilson, J. M., Khabazian, I., Wong, M. C., Seyedalikhani, A., Bains, J. S., Pasqualotto, B. A., Williams, D. E., Andersen, R. J., Simpson, R. J., Smith, R., Craig, U. K., Kurland, L. T., and Shaw, C. A. "Behavioral and neurological correlates of ALS-parkinsonism dementia complex in adult mice fed washed cycad flour." *Neuromolecular Med*, vol. 1, no. 3 (2002): 207–21. doi:10.1385/NMM:1:3:207 PMID: 12095162.

24. Yes, there are a lot of alcohol-related analogies in this book … but we are in the age of COVID-19, where frequent smell and taste testing are important.

25. I tend to use bar analogies because a lot of my writing is done in such places.

26. Taylor, L. E., Swerdfeger, A. L., and Eslick, G. D. "Vaccines are not associated with autism: an evidence-based meta-analysis of case-control and cohort studies." *Vaccine*, vol. 32, no. 29 (Jun 17 2014): 3623–9. doi:10.1016/j.vaccine.2014.04.085 PMID: 24814559.; Iqbal, S., Barile, J. P., Thompson, W. W., and DeStefano, F. "Number of antigens in early childhood vaccines and neuropsychological outcomes at age 7–10 years." *Pharmacoepidemiol Drug Saf*, vol. 22, no. 12 (Dec 2013): 1263–70. doi:10.1002/ pds.3482 PMID: 23847024.; DeStefano, F., Price, C. S., and Weintraub, E. S. "Increasing exposure to antibody-stimulating proteins and polysaccharides in vaccines is not associated with risk of autism." *J Pediatr*, vol. 163, no. 2 (Aug 2013): 561–7. doi:10.1016/j.jpeds.2013.02.001 PMID: 23545349.; Smith, M. J., and Woods, C. R.

"On-time vaccine receipt in the first year does not adversely affect neuropsychological outcomes." *Pediatrics,* vol. 125, no. 6 (Jun 2010): 1134–41. doi:10.1542/peds.2009–2489 PMID: 20498176.

27. Jain, A., Marshall, J., Buikema, A., Bancroft, T., Kelly, J. P., and Newschaffer, C. J. "Autism occurrence by MMR vaccine status among US children with older siblings with and without autism." *JAMA,* vol. 313, no. 15 (Apr 21 2015): 1534–40. doi:10.1001/jama.2015.3077 PMID: 25898051; Klein, N. P. "Vaccine safety in special populations." *Hum Vaccin,* vol. 7, no. 2 (Feb 2011): 269–71. doi:10.4161/hv.7.2.13860 PMID: 21307654.; Mrozek-Budzyn, D., Kieltyka, A., and Majewska, R. "Lack of association between measles-mumps-rubella vaccination and autism in children: a case-control study." *Pediatr Infect Dis J,* vol. 29, no. 5 (May 2010): 397–400. doi:10.1097/INF.0b013e3181c40a8a PMID: 19952979.; Fombonne, E., Zakarian, R., Bennett, A., Meng, L., and Mclean-Heywood, D. "Pervasive developmental disorders in Montreal, Quebec, Canada: prevalence and links with immunizations." *Pediatrics,* vol. 118, no. 1 (Jul 2006): e139-50. doi:10.1542/peds.2005-2993 PMID: 16818529.; Makela, A., Nuorti, J. P., and Peltola, H. "Neurologic disorders after measles-mumps-rubella vaccination." *Pediatrics,* vol. 110, no. 5 (Nov 2002): 957–63. doi:10.1542/peds.110.5.957 PMID: 12415036.; Chakrabarti, S., and Fombonne, E. "Pervasive developmental disorders in preschool children." *JAMA,* vol. 285, no. 24 (Jun 27 2001): 3093–9. doi:10.1001/jama.285.24.3093 PMID: 11427137.; Kaye, J. A., Del Mar Melero-Montes, M., and Jick, H. "Mumps, measles, and rubella vaccine and the incidence of autism recorded by general practitioners: a time trend analysis." *BMJ,* vol. 322, no. 7284 (Feb 24 2001): 460–3. doi:10.1136/bmj.322.7284.460 PMID: 11222420.; Committee to Review Adverse Effects of Vaccines, and Institute of Medicine. *Adverse Effects of Vaccines: Evidence and Causality.* Edited by K. Stratton, A. Ford, E. Rusch and E. W. Clayton. Washington, DC: National Academies Press, 2011. doi:10.17226/13164.

28. Smith, M. J., and Woods, C. R. "On-time vaccine receipt in the first year does not adversely affect neuropsychological outcomes." *Pediatrics.*

29. DeStefano, F., Price, C. S., and Weintraub, E. S. "Increasing exposure to antibody-stimulating proteins and polysaccharides in vaccines is not associated with risk of autism." *J Pediatr.*

30. Iqbal, S., Barile, J. P., Thompson, W. W., and DeStefano, F. "Number of antigens in early childhood vaccines and neuropsychological outcomes at age 7–10 years." *Pharmacoepidemiol Drug Saf.*

31. Taylor, L. E., Swerdfeger, A. L., and Eslick, G. D. "Vaccines are not associated with autism: an evidence-based meta-analysis of case-control and cohort studies." *Vaccine.*

32. Smeeth, L., Cook, C., Fombonne, E., Heavey, L., Rodrigues, L. C., Smith, P. G., and Hall, A. J. "MMR vaccination and pervasive developmental disorders: a case-control study." *Lancet,* vol. 364, no. 9438 (Sep 11–17 2004): 963–9. doi:10.1016/S0140-6736(04)17020-7 PMID: 15364187.

33. *Tic.* Retrieved from Wikipedia, https://en.wikipedia.org/wiki/Tic#:~:text=A%20tic%20is%20a%20sudden,eye%20blinking%20and%20throat%20clearing.

34. Andrews, N., Miller, E., Grant, A., Stowe, J., Osborne, V., and Taylor, B. "Thimerosal exposure in infants and developmental disorders: a retrospective cohort study in the United kingdom does not support a causal association." *Pediatrics,* vol. 114, no. 3 (Sep 2004): 584–91. doi:10.1542/peds.2003-1177-L PMID: 15342825.

35. Ibid.

36. Smeeth, L., Cook, C., Fombonne, E., Heavey, L., Rodrigues, L. C., Smith, P. G., and Hall, A. J. "MMR vaccination and pervasive developmental disorders: a case-control study." *Lancet.*

37. Wakefield, A. J., Murch, S. H., Anthony, A., Linnell, J., Casson, D. M., Malik, M., Berelowitz, M., Dhillon, A. P., Thomson, M. A., Harvey, P., Valentine, A., Davies, S. E., and Walker-Smith, J. A. "Ileal-lymphoid-nodular hyperplasia, non-specific colitis, and pervasive developmental disorder in children." *Lancet,* vol. 351, no. 9103 (Feb 28 1998): 637–41. doi:10.1016/s0140-6736(97)11096–0 PMID: 9500320.

38. Taylor, B., Miller, E., Farrington, C. P., Petropoulos, M. C., Favot-Mayaud, I., Li, J., and Waight, P. A. "Autism and measles, mumps, and rubella vaccine: no epidemiological evidence for a causal association." *Lancet,* vol. 353, no. 9169 (Jun 12 1999): 2026–9. doi:10.1016/s0140-6736(99)01239–8 PMID: 10376617.

39. Ibid.

40. Madsen, K. M., Hviid, A., Vestergaard, M., Schendel, D., Wohlfahrt, J., Thorsen, P., Olsen, J., and Melbye, M. "A population-based study of measles, mumps, and rubella vaccination and autism." *N Engl J Med,* vol. 347, no. 19 (Nov 7 2002): 1477–82. doi:10.1056/NEJMoa021134 PMID: 12421889.

41. In Denmark, children only receive the second MMR at age twelve, thus out of the study period.

42. Wilson, K., Mills, E., Ross, C., McGowan, J., and Jadad, A. "Association of autistic spectrum disorder and the measles, mumps, and rubella vaccine: a systematic review of current epidemiological evidence." *Arch Pediatr Adolesc Med,* vol. 157, no. 7 (Jul 2003): 628–34. doi:10.1001/archpedi.157.7.628 PMID: 12860782.

43. Committee to Review Adverse Effects of Vaccines, and Institute of Medicine. *Adverse Effects of Vaccines: Evidence and Causality.* Edited by K. Stratton, A. Ford, E. Rusch and E. W. Clayton.

44. Committee to Review Adverse Effects of Vaccines, and Institute of Medicine. "Preface." In *Adverse Effects of Vaccines: Evidence and Causality.* edited by K. Stratton, A. Ford, E. Rusch and E. W. Clayton.pg. ix. Washington, DC: National Academies Press, 2011. https://www.ncbi.nlm.nih.gov/books/NBK190024/.

45. Ibid.

46. Ibid.

47. Committee to Review Adverse Effects of Vaccines, and Institute of Medicine. "Evaluating Biological Mechanisms of Adverse Events." In *Adverse Effects of Vaccines: Evidence and Causality.* edited by K. Stratton, A. Ford, E. Rusch and E. W. Clayton.pg. 71. Washington, DC: National Academies Press, 2011. https://www.ncbi.nlm.nih.gov /books/NBK190024/.

48. Committee to Review Adverse Effects of Vaccines, and Institute of Medicine. "Measles, Mumps, and Rubella Vaccine." In *Adverse Effects of Vaccines: Evidence and Causality.* edited by K. Stratton, A. Ford, E. Rusch and E. W. Clayton.pg. 153. Washington, DC: National Academies Press, 2011. https://www.ncbi.nlm.nih.gov/books/NBK190024/.

49. Lee, S. H. *From Pap smear to HPV vaccine: the cervical cancer prevention industry.*pages cm?. Obstetrics and gynecology advances women's issues. New York: Nova Science Publishers, 2021.; Inbar, R., Weiss, R., Tomljenovic, L., Arango, M. T., Deri, Y., Shaw, C. A., Chapman, J., Blank, M., and Shoenfeld, Y. "Behavioral abnormalities in female mice following administration of aluminum adjuvants and the human papillomavirus (HPV) vaccine Gardasil." *Immunol Res,* vol. 65, no. 1 (Feb 2017): 136–49. doi:10.1007/ s12026-016-8826-6 PMID: 27421722.; Tomljenovic, L., Wilyman, J., Vanamee, E.,

Bark, T., and Shaw, C. A. "HPV vaccines and cancer prevention, science versus activism." *Infect Agent Cancer,* vol. 8, no. 1 (Feb 1 2013): 6. doi:10.1186/1750-9378-8-6 PMID: 23369430.; Tomljenovic, L., and Shaw, C. A. "Too fast or not too fast: the FDA's approval of Merck's HPV vaccine Gardasil." *J Law Med Ethics,* vol. 40, no. 3 (Fall 2012): 673–81. doi:10.1111/j.1748-720X.2012.00698.x PMID: 23061593.; Tomljenovic, L., Spinosa, J. P., and Shaw, C. A. "Human papillomavirus (HPV) vaccines as an option for preventing cervical malignancies: (how) effective and safe?". *Curr Pharm Des,* vol. 19, no. 8 (2013): 1466–87. https://www.ncbi.nlm.nih.gov/pubmed/23016780 -PMID: 23016780.; Tomljenovic, L., and Shaw, C. A. "No autoimmune safety signal after vaccination with quadrivalent HPV vaccine Gardasil?". *J Intern Med,* vol. 272, no. 5 (Nov 2012): 514–5; author reply 16. doi:10.1111/j.1365–2796.2012.02551.x PMID: 22540172.; Tomljenovic, L., and Shaw, C. A. "Human papillomavirus (HPV) vaccine policy and evidence-based medicine: are they at odds?". *Ann Med,* vol. 45, no. 2 (Mar 2013): 182–93. doi:10.3109/07853890.2011.645353 PMID: 22188159.; Tomljenovic, L., and Shaw, C. A. "Mandatory HPV Vaccination." *JAMA,* vol. 307, no. 3 (2012): 252–55. doi:10.1001/jama.2011.2020 %J JAMA.; Tomljenovic, L., and Shaw, C. A. "Death after Quadrivalent Human Papillomavirus (HPV) Vaccination: Causal or Coincidental?". *Pharmaceutical Regulatory Affairs,* vol. S12, no. 1 (2012). doi:10.4172/2167–7689.S12-001 PMID: 22188159.

50. Committee to Review Adverse Effects of Vaccines, and Institute of Medicine. "Diphtheria Toxoid–, Tetanus Toxoid–, and Acellular Pertussis– Containing Vaccines." In *Adverse Effects of Vaccines: Evidence and Causality.* edited by K. Stratton, A. Ford, E. Rusch and E. W. Clayton. pg. 525. Washington, DC: National Academies Press, 2011. https://www.ncbi.nlm.nih.gov/books/NBK190024/, https://www.ncbi.nlm.nih.gov/books/NBK -190024/pdf/Bookshelf_NBK190024.pdf.

51. Committee to Review Adverse Effects of Vaccines, and Institute of Medicine. "Increased Susceptibility." In *Adverse Effects of Vaccines: Evidence and Causality.* edited by K. Stratton, A. Ford, E. Rusch and E. W. Clayton. pg. 83. Washington, DC: National Academies Press, 2011. https://www.ncbi.nlm.nih.gov/pubmed/24624471.

52. *Vaccine Safety Publications.* (2020). From "Vaccine Safety". Retrieved from US Centers for Disease Control and Prevention National Center for Emerging and Zoonotic Infectious Diseases (NCEZID), Division of Healthcare Quality Promotion (DHQP), US Department of Health & Human Services, https://www.cdc.gov/vaccinesafety /research/publications/index.html.

53. Before going into this list of publications, it may be important to keep in mind that the CDC is a branch of the Department of Health and Human Services (HHS), which holds vaccine-related patents that may number in the dozens or more. Many of these patents relate to methods. Mark Blaxill, cited in GreenMedInfo (https: //www.greenmedinfo.com/blog/herbal-cream-clears-877-hpv-cases-naturallyclaims), claims that HHS had fifty-seven such patents, but that number may be dated by any newer ones associated with COVID-19 (https://www.google.com/search?tbo=p&tb -m=pts&hl=en&q=vaccine+inassignee:centers+inassignee:for+inassignee:disease+i -nassignee:control&tbs=,ptss:g&num=100As). Further, individual members of ACIP, the advisory body for the CDC that approves vaccines for use in the US, hold many more (https://www.lawfirms.com/resources/environment/environment-health/cdc-mem- bers-own-more-50-patents-connected-vaccinations). Even a very pro-vaccine website notes CDC/HHS patent holdings in this regard (https://vaxopedia.org/2018/05/19 /does-the-cdc-own-any-patents-on-vaccines/).

54. Ironically, VAERS is the same system that some bloggers tend to mock when independent scientists use it. For example, Dr. David Gorski, whom we will meet in a later chapter, terms the use of VAERS "dumpster diving."

55. CDC. (2021). From "Morbidity and Mortality Weekly Report (MMWR)". Retrieved from US Centers for Disease Control and Prevention, U.S. Department of Health & Human Services, https://www.cdc.gov/mmwr/index.html.

56. Darwish, A. A. "Presenting the case for an immunization safety surveillance system." *Bull World Health Organ,* vol. 78, no. 2 (2000): 216. https://www.ncbi.nlm.nih.gov/pubmed/10743287 PMID: 10743287.

57. Ward, B. J. "Vaccine adverse events in the new millennium: is there reason for concern?". *Bull World Health Organ,* vol. 78, no. 2 (2000): 205–15. https://www.ncbi.nlm.nih.gov/pubmed/10743286 PMID: 10743286.

58. Ibid, 212.

59. Ibid, 216.

60. The Institute was founded by the UN Development Program. Clemens was soon to more formally join the WHO.

61. Clemens, J. "Evaluating vaccine safety before and after licensure." *Bull World Health Organ,* vol. 78, no. 2 (2000): 218–9. https://www.ncbi.nlm.nih.gov/pubmed/10743289 PMID: 10743289.

62. Brown, P. "A view from the media on vaccine safety : round table discussion / Phyllida Brown." *Bulletin of the World Health Organization : the International Journal of Public Health 2000 ; 78(2) : 216–218,* (2000 2000): 217. https://apps.who.int/iris/handle/10665/57199.

63. Ivinson, A. J. "Concern, but not with surveillance." *Bull World Health Organ,* vol. 78, no. 2 (2000): 222–3. https://www.ncbi.nlm.nih.gov/pubmed/10743292 PMID:10743292.

64. Takahashi, H., Arai, S., Tanaka-Taya, K., and Okabe, N. "Autism and infection/immunization episodes in Japan." *Jpn J Infect Dis,* vol. 54, no. 2 (Apr 2001): 78–9. https://www.ncbi.nlm.nih.gov/pubmed/11427748 PMID: 11427748.

65. Chen, R. T., DeStefano, F., Davis, R. L., Jackson, L. A., Thompson, R. S., Mullooly, J. P., Black, S. B., Shinefield, H. R., Vadheim, C. M., Ward, J. I., and Marcy, S. M. "The Vaccine Safety Datalink: immunization research in health maintenance organizations in the USA." *Bull World Health Organ,* vol. 78, no. 2 (2000): 186–94. https://www.ncbi.nlm.nih.gov/pubmed/10743283 PMID: 10743283.

66. Lindsey, N. P., Rabe, I. B., Miller, E. R., Fischer, M., and Staples, J. E. "Adverse event reports following yellow fever vaccination, 2007–13." *J Travel Med,* vol. 23, no. 5 (May 2016). doi:10.1093/jtm/taw045 PMID: 27378369.

67. Loughlin, A. M., Marchant, C. D., Adams, W., Barnett, E., Baxter, R., Black, S., Casey, C., Dekker, C., Edwards, K. M., Klein, J., Klein, N. P., Larussa, P., Sparks, R., and Jakob, K. "Causality assessment of adverse events reported to the Vaccine Adverse Event Reporting System (VAERS)." *Vaccine,* vol. 30, no. 50 (Nov 26 2012): 7253–9. doi:10.1016/j.vaccine.2012.09.074 PMID: 23063829.

68. Barile, J. P., Kuperminc, G. P., Weintraub, E. S., Mink, J. W., and Thompson, W. W. "Thimerosal exposure in early life and neuropsychological outcomes 7–10 years later." *J Pediatr Psychol,* vol. 37, no. 1 (Jan-Feb 2012): 106–18. doi:10.1093/jpepsy/jsr048 PMID: 21785120.

69. Donahue, J. G., Kieke, B. A., King, J. P., DeStefano, F., Mascola, M. A., Irving, S. A., Cheetham, T. C., Glanz, J. M., Jackson, L. A., Klein, N. P., Naleway, A. L., Weintraub,

E., and Belongia, E. A. "Association of spontaneous abortion with receipt of inactivated influenza vaccine containing H1N1pdm09 in 2010–11 and 2011–12." *Vaccine,* vol. 35, no. 40 (Sep 25 2017): 5314–22. doi:10.1016/j.vaccine.2017.06.069 PMID: 28917295.

70. Taylor, L. E., Swerdfeger, A. L., and Eslick, G. D. "Vaccines are not associated with autism: an evidence-based meta-analysis of case-control and cohort studies." *Vaccine.*

71. Yamamoto-Hanada, K., Pak, K., Saito-Abe, M., Yang, L., Sato, M., Mezawa, H., Sasaki, H., Nishizato, M., Konishi, M., Ishitsuka, K., Matsumoto, K., Saito, H., Ohya, Y., Japan, E., and Children's Study, G. "Cumulative inactivated vaccine exposure and allergy development among children: a birth cohort from Japan." *Environ Health Prev Med,* vol. 25, no. 1 (Jul 7 2020): 27. doi:10.1186/s12199-020-00864-7 PMID: 32635895.

72. Ibid, p. 37.

73. Lyons-Weiler, J., and Ricketson, R. "Reconsideration of the immunotherapeutic pediatric safe dose levels of aluminum." *Journal of Trace Elements in Medicine and Biology,* vol. 48 (2018/07/01/ 2018): 67–73. doi:https://doi.org/10.1016/j.jtemb.2018.02.025.

74. Lyons-Weiler, J. "Systematic review of historical epidemiologic studies influencing public health policies on vaccination." *Institute for Pure and Applied Knowledge,* (2018). http://ipaknowledge.org/resources/LYONSWEILERSYSTREVIEW.pdf.

75. Many of the studies on the toxicity of ethyl mercury and Thimerosal can be found in: Kennedy Jr., R. F. (Ed.). (2015). *Thimerosal: Let the Science Speak.* New York: Skyhorse Publishing.

76. A definition of a "cluster fuck": a disastrously mishandled situation or undertaking. It's a common military saying.

77. *MMR Vaccine and Thimerosal-Containing Vaccines Are Not Associated With Autism, IOM Report Says.* (2004). Retrieved from National Academies of Sciences, Engineering, Medicine, https://www.nationalacademies.org/news/2004/05/mmr-vaccine-and-thimerosal-containing-vaccines-are-not-associated-with-autism-iom-report-says; https://www.nationalacademies.org/news/2004/05/mmr-vaccine-and-thimerosal-containing-vaccines-are-not-associated-with-autism-iom-report-says; https://www.nap.edu/catalog/10997/immunization-safety-review-vaccines-and-autism?onpi_newsdoc05182004=.

78. Lyons-Weiler, J. "Balance of Risk in COVID-19 Reveals the Extreme Cost of False Positives." *International Journal of Vaccine Theory, Practice, and Research,* vol. 1, no. 2 (January 05 2021): 209–22. https://ijvtpr.com/index.php/IJVTPR/article/view/15/24.

79. Lyons-Weiler, J. "Plan B Public Health Infrastructure and Operations Oversight Reform for America." *International Journal of Vaccine Theory, Practice, and Research,* vol. 1, no. 2 (January 05 2021): 283–94. https://ijvtpr.com/index.php/IJVTPR/article/view/19/20.

80. Hooker, B., Kern, J., Geier, D., Haley, B., Sykes, L., King, P., and Geier, M. "Methodological issues and evidence of malfeasance in research purporting to show thimerosal in vaccines is safe." *Biomed Res Int,* vol. 2014 (2014): 247218. doi:10.1155/2014/247218 PMID: 24995277.

81. Goldman, G., and Yazbak, F. "An investigation of the association between MMR vaccination and autism in Denmark." *Journal of American Physicians Surgeons,* vol. 9 (2004): 70–75. https://jpands.org/vol9no3/goldman.pdf.

82. Skowronski, D. M., Hamelin, M. E., De Serres, G., Janjua, N. Z., Li, G., Sabaiduc, S., Bouhy, X., Couture, C., Leung, A., Kobasa, D., Embury-Hyatt, C., De Bruin, E., Balshaw, R., Lavigne, S., Petric, M., Koopmans, M., and Boivin, G. "Randomized controlled ferret study to assess the direct impact of 2008–09 trivalent inactivated influenza vaccine on A(H1N1)pdm09 disease risk." *PLoS One,* vol. 9, no. 1 (2014): e86555. doi:10.1371/journal.pone.0086555 PMID: 24475142.

83. Kennedy, S. *The Tuskegee Syphilis Study (1932–1972).* (2014). Retrieved from IMARC, https://www.imarcresearch.com/blog/bid/351998/The-Tuskegee-Syphilis-Study -1932-1972.

84. Callaway, E. "Dozens to be deliberately infected with coronavirus in UK 'human challenge' trials." *Nature,* vol. 586, no. 7831 (Oct 2020): 651–52. doi:10.1038/d41586-020-02821-4 PMID: 33082550.

85. *CDC Works With Global Partners to End Cholera* (2020). From "CDC at Work: Cholera". Retrieved from US Centers for Disease Control and Prevention, U.S. Department of Health & Human Services, https://www.cdc.gov/cholera/ending-cholera.html.

86. Van Panhuis, W. G., Grefenstette, J., Jung, S. Y., Chok, N. S., Cross, A., Eng, H., Lee, B. Y., Zadorozhny, V., Brown, S., Cummings, D., and Burke, D. S. "Contagious diseases in the United States from 1888 to the present." *N Engl J Med,* vol. 369, no. 22 (Nov 28 2013): 2152–8. doi:10.1056/NEJMms1215400 PMID: 24283231.

87. Leslie, S. *Project Tycho™: Data for Health.* (2013). From "Library Blog". Retrieved from Georgia State University, https://blog.library.gsu.edu/2013/12/05/project-tycho %e2%84%a2-data-for-health.

88. *Death Rates for Selected Causes by 10-Year Age Groups, Race, and Sex: Death Registration States, 1900–32, and United States, 1933–98.* (2015). Edited by National Center for Health Statistics. From "National Vital Statistics System Mortality Tables HIST290". Retrieved from US Centers for Disease Control and Prevention, U.S. Department of Health & Human Services, https://www.cdc.gov/nchs/nvss/mortality/hist290.htm.

89. https://www.census.gov/library/publications/1975/compendia/hist_stats_colonial -1970.html.

90. We also realized at this point that we were likely to annoy both anti- and pro-vaccine camps, since we had demonstrated that some vaccines work to some extent, just not all.

91. Personal communication.

92. Shaw, C. A. *Neural Dynamics of Neurological Disease.* Boston: John Wiley & Sons, 2017. doi:https://doi.org/10.1002/9781118634523.ch1.

93. Ioannidis, J. P. "Why most published research findings are false." *PLoS Med,* vol. 2, no. 8 (Aug 2005): e124. doi:10.1371/journal.pmed.0020124 PMID: 16060722.

94. Mawson, A., Ray, B., Bhuiyan, A. R., and Jacob, B. J. F. I. P. H. "Vaccination and Health Outcomes: A Survey of 6- to 12-year-old Vaccinated and Unvaccinated Children based on Mothers Reports." vol. 4 (2016).

95. Hooker, B. S., and Miller, N. Z. "Analysis of health outcomes in vaccinated and unvaccinated children: Developmental delays, asthma, ear infections and gastrointestinal disorders." *SAGE Open Med,* vol. 8 (2020): 2050312120925344. doi:10.1177/2050312120925344 PMID: 32537156.

96. Lyons-Weiler, J., and Thomas, P. "Relative Incidence of Office Visits and Cumulative Rates of Billed Diagnoses Along the Axis of Vaccination." *Int J Environ Res Public Health,* vol. 17, no. 22 (Nov 22 2020). doi:10.3390/ijerph17228674 PMID: 33266457.

97. J. Lyons-Weiler, personal communication.

98. *Vaccines.* (2021). From "Vaccines, Blood & Biologics". Retrieved from US Food and Drug Administration, U.S. Department of Health & Human Services, https://www.fda .gov/vaccines-blood-biologics/vaccines.

99. *Vaccine Safety.* (2020). From "Vaccine Basics". Retrieved from Vaccines.gov, U.S. Department of Health & Human Services, https://www.vaccines.gov/basics/safety.

100. https://www.cnbc.com/2021/04/19/fda-asks-emergent-plant-to-pause-manufacturing -while-it-investigates-botched-covid-vaccines.html#:~:text=The%20Food%20and%20 -Drug%20Administration,in%20a%20regulatory%20filing%20Monday.

101. Gatti, A. M., and Montanari, S. "New quality-control investigations on vaccines: micro-and nanocontamination." *Int J Vaccines Vaccin*, vol. 4, no. 1 (2017): 00072. https://siksik.org/wp-content/uploads/vaccins/02-2017-Medcrave-Nanocontamina-tion.pdf.; *Summary of data confirmations through interlaboratory analysis.* (2019). From "Vaccinegate". Retrieved from Corvelva, https://www.corvelva.it/en/speciale-corvelva /vaccinegate/sommario-delle-conferme-dei-dati-tramite-analisi-interlaboratorio.html.

102. Lyons-Weiler, J., and Ricketson, R. "Reconsideration of the immunotherapeutic pedi-atric safe dose levels of aluminum." *J Trace Elem Med Biol*, vol. 48 (Jul 2018): 67–73. doi:10.1016/j.jtemb.2018.02.025 PMID: 29773196.

103. Haseltine, W. A. *The FDA Will Not Inspect Vaccine Production Plants.* (2020). From "Healthcare". Retrieved from Forbes, https://www.forbes.com/sites/williamhasel -tine/2020/10/29/fda-will-not-inspect-vaccine-production-plants/?sh=564d82794616.

104. *CBER Offices & Divisions.* (2018). From "Vaccines, Blood & Biologics". Retrieved from US Food and Drug Administration, Center for Biologics Evaluation and Research (CBER), https://www.fda.gov/about-fda/fda-organization/center-biologics-evaluation -and-research-cber.

105. *CLASSIFICATION OF AEFIS.* From "MODULE 3: Adverse events following immu-nization". Retrieved from World Health Organization, Vaccine Safety Basics, https: //vaccine-safety-training.org/classification-of-aefis.html.

106. *Vaccine Safety.* (2020). From "Vaccine Basics". Retrieved from Vaccines.gov, U.S. Department of Health & Human Services, https://www.vaccines.gov/basics/safety.

107. On July 1, 1946, the then-called Communicable Disease Center (CDC) opened its doors and occupied one floor of a small building in Atlanta. It later changed its name to the Centers for Disease Control and Prevention.

108. *Ensuring the Safety of Vaccines in the United States.* (2018). From "Understanding Vaccines and Vaccine Safety". Retrieved from US Centers for Disease Control and Prevention, National Center for Immunization and Respiratory Diseases, https://www .fda.gov/files/vaccines,%20blood%20&%20biologics/published/Ensuring-the-Safety -of-Vaccines-in-the-United-States.pdf, https://www.cdc.gov/vaccines/hcp/conversations /ensuring-safe-vaccines.html.

109. Tomljenovic, L., Tarsell, E., Garrett, J., Shaw, C. A., and Holland, M. S. "'Serious' under-reporting? An examination of vaccine adverse event reporting after quadrivalent human papillomavirus (QHPV) vaccination." (2021).

110. *Electronic Support for Public Health - Vaccine Adverse Event Reporting System (ESP:VAERS) (Massachusetts).* (2007). Edited by Harvard Pilgrim Healthcare Inc. From "Digital Healthcare Research". Retrieved from US Department of Health & Human Services, Agency for Healthcare Research and Quality, https://digital.ahrq.gov/ahrq-funded-projects /electronic-support-public-health-vaccine-adverse-event-reporting-system.

111. *Ensuring the Safety of Vaccines in the United States.* (2018). From "Understanding Vaccines and Vaccine Safety". Retrieved from US Centers for Disease Control and Prevention, National Center for Immunization and Respiratory Diseases, https://www.fda.gov/files /vaccines,%20blood%20&%20biologics/published/Ensuring-the-Safety-of-Vaccines -in-the-United-States.pdf, https://www.cdc.gov/vaccines/hcp/conversations/ensuring-safe -vaccines.html.

112. Ibid.

113. Ibid.

114. Ibid.

115. *Advisory Committee on Immunization Practices (ACIP).* (2021). Retrieved from US Centers for Disease Control and Prevention, https://www.cdc.gov/vaccines/acip/index .html.

116. *Ensuring the Safety of Vaccines in the United States.* (2018). From "Understanding Vaccines and Vaccine Safety". Retrieved from US Centers for Disease Control and Prevention, National Center for Immunization and Respiratory Diseases, https://www.fda.gov/files /vaccines,%20blood%20&%20biologics/published/Ensuring-the-Safety-of-Vaccines -in-the-United-States.pdf, https://www.cdc.gov/vaccines/hcp/conversations/ensuring-safe -vaccines.html.

117. Ibid.

118. Ibid.

119. Macdonald, N. E., and Law, B. J. "Canada's eight-component vaccine safety system: A primer for health care workers." *Paediatr Child Health,* vol. 22, no. 4 (Jul 2017): e13- e16. doi:10.1093/pch/pxx073 PMID: 29507505.

120. *Testing, approval and monitoring.* (2020). From "ImmunizeBC - Safety". Retrieved from BC Centre for Disease Control, https://immunizebc.ca/testing-approval-and-monitoring.

121. Ibid.

122. *Vaccine safety and pharmacovigilance: Canadian Immunization Guide.* (2020). From "Canadian Immunization Guide: Part 2 - Vaccine Safety". Retrieved from Government of Canada, https://www.canada.ca/en/public-health/services/publications/healthy-living /canadian-immunization-guide-part-2-vaccine-safety/page-2-vaccine-safety.html.

123. Ismail, S. J., Langley, J. M., Harris, T. M., Warshawsky, B. F., Desai, S., and Farhangmehr, M. "Canada's National Advisory Committee on Immunization (NACI): evidence-based decision-making on vaccines and immunization." *Vaccine,* vol. 28 Suppl 1 (Apr 19 2010): A58-63. doi:10.1016/j.vaccine.2010.02.035 PMID: 20412999.

124. *The National Advisory Committee on Immunization (NACI).* (2020). From "About Us". Retrieved from Immunize Canada, https://immunize.ca/about-us.

125. *Vaccine safety and pharmacovigilance: Canadian Immunization Guide.* (2020). From "Canadian Immunization Guide: Part 2 - Vaccine Safety". Retrieved from Government of Canada, https://www.canada.ca/en/public-health/services/publications/healthy-living /canadian-immunization-guide-part-2-vaccine-safety/page-2-vaccine-safety.html.

126. *Canadian Immunization Guide: Table of updates.* (2020). From "Vaccines and immuniza- tion". Retrieved from Government of Canada, https://www.canada.ca/en/public-health /services/canadian-immunization-guide/updates.html.

127. "CIOMS is to advance public health through guidance on health research including ethics, medical product development and safety." CIOMS is an international nongov- ernmental organization established jointly by World Health Organization (WHO) and United Nations Educational, Scientific and Cultural Organization (UNESCO).

128. Kohl, K. S., Bonhoeffer, J., Braun, M. M., Chen, R. T., Duclos, P., Heijbel, H., Heininger, U., Loupi, E., and Marcy, S. M. "The Brighton Collaboration: Creating a Global Standard for Case Definitions (and Guidelines) for Adverse Events Following Immunization." In *Advances in Patient Safety: From Research to Implementation (Volume 2: Concepts and Methodology).* edited by K. Henriksen, J. B. Battles, E. S. Marks and D. I. Lewin. Advances in Patient Safety. Rockville (MD), 2005. https://www.ncbi.nlm.nih .gov/pubmed/21249832.

129. *Institute of Medicine (IOM) Reports on Vaccine Safety.* (2020). From "Vaccine Safety". Retrieved from US Centers for Disease Control and Prevention National Center for Emerging and Zoonotic Infectious Diseases (NCEZID) Division of Healthcare Quality Promotion (DHQP), US Department of Health & Human Services, https://www.cdc.gov/vaccinesafety/research/iomreports/index.html.

130. *Robert F. Kennedy Jr. vs Alan Dershowitz: The Great Vaccine Debate!* (2020). Retrieved from Children's Health Defense, https://childrenshealthdefense.org/transcripts/robert-f-kennedy-jr-vs-alan-dershowitz-the-great-vaccine-debate/.

131. Gatti, A. M., and Montanari, S. "New quality-control investigations on vaccines: micro- and nanocontamination." *Int J Vaccines Vaccin.; Summary of data confirmations through interlaboratory analysis.* (2019). From "Vaccinegate". Retrieved from Corvelva.

132. *What You Need to Know About the National Vaccine Injury Compensation Program (VICP).* (2019). Retrieved from Health Resources & Services Administration, https://www.hrsa.gov/sites/default/files/hrsa/vaccine-compensation/resources/about-vaccine-injury-compensation-program-booklet.pdf.

133. *Covered Vaccines.* (2020). From "National Vaccine Injury Compensation Program". Retrieved from Health Resources & Services Administration, https://www.hrsa.gov/vaccine-compensation/covered-vaccines/index.html.

134. Mikovits, J., Heckenlively, K., and Kennedy, R. J. F. *Plague of Corruption: Restoring Faith in the Promise of Science.* Children's Health Defense. Skyhorse Publishing, 2020. https://www.skyhorsepublishing.com/9781510752245/plague-of-corruption/.

135. Ibid, p. 203.

136. *Government of Canada Announces pan-Canadian Vaccine Injury Support Program.* (2020). From "Public Health Agency of Canada". Retrieved from Government of Canada, https://www.canada.ca/en/public-health/news/2020/12/government-of-canada-announces-pan-canadian-vaccine-injury-support-program.html.

Chapter 4

1. Crichton, M (1942–2008). *Aliens Cause Global Warming.* (2003). Retrieved from Caltech Michelin Lecture, https://stephenschneider.stanford.edu/Publications/PDF_Papers/Crichton2003.pdf.; Crichton was a popular science fiction author. Before this, he had been a practicing physician. His comments on consensus and science were written in the context of climate change. He might have been wrong about this, but he was not wrong about his contention that claiming the "science is settled" about climate change, vaccines, or anything else is a profoundly nonscientific thing to argue, and one that pretty much shows that the speaker has no real idea how science is supposed to work. There are, no doubt, some who will see this quote and be instantly convinced that citing Crichton for an unpopular opinion on climate change means that those skeptical of vaccines are also climate change "denialists" and/or part of the alt-right. Such a position is merely part of the collection of ad hominem slurs that are used to discredit people, not based on their actual views or knowledge, but based on some perceived positions or associations. My only advice to such people is to grow a skeptical brain and stop acting like everything anybody has ever said about any subject has to match your own narrow interpretations of science and history.

2. Key concerns in Jenner's day in the resistance to vaccines.

3.   *Jonas Salk - American physician and medical researcher.* Retrieved from Encyclopaedia Britannica, https://www.britannica.com/biography/Jonas-Salk.

4.   *Albert Bruce Sabin - American physician and microbiologist.* Retrieved from Encyclopaedia Britannica, https://www.britannica.com/biography/Albert-Bruce-Sabin.

5.   Hansen, K. *Faces of Healthcare: What's a Gastroenterologist?* Edited by Timothy J. Legg. Retrieved from Healthline, https://www.healthline.com/find-care/articles/gastroenterologists /what-is-a-gastroenterologist.

6.   A "perfect storm" is one in which a series of minor events combines to become something far more dangerous than any of the initial parts.

7.   A frequent insult aimed at vaccine safety researchers is to call them something like "the new Wakefield." I've had this epithet tossed at me numerous times, and each time I have to stop and think whether I should be offended or simply say, "Thank you."

8.   Alli, R. A. *What Does the Word 'Autism' Mean?* (2019). From "Autism". Retrieved from WebMD, https://www.webmd.com/brain/autism/what-does-autism-mean.

9.   Kanner, L. "Autistic disturbances of affective contact." *Nervous Child,* vol. 2, no. 3 (1943): 217–50.

10. Manouilenko, I., and Bejerot, S. "Sukhareva-Prior to Asperger and Kanner." *Nord J Psychiatry,* vol. 69, no. 6 (Aug 2015): 479–82. doi:10.3109/08039488.2015.1005022 PMID: 25826582.

11. This improvement may not be particularly surprising given the innate "neuroplasticity" of the nervous system, particularly in young children or animals. As discussed in some detail in my 2017 book, *Neural Dynamics of Neurological Disease,* cited elsewhere in this book, plasticity to neuronal injury can occur in several ways, depending on both the type of insult and the age at which such insults occur. For example, early gene mutations, if not fatal, may be compensated for by redundant functions of other genes; toxin damage, if not too severe, may allow other regions of the CNS to compensate by processes described as restitution and substitution. Plasticity in the CNS is not, however, a simple process, and while the impacts of sensory deprivation or unilateral stroke can be overcome by system redundancy if caught early, disruption of developmental programs such as the orderly sequence of events of neuronal migration, differentiation, and synaptic pruning are less likely to be adequately compensated for. In other words, in the latter case, as in autism, while the neuroplasticity capacity is there, recovery to reset the disruption on the programmed developmental sequence may not fully occur. In my book, I cited references to a "recovery continuum," in which the variables contributing to the different levels of recovery include the following: the site, nature, and severity of the injury; genetic susceptibility factors; sex; and, most important of all, age at insult in relation to the stage of CNS development. The fact that most of the children observed by Kanner came from relatively well-off families in which the parents were highly educated may suggest that the prospects for autism recovery, at least in part, were more likely. Much of what Kanner describes in his article would seem to fit into a form of compensatory neuroplasticity.

12. Kanner, L. "Autistic disturbances of affective contact." *Nervous Child,* vol. 2, no. 3 (1943): 249.

13. Oller, J.W, Oller S.D., Autism: the diagnosis, treatment, and etiology of the undeniable epidemic. James and Bartlett Learning. 2009.

14. *Asperger Syndrome.* Retrieved from Autism Speaks Canada, https://www.autismspeaks .ca/about-autism/what-is-autism/asperger-syndrome/#:~:text=What%20is%20Asperger -%20Syndrome%3F,interests%20and%2For%20repetitive%20behaviors.

15. Al Backer, N. B. "Developmental regression in autism spectrum disorder." *Sudan J Paediatr,* vol. 15, no. 1 (2015): 21–6. https://www.ncbi.nlm.nih.gov/pubmed/27493417 -PMID: 27493417.

16. *What is Rett syndrome?* From "About Rett Syndrome". Retrieved from Rettsyndrome. org, https://www.rettsyndrome.org/about-rett-syndrome/what-is-rett-syndrome/#:~: -text=Rett%20syndrome%20is%20a%20rare,near%20constant%20repetitive%20 -hand%20movements.

17. Rapin, I. "Autism." *N Engl J Med,* vol. 337, no. 2 (Jul 10 1997): 97–104. doi:10.1056 /NEJM199707103370206 PMID: 9211680.

18. Fombonne, E. "Epidemiological surveys of autism and other pervasive developmental disorders: an update." *J Autism Dev Disord,* vol. 33, no. 4 (Aug 2003): 365–82. doi:10.1023/a:1025054610557 PMID: 12959416.; Fombonne, E. "The epidemiology of autism: a review." *Psychol Med,* vol. 29, no. 4 (Jul 1999): 769–86. doi:10.1017/ s0033291799008508 PMID: 10473304

19. Tuchman, R., and Rapin, I. "Epilepsy in autism." *Lancet Neurol,* vol. 1, no. 6 (Oct 2002): 352–8. doi:10.1016/s1474-4422(02)00160-6 PMID: 12849396.

20. Ashwood, P., Wills, S., and Van De Water, J. "The immune response in autism: a new frontier for autism research." *J Leukoc Biol,* vol. 80, no. 1 (Jul 2006): 1–15. doi:10.1189/jlb.1205707 PMID: 16698940.; Jass, J. R. "The intestinal lesion of autistic spectrum disorder." *Eur J Gastroenterol Hepatol,* vol. 17, no. 8 (Aug 2005): 821–2. doi:10.1097/00042737-200508000-00007 PMID: 16003130.; Balzola, F., Barbon, V., Repici, A., Rizzetto, M., Clauser, D., Gandione, M., and Sapino, A. "Panenteric IBD-like disease in a patient with regressive autism shown for the first time by the wireless capsule enteroscopy: another piece in the jigsaw of this gut-brain syndrome?". *Am J Gastroenterol,* vol. 100, no. 4 (Apr 2005): 979–81. doi:10.1111/j.1572–0241.2005.41202_4.x PMID: 15784047.; White, J. F. "Intestinal pathophysiology in autism." *Exp Biol Med (Maywood),* vol. 228, no. 6 (Jun 2003): 639–49. doi:10.1177/153537020322800601 PMID: 12773694.; Torrente, F., Ashwood, P., Day, R., Machado, N., Furlano, R. I., Anthony, A., Davies, S. E., Wakefield, A. J., Thomson, M. A., Walker-Smith, J. A., and Murch, S. H. "Small intestinal enteropathy with epithelial IgG and complement deposition in children with regressive autism." *Mol Psychiatry,* vol. 7, no. 4 (2002): 375–82, 34. doi:10.1038/sj.mp.4001077 PMID: 11986981.; Chen, J. A., Penagarikano, O., Belgard, T. G., Swarup, V., and Geschwind, D. H. "The emerging picture of autism spectrum disorder: genetics and pathology." *Annu Rev Pathol,* vol. 10 (2015): 111–44. doi:10.1146/annurev-pathol-012414-040405 PMID: 25621659.

21. Newschaffer, C. J., Croen, L. A., Daniels, J., Giarelli, E., Grether, J. K., Levy, S. E., Mandell, D. S., Miller, L. A., Pinto-Martin, J., Reaven, J., Reynolds, A. M., Rice, C. E., Schendel, D., and Windham, G. C. "The epidemiology of autism spectrum disorders." *Annu Rev Public Health,* vol. 28 (2007): 235–58. doi:10.1146/annurev. publhealth.28.021406.144007 PMID: 17367287.

22. For details, see Shaw, 2017.

23. Zablotsky, B., Black, L. I., and Blumberg, S. J. "Estimated Prevalence of Children With Diagnosed Developmental Disabilities in the United States, 2014–2016." *NCHS Data Brief,* no. 291 (Nov 2017): 1–8. https://www.ncbi.nlm.nih.gov/pubmed/29235982 PMID: 29235982.

24. *Data & Statistics on Autism Spectrum Disorder.* (2020). From "Autism Spectrum Disorder (ASD)". Retrieved from US Centers for Disease Control and Prevention, https://www .cdc.gov/ncbddd/autism/data.html.

25. Rogers, T. M. "The Political Economy of Autism." pg. 12. Doctor of Philosophy Ph.D. thesis for University of Sydney, 2018. Retrieved from http://hdl.handle.net/2123/20198.

26. Edwards, A. *Global forced displacement hits record high.* (2016). Retrieved from UNHCR, http://www.unhcr.org/en-us/news/latest/2016/6/5763b65a4/global-forced-displacement -hits-record-high.html.

27. SNP: single nucleotide polymorphism: basically a replacement of one nucleotide by another.

28. *Genetic drift.* (2020). Edited by BiologyOnline Editors. Retrieved from Biology Online, https://www.biologyonline.com/dictionary/genetic-drift#:~:text=Genetic%20drift%20 -(biology%20definition)%3A,the%20so%2Dcalled%20bottleneck%20effect.

29. Bacchelli, E., and Maestrini, E. "Autism spectrum disorders: molecular genetic advances." *Am J Med Genet C Semin Med Genet,* vol. 142C, no. 1 (Feb 15 2006): 13–23. doi:10.1002/ajmg.c.30078 PMID: 16419096.; Lauritsen, M. B., Pedersen, C. B., and Mortensen, P. B. "Effects of familial risk factors and place of birth on the risk of autism: a nationwide register-based study." *J Child Psychol Psychiatry,* vol. 46, no. 9 (Sep 2005): 963–71. doi:10.1111/j.1469–7610.2004.00391.x PMID: 16108999.; Klauck, S. M. "Genetics of autism spectrum disorder." *Eur J Hum Genet,* vol. 14, no. 6 (Jun 2006): 714–20. doi:10.1038/sj.ejhg.5201610 PMID: 16721407.

30. Yuen, R. K., Thiruvahindrapuram, B., Merico, D., Walker, S., Tammimies, K., Hoang, N., Chrysler, C., Nalpathamkalam, T., Pellecchia, G., Liu, Y., Gazzellone, M. J., D'abate, L., Deneault, E., Howe, J. L., Liu, R. S., Thompson, A., Zarrei, M., Uddin, M., Marshall, C. R., Ring, R. H., Zwaigenbaum, L., Ray, P. N., Weksberg, R., Carter, M. T., Fernandez, B. A., Roberts, W., Szatmari, P., and Scherer, S. W. "Whole-genome sequencing of quartet families with autism spectrum disorder." *Nat Med,* vol. 21, no. 2 (Feb 2015): 185–91. doi:10.1038/nm.3792 PMID: 25621899.

31. Vanzo, R. J., Prasad, A., Staunch, L., Hensel, C. H., Serrano, M. A., Wassman, E. R., Kaplun, A., Grandin, T., and Boles, R. G. "The Temple Grandin Genome: Comprehensive Analysis in a Scientist with High-Functioning Autism." *J Pers Med,* vol. 11, no. 1 (Dec 29 2020). doi:10.3390/jpm11010021 PMID: 33383702.

32. Hunter, J. W., Mullen, G. P., Mcmanus, J. R., Heatherly, J. M., Duke, A., and Rand, J. B. "Neuroligin-deficient mutants of C. elegans have sensory processing deficits and are hypersensitive to oxidative stress and mercury toxicity." *Dis Model Mech,* vol. 3, no. 5–6 (May-Jun 2010): 366–76. doi:10.1242/dmm.003442 PMID: 20083577.

33. Giulivi, C., Zhang, Y. F., Omanska-Klusek, A., Ross-Inta, C., Wong, S., Hertz-Picciotto, I., Tassone, F., and Pessah, I. N. "Mitochondrial dysfunction in autism." *JAMA,* vol. 304, no. 21 (Dec 1 2010): 2389–96. doi:10.1001/jama.2010.1706 PMID: 21119085.

34. Shaw, C. A. *Neural Dynamics of Neurological Disease.* Boston: John Wiley & Sons, 2017. doi:https://doi.org/10.1002/9781118634523.ch1.

35. Ibid.

36. Lewis, D. L. "The Autism Biosolids Conundrum." *International Journal of Vaccine Theory, Practice, and Research,* vol. 1, no. 1 (July 15 2020): 51–74. https://ijvtpr.com/index.php/IJVTPR /article/view/4/12.

37. Shaw, C. A. *Neural Dynamics of Neurological Disease.*

38. Summarized in Shaw, 2017, ibid

39. *Diagnostic and Statistical Manual of Mental Disorders (DSM–5).* (2020). Retrieved from American Psychiatric Association, https://www.psychiatry.org/psychiatrists/practice/dsm.

40. Adapted from Lord, C. and Jones, R. M., "Annual research review: re-thinking the classification of autism spectrum disorders." *J Child Psychol Psychiatry*, vol. 53, no. 5 (2012): 490–509. doi:10.1111/j.1469–7610.2012.02547.x. PMID: 22486486.

41. Rogers, T. M. "The Political Economy of Autism." pg. 11. Doctor of Philosophy Ph.D. thesis for University of Sydney, 2018. Retrieved from http://hdl.handle.net/2123/20198.

42. Hansen, S. N., Schendel, D. E., and Parner, E. T. "Explaining the increase in the prevalence of autism spectrum disorders: the proportion attributable to changes in reporting practices." *JAMA Pediatr,* vol. 169, no. 1 (Jan 2015): 56–62. doi:10.1001/jamapediatrics.2014.1893 PMID: 25365033.

43. Angelidou, A., Asadi, S., Alysandratos, K. D., Karagkouni, A., Kourembanas, S., and Theoharides, T. C. "Perinatal stress, brain inflammation and risk of autism-review and proposal." *BMC Pediatr,* vol. 12 (Jul 2 2012): 89. doi:10.1186/1471-2431-12-89 PMID: 22747567.; Theoharides, T. C., and Zhang, B. "Neuro-inflammation, blood-brain barrier, seizures and autism." *J Neuroinflammation,* vol. 8 (Nov 30 2011): 168. doi:10.1186/1742-2094-8-168 PMID: 22129087.; Pardo, C. A., Vargas, D. L., and Zimmerman, A. W. "Immunity, neuroglia and neuroinflammation in autism." *Int Rev Psychiatry,* vol. 17, no. 6 (Dec 2005): 485–95. doi:10.1080/02646830500381930 PMID: 16401547.; Vargas, D. L., Nascimbene, C., Krishnan, C., Zimmerman, A. W., and Pardo, C. A. "Neuroglial activation and neuroinflammation in the brain of patients with autism." *Ann Neurol,* vol. 57, no. 1 (Jan 2005): 67–81. doi:10.1002/ana.20315 PMID: 15546155.

44. Bennie, M. *What is Neurodiversity?* (2016). From "Articles & Blogs". Retrieved from Autism Awareness Center Inc., https://autismawarenesscentre.com/un-adopts-new-goals -disabilities/.

45. *Advisory Committee on Immunization Practices (ACIP).* (2021). Retrieved from US Centers for Disease Control and Prevention, https://www.cdc.gov/vaccines/acip/index .html.

46. Miller, N. Z., and Goldman, G. S. "Infant mortality rates regressed against number of vaccine doses routinely given: is there a biochemical or synergistic toxicity?". *Hum Exp Toxicol,* vol. 30, no. 9 (Sep 2011): 1420–8. doi:10.1177/0960327111407644 PMID: 21543527.

47. *COVID-19 ACIP Vaccine Recommendations.* (2020). From "Vaccine-Specific Recommendations". Retrieved from US Centers for Disease Control and Prevention, https://www.cdc.gov/vaccines/hcp/acip-recs/vacc-specific/covid-19.html.

48. Petrik, M. S., Wong, M. C., Tabata, R. C., Garry, R. F., and Shaw, C. A. "Aluminum adjuvant linked to Gulf War illness induces motor neuron death in mice." *Neuromolecular Med,* vol. 9, no. 1 (2007): 83–100. doi:10.1385/nmm:9:1:83 PMID: 17114826.

49. Admin. *Thoughtful House changes name and focus.* (2011). From "Autism News Science and Opinion". Retrieved from Left Brain/Right Brain, https://leftbrainrightbrain.co.uk /2011/07/21/thoughtful-house-changes-name-focus/.

50. Apoptosis is one of the forms of cell death, termed programmed cell death, and it has a number of distinguishing features from other forms, the latter including necrosis and a hybrid form called necroptosis. All of these forms of cell death are common in neurological diseases, although which one is primary in any of these is still a subject of debate. See: Morrice, J. R., Gregory-Evans, C. Y., & Shaw, C. A. (2017) Necroptosis in amyotrophic lateral sclerosis and other neurological disorders. *Biochim Biophys Acta Mol Basis Dis,* 1863(2), 347–53.

51. I'm guessing the investigator who left was Dr. Laura Hewitson, the lead author on two subsequent papers about this experiment.

52. Dwoskin Family Foundation, a former philanthropic organization founded by Claire Dwoskin and her ex-husband, Al Dwoskin. The Foundation funded critical vaccine studies, particularly on the subjects of aluminum toxicity and autoimmune reactions, both potentially linked to vaccines.

53. Prof. Yehuda Shonefeld is a physician and autoimmune specialist who runs a clinic in Tel Aviv, the Zabludowicz Center for Autoimmune Diseases, at the Sheba Medical Center. He is the author of well over 1,900 publications (at last count). Prof. Roman Gherardi is at the Université de Paris Est, where he deals with patients suffering from neuromuscular disorders, as well as a disorder he first identified, macrophagic myofasciitis (MMF). He works closely with a senior colleague, Prof. Jerome Authier.

54. AutismOne, https://www.autismone.org/content/autismone-0.

55. Most of the academic units based at the Royal Free campus were previously part of the Royal Free Hospital School of Medicine, which, in August 1998, merged with University College London (UCL). The Royal Free Campus is now one of the UCL Medical School's main teaching and research sites.

56. Deer, B. "How the case against the MMR vaccine was fixed." BMJ, vol. 342 (Jan 5 2011): c5347. doi:10.1136/bmj.c5347 PMID: 21209059.

57. The stages in university faculty rankings in the United Kingdom differ from those in the United States and Canada. Here is the equivalence: professor (same in both), reader (more like an associate professor), senior lecturer (somewhere between an associate and assistant professor), lecturer (assistant professor), and assistant lecturer (no real equivalent rank).

58. Wakefield, A. J., Murch, S. H., Anthony, A., Linnell, J., Casson, D. M., Malik, M., Berelowitz, M., Dhillon, A. P., Thomson, M. A., Harvey, P., Valentine, A., Davies, S. E., and Walker-Smith, J. A. "Ileal-lymphoid-nodular hyperplasia, non-specific colitis, and pervasive developmental disorder in children." Lancet, vol. 351, no. 9103 (Feb 28 1998): 637–41. doi:10.1016/s0140-6736(97)11096-0 PMID: 9500320.

59. Thompson, N. P., Montgomery, S. M., Pounder, R. E., and Wakefield, A. J. "Is measles vaccination a risk factor for inflammatory bowel disease?". Lancet, vol. 345, no. 8957 (Apr 29 1995): 1071–4. doi:10.1016/s0140-6736(95)90816-1 PMID: 7715338.

60. Prof. John Walker-Smith of the Royal Free Hospital was, at the time, probably the most famous specialist in GI pathologies in the United Kingdom.

61. Enterocolitis, an inflammation of the small intestine and colon.

62. Hall, C. Consultant's theory sparked the debate over separate jabs. (2002). Retrieved from The Telegraph, https://www.telegraph.co.uk/news/uknews/1384004/Consultants-theory-sparked-the-debate-over-separate-jabs.html.

63. New Research links Autism and Bowel Disease. (1998). Retrieved from The Royal Free Hospital School of Medicine, https://briandeer.com/mmr/royal-free-press-1998.pdf.

64. Shakespeare, W., Mowat, B. A., and Werstine, P. The tragedy of Julius Caesar. Act 3, Scene 1, Line 273. The New Folger Library Shakespeare. Simon & Schuster pbk. ed. New York: Simon & Schuster Paperbacks, 2011.

65. D'Eufemia, P., Celli, M., Finocchiaro, R., Pacifico, L., Viozzi, L., Zaccagnini, M., Cardi, E., and Giardini, O. "Abnormal intestinal permeability in children with autism." Acta Paediatr, vol. 85, no. 9 (Sep 1996): 1076–9. doi:10.1111/j.1651-2227.1996.tb14220.x PMID: 8888921.; Goodwin, M. S., Cowen, M. A., and Goodwin, T. C. "Malabsorption and cerebral dysfunction: a multivariate and comparative study of autistic children." J

Autism Child Schizophr, vol. 1, no. 1 (Jan-Mar 1971): 48–62. doi:10.1007/BF01537742 PMID: 5172439.

66. Wakefield, A. J., Anthony, A., Murch, S. H., Thomson, M., Montgomery, S. M., Davies, S., O'Leary, J. J., Berelowitz, M., and Walker-Smith, J. A. "Enterocolitis in children with developmental disorders." *Am J Gastroenterol,* vol. 95, no. 9 (Sep 2000): 2285–95. doi:10.1111/j.1572–0241.2000.03248.x PMID: 11007230.

67. Uhlmann, V., Martin, C. M., Sheils, O., Pilkington, L., Silva, I., Killalea, A., Murch, S. B., Walker-Smith, J., Thomson, M., Wakefield, A. J., and O'Leary, J. J. "Potential viral pathogenic mechanism for new variant inflammatory bowel disease." *Mol Pathol,* vol. 55, no. 2 (Apr 2002): 84–90. doi:10.1136/mp.55.2.84 PMID: 11950955.

68. Torrente, F., Ashwood, P., Day, R., Machado, N., Furlano, R. I., Anthony, A., Davies, S. E., Wakefield, A. J., Thomson, M. A., Walker-Smith, J. A., and Murch, S. H. "Small intestinal enteropathy with epithelial IgG and complement deposition in children with regressive autism." *Mol Psychiatry.*; Uhlmann, V., Martin, C. M., Sheils, O., Pilkington, L., Silva, I., Killalea, A., Murch, S. B., Walker-Smith, J., Thomson, M., Wakefield, A. J., and O'Leary, J. J. "Potential viral pathogenic mechanism for new variant inflammatory bowel disease." *Mol Pathol.*; Ashwood, P., Anthony, A., Torrente, F., and Wakefield, A. J. "Spontaneous mucosal lymphocyte cytokine profiles in children with autism and gastrointestinal symptoms: mucosal immune activation and reduced counter regulatory interleukin-10." *J Clin Immunol,* vol. 24, no. 6 (Nov 2004): 664–73. doi:10.1007/s10875-004-6241-6 PMID: 15622451.; Ashwood, P., Anthony, A., Pellicer, A. A., Torrente, F., Walker-Smith, J. A., and Wakefield, A. J. "Intestinal lymphocyte populations in children with regressive autism: evidence for extensive mucosal immunopathology." *J Clin Immunol,* vol. 23, no. 6 (Nov 2003): 504–17. doi:10.1023/b:joci.0000010427.05143. bb PMID: 15031638.; Horvath, K., Papadimitriou, J. C., Rabsztyn, A., Drachenberg, C., and Tildon, J. T. "Gastrointestinal abnormalities in children with autistic disorder." *J Pediatr,* vol. 135, no. 5 (Nov 1999): 559–63. doi:10.1016/s0022-3476(99)70052–1 PMID: 10547242.

69. A short retraction of the interpretation of the original data by 10 of the 12 coauthors of the paper was on March 6th, 2004. Interestingly, the lead on this "retraction" was none other than Simon Murch, one of those investigated by the GMC. See: Sathyanarayana Rao, T. S., & Andrade, C. (2011). The MMR vaccine and autism: Sensation, refutation, retraction, and fraud. Indian J Psychiatry, 53(2), 95–96.

70. *Vaccine Safety: Evaluating The Science Conference.* (2011). Retrieved from http://www.vaccinesafetyconference.com/index.html.

71. Dr. Chris Exley is one of the most highly regarded aluminum scientists in the world with many publications in the peer-reviewed literature.; https://www.keele.ac.uk/aluminium/groupmembers/chrisexley/.

72. Wakefield, A. J., dir. *Vaxxed: From Cover-Up to Catastrophe.* (2016; GathrFilms, 2017), Documentary. IMDB: https://www.imdb.com/title/tt5562652/.; Burrowes, B., dir. *Vaxxed II: The People's Truth.* (2019; Busch Media Group, 2020), Documentary. IMDB: https://www.imdb.com/title/tt11137248/.; Bailey, M., dir. *The Pathological Optimist.* (2017; GathrFilms, 20174), Documentary. IMDB: https://www.imdb.com/title/tt6391732/.

73. Fraser, L. *Anti-MMR doctor is forced out.* (2001). Retrieved from The Telegraph, https://www.telegraph.co.uk/news/uknews/1364080/Anti-MMR-doctor-is-forced-out.html.

74. Deer, B. *About Brian Deer.* (2021). Retrieved from Brian Deer–Award Winning Investigations, https://briandeer.com/about-us.htm.

75. Cassell, W., Jr. *How Rupert Murdoch Became a Media Tycoon.* (2020). From "RICH & POWERFUL". Retrieved from Investopedia, https://www.investopedia.com/articles/investing/083115/how-rupert-murdoch-became-media-tycoon.asp#:~:text=Murdoch's-%20media%20empire%20includes%20Fox,Disney%20Company%20for%20-%2471.3%20billion.

76. Allegedly, according to a Canary Party article, *Sunday Times* Editor Paul Nuki told Brian Deer: "Find something big." Deer responded: "About what?" "MMR?" was the reply. This exchange was later confirmed by Deer in his article in the *BMJ* 340 in 2010, "Reflections on investigating Wakefield."

According to this post, Nuki had a direct family tie to a government employee responsible for MMR safety. The article went on to claim that "Paul Nuki is the son of Professor George Nuki who sat on the Committee on Safety in Medicines when it passed [the] Pluserix MMR vaccine as safe for use in 1987." See: VanDerHorst-Larson, J. (2012). Canary Party letter to officials at La Crosse University, WI, regarding Brian Deer. Retrieved from The Canary Party: https://us2.campaign-archive.com/?u=b62698a50aececa2aded9f56b&id=0a8ef0cb42&e=9d349a3300.

Is this story true? It is hard to be sure, but it does seem to fit with other allegations of an agenda by the *Sunday Times*. However, the story is understandably denied by Brian Deer himself: Deer, B. (2012). "Jennifer Larson: food for the birds." Retrieved from Brian Deer: Award-Winning Investigations: https://briandeer.com/solved/vanderhorst-larson.htm.

77. Dr. David Lewis and Mary Holland: Lewis was a senior scientist with the Environmental Protection Agency (EPA) who was initially fired by the EPA following a dispute about the safety of what is euphemistically called "biosludge," also known as biosolids, a dismissal that was later overturned in the courts – Lewis v. EPA, U.S. Department of Labor, Office of Administrative Law Judges, Case No. 97-CAA-7. "In retaliation over my 1996 Nature commentary, which discussed gaps in the science EPA uses to support a number of regulations, the DOL found: 'Dr. David Lewis was discriminated against by EPA's inquiry into ethics violations at the highest levels and communicating these allegations to members of Congress.' Settlement: EPA paid $115,000 ($75,000 in attorneys' fees, $40,000 to plaintiff), and issued a letter stating that I did not violate the Hatch Act or any ethics rules as EPA had falsely claimed." Lewis was, and remains, of the view that biosludge is a major health hazard. Lewis had been put through the wringer by EPA until lawsuits launched by him forced EPA to reinstate him.

Mary Holland is a New York lawyer, now with Children's Health Defense, who has written on the Deer accusations against Wakefield.

78. Deer, B. "How the case against the MMR vaccine was fixed." *BMJ.*

79. Deer, B. "Secrets of the MMR scare. How the vaccine crisis was meant to make money." *BMJ*, vol. 342 (Jan 11 2011): c5258. doi:10.1136/bmj.c5258 PMID: 21224310.

80. Deer, B. "Secrets of the MMR scare. The Lancet's two days to bury bad news." *BMJ*, vol. 342 (Jan 18 2011): c7001. doi:10.1136/bmj.c7001 PMID: 21245118.

81. Deer, B. "Pathology reports solve 'new bowel disease' riddle." *BMJ*, vol. 343 (Nov 9 2011): d6823. doi:10.1136/bmj.d6823 PMID: 22077090.; Deer, B. *Andrew Wakefield concocts a conspiracy.* (2010). From "Andrew Wakefield: the fraud investigation". Retrieved from Brian Deer–Award Winning Investigations, https://briandeer.com/solved/tall-story.htm.

82. Godlee, F. "Institutional and editorial misconduct in the MMR scare." *BMJ*, vol. 342 (2011): d378. doi:10.1136/bmj.d378 %J BMJ.; Godlee, F., Smith, J., and Marcovitch,

H. "Wakefield's article linking MMR vaccine and autism was fraudulent." *BMJ*, vol. 342 (Jan 5 2011): c7452. doi:10.1136/bmj.c7452 PMID: 21209060.

83. I experienced the "extra authors" phenomenon in several publications I had written about my own work on glutathione in collaboration with a Hungarian neuroscientist, Dr. Reka Janáky. One of the final revisions of the manuscript with her came back with two additional authors whom I didn't know: one was the head of the laboratory where Janáky worked; the other was the laboratory head's secretary and mistress. Neither had done anything for the manuscript besides add their names. Hence, is it possible that many of the authors on Wakefield et al. simply signed off on the manuscript just to get their name on a paper that might gain a lot of attention? Yes, at that time certainly. Is that what happened? There is no way to know.

84. Wakefield, A. J., and Fudenberg, H. H. 1998. "Pharmaceutical composition for treatment of MMR virus mediated disease comprising a transfer factor obtained from the dialysis of virus-specific lymphocytes." United Kingdom. Assignee: Royal Free Hospital School of Medicine. Patent: GB 2 325 856 A. Filed (June 04, 1998) and issued (December 09, 1998). URL: https://briandeer.com/mmr/1998-vaccine-patent.pdf, https://patentim -ages.storage.googleapis.com/43/4a/98/cd9945c9084be1/GB2325856A.pdf.

85. Hall, C. *Consultant's theory sparked the debate over separate jabs.* (2002). Retrieved from *The Telegraph.*; Holland, M. "Who is Dr. Andrew Wakefield?". Chap. 25 In *Vaccine epidemic: how corporate greed, biased science, and coercive government threaten our human rights, our health, and our children.* edited by Louise Kuo Habakus and Mary Holland. pg. 311–20. New York: Skyhorse, 2011. https://vaxxedthemovie.com/wp-content /uploads/2016/04/Who-is-Dr.-Andrew-Wakefield-by-Mary-Holland-JD.pdf.

86. Deer, B. *Transfer factor scam in autism.* (2017). Retrieved from Brian Deer–Award Winning Investigations, https://briandeer.com/wakefield/transfer-factor.htm.

87. Deer, B. *The Doctor Who Fooled the World: Science, Deception, and the War on Vaccines.* 408. Baltimore: Johns Hopkins University Press, 2020. text. https://jhupbooks.press.jhu .edu/title/doctor-who-fooled-world.

88. From their website: https://www.gmc-uk.org/; "…formed in 1858. We work to protect patient safety and support medical education and practice across the UK. We do this by working with doctors, employers, educators, patients and other key stakeholders in the UK's healthcare systems… Before we were formed under the Medical Act, there were 19 bodies regulating the UK medical profession."

89. *Fitness to Practice Hearing* before Fitness to Practise Panel applying the General Medical Council's Preliminary Proceedings Committee and Professional Conduct Committee (Procedure) Rules 1988. (2010), January 28, 2010. (Statement of Dr. Andrew J. Wakefield, Prof. John A. Walker-Smith, Prof. Simon H. Murch). pg 2.

90. Ibid.

91. Boseley, S. *Andrew Wakefield struck off register by General Medical Council.* (2010). Retrieved from The Guardian, https://www.theguardian.com/society/2010/may/24 /andrew-wakefield-struck-off-gmc.

92. From Wikipedia: "The High Court of Justice in London, together with the Court of Appeal and the Crown Court, are the Senior Courts of England and Wales. The High Court deals at first instance with all high value and high importance civil law (non-criminal) cases, and also has a supervisory jurisdiction over all subordinate courts and tribunals, with a few statutory exceptions." https://en.wikipedia.org/wiki/High_Court_of_Justice.

93. *Walker-Smith v. General Medical Council* Attorneys: Mr. Stephen Miller QC and Ms. Andrea Lindsay-Strugo. EWHC 503 (England and Wales High Court

(Administrative Court), 2012). Retrieved from https://www.casemine.com/judgement/uk/5a8ff7d760d03e7f57eb269c.

94. Ibid.

95. Ibid.

96. Ibid.

97. *Appendix 1: The UK High Court Decision in the appeal by Professor John Walker-Smith.* (2018). From "Current Controversies". Retrieved from Alliance for Human Research Protection, https://ahrp.org/laffaire-wakefield-appendix-1-the-uk-high-court-decision-in-the-appeal-by-professor-john-walker-smith/.

98. Cirstea, M. S., et al. (2020). 'Microbiota Composition and Metabolism are Associated With Gut Function in Parkinson's Disease.' *Mov Disord*, 35(7), 1208–17; Dysbiosis and microbiome, definitions: Dysbiosis is a term for a microbial imbalance or maladaptation on or inside the body, such as an impaired microbiota; Microbiome is a term for the collection of beneficial and pathological microbiota whose actiosn lead to either healthy outcomes or a microbial imbalance or maladaptation on or inside the body leading to some disease state.

99. Parkinson, J. "An essay on the shaking palsy. 1817." *J Neuropsychiatry Clin Neurosci*, vol. 14, no. 2 (Spring 2002): 223–36; discussion 22. doi:10.1176/jnp.14.2.223 PMID: 11983801.

100. Wakefield, A. J., and Mccarthy, J. *Callous Disregard: Autism and Vaccines- The Truth Behind a Tragedy.* New York: Skyhorse, 2017. https://books.google.ca/books?id=uGeCDwAAQBAJ.

101. Deer, B. *The Doctor Who Fooled the World: Science, Deception, and the War on Vaccines.*

102. Ellis, R., Levs, J., and Hamasaki, S. *Outbreak of 51 measles cases linked to Disneyland.* (2015). Retrieved from CNN, https://www.cnn.com/2015/01/21/health/disneyland-measles/index.html.

103. *Measles.* (2019). From "Newsroom-Fact sheets". Retrieved from World Health Organization, https://www.who.int/news-room/fact-sheets/detail/measles.

104. Pager, T., and Mays, J. C. *New York Declares Measles Emergency, Requiring Vaccinations in Parts of Brooklyn.* (2019). Retrieved from the New York Times, https://www.nytimes.com/2019/04/09/nyregion/measles-vaccination-williamsburg.html.

105. Bowler, J. *The Latest Reported Measles Outbreak Could Basically Be Blamed on Andrew Wakefield.* (2019). From "Health". Retrieved from Science Alert, https://www.sciencealert.com/latest-measles-outbreak-in-canada-could-be-traced-to-andrew-wakefield.

106. Green, E. *Measles Can Be Contained. Anti-Semitism Cannot.* (2019). From "Politics". Retrieved from The Atlantic, https://www.theatlantic.com/politics/archive/2019/05/orthodox-jews-face-anti-semitism-after-measles-outbreak/590311/.

107. *Problem Reaction Solution.* Retrieved from Ethics Wiki, https://ethics.wikia.org/wiki/Problem_Reaction_Solution.

108. *Measles morbillivirus.* (2021). Retrieved from Wikipedia, https://en.wikipedia.org/wiki/Measles_morbillivirus.; *Measles.* (2013). From "Biologicals >Vaccine Standardization". Retrieved from World Health Organization, https://www.who.int/biologicals/vaccines/measles/en/#:~:text=Measles%20vaccines&text=Many%20of%20the%20-attenuated%20strains,(Ji%2D191)%20strains.

109. Rosenwald, M. S. *Columbus brought measles to the New World. It was a disaster for Native Americans.* (2019). From "Retropolis". Retrieved from the Washington Post, https://www.washingtonpost.com/history/2019/05/05/columbus-brought-measles-new-world-it-was-disaster-native-americans/.

110. *Measles.* (2020). From "Diseases & Conditions". Retrieved from Mayo Clinic, https://www.mayoclinic.org/diseases-conditions/measles/symptoms-causes/syc-20374857.

111. *Ebola virus disease.* From "Health Topics". Retrieved from World Health Organization, https://www.who.int/health-topics/ebola/.; *Plague.* From "Health Topics". Retrieved from World Health Organization, https://www.who.int/health-topics/plague#tab=tab_1.

112. *Measles Epidemiological Summary, British Columbia, 2019 year to date – July 26th.* (2019). Retrieved from BC Centre for Disease Control, http://www.bccdc.ca/resource-gallery/Documents/Measles%20BC%20epi%20summary%202019%20YTD.pdf.

113. The likely reason for this was that the province's attorney general, David Eby, is a former head of the BC Civil Liberties Association and must have realized that a mandate would backfire with endless lawsuits.

114. Dr. Bonnie Henry, personal communication.

115. Data not available, but for comparison, see the numbers for the US for the same period, still significant at, about 11%: Patel, M., et al. (2019). Increase in Measles Cases—United States, January 1–April 26, 2019. MMWR Morb Mortal Wkly Rep, 68, 402–04. Retrieved from Centers for Disease Control & Prevention: https://www.cdc.gov/mmwr/volumes/68/wr/pdfs/mm6817e1-H.pdf.

116. Aliferis, L. *Disneyland Measles Outbreak Hits 59 Cases And Counting.* (2015). From "Public Health". Retrieved from NPR, https://www.npr.org/sections/health-shots/2015/01/22/379072061/disneyland-measles-outbreak-hits-59-cases-and-counting.

117. *Vital Statistics of the United States* (1964). Chapter "Volume II, Mortality, Part A". Retrieved from US Centers for Disease Control and Prevention, National Center for Health Statistics, https://www.cdc.gov/nchs/data/vsus/VSUS_1962_2A.pdf.

118. CDC. (2020). Questions About Measles. Retrieved from Centers for Disease Control and Prevention: https://www.cdc.gov/measles/about/faqs.html—in the FAQ section, "How common was measles before the vaccine?" it states that four million caught the disease each year, and four hundred died—thus a fatality rate of one in ten thousand.

119. Dr. Bob Sears, personal communication.

120. US Department of Health and Human Services. (2021). VAERS Data. Retrieved from Vaccine Adverse Event Reporting System: https://vaers.hhs.gov/data.html; M. Kuo, personal communication; The CDC data list 111 deaths due to measles complications from 1990 to 2018. In the same time period, VAERS data show 148 deaths attributed to the measles vaccines. Adjustments have to be made for two factors in trying to determine the risk from either the disease or the MMR vaccine. The first adjustment is that MMR uptake depends on state and county, but the number of children so vaccinated is routinely much higher than those who are not vaccinated. The second factor is that the data from VAERS are underreported by a factor of approximately one hundred. See Harvard Pilgrim Healthcare Inc. (2010). Electronic Support for Public Health – Vaccine Adverse Event Reporting System (ESP:VAERS). Massachusetts: Agency for Healthcare Research and Quality. Retrieved from https://digital.ahrq.gov/ahrq-funded-projects/electronic-support-public-health-vaccine-adverse-event-reporting-system.

121. Paunio, M., Hedman, K., Davidkin, I., Valle, M., Heinonen, O. P., Leinikki, P., Salmi, A., and Peltola, H. "Secondary measles vaccine failures identified by measurement of IgG avidity: high occurrence among teenagers vaccinated at a young age." *Epidemiol Infect,* vol. 124, no. 2 (Apr 2000): 263–71. doi:10.1017/s0950268899003222 PMID: 10813152.

122. Naniche, D., Garenne, M., Rae, C., Manchester, M., Buchta, R., Brodine, S. K., and Oldstone, M. B. "Decrease in measles virus-specific CD4 T cell memory in vaccinated

subjects." *J Infect Dis,* vol. 190, no. 8 (Oct 15 2004): 1387–95. doi:10.1086/424571 PMID: 15378430.

123. Di Pietrantoni, C., Rivetti, A., Marchione, P., Debalini, M. G., and Demicheli, V. "Vaccines for measles, mumps, rubella, and varicella in children." *The Cochrane Database of Systematic Reviews,* vol. 4 (Apr 20 2020): CD004407. doi:10.1002/14651858. CD004407.pub4 PMID: 32309885.

124. T. Bark, personal communication.

125. De Vries, R. D., McQuaid, S., Van Amerongen, G., Yuksel, S., Verburgh, R. J., Osterhaus, A. D., Duprex, W. P., and De Swart, R. L. "Measles immune suppression: lessons from the macaque model." *PLoS Pathog,* vol. 8, no. 8 (2012): e1002885. doi:10.1371/journal. ppat.1002885 PMID: 22952446.

126. Mina, M. J., Metcalf, C. J., De Swart, R. L., Osterhaus, A. D., and Grenfell, B. T. "Long-term measles-induced immunomodulation increases overall childhood infectious disease mortality." *Science,* vol. 348, no. 6235 (May 8 2015): 694–9. doi:10.1126/science .aaa3662 PMID: 25954009.

127. Measles leads to immune suppression generally.

128. Jennewein, M. F., Goldfarb, I., Dolatshahi, S., Cosgrove, C., Noelette, F. J., Krykbaeva, M., Das, J., Sarkar, A., Gorman, M. J., Fischinger, S., Boudreau, C. M., Brown, J., Cooperrider, J. H., Aneja, J., Suscovich, T. J., Graham, B. S., Lauer, G. M., Goetghebuer, T., Marchant, A., Lauffenburger, D., Kim, A. Y., Riley, L. E., and Alter, G. "Fc Glycan-Mediated Regulation of Placental Antibody Transfer." *Cell,* vol. 178, no. 1 (Jun 27 2019): 202–15 e14. doi:10.1016/j.cell.2019.05.044 PMID: 31204102.; Waaijenborg, S., Hahne, S. J., Mollema, L., Smits, G. P., Berbers, G. A., Van Der Klis, F. R., De Melker, H. E., and Wallinga, J. "Waning of maternal antibodies against measles, mumps, rubella, and varicella in communities with contrasting vaccination coverage." *J Infect Dis,* vol. 208, no. 1 (Jul 2013): 10–6. doi:10.1093/infdis/jit143 PMID: 23661802.; Leuridan, E., Hens, N., Hutse, V., Ieven, M., Aerts, M., and Van Damme, P. "Early waning of maternal measles antibodies in era of measles elimination: longitudinal study." *BMJ,* vol. 340 (May 18 2010): c1626. doi:10.1136/bmj.c1626 PMID: 20483946.; Ohsaki, M., Tsutsumi, H., Takeuchi, R., Kuniya, Y., and Chiba, S. "Reduced passive measles immunity in infants of mothers who have not been exposed to measles outbreaks." *Scand J Infect Dis,* vol. 31, no. 1 (1999): 17–9. doi:10.1080/00365549950161826 PMID: 10381212.; Jenks, P. J., Caul, E. O., and Roome, A. P. "Maternally derived measles immunity in children of naturally infected and vaccinated mothers." *Epidemiol Infect,* vol. 101, no. 2 (Oct 1988): 473–6. doi:10.1017/s095026880005442x PMID: 3053223.

129. *Brian Hooker (bioengineer).* Retrieved from Wikipedia, https://en.wikipedia.org/wiki /Brian_Hooker_(bioengineer).

130. Fine, P. E. "Herd immunity: history, theory, practice." *Epidemiol Rev,* vol. 15, no. 2 (1993): 265–302. doi:10.1093/oxfordjournals.epirev.a036121 PMID: 8174658.

131. Hussain, A., Ali, S., Ahmed, M., and Hussain, S. "The Anti-vaccination Movement: A Regression in Modern Medicine." *Cureus,* vol. 10, no. 7 (Jul 3 2018): e2919. doi:10.7759/cureus.2919 PMID: 30186724.

132. BC Centre for Disease Control, personal communication.

133. Fine, P. E. "Herd immunity: history, theory, practice." *Epidemiol Rev.*

134. For a general view on this, see Mathias, R. G., et al. (1989). The role of secondary vaccine failures in measles outbreaks. Am J Public Health, 79(4), 475–8.

135. Naniche, D., Garenne, M., Rae, C., Manchester, M., Buchta, R., Brodine, S. K., and Oldstone, M. B. "Decrease in measles virus-specific CD4 T cell memory in vaccinated subjects." *J Infect Di.*

136. Klein, N. P., Bartlett, J., Fireman, B., and Baxter, R. "Waning Tdap Effectiveness in Adolescents." *Pediatrics,* vol. 137, no. 3 (Mar 2016): e20153326. doi:10.1542/peds. 2015–3326 PMID: 26908667.

137. *QuickFacts - United States.* Retrieved from US Census Bureau, https://www.census.gov /quickfacts/fact/table/US/PST045219.

138. *Resident population of the United States by sex and age as of July 1, 2019 (in millions).* (2021). From "Demographics". Retrieved from Statista, https://www.statista.com /statistics/241488/population-of-the-us-by-sex-and-age/.

139. Haseltine, W. A. *The Moderna Vaccine's Antibodies May Not Last As Long As We Hoped.* (2020). From "Healthcare". Retrieved from Forbes, https://www.forbes.com/sites /williamhaseltine/2020/12/22/the-moderna-vaccines-antibodies-may-not-last-as-long -as-we-hoped/?sh=697c93624567.

140. Mello, M. M., Silverman, R. D., and Omer, S. B. "Ensuring Uptake of Vaccines against SARS-CoV-2." *N Engl J Med,* vol. 383, no. 14 (Oct 1 2020): 1296–99. doi:10.1056/ NEJMp2020926 PMID: 32589371.; Goodwin, L. *Eventually, getting the COVID-19 vaccine could be required for many.* (2020). From "Nation". Retrieved from Boston Globe, https: //www.bostonglobe.com/2020/12/15/nation/will-covid-19-vaccine-be-mandatory -future/.; Braley-Rattai, A. *Can COVID-19 vaccinations be mandated? Short answer: Yes.* (2020). From "PMN News". Retrieved from National Post, https://nationalpost.com /pmn/news-pmn/can-covid-19-vaccinations-be-mandated-short-answer-yes.

141. Personal communication concerning changing rate of ASD after removal of Thimerosal, see also: Nevison, C. D. (2014). A comparison of temporal trends in the United States autism prevalence to trends in suspected environmental factors. J Environ Health, 13, 73.

142. *Exposome and Exposomics.* (2014). From "Workplace Safety and Health Topics". Retrieved from US Centers for Disease Control and Prevention, https://www.cdc.gov /niosh/topics/exposome/default.html#:~:text=The%20exposome%20can%20be%20 -defined,from%20environmental%20and%20occupational%20sources.

143. Lewis, D. L. *Science for Sale: How the US Government Uses Powerful Corporations and Leading Universities to Support Government Policies, Silence Top Scientists, Jeopardize Our Health, and Protect Corporate Profits.* New York: Skyhorse, 2014. https://books.google .ca/books?id=UFyCDwAAQBAJ.

144. Shaw, C. A. *Neural Dynamics of Neurological Disease.* Boston: John Wiley & Sons, 2017. doi:https://doi.org/10.1002/9781118634523.ch1.

145. Eidi, H., Yoo, J., Bairwa, S. C., Kuo, M., Sayre, E. C., Tomljenovic, L., and Shaw, C. A. "Early postnatal injections of whole vaccines compared to placebo controls: Differential behavioural outcomes in mice." *J Inorg Biochem,* vol. 212 (Nov 2020): 111200. doi:10.1016/j.jinorgbio.2020.111200 PMID: 33039918

Chapter 5

1. Ganrot, P. O. "Metabolism and possible health effects of aluminum." *Environ Health Perspect,* vol. 65 (Mar 1986): 363–441. doi:10.1289/ehp.8665363 PMID: 2940082.

2. Shaw, C. A. *Neural Dynamics of Neurological Disease.* Boston: John Wiley & Sons, 2017. doi:https://doi.org/10.1002/9781118634523.ch1.; Shaw, C. A., and Marler, T. E. "The lessons of ALS-PDC: Environmental factors in ALS etiology." In *Spectrums of Amyotrophic Lateral Sclerosis: Heterogeneity, Pathogenesis and Therapeutic Directions.* edited by C. A. Shaw and Jessica R. Morrice. New Jersey: John Wiley & Sons, 2021.

3. Cluster definition: a large number of similar cases of disease presentation in a relatively small geographical space and time.

4. Rosetta Stone: A stone stele found by Napoleon's soldiers in Rashid (Rosetta), Egypt, during the French campaign in Egypt in the late eighteenth century. The stone contained texts of the same decree written in 196 B.C. in three languages: Egyptian hieroglyphics, demotic (another form of Egyptian), and ancient Greek. The fact that ancient Greek was understood allowed the French linguist Jean-François Champollion to translate the previously indecipherable hieroglyphics.

5. For a full description of the search for ALS-PDC etiology, see Shaw, C. A., & Marler, T. E. (2021). The Lessons of ALS-PDC: Environmental Factors in ALS Etiology. In C. A. Shaw, & J. R. Morrice (Eds.), Spectrums of Amyotrophic Lateral Sclerosis: Heterogeneity, Pathogenesis and Therapeutic Directions. New Jersey: John Wiley & Sons.

6. BMAA: beta-N-methylamino-L-alanine; BOAA: beta-N-oxalylamino-L-alanine.

7. Khabazian, I., Bains, J. S., Williams, D. E., Cheung, J., Wilson, J. M., Pasqualotto, B. A., Pelech, S. L., Andersen, R. J., Wang, Y. T., Liu, L., Nagai, A., Kim, S. U., Craig, U. K., and Shaw, C. A. "Isolation of various forms of sterol beta-D-glucoside from the seed of Cycas circinalis: neurotoxicity and implications for ALS-parkinsonism dementia complex." *J Neurochem,* vol. 82, no. 3 (Aug 2002): 516–28. doi:10.1046/j.1471-4159.2002.00976.x PMID: 12153476.; Wilson, J. M., Khabazian, I., Wong, M. C., Seyedalikhani, A., Bains, J. S., Pasqualotto, B. A., Williams, D. E., Andersen, R. J., Simpson, R. J., Smith, R., Craig, U. K., Kurland, L. T., and Shaw, C. A. "Behavioral and neurological correlates of ALS-parkinsonism dementia complex in adult mice fed washed cycad flour." *Neuromolecular Med,* vol. 1, no. 3 (2002): 207–21. doi:10.1385/NMM:1:3:207 PMID: 12095162.

8. Calvin ball was a game played by Calvin and his tiger Teddy in Bill Waterson's cartoon series, *Calvin and Hobbes.* The joke about Calvin ball in the series was that the rules could continually be changed at any stage of the game.

9. *Projection.* (2020). From "APA Dictionary of Psychology". Retrieved from American Psychological Association, https://dictionary.apa.org/projection.

10. *Thimerosal and Vaccines.* (2020). From "Vaccine Safety—Questions and Concerns". Retrieved from US Centers for Disease Control and Prevention, https://www.cdc.gov/vaccinesafety/concerns/thimerosal/index.html.

11. Zambon, V. *What is mad hatter's disease?* (2020). Retrieved from Medical News Today, https://www.medicalnewstoday.com/articles/mad-hatters-disease.

12. *Merbromin.* Retrieved from Wikipedia, https://en.wikipedia.org/wiki/Merbromin.

13. *Quantitative and Qualitative Analysis of Mercury Compounds in the List.* (2018). From "Food and Drug Administration Modernization Act (FDAMA) of 1997". Retrieved from US Food and Drug Administration, https://www.fda.gov/regulatory-information/food-and-drug-administration-modernization-act-fdama-1997/quantitative-and-qualitative-analysis-mercury-compounds-list.

14. Risher, J. F., Murray, H. E., and Prince, G. R. "Organic mercury compounds: human exposure and its relevance to public health." *Toxicol Ind Health,* vol. 18, no. 3 (Apr 2002): 109–60. doi:10.1191/0748233702th138oa PMID: 12974562

15. *Thimerosal and Vaccines.* (2020). From "Vaccine Safety—Questions and Concerns". Retrieved from US Centers for Disease Control and Prevention, https://www.cdc.gov /vaccinesafety/concerns/thimerosal/index.html.; Tosti, A., and Tosti, G. "Thimerosal: a hidden allergen in ophthalmology." *Contact Dermatitis,* vol. 18, no. 5 (May 1988): 268–73. doi:10.1111/j.1600–0536.1988.tb02831.x PMID: 3416589.

16. *Understanding Thimerosal, Mercury, and Vaccine Safety.* (2011). Retrieved from US Food and Drug Administration, https://www.fda.gov/media/83535/download.

17. *Timeline: Thimerosal in Vaccines (1999–2010).* (2020). From "Vaccine Safety - Questions and Concerns". Retrieved from US Food and Drug Administration, https://www .cdc.gov/vaccinesafety/concerns/thimerosal/timeline.html#:~:text=The%20FDA%20 reviews%20the%20use,November%205.

18. Kennedy, R. F., Hyman, M., and Herbert, M. R. *Thimerosal: Let the Science Speak: The Evidence Supporting the Immediate Removal of Mercury-a Known Neurotoxin-from Vaccines.* New York: Skyhorse, 2014. https://www.skyhorsepublishing.com/9781634504423 /thimerosal-let-the-science-speak/.

19. *Scientific Review of Vaccine Safety Datalink Information.* (2000). Retrieved from Simpson-wood Retreat Center, Norcross, Georgia, https://skeptico.blogs.com/Simpsonwood _Transcript.pdf.

20. Shaw, C. A. *Neural Dynamics of Neurological Disease.* Boston: John Wiley & Sons, 2017. doi:https://doi.org/10.1002/9781118634523.ch1.

21. Tomljenovic, L., and Shaw, C. A. "Do aluminum vaccine adjuvants contribute to the rising prevalence of autism?". *J Inorg Biochem,* vol. 105, no. 11 (Nov 2011): 1489–99. doi:10.1016/j.jinorgbio.2011.08.008 PMID: 22099159.

22. HMO (health maintenance organization): a network or organization that providing health insurance coverage for a monthly or annual fee. It is made up of medical insur-ance providers that limit coverage to medical care provided through doctors and other providers who are under contract to the HMO.

23. Institute of Medicine, Board on Health Promotion and Disease Prevention, and Immunization Safety Review Committee. *Immunization Safety Review: Thimerosal-Containing Vaccines and Neurodevelopmental Disorders.* Edited by M.C. McCormick, A. Gable and K. Stratton. Washington, DC: National Academies Press, 2001. https://books .google.ca/books?id=IHSwG1mw6ukC.

24. Committee to Review Adverse Effects of Vaccines, and Institute of Medicine. *Adverse Effects of Vaccines: Evidence and Causality.* Edited by K. Stratton, A. Ford, E. Rusch and E. W. Clayton. Washington, DC: National Academies Press, 2011. doi:10.17226/13164.

25. Burbacher, T. M., Shen, D. D., Liberato, N., Grant, K. S., Cernichiari, E., and Clarkson, T. "Comparison of blood and brain mercury levels in infant monkeys exposed to meth-ylmercury or vaccines containing thimerosal." *Environ Health Perspect,* vol. 113, no. 8 (Aug 2005): 1015–21. doi:10.1289/ehp.7712 PMID: 16079072.

26. Kirby, D. *Evidence of Harm: Mercury in Vaccines and the Autism Epidemic: A Medical Controversy.* New York: St. Martin's Press, 2006. https://books.google.ca/books?id=w2PwV -MgCK1UC.

27. Geier, D. A., and Geier, M. R. "An assessment of the impact of thimerosal on childhood neurodevelopmental disorders." *Pediatr Rehabil,* vol. 6, no. 2 (Apr–Jun 2003): 97–102. doi:10.1080/1363849031000139315 PMID: 14534046.

28. Geier, D. A., Kern, J. K., Homme, K. G., and Geier, M. R. "The risk of neurode-velopmental disorders following Thimerosal-containing Hib vaccine in compar-ison to Thimerosal-free Hib vaccine administered from 1995 to 1999 in the United

States." *Int J Hyg Environ Health,* vol. 221, no. 4 (May 2018): 677–83. doi:10.1016/j. ijheh.2018.03.004 PMID: 29573974.

29. Verstraeten, T., Davis, R. L., DeStefano, F., Lieu, T. A., Rhodes, P. H., Black, S. B., Shinefield, H., Chen, R. T., and Vaccine Safety Datalink, T. "Safety of thimerosal-containing vaccines: a two-phased study of computerized health maintenance organization databases." *Pediatrics,* vol. 112, no. 5 (Nov 2003): 1039–48. https://www.ncbi.nlm.nih .gov/pubmed/14595043 PMID: 14595043.

30. Hooker, B. *Dr. Brian Hooker's Official Statement Regarding Vaccine Whistleblower William Thompson.* (2016). Retrieved from https://childrenshealthdefense.org/wp-content /uploads/2016/11/Dr_BrianHooker_statement_regarding_Vaccine_Whistleblower _William_Thompson.pdf.

31. Wakefield, A. J., dir. *Vaxxed: From Cover-Up to Catastrophe.* (2016; GathrFilms, 2017), Documentary. IMDB: https://www.imdb.com/title/tt5562652/.; Yuhas, A. *Robert De Niro pulls anti-vaccination film from Tribeca film festival.* (2016). From "Culture". Retrieved from The Guardian, https://www.theguardian.com/film/2016/mar/27/robert -de-niro-pulls-vaccination-film-tribeca-vaxxed.

32. Lauerman, K. *Correcting our record: We've removed an explosive 2005 report by Robert F. Kennedy Jr. about autism and vaccines. Here's why.* (2011). Retrieved from Salon, https ://www.salon.com/2011/01/16/dangerous_immunity/.

33. Nevison, C. D. "A comparison of temporal trends in United States autism prevalence to trends in suspected environmental factors." *Environ Health,* vol. 13 (Sep 5 2014): 73. doi:10.1186/1476-069X-13-73 PMID: 25189402

34. Nevison, C. D. *The 2020 ADDM Report on U.S. Autism Prevalence: Three Reasons Why The Popular Narrative Was Misleading.* (2020). Retrieved from Children's Health Defense, https://childrenshealthdefense.org/news/the-2020-addm-report-on-u-s-autism -prevalence-three-reasons-why-the-popular-narrative-was-misleading/.

35. C. D. Nevison, personal communication.

36. Whitehouse, A. J., Cooper, M. N., Bebbington, K., Alvares, G., Lin, A., Wray, J., and Glasson, E. J. "Evidence of a reduction over time in the behavioral severity of autistic disorder diagnoses." *Autism Res,* vol. 10, no. 1 (Jan 2017): 179–87. doi:10.1002/aur.1740 PMID: 28102641.

37. Eickhoff, T. C., and Myers, M. "Workshop summary. Aluminum in vaccines." *Vaccine,* vol. 20 Suppl 3 (May 31 2002): S1-4. doi:10.1016/s0264-410x(02)00163–9 PMID: 12184358.

38. *Aluminum: chemical element.* (2020). Retrieved from Britannica, https://www.britannica .com/science/aluminum.

39. *Hans Christian Oersted.* (2015). Retrieved from Famous Scientists: The Art of Genius, https://www.famousscientists.org/hans-christian-oersted/.

40. *Aluminum.* Retrieved from Lenntech, https://www.lenntech.com/periodic/elements /al.htm.

41. *Vaccine Ingredients - Aluminum.* From "Vaccine Education Center". Retrieved from Children's Hospital of Philadelphia, https://www.chop.edu/centers-programs/vaccine -education-center/vaccine-ingredients/aluminum.

42. Exley, C. "A biogeochemical cycle for aluminium?". *J Inorg Biochem,* vol. 97, no. 1 (Sep 15 2003): 1–7. doi:10.1016/s0162-0134(03)00274–5 PMID: 14507454.

43. Ehgartner, B., and Exley, C. *The Age of Aluminum: The Dark Side of the Shiny Metal.* Publishing259 Seiten, 2019. https://books.google.ca/books?id=IoBUugEACAAJ.; Exley,

C. *Imagine You Are An Aluminum Atom: Discussions With Mr. Aluminum*. New York: Skyhorse, 2020. https://books.google.ca/books?id=SDv6DwAAQBAJ.

44. Saiyed, S. M., and Yokel, R. A. "Aluminium content of some foods and food products in the USA, with aluminium food additives." *Food Addit Contam*, vol. 22, no. 3 (Mar 2005): 234–44. doi:10.1080/02652030500073584 PMID: 16019791.

45. *Aluminum salts: Aluminum chloride, Aluminum nitrate, Aluminum sulphate*. (2010). From "Chemicals at a glance". Retrieved from Government of Canada, https://www.canada.ca/en/health-canada/services/chemical-substances/fact-sheets/chemicals-glance/aluminum-salts.html.

46. Shaw, C. A., Seneff, S., Kette, S. D., Tomljenovic, L., Oller, J. W., Jr., and Davidson, R. M. "Aluminum-induced entropy in biological systems: implications for neurological disease." *J Toxicol*, vol. 2014 (2014): 491316. doi:10.1155/2014/491316 PMID: 25349607.; Yokel, R. A., and Mcnamara, P. J. "Aluminium toxicokinetics: an updated minireview." *Pharmacol Toxicol*, vol. 88, no. 4 (Apr 2001): 159–67. doi:10.1034/j.1600–0773.2001.d01-98.x PMID: 11322172

47. Strunecka, A., Strunecky, O., and Patocka, J. "Fluoride plus aluminum: useful tools in laboratory investigations, but messengers of false information." *Physiol Res*, vol. 51, no. 6 (2002): 557–64. https://www.ncbi.nlm.nih.gov/pubmed/12511178 PMID: 12511178.

48. Minshall, C., Nadal, J., and Exley, C. "Aluminium in human sweat." *J Trace Elem Med Biol*, vol. 28, no. 1 (Jan 2014): 87–8. doi:10.1016/j.jtemb.2013.10.002 PMID: 24239230.

49. Yokel, R. A., and McNamara, P. J. "Aluminium toxicokinetics: an updated minireview." *Pharmacol Toxicol*.

50. Iheozor-Ejiofor, Z., Worthington, H. V., Walsh, T., O'Malley, L., Clarkson, J. E., Macey, R., Alam, R., Tugwell, P., Welch, V., and Glenny, A. M. "Water fluoridation for the prevention of dental caries." *The Cochrane Database of Systematic Reviews*, no. 6 (Jun 18 2015): CD010856. doi:10.1002/14651858.CD010856.pub2 PMID: 26092033.

51. Das, T. K., Susheela, A. K., Gupta, I. P., Dasarathy, S., and Tandon, R. K. "Toxic effects of chronic fluoride ingestion on the upper gastrointestinal tract." *J Clin Gastroenterol*, vol. 18, no. 3 (Apr 1994): 194–9. doi:10.1097/00004836-199404000-00004 PMID: 8034913.

52. Strunecka, A., Strunecky, O., and Patocka, J. "Fluoride plus aluminum: useful tools in laboratory investigations, but messengers of false information." *Physiol Res. [another duplicate?]*

53. Crisponi, G., Fanni, D., Gerosa, C., Nemolato, S., Nurchi, V. M., Crespo-Alonso, M., Lachowicz, J. I., and Faa, G. "The meaning of aluminium exposure on human health and aluminium-related diseases." *Biomol Concepts*, vol. 4, no. 1 (Feb 2013): 77–87. doi:10.1515/bmc-2012-0045 PMID: 25436567.; Riihimaki, V., and Aitio, A. "Occupational exposure to aluminum and its biomonitoring in perspective." *Crit Rev Toxicol*, vol. 42, no. 10 (Nov 2012): 827–53. doi:10.3109/10408444.2012.725027 PMID: 23013241.; Krewski, D., Yokel, R. A., Nieboer, E., Borchelt, D., Cohen, J., Harry, J., Kacew, S., Lindsay, J., Mahfouz, A. M., and Rondeau, V. "Human health risk assessment for aluminium, aluminium oxide, and aluminium hydroxide." *J Toxicol Environ Health B Crit Rev*, vol. 10 Suppl 1 (2007): 1–269. doi:10.1080/10937400701597766 PMID: 18085482.; Gitelman, H. J. "Aluminum exposure and excretion." *Sci Total Environ*, vol. 163, no. 1–3 (Feb 24 1995): 129–35. doi:10.1016/0048–9697(95)04483-h PMID: 7716490.; Elinder, C. G., Ahrengart, L., Lidums, V., Pettersson, E., and Sjögren, B. "Evidence of aluminium accumulation in aluminium welders." *Br J Ind Med*, vol. 48,

no. 11 (Nov 1991): 735–8. doi:10.1136/oem.48.11.735 PMID: 1954151.; Ljunggren, K. G., Lidums, V., and Sjögren, B. "Blood and urine concentrations of aluminium among workers exposed to aluminium flake powders." *Br J Ind Med,* vol. 48, no. 2 (Feb 1991): 106–9. doi:10.1136/oem.48.2.106 PMID: 1998604.

54. *Adjuvare,* to help, from the Latin. An adjuvant helps activate the immune response.

55. Adapted from Tomljenovic, L., "Aluminum and Alzheimer's disease: after a century of controversy, is there a plausible link?" *J Alzheimers Dis,* vol. 23, no. 4 (2011): 567–98. doi:10.3233/JAD-2010-101494. PMID: 21157018.

56. Gies, W. J. "Some objections to the use of alum baking-powder." *Journal of the American Medical Association,* vol. LVII, no. 10 (1911): 816. doi:10.1001/jama.1911.04260090038015 %J Journal of the American Medical Association.

57. The McIntrye Research Foundation (1930–1992). Their archived website states that "The McIntyre Research Foundation was a non-profit corporation formed to carry on research and investigation in connection with the prevention, mitigation, and eradication of industrial diseases." https://www.tandfonline.com/doi/full/10.1080/15459624.2019.1657581.

58. *Silicosis Symptoms and Diagnosis.* (2020). From "Lung Health & Diseases". Retrieved from American Lung Association, https://www.lung.org/lung-health-diseases/lung-disease-lookup/silicosis/symptoms-diagnosis.

59. Martell, J. *Aluminum Dust Exposure & Health Issues.* (2013). Retrieved from McIntyre Powder Project, http://www.mcintyrepowderproject.com/.; Occupational Cancer Research Centre, and Ontario Health (Cancer Care Ontario). *Investigation of McIntyre Powder Exposure and Neurological Outcomes in the Mining Master File Cohort: Final Report.* (2020). Retrieved from Occupational Cancer Research Centre, http://www.occupationalcancer.ca/wp-content/uploads/2020/04/OCRC_McIntyrePowder_FinalReport_2020.pdf.

60. Shaw, C. A. *Neural Dynamics of Neurological Disease.* Boston: John Wiley & Sons, 2017. doi:https://doi.org/10.1002/9781118634523.ch1.

61. Reusche, E., Koch, V., Friedrich, H. J., Nunninghoff, D., Stein, P., and Rob, P. M. "Correlation of drug-related aluminum intake and dialysis treatment with deposition of argyrophilic aluminum-containing inclusions in CNS and in organ systems of patients with dialysis-associated encephalopathy." *Clin Neuropathol,* vol. 15, no. 6 (Nov-Dec 1996): 342–7. https://www.ncbi.nlm.nih.gov/pubmed/8937781 PMID: 8937781.

62. Killin, L. O., Starr, J. M., Shiue, I. J., and Russ, T. C. "Environmental risk factors for dementia: a systematic review." *BMC Geriatr,* vol. 16, no. 1 (Oct 12 2016): 175. doi:10.1186/s12877-016-0342-y PMID: 27729011/; Andrasi, E., Pali, N., Molnar, Z., and Kosel, S. "Brain aluminum, magnesium and phosphorus contents of control and Alzheimer-diseased patients." *J Alzheimers Dis,* vol. 7, no. 4 (Aug 2005): 273–84. doi:10.3233/jad-2005-7402 PMID: 16131728.

63. Exley, C., and House, E. R. "Aluminium in the human brain." *Monatshefte für Chemie –Chemical Monthly,* vol. 142, no. 4 (2011/04/01 2011): 357–63. doi:10.1007/s00706-010-0417-y .

64. *Camelford water pollution incident.* (2016). Retrieved from Wikipedia, https://en.wikipedia.org/wiki/Camelford_water_pollution_incident.; Lean, G. *Poisoned: The Camelford scandal.* (2013). From "Environment". Retrieved from Independent, https://www.independent.co.uk/environment/poisoned-the-camelford-scandal-358010.html.

65. Exley, C., and Esiri, M. M. "Severe cerebral congophilic angiopathy coincident with increased brain aluminium in a resident of Camelford, Cornwall, UK." *J Neurol*

Neurosurg Psychiatry, vol. 77, no. 7 (Jul 2006): 877–9. doi:10.1136/jnnp.2005.086553 PMID: 16627535.

66. Bolognin, S., Messori, L., Drago, D., Gabbiani, C., Cendron, L., and Zatta, P. "Aluminum, copper, iron and zinc differentially alter amyloid-Abeta(1–42) aggregation and toxicity." *Int J Biochem Cell Biol,* vol. 43, no. 6 (Jun 2011): 877–85. doi:10.1016/j. biocel.2011.02.009 PMID: 21376832.

67. Reusche, E., Koch, V., Friedrich, H. J., Nunninghoff, D., Stein, P., and Rob, P. M. "Correlation of drug-related aluminum intake and dialysis treatment with deposition of argyrophilic aluminum-containing inclusions in CNS and in organ systems of patients with dialysis-associated encephalopathy." *Clin Neuropathol.;* Aremu, D. A., and Meshitsuka, S. "Accumulation of aluminum by primary cultured astrocytes from aluminum amino acid complex and its apoptotic effect." *Brain Res,* vol. 1031, no. 2 (Jan 21 2005): 284–96. doi:10.1016/j.brainres.2004.06.090 PMID: 15649454.; Levesque, L., Mizzen, C. A., McLachlan, D. R., and Fraser, P. E. "Ligand specific effects on aluminum incorporation and toxicity in neurons and astrocytes." *Brain Res,* vol. 877, no. 2 (Sep 22 2000): 191–202. doi:10.1016/s0006-8993(00)02637–8 PMID: 10986332.

68. Walton, J. R. "Cognitive deterioration and associated pathology induced by chronic low-level aluminum ingestion in a translational rat model provides an explanation of Alzheimer's disease, tests for susceptibility and avenues for treatment." *Int J Alzheimers Dis,* vol. 2012 (2012): 914947. doi:10.1155/2012/914947 PMID: 22928148.; Walton, J. R., and Wang, M. X. "APP expression, distribution and accumulation are altered by aluminum in a rodent model for Alzheimer's disease." *J Inorg Biochem,* vol. 103, no. 11 (Nov 2009): 1548–54. doi:10.1016/j.jinorgbio.2009.07.027 PMID: 19818510.; Walton, J. R. "Functional impairment in aged rats chronically exposed to human range dietary aluminum equivalents." *Neurotoxicology,* vol. 30, no. 2 (Mar 2009): 182–93. doi:10.1016/j.neuro.2008.11.012 PMID: 19109991.; Walton, J. R. "An aluminum-based rat model for Alzheimer's disease exhibits oxidative damage, inhibition of PP2A activity, hyperphosphorylated tau, and granulovacuolar degeneration." *J Inorg Biochem,* vol. 101, no. 9 (Sep 2007): 1275–84. doi:10.1016/j.jinorgbio.2007.06.001 PMID: 17662457.; Walton, J. R. "A longitudinal study of rats chronically exposed to aluminum at human dietary levels." *Neurosci Lett,* vol. 412, no. 1 (Jan 22 2007): 29–33. doi:10.1016/j.neulet.2006.08.093 PMID: 17156917.

69. McLachlan, D. R., Kruck, T. P., Lukiw, W. J., and Krishnan, S. S. "Would decreased aluminum ingestion reduce the incidence of Alzheimer's disease?". *CMAJ,* vol. 145, no. 7 (Oct 1 1991): 793–804. https://www.ncbi.nlm.nih.gov/pubmed/1822096 PMID: 1822096.

70. Anodize (a metal, especially aluminum) with a protective oxide layer by an electrolytic process in which the metal forms the anode.

71. Tomljenovic, L. "Aluminum and Alzheimer's disease: after a century of controversy, is there a plausible link?". *J Alzheimers Dis,* vol. 23, no. 4 (2011): 567–98. doi:10.3233 /JAD-2010-101494 PMID: 21157018.

72. Brunner, R., Jensen-Jarolim, E., and Pali-Scholl, I. "The ABC of clinical and experimental adjuvants-a brief overview." *Immunol Lett,* vol. 128, no. 1 (Jan 18 2010): 29–35. doi:10.1016/j.imlet.2009.10.005 PMID: 19895847.; Aluminum compounds such as aluminum hydroxide and aluminum potassium are still the main adjuvant molecules used in vaccines, in North America. Aluminum hydroxide and aluminum phosphate are the most commonly used adjuvants. As cited in the above reference, they are classified as Type B adjuvants (see the article for a more complete discussion of the other

types). Basically, the antigen is adsorbed in aluminum nanoparticles of the compound by various chemical forces: electrostatic, hydrophobic, and ligand exchange as described in this article. More properly, the antigen is actually considered to be adsorbed onto the aluminum nanoparticles as a thin sheet that is released slowly, this being the basis of the so-called "depot" effect.

73. Park, C. "Body burden." In *A Dictionary of Environment and Conservation*. Oxford, UK: Oxford University Press, 2007. https://www.oxfordreference.com/view/10.1093/oi/authority.20110803095514996.

74. Crisponi, G., Fanni, D., Gerosa, C., Nemolato, S., Nurchi, V. M., Crespo-Alonso, M., Lachowicz, J. I., and Faa, G. "The meaning of aluminium exposure on human health and aluminium-related diseases." *Biomol Concepts*.

75. In chemistry, a salt is a chemical compound consisting of an ionic assembly of cations and anions. Salts are composed of related numbers of cations (positively charged ions) and anions (negatively charged ions) so that the product is electrically neutral (without a net charge).

76. Glenny, A. T., Pope, C. G., Waddington, H., and Wallace, U. "Immunological notes. XVII: The antigenic value of toxoid precipitated by potassium alum." *The Journal of Pathology and Bacteriology*, vol. 29, no. 1 (1926): 31–40. doi:https://doi.org/10.1002/path.1700290106.

77. *Alum (1926)*. Retrieved from British Society for Immunology, https://www.immunology.org/alum-1926#:~:text=Sometimes%20in%20medical%20science%20it,without%20-anyone%20really%20knowing%20why.

78. Lee, S. H. "Detection of human papillomavirus (HPV) L1 gene DNA possibly bound to particulate aluminum adjuvant in the HPV vaccine Gardasil." *J Inorg Biochem*, vol. 117 (Dec 2012): 85–92. doi:10.1016/j.jinorgbio.2012.08.015 PMID: 23078778.

79. Shardlow, E., Mold, M., and Exley, C. "Unraveling the enigma: elucidating the relationship between the physicochemical properties of aluminium-based adjuvants and their immunological mechanisms of action." *Allergy Asthma Clin Immunol*, vol. 14 (2018): 80. doi:10.1186/s13223-018-0305-2 PMID: 30455719.

80. Gherardi, R. K., Eidi, H., Crépeaux, G., Authier, F. J., and Cadusseau, J. "Biopersistence and brain translocation of aluminum adjuvants of vaccines." *Front Neurol*, vol. 6 (2015): 4. doi:10.3389/fneur.2015.00004 PMID: 25699008.; Khan, Z., Combadiére, C., Authier, F. J., Itier, V., Lux, F., Exley, C., Mahrouf-Yorgov, M., Decrouy, X., Moretto, P., Tillement, O., Gherardi, R. K., and Cadusseau, J. "Slow CCL2-dependent translocation of biopersistent particles from muscle to brain." *BMC Med*, vol. 11 (Apr 4 2013): 99. doi:10.1186/1741-7015-11-99 PMID: 23557144.

81. Ciba-Geiby and Chiron developed M59 before the company was acquired by Novartis, which was bought by CSL Bering, which became Seqiris.

82. Petrik, M. S., Wong, M. C., Tabata, R. C., Garry, R. F., and Shaw, C. A. "Aluminum adjuvant linked to Gulf War illness induces motor neuron death in mice." *Journal of Neuromolecular Medicine*, vol. 9, no. 1 (2007): 83–100. doi:10.1385/nmm:9:1:83 PMID: 17114826.

83. Garcon, N., and Di Pasquale, A. "From discovery to licensure, the Adjuvant System story." *Hum Vaccin Immunother*, vol. 13, no. 1 (Jan 2 2017): 19–33. doi:10.1080/21645515.2016.1225635 PMID: 27636098.; AS01 A combination of QS-21 (immunostimulants Quillaja saponaria Molina: fraction 21) and MPL (3-deacylated monophosphoryl lipid) with liposomes AS02 A combination of immunostimulants of QS-21 (immunostimulants Quillaja Saponaria Molina: fraction 21) and MPL (3-deacylated

monophosphoryl lipid) with an oil in water Emulsion AS03 A combination of an oil in water emulsion with alpha-tocopherol (Vitamin E) as immuno-enhancing component AS04 MPL (3-deacylated monophosphoryl lipid) is adsorbed onto aluminum hydroxide or aluminum phosphate, depending on the vaccine with which it is used.

84. Petrik, M. S., Wong, M. C., Tabata, R. C., Garry, R. F., and Shaw, C. A. "Aluminum adjuvant linked to Gulf War illness induces motor neuron death in mice." *Neuromolecular Med.;[noted above already]* Sheth, S. K. S., Li, Y., and Shaw, C. A. "Is exposure to aluminium adjuvants associated with social impairments in mice? A pilot study." *J Inorg Biochem,* vol. 181 (Apr 2018): 96–103. doi:10.1016/j.jinorgbio.2017.11.012 PMID: 29221615.; Shaw, C. A., Li, Y., and Tomljenovic, L. "Administration of aluminium to neonatal mice in vaccine-relevant amounts is associated with adverse long term neurological outcomes." *J Inorg Biochem,* vol. 128 (Nov 2013): 237–44. doi:10.1016/j.jinorgbio.2013.07.022 PMID: 23932735.

85. Conklin, L., Hviid, A., Orenstein, W., Pollard, A., Wharton, M., and Zuber, P. *Vaccine safety issues at the turn of the 21st century.* From "Vaccine Safety". Retrieved from World Health Organization, https://www.who.int/vaccine_safety/GACVSsymposiumTrack1-Safety-issues-reviewed-during-early21st-centuryRev2.pdf.

86. Holiday, R., and Hanselman, S. "Epictetus the Free Man." In *Lives of the Stoics: The Art of Living from Zeno to Marcus Aurelius.* pg. 251–86: New York: Penguin Publishing Group, 2020. https://books.google.ca/books?id=xfvHDwAAQBAJ.

87. Sheth, S. K. S., Li, Y., and Shaw, C. A. "Is exposure to aluminium adjuvants associated with social impairments in mice? A pilot study." *J Inorg Biochem.;* Shaw, C. A., Li, Y., and Tomljenovic, L. "Administration of aluminium to neonatal mice in vaccine-relevant amounts is associated with adverse long term neurological outcomes." *J Inorg Biochem.*

88. Eidi, H., Yoo, J., Bairwa, S. C., Kuo, M., Sayre, E. C., Tomljenovic, L., and Shaw, C. A. "Early postnatal injections of whole vaccines compared to placebo controls: Differential behavioural outcomes in mice." *J Inorg Biochem,* vol. 212 (Nov 2020): 111200. doi:10.1016/j.jinorgbio.2020.111200 PMID: 33039918.

89. Asin, J., Pascual-Alonso, M., Pinczowski, P., Gimeno, M., Perez, M., Muniesa, A., De Pablo-Maiso, L., De Blas, I., Lacasta, D., Fernandez, A., De Andres, D., Reina, R., and Luján, L. "Cognition and behavior in sheep repetitively inoculated with aluminum adjuvant-containing vaccines or aluminum adjuvant only." *J Inorg Biochem,* vol. 203 (Feb 2020): 110934. doi:10.1016/j.jinorgbio.2019.110934 PMID: 31783216.; Luján, L., Perez, M., Salazar, E., Alvarez, N., Gimeno, M., Pinczowski, P., Irusta, S., Santamaria, J., Insausti, N., Cortes, Y., Figueras, L., Cuartielles, I., Vila, M., Fantova, E., and Chapulle, J. L. "Autoimmune/autoinflammatory syndrome induced by adjuvants (ASIA syndrome) in commercial sheep." *Immunol Res,* vol. 56, no. 2–3 (Jul 2013): 317–24. doi:10.1007/s12026-013-8404-0 PMID: 23579772.

90. Khan, Z., Combadière, C., Authier, F. J., Itier, V., Lux, F., Exley, C., Mahrouf-Yorgov, M., Decrouy, X., Moretto, P., Tillement, O., Gherardi, R. K., and Cadusseau, J. "Slow CCL2-dependent translocation of biopersistent particles from muscle to brain." *BMC Med.;* Crèpeaux, G., Eidi, H., David, M. O., Tzavara, E., Giros, B., Exley, C., Curmi, P. A., Shaw, C. A., Gherardi, R. K., and Cadusseau, J. "Highly delayed systemic translocation of aluminum-based adjuvant in CD1 mice following intramuscular injections." *J Inorg Biochem,* vol. 152 (Nov 2015): 199–205. doi:10.1016/j.jinorgbio.2015.07.004 PMID: 26384437.

91. Feynman, R. P. *"It doesn't matter how beautiful your theory is, it doesn't matter how smart you are. If it doesn't agree with experiment, it's wrong.".* Retrieved from

Goodreads Quotes, https://www.goodreads.com/quotes/625767-it-doesn-t-matter-how -beautiful-your-theory-is-it-doesn-t.

92. Pharmacodynamics, in contrast, refers to what the drug does to the body.

93. Agency for Toxic Substances and Disease Registry. *Toxicological Profile for Aluminum.* (2008). Retrieved from US Centers for Disease Control and Prevention, https://www .atsdr.cdc.gov/toxprofiles/tp.asp?id=191&tid=34.

94. Priest, N. D., Newton, D., Day, J. P., Talbot, R. J., and Warner, A. J. "Human metabolism of aluminium-26 and gallium-67 injected as citrates." *Hum Exp Toxicol,* vol. 14, no. 3 (Mar 1995): 287–93. doi:10.1177/096032719501400309 PMID: 7779460.

95. Flarend, R. E., Hem, S. L., White, J. L., Elmore, D., Suckow, M. A., Rudy, A. C., and Dandashli, E. A. "In vivo absorption of aluminium-containing vaccine adjuvants using 26Al." *Vaccine,* vol. 15, no. 12–13 (Aug–Sep 1997): 1314–8. doi:10.1016/s0264-410x(97)00041–8 PMID: 9302736.

96. Keith, L. S., Jones, D. E., and Chou, C. H. "Aluminum toxicokinetics regarding infant diet and vaccinations." *Vaccine,* vol. 20 Suppl 3 (May 31 2002): S13-7. doi:10.1016/ s0264-410x(02)00165–2 PMID: 12184359.

97. Mitkus, R. J., King, D. B., Hess, M. A., Forshee, R. A., and Walderhaug, M. O. "Updated aluminum pharmacokinetics following infant exposures through diet and vaccination." *Vaccine,* vol. 29, no. 51 (Nov 28 2011): 9538–43. doi:10.1016/j.vaccine.2011.09.124 PMID: 22001122.

98. Masson, J. D., Crépeaux, G., Authier, F. J., Exley, C., and Gherardi, R. K. "Critical analysis of reference studies on the toxicokinetics of aluminum-based adjuvants." *J Inorg Biochem,* vol. 181 (Apr 2018): 87–95. doi:10.1016/j.jinorgbio.2017.12.015 PMID: 29307441.

99. McFarland, G., La Joie, E., Thomas, P., and Lyons-Weiler, J. "Acute exposure and chronic retention of aluminum in three vaccine schedules and effects of genetic and environmental variation." *J Trace Elem Med Biol,* vol. 58 (Mar 2020): 126444. doi:10.1016/j. jtemb.2019.126444 PMID: 31846784.

100. Dr. Paul Thomas is a Portland, Oregon-based pediatrician who has designed what he terms a "vaccine-friendly plan" designed to reduce exposure to tovaccine aluminum adjuvants in early periods of child development.; Thomas, P. *The Dr. Paul Approved Vaccine Plan.* Retrieved from Integrative Pediatrics, https://www.integrativepediatricsonline.com/uploads/1/0/9/2/109222957/the_vaccine-friendly_plan.pdf.

101. Tomljenovic, L., and Shaw, C. A. "Aluminum vaccine adjuvants: are they safe?". *Curr Med Chem,* vol. 18, no. 17 (2011): 2630–7. doi:10.2174/092986711795933740 PMID: 21568886.

102. Dorea, J. G., and Marques, R. C. "Infants' exposure to aluminum from vaccines and breast milk during the first 6 months." *J Expo Sci Environ Epidemiol,* vol. 20, no. 7 (Nov 2010): 598–601. doi:10.1038/jes.2009.64 PMID: 20010978.

103. *Physicians for Informed Consent.* (2020). Retrieved from Physicians for Informed Consent, https://physiciansforinformedconsent.org/.

104. Exley, C. "Aluminium adjuvants and adverse events in sub-cutaneous allergy immunotherapy." *Allergy Asthma Clin Immunol,* vol. 10, no. 1 (Jan 20 2014): 4. doi:10.1186/1710-1492-10-4 PMID: 24444186.

105. https://www.healthlinkbc.ca/sites/default/files/pdf/immunization-infants-children.pdf.

106. *Paul A. Offit, MD.* From "Doctors". Retrieved from Children's Hospital of Philadelphia, https://www.chop.edu/doctors/offit-paul-a.

107. *12th International Congress on Autoimmunity.* (2021). Retrieved from https://autoimmunity .kenes.com/.

108. Multiple sclerosis is a disorder of the brain in which the insulating covering of some neuronal axons, myelin, is destroyed by the immune system. See: Multiple Sclerosis. (2020). Retrieved from Mayo Clinic: https://www.mayoclinic.org/diseases-conditions /multiple-sclerosis/symptoms-causes/syc-20350269.

109. Myasthenia gravis is also an autoimmune disorder in which immune system antibodies target the neuromuscular junction on muscle fibers, resulting in the loss of neurotrans-mitter receptors. The result is that neurotransmission fails. See: Myasthenia Gravis. (2021). Retrieved from Johns Hopkins Medicine: https://www.hopkinsmedicine.org /health/conditions-and-diseases/myasthenia-gravis#:~:text=Myasthenia%20gravis%20 -(MG)%20is%20a,%2C%20mouth%2C%20throat%20and%20limbs.

110. *Guillain-Barré Syndrome and Vaccines.* (2020). From "Vaccine Safety - Questions and Concerns". Retrieved from US Centers for Disease Control and Prevention, https://www .cdc.gov/vaccinesafety/concerns/guillain-barre-syndrome.html.

111. Shaw, C. A. *Neural Dynamics of Neurological Disease.* Boston: John Wiley & Sons, 2017. doi:https://doi.org/10.1002/9781118634523.ch1.

112. Gherardi, R. K., and Authier, F. J. "Macrophagic myofasciitis: characterization and patho-physiology." *Lupus,* vol. 21, no. 2 (Feb 2012): 184–9. doi:10.1177/0961203311429557 PMID: 22235051.; Gherardi, R. K., Coquet, M., Cherin, P., Belec, L., Moretto, P., Dreyfus, P. A., Pellissier, J. F., Chariot, P., and Authier, F. J. "Macrophagic myofasciitis lesions assess long-term persistence of vaccine-derived aluminium hydroxide in muscle." *Brain,* vol. 124, no. Pt 9 (Sep 2001): 1821–31. doi:10.1093/brain/124.9.1821 PMID: 11522584.

113. Gherardi, R. K., and Authier, F. J. "Macrophagic myofasciitis: characterization and pathophysiology." *Lupus.*

114. Rigolet, M., Aouizerate, J., Couette, M., Ragunathan-Thangarajah, N., Aoun-Sebaiti, M., Gherardi, R. K., Cadusseau, J., and Authier, F. J. "Clinical features in patients with long-lasting macrophagic myofasciitis." *Front Neurol,* vol. 5 (2014): 230. doi:10.3389/ fneur.2014.00230 PMID: 25506338.

115. Tomljenovic, L., and Shaw, C. A. "Do aluminum vaccine adjuvants contribute to the rising prevalence of autism?". *J Inorg Biochem.;* Tomljenovic, L., and Shaw, C. A. "Aluminum vaccine adjuvants: are they safe?". *Curr Med Chem.*

116. *Questions and Answers about macrophagic myofasciitis (MMF).* (2008). From "The Global Advisory Committee on Vaccine Safety". Retrieved from World Health Organization, https://www.who.int/vaccine_safety/committee/topics/aluminium/questions/en/.

117. Meroni, P. L. "Autoimmune or auto-inflammatory syndrome induced by adjuvants (ASIA): old truths and a new syndrome?". *J Autoimmun,* vol. 36, no. 1 (Feb 2011): 1–3. doi:10.1016/j.jaut.2010.10.004 PMID: 21051205.; Shoenfeld, Y., and Agmon-Levin, N. "'ASIA' - autoimmune/inflammatory syndrome induced by adjuvants." *J Autoimmun,* vol. 36, no. 1 (Feb 2011): 4–8. doi:10.1016/j.jaut.2010.07.003 PMID: 20708902.

118. Ameratunga, R., Gillis, D., Gold, M., Linneberg, A., and Elwood, J. M. "Evidence Refuting the Existence of Autoimmune/Autoinflammatory Syndrome Induced by Adjuvants (ASIA)." *J Allergy Clin Immunol Pract,* vol. 5, no. 6 (Nov - Dec 2017): 1551–55 e1. doi:10.1016/j.jaip.2017.06.033 PMID: 28888842.

119. Crépeaux, G., Gherardi, R. K., and Authier, F. J. "ASIA, chronic fatigue syndrome, and selective low dose neurotoxicity of aluminum adjuvants." *J Allergy Clin Immunol Pract,* vol. 6, no. 2 (Mar - Apr 2018): 707. doi:10.1016/j.jaip.2017.10.039 PMID: 29525002.

120. Galic, M. A., Spencer, S. J., Mouihate, A., and Pittman, Q. J. "Postnatal programming of the innate immune response." *Integr Comp Biol,* vol. 49, no. 3 (Sep 2009): 237–45. doi:10.1093/icb/icp025 PMID: 21665816.; Boisse, L., Mouihate, A., Ellis, S., and Pittman, Q. J. "Long-term alterations in neuroimmune responses after neonatal exposure to lipopolysaccharide." *J Neurosci,* vol. 24, no. 21 (May 26 2004): 4928–34. doi:10.1523/JNEUROSCI.1077–04.2004 PMID: 15163684.

121. Hsiao, E. Y., and Patterson, P. H. "Activation of the maternal immune system induces endocrine changes in the placenta via IL-6." *Brain Behav Immun,* vol. 25, no. 4 (May 2011): 604–15. doi:10.1016/j.bbi.2010.12.017 PMID: 21195166.; Smith, S. E., Li, J., Garbett, K., Mirnics, K., and Patterson, P. H. "Maternal immune activation alters fetal brain development through interleukin-6." *J Neurosci,* vol. 27, no. 40 (Oct 3 2007): 10695–702. doi:10.1523/JNEUROSCI.2178–07.2007 PMID: 17913903.; Smith, S. E. P., and Patterson, P. H. "Alteration of Neurodevelopment and behavior by Maternal Immune Activation." In *The Neuroimmunological Basis of Behavior and Mental Disorders.* edited by Allan Siegel and Steven S. Zalcman.pg. 111–30. Boston, MA: Springer US, 2009. https://doi.org/10.1007/978-0-387-84851-8_7.

122. Dinarello, C. A. "Cytokines as endogenous pyrogens." *J Infect Dis,* vol. 179 Suppl 2 (Mar 1999): S294-304. doi:10.1086/513856 PMID: 10081499.

123. Louveau, A., Smirnov, I., Keyes, T. J., Eccles, J. D., Rouhani, S. J., Peske, J. D., Derecki, N. C., Castle, D., Mandell, J. W., Lee, K. S., Harris, T. H., and Kipnis, J. "Structural and functional features of central nervous system lymphatic vessels." *Nature,* vol. 523, no. 7560 (Jul 16 2015): 337–41. doi:10.1038/nature14432 PMID: 26030524.

124. Akiyama, H., Barger, S., Barnum, S., Bradt, B., Bauer, J., Cole, G. M., Cooper, N. R., Eikelenboom, P., Emmerling, M., Fiebich, B. L., Finch, C. E., Frautschy, S., Griffin, W. S., Hampel, H., Hull, M., Landreth, G., Lue, L., Mrak, R., Mackenzie, I. R., Mcgeer, P. L., O'banion, M. K., Pachter, J., Pasinetti, G., Plata-Salaman, C., Rogers, J., Rydel, R., Shen, Y., Streit, W., Strohmeyer, R., Tooyoma, I., Van Muiswinkel, F. L., Veerhuis, R., Walker, D., Webster, S., Wegrzyniak, B., Wenk, G., and Wyss-Coray, T. "Inflammation and Alzheimer's disease." *Neurobiol Aging,* vol. 21, no. 3 (May-Jun 2000): 383–421. doi:10.1016/s0197-4580(00)00124-x PMID: 10858586.; Pardo, C. A., Vargas, D. L., and Zimmerman, A. W. "Immunity, neuroglia and neuroinflammation in autism." *Int Rev Psychiatry,* vol. 17, no. 6 (Dec 2005): 485–95. doi:10.1080/02646830500381930 PMID: 16401547.; Vargas, D. L., Nascimbene, C., Krishnan, C., Zimmerman, A. W., and Pardo, C. A. "Neuroglial activation and neuroinflammation in the brain of patients with autism." *Ann Neurol,* vol. 57, no. 1 (Jan 2005): 67–81. doi:10.1002/ana.20315 PMID: 15546155.

125. Shaw, C. A., Li, D., and Tomljenovic, L. "Are there negative CNS impacts of aluminum adjuvants used in vaccines and immunotherapy?". *Immunotherapy,* vol. 6, no. 10 (2014): 1055–71. doi:10.2217/imt.14.81 PMID: 25428645.

126. Asin, J., Pascual-Alonso, M., Pinczowski, P., Gimeno, M., Perez, M., Muniesa, A., De Pablo-Maiso, L., De Blas, I., Lacasta, D., Fernandez, A., De Andres, D., Reina, R., and Luján, L. "Cognition and behavior in sheep repetitively inoculated with aluminum adjuvant-containing vaccines or aluminum adjuvant only." *J Inorg Biochem.;* Luján, L., Perez, M., Salazar, E., Alvarez, N., Gimeno, M., Pinczowski, P., Irusta, S., Santamaria, J., Insausti, N., Cortes, Y., Figueras, L., Cuartielles, I., Vila, M., Fantova, E., and Chapulle, J. L. "Autoimmune/autoinflammatory syndrome induced by adjuvants (ASIA syndrome) in commercial sheep." *Immunol Res.*

127. Masson, J. D., Thibaudon, M., Bélec, L., and Crépeaux, G. "Calcium phosphate: a substitute for aluminum adjuvants?". *Expert Rev Vaccines,* vol. 16, no. 3 (Mar 2017): 289–99. doi:10.1080/14760584.2017.1244484 PMID: 27690701.

128. Vande Velde, V. 2016. "Calcium Fluoride Compositions." US. Assignee: GLAXOSMITHKLINE BIOLOGICALS S.A. (Rixensart, BE). Filed (October 23, 2014), and issued (September 15, 2016). URL: https://www.freepatentsonline.com/y2016/0263214.html.

129. Ledford, H. "Super-precise new CRISPR tool could tackle a plethora of genetic diseases." *Nature,* vol. 574, no. 7779 (Oct 2019): 464–65. doi:10.1038/d41586-019-03164-5 PMID: 31641267.; Neldeborg, S., Lin, L., Stougaard, M., and Luo, Y. "Rapid and Efficient Gene Deletion by CRISPR/Cas9." *Methods Mol Biol,* vol. 1961 (2019): 233–47. doi:10.1007/978-1-4939-9170-9_14 PMID: 30912049.

130. Kovac, J., Macedoni Luksic, M., Trebusak Podkrajsek, K., Klancar, G., and Battelino, T. "Rare single nucleotide polymorphisms in the regulatory regions of the superoxide dismutase genes in autism spectrum disorder." *Autism Res,* vol. 7, no. 1 (Feb 2014): 138–44. doi:10.1002/aur.1345 PMID: 24155217.

131. Oller, J. W., Jr. "Pragmatic Information." In *Biological Information: New Perspectives.* edited by R.J. Marks, M.J. Behe, W.A. Dembski, B.L. Gordon and J.C. Sanford.pg. 64–86: Singapore: World Scientific Publishing Company, 2013. https://books.google.ca/books?id=xYtEDwAAQBAJ.

132. Oller, J. W., Jr., and Shaw, C. A. "From Superficial Damage to Invasion of the Nucleosome: Ranking of Morbidities by the Biosemiotic Depth Hypothesis." *International Journal of Sciences,* vol. 8, no. 06 (2019): 51–73. doi:10.18483/ijSci.2069.

133. Ibid.

134. *Mary Holland.* (2020). Retrieved from Health Choice, https://healthchoice.org/leader/mary-holland/.

135. Crepeaux, G., Eidi, H., David, M. O., Tzavara, E., Giros, B., Exley, C., Curmi, P. A., Shaw, C. A., Gherardi, R. K., and Cadusseau, J. "Highly delayed systemic translocation of aluminum-based adjuvant in CD1 mice following intramuscular injections." *J Inorg Biochem.*; Crépeaux, G., Eidi, H., David, M. O., Baba-Amer, Y., Tzavara, E., Giros, B., Authier, F. J., Exley, C., Shaw, C. A., Cadusseau, J., and Gherardi, R. K. "Non-linear dose-response of aluminium hydroxide adjuvant particles: Selective low dose neurotoxicity." *Toxicology,* vol. 375 (Jan 15 2017): 48–57. doi:10.1016/j.tox.2016.11.018 PMID: 27908630.

136. Exley, C. *Imagine You Are An Aluminum Atom: Discussions With Mr. Aluminum.* New York: Skyhorse, 2020. https://books.google.ca/books?id=SDv6DwAAQBAJ.

137. Exley, C. *Imagine you are an Aluminum Atom.* (2020). Retrieved from The Hippocratic Post, https://www.hippocraticpost.com/global-reach/imagine-you-are-an-aluminum-atom/. [multiple citations]

138. Willhite, C. C., Karyakina, N. A., Yokel, R. A., Yenugadhati, N., Wisniewski, T. M., Arnold, I. M., Momoli, F., and Krewski, D. "Systematic review of potential health risks posed by pharmaceutical, occupational and consumer exposures to metallic and nanoscale aluminum, aluminum oxides, aluminum hydroxide and its soluble salts." *Crit Rev Toxicol,* vol. 44 Suppl 4 (Oct 2014): 1–80. doi:10.3109/10408444.2014.934439 PMID: 25233067.

139. Corkins, M. R., and Committee On, N. "Aluminum Effects in Infants and Children." *Pediatrics,* vol. 144, no. 6 (Dec 2019). doi:10.1542/peds.2019–3148 PMID: 31767714.

140. Boretti, A., 2021. Reviewing the association between aluminum adjuvants in the vaccines and autism spectrum disorder. *J. Trace Elements in Medicine and Biology*. 66, 126764.

Chapter 6

1. Howard Zinn (1923–2012) was an American historian known for a socialist interpretation of history, particularly American history. The quote is from On War, pg. 171, New York: Seven Stories Press, 2001.
2. Abad-Santos, A. *Lizard people: the greatest political conspiracy ever created.* (2015). Retrieved from Vox, https://www.vox.com/2014/11/5/7158371/lizard-people-conspiracy -theory-explainer.; Edwards, P. *9 questions about the Illuminati you were too afraid to ask.* (2015). Retrieved from Vox, https://www.vox.com/2015/5/19/8624675 /what-is-illuminati-meaning-conspiracy-beyonce.
3. Hammond, J. R. *The Rotavirus Vaccine: A Case Study in Government Corruption and Malfeasance.* (2019). From "News". Retrieved from Children's Health Defense, https://childrenshealthdefense.org/news/the-rotavirus-vaccine-a-case-study-in -government-corruption-and-malfeasance/.; Olmsted, D., and Blaxill, M. *Counting Offit's Millions: More on How Merck's Rotateq Vaccine Made Paul Offit Wealthy.* (2011). Retrieved from Age of Autism, https://www.ageofautism.com/2011/01/counting-offits -millions-more-on-how-mercks-rotateq-vaccine-made-paul-offit-wealthy.html.; *Vaccine (Drops) for Rotavirus.* (2019). From "Diseases that Vaccines Prevent". Retrieved from US Centers for Disease Control and Prevention, https://www.cdc.gov/vaccines/parents /diseases/rotavirus.html.
4. *Maurice Hilleman.* (2016). Retrieved from Famous Scientists: The Art of Genius, https: //www.famousscientists.org/maurice-hilleman/.
5. *About the National Academy of Medicine.* Retrieved from National Academy of Medicine, https://nam.edu/about-the-nam/.
6. *Advisory Committee on Immunization Practices (ACIP).* (2021). From "Vaccines". Retrieved from US Centers for Disease Control and Prevention, https://www.cdc.gov /vaccines/acip/index.html.
7. *The Center for Vaccine Ethics and Policy.* (2010). Retrieved from Penn Center for Bioethics of the University of Pennsylvania, https://centerforvaccineethicsandpolicy.files .wordpress.com/2010/01/center-for-vaccine-ethics-and-policy_overview_february-2010 .pdf.
8. GE2P2 stands for Governance, Ethics, Evidence, Policy and Practice. GE2P2 conducts its work through a "centers of excellence" structure, and it currently has the following: Center for Disaster and Humanitarian Ethics, Center for Vaccine Ethics and Policy (CVEP), Center for Ethics and Policy on Access to Medicines, Center for Informed Consent Integrity, and Independent Bioethics Advisory Committee. See: http://www .ge2p2.org/.
9. PATH is an international nonprofit organization founded in 1977 and based in Seattle with more than seventy offices around the world. PATH focuses on five platforms— vaccines, drugs, diagnostics, devices, and system and service innovations. PATH's partners include GSK, Johnson & Johnson, Pfizer, the Bill and Melinda Gates Foundation, and others. Its best-known technology is the vaccine vial monitor, a small sticker that adheres to a vaccine vial and changes color as the vaccine is exposed to heat over time. Also, PATH has worked with pharmaceutical companies to support the development of newer vaccines (meningitis and pneumonia) and to introduce vaccines, such as rotavirus

and Japanese encephalitis. An HPV vaccine study conducted by PATH in 2009 caused controversy in India when seven of the girls participating in the trial died after receiving the vaccines, but state investigations later considered that those deaths were unrelated to the vaccinations. See: https://www.path.org/.

10. Valera LLC is the second venture company formed by Moderna in 2015. Valera focused exclusively on developing prophylactic and therapeutic vaccines. In 2017, Moderna ditched the venture-based R&D model and brought all four units including Valera under one single corporation roof.

11. *Center for Vaccine Ethics and Policy (CVEP): Statement of Independence.* (2011). Retrieved from Penn Center for Bioethics of the University of Pennsylvania, https://globalvaccines 202xsymposium.wordpress.com/independence/.

12. COI declared from Offit's Publications (2000–2020)
    - 2013: Supported in part funding from Vanderbilt Institute for Clinical and Translational Research grant support and Agency for Healthcare Research Quality (https://www.sciencedirect.com/science/article/abs/pii/S1876285913000661)
    - 2011: The financial support from CHOP Center for Pediatric Clinical Effectiveness Pilot Grant Program and the Center for Bioethics at the U of Penn School of Med (https://www.sciencedirect.com/science/article/pii/S0264410X1001902X)
    - 2009: PAO has served on a scientific advisory board to Merck (https://www.ncbi .nlm.nih.gov/pmc/articles/PMC2908388/pdf/nihms-212222.pdf)
    - 2006: Other than employees of Merck, all authors are investigators and/or consultants for the sponsor. Merck employees may hold stock and/or stock options in the company. H Fred Clark and Paul Offit are coholders of the patent on RotaTeq (https://www.sciencedirect.com/science/article/pii/S0264410X0600288X)
    - 2006: Many of the coauthors reported the receipt of consulting fees, lecture fees, and grant support from Merck and GSK (https://www.nejm.org/doi/full/10.1056 /nejmoa052664)
    - 2004: Supported by a grant from Merck (https://www.sciencedirect.com/science /article/abs/pii/S0022347603007741)
    - 2003: Some vaccines discussed in this article are manufactured by Merck and Co. Dr. Offit is the coholder of a patent on a bovine-human reassortant rotavirus vaccine that is being developed by Merck. Dr. Offit's laboratory support comes from the NIH, and he does not receive personal support or honoraria from Merck and does not have a financial interest in the company. (https://pediatrics.aappublications.org /content/pediatrics/112/6/1394.full.pdf).

13. *The Foundation for Vaccine Research.* (2021). Retrieved from The Foundation for Vaccine Research, https://www.vaccinefoundation.org/.

14. *Autism Science Foundation.* (2021). Retrieved from https://autismsciencefoundation.org/.

15. Kroll, A., and Schulman, J. *Leaked Documents Reveal the Secret Finances of a Pro-Industry Science Group.* (2013). Retrieved from Mother Jones, https://www.motherjones.com /politics/2013/10/american-council-science-health-leaked-documents-fundraising/.

16. Attkisson, S. *How Independent Are Vaccine Defenders?* (2008). Retrieved from CBS News, https://www.cbsnews.com/news/how-independent-are-vaccine-defenders/.

17. *Funding & Annual Reports.* (2020). Retrieved from Vaccinate Your Family, https: //vaccinateyourfamily.org/about-us/funding/.

18. *Our Alliance.* (2020). Retrieved from Gavi: The Vaccine Alliance, https://www.gavi.org /our-alliance.

19. *Gavi.* (2020). Retrieved from Wikipedia, https://en.wikipedia.org/wiki/GAVI.; *Gavi must ensure more children get new, more affordable pneumonia vaccine.* (2020). From "Press Release". Retrieved from Medecins Sans Frontieres, https://www.msf.org/gavi -must-work-ensure-more-children-get-new-more-affordable-pneumonia-vaccine.

20. Holland, S., and Nichols, M. *Trump cutting U.S. ties with World Health Organization over virus.* (2020). From "Healthcare & Pharma". Retrieved from Reuters, https://www.reuters.com/article/us-health-coronavirus-trump-who-idUSKB-N2352YJ.; Ren, G., and Lustig Vijay, S. *US $8.8 Billion Pledged For Gavi, The Vaccine Alliance – Smashing US $7.4 Billion Goal.* (2020). From "Gavi: the Vaccine Alliance". Retrieved from Health Policy Watch, https://healthpolicy-watch.news /us-8-8-billion-pledged-for-gavi-the-vaccine-alliance-smashing-us-7-4-billion-goal/.

21. A mortal sin is defined as a grave action that is committed in full knowledge of its gravity and with the full consent of the sinner's will. Such a sin cuts the sinner off from God's sanctifying grace until it is repented, usually in confession with a priest—Mortal sin. (2021). Retrieved from Britannica: https://www.britannica.com/topic/cardinal-sin.
    I would be inclined to argue that it is a mortal sin.

22. *Paul A. Offit, MD.* From "Doctors". Retrieved from Children's Hospital of Philadelphia, https://www.chop.edu/doctors/offit-paul-a.; *Paul A. Offit, MD, Offical Author Page.* (2020). Retrieved from https://www.paul-offit.com/.

23. Offit, P. A. *Autism's False Prophets: Bad Science, Risky Medicine, and the Search for a Cure.* New York: Columbia University Press, 2008. https://books.google.ca/books?id =_epul7PlKccC.

24. Offit, P. A. *Deadly Choices (UK edition): How the Anti-Vaccine Movement Threatens Us All.* New York: Basic Books, 2011. https://books.google.ca/books?id=yjoBfAEACAAJ.

25. Offit, P. A. *Bad Faith: When Religious Belief Undermines Modern Medicine.* New York: Basic Books, 2015. https://books.google.ca/books?id=472DBQAAQBAJ.

26. https://en.wikipedia.org/wiki/Paul_Offit#/media/File:Paul_Offit.jpg; https://en.wikipedia .org/wiki/Stanley_Plotkin#/media/File:Photo_Plotkin1.jpg; https://en.wikipedia.org/wiki /Peter_Hotez#/media/File:Peter_Hotez_2019_Texas_Book_Festival.jpg.

27. *Dosis sola facit venenum.* See: The dose makes the poison. (2020). Retrieved from Wikipedia: https://en.wikipedia.org/wiki/The_dose_makes_the_poison.

28. Offit, P. A., and Jew, R. K. "Addressing parents' concerns: do vaccines contain harmful preservatives, adjuvants, additives, or residuals?". *Pediatrics,* vol. 112, no. 6 Pt 1 (Dec 2003): 1394–7. doi:10.1542/peds.112.6.1394 PMID: 14654615.

29. Golub, M. S., Donald, J. M., Gershwin, M. E., and Keen, C. L. "Effects of aluminum ingestion on spontaneous motor activity of mice." *Neurotoxicol Teratol,* vol. 11, no. 3 (May-Jun 1989): 231–5. doi:10.1016/0892–0362(89)90064–0 PMID: 2755419.

30. Golub, M. S., Han, B., Keen, C. L., and Gershwin, M. E. "Auditory startle in Swiss Webster mice fed excess aluminum in diet." *Neurotoxicol Teratol,* vol. 16, no. 4 (Jul-Aug 1994): 423–5. doi:10.1016/0892–0362(94)90031–0 PMID: 7968944.

31. Golub, M. S., Han, B., Keen, C. L., Gershwin, M. E., and Tarara, R. P. "Behavioral performance of Swiss Webster mice exposed to excess dietary aluminum during development or during development and as adults." *Toxicol Appl Pharmacol,* vol. 133, no. 1 (Jul 1995): 64–72. doi:10.1006/taap.1995.1127 PMID: 7597711.

32. Golub, M. S., Germann, S. L., and Keen, C. L. "Developmental aluminum toxicity in mice can be modulated by low concentrations of minerals (Fe, Zn, P, Ca, Mg) in the diet." *Biol Trace Elem Res,* vol. 93, no. 1–3 (Summer 2003): 213–26. doi:10.1385/ BTER:93:1–3:213 PMID: 12835503.; Golub, M. S., and Germann, S. L. "Long-term

consequences of developmental exposure to aluminum in a suboptimal diet for growth
and behavior of Swiss Webster mice." *Neurotoxicol Teratol,* vol. 23, no. 4 (Jul–Aug
2001): 365–72. doi:10.1016/s0892-0362(01)00144–1 PMID: 11485839.; Golub, M.
S., Germann, S. L., Han, B., and Keen, C. L. "Lifelong feeding of a high aluminum
diet to mice." *Toxicology,* vol. 150, no. 1–3 (Sep 7 2000): 107–17. doi:10.1016/s0300-
483x(00)00251–1 PMID: 10996667.

33. Courchesne, E., Mouton, P. R., Calhoun, M. E., Semendeferi, K., Ahrens-Barbeau, C.,
Hallet, M. J., Barnes, C. C., and Pierce, K. "Neuron number and size in prefrontal cortex
of children with autism." *JAMA,* vol. 306, no. 18 (Nov 9 2011): 2001–10. doi:10.1001/
jama.2011.1638 PMID: 22068992.

34. Some of the former rise to the level of "mortal sins" in the eyes of the Catholic Church
as cited earlier: A mortal sin in Catholic theology is a gravely sinful act, which can lead
to damnation if a person does not repent of the sin before death. In order for a sin to be
mortal, it must meet three conditions: 1. Mortal sin is a sin of grave matter; 2. Mortal
sin is committed with full knowledge of the sinner; 3. Mortal sin is committed with
deliberate consent of the sinner. https://www.vatican.va/archive/ENG0015/__P6C.
HTM#$21R.

35. Bishop, N. J., Morley, R., Day, J. P., and Lucas, A. "Aluminum neurotoxicity in preterm
infants receiving intravenous-feeding solutions." *N Engl J Med,* vol. 336, no. 22 (May 29
1997): 1557–61. doi:10.1056/NEJM199705293362203 PMID: 9164811.

36. This is the sort of thing that would normally land an academic in hot water at most universities. It's one thing to disagree with published literature, and that is fine. It's another
thing to cherry-pick studies that support one's views, and while reprehensible scholarship, that happens. It is, however, an entirely different category of offense to factually
misrepresent the actual literature for some purpose. The last would fall squarely into the
box of academic malfeasance. Had I done this, my university would have had my behind
hauled before an investigation panel at the earliest opportunity. But maybe it works differently at CHOP and the University of Pennsylvania. Maybe all the royalty dollars that
RotaTeq generated help make academic misconduct simply invisible?

37. *Vaccine- and Vaccine Safety-Related Q&A Sheets.* (2020). From "Vaccine Education
Center". Retrieved from Children's Hospital of Philadelphia, https://www.chop.edu
/centers-programs/vaccine-education-center/resources/vaccine-and-vaccine-safety
-related-qa-sheets.

38. https://www.chop.edu/centers-programs/vaccine-education-center.

39. Velazquez, F. R., Colindres, R. E., Grajales, C., Hernandez, M. T., Mercadillo, M. G.,
Torres, F. J., Cervantes-Apolinar, M., Deantonio-Suarez, R., Ortega-Barria, E., Blum,
M., Breuer, T., and Verstraeten, T. "Postmarketing surveillance of intussusception fol-
lowing mass introduction of the attenuated human rotavirus vaccine in Mexico." *Pediatr
Infect Dis J,* vol. 31, no. 7 (Jul 2012): 736–44. doi:10.1097/INF.0b013e318253add3
PMID: 22695189.; Patel, M. M., Lopez-Collada, V. R., Bulhoes, M. M., De Oliveira, L.
H., Bautista Marquez, A., Flannery, B., Esparza-Aguilar, M., Montenegro Renoiner, E.
I., Luna-Cruz, M. E., Sato, H. K., Hernandez-Hernandez Ldel, C., Toledo-Cortina, G.,
Ceron-Rodriguez, M., Osnaya-Romero, N., Martinez-Alcazar, M., Aguinaga-Villasenor,
R. G., Plasencia-Hernandez, A., Fojaco-Gonzalez, F., Hernandez-Peredo Rezk, G.,
Gutierrez-Ramirez, S. F., Dorame-Castillo, R., Tinajero-Pizano, R., Mercado-Villegas,
B., Barbosa, M. R., Maluf, E. M., Ferreira, L. B., De Carvalho, F. M., Dos Santos, A.
R., Cesar, E. D., De Oliveira, M. E., Silva, C. L., De Los Angeles Cortes, M., Ruiz
Matus, C., Tate, J., Gargiullo, P., and Parashar, U. D. "Intussusception risk and health

benefits of rotavirus vaccination in Mexico and Brazil." *N Engl J Med,* vol. 364, no. 24 (Jun 16 2011): 2283–92. doi:10.1056/NEJMoa1012952 PMID: 21675888.; Buttery, J. P., Danchin, M. H., Lee, K. J., Carlin, J. B., Mcintyre, P. B., Elliott, E. J., Booy, R., Bines, J. E., and Group, P. a. S. "Intussusception following rotavirus vaccine administration: post-marketing surveillance in the National Immunization Program in Australia." *Vaccine,* vol. 29, no. 16 (Apr 5 2011): 3061–6. doi:10.1016/j.vaccine.2011.01.088 PMID: 21316503.; In response, the Center for Biologics Evaluation and Research of the FDA initiated the study of RotaTeq and Rotarix in the Post-Licensure Rapid Immunization Safety Monitoring (PRISM) program and reported that RotaTeq was associated with ~1.5 excess cases per 100,000 recipients of the first dose (Yih, W. K., et al., 2014). Intussusception risk after rotavirus vaccination in U.S. infants. N Engl J Med, 370(6), 503–12. The risk is about one in twenty thousand US infants to one in a hundred thousand US infants within a week of getting the vaccine, and this means that between forty and 120 US infants might develop intussusception associated with rotavirus vaccine each year.

40. *Our Scientists.* (2017). Retrieved from The Wistar Institute, https://wistar.org/our -scientists.; How they describe themselves: "Wistar is a world leader in early stage discovery science in the areas of cancer, immunology and infectious disease." And they do vaccine research.

41. Olmsted, D., and Blaxill, M. *Counting Offit's Millions: More on How Merck's Rotateq Vaccine Made Paul Offit Wealthy.* (2011). Retrieved from Age of Autism, https://www .ageofautism.com/2011/01/counting-offits-millions-more-on-how-mercks-rotateq -vaccine-made-paul-offit-wealthy.html.; Kelly, M. *Dr Paul "for profit" Offit, measles and the BBC.* (2013). From "OPENDEMOCRACYUK". Retrieved from Open Democracy, https://www.opendemocracy.net/en/opendemocracyuk/dr-paul-for-profit-offit-measles -and-bbc/.; One should note that *Age of Autism* is considered to be quite anti-vaccine, as are the authors of this article. The reader can decide if this is a fair evaluation.

42. Olmsted, D., and Blaxill, M. *Voting Himself Rich: CDC Vaccine Adviser Made $29 Million Or More After Using Role to Create Market.* (2009). Retrieved from Age of Autism, https://www.ageofautism.com/2009/02/voting-himself-rich-cdc-vaccine-adviser-made- 29-million-or-more-after-using-role-to-create-market.html.; Offit seems to have recused himself from the vote on his own rotavirus vaccine but did vote on that of a competitor.

43. Boghani, P. *Dr. Paul Offit: "A Choice Not To Get a Vaccine Is Not a Risk-Free Choice".* (2015). Retrieved from Frontline, https://www.pbs.org/wgbh/frontline/article/paul-offit -a-choice-not-to-get-a-vaccine-is-not-a-risk-free-choice/.

44. We have a prime minister in Canada at present who does this all of the time each time he gets caught in an ongoing series of scandals.; Gollom, M. *What you need to know about the SNC-Lavalin affair.* (2019). From "Politics·CBC EXPLAINS". Retrieved from CBC News, https://www.cbc.ca/news/politics/trudeau-wilson-raybould-attorney-general -snc-lavalin-1.5014271.; Ivison, J. *John Ivison: PM's defence to ethics czar reveals his nasty political side.* (2019). From "Opinion". Retrieved from National Post, https://nationalpost .com/opinion/john-ivison-pms-defence-to-ethics-czar-reveals-his-nasty-political-side.

45. Brunk, D. *Tide beginning to turn on vaccine hesitancy.* (2019). From "Conference Converage". Retrieved from MDedge, https://www.mdedge.com/pediatrics/article/211966/vaccines /tide-beginning-turn-vaccine-hesitancy?sso=true.

46. *"concommitant use studies of vaccines".* From "Search Results". Retrieved from US Food & Drug Administration, https://search.usa.gov/search?utf8=%E2%9C%93&affiliate =fda1&sort_by=&query=concommitant+use+studies+of+vaccines&commit=Search.

47. Mullersman, J. *Required Postmarketing Studies.* (2012). From "Application of Pharmacovigilance to U.S. FDA Regulatory Decisions for Vaccines". Retrieved from US Food & Drug Administration, https://www.fda.gov/media/93833/download.

48. Patel, D. *10 Telltale Phrases That Indicate Somebody Isn't Telling the Truth.* (2018). From "Communication". Retrieved from Entrepreneur, https://www.entrepreneur.com/article/321282.

49. Dang-Tan, T., Mahmud, S. M., Puntoni, R., and Franco, E. L. "Polio vaccines, Simian Virus 40, and human cancer: the epidemiologic evidence for a causal association." *Oncogene,* vol. 23, no. 38 (Aug 23 2004): 6535–40. doi:10.1038/sj.onc.1207877 PMID: 15322523.; Butel, J. S. "Simian virus 40, poliovirus vaccines, and human cancer: research progress versus media and public interests." *Bull World Health Organ,* vol. 78, no. 2 (2000): 195–8. https://www.ncbi.nlm.nih.gov/pubmed/10743284 PMID:10743284.

50. Keim, B. *Did Merck Bring AIDS to America? No.* (2007). From "Science". Retrieved from Wired, https://www.wired.com/2007/09/did-merck-bring/.

51. Caplan, A. L. *Revoke the license of any doctor who opposes vaccination.* (2015). From "Opinions". Retrieved from the Washington Post, https://www.washingtonpost.com/opinions/revoke-the-license-of-any-doctor-who-opposes-vaccination/2015/02/06/11a05e50-ad7f-11e4-9c91-e9d2f9fde644_story.html.

52. Irving, D. N. "What Is Bioethics? (Quid Est 'Bioethics'?)." In *Life and Learning Conference.* edited by Joseph W. Koterski.pg. 1–84. Georgetown University: University Faculty for Life, 2000. https://www.catholicculture.org/culture/library/view.cfm?recnum=3320, https://books.google.ca/books/about/Life_and_Learning_Ten.html?id=55biRQAACAAJ&hl=en&output=html_text&redir_esc=y.

53. Lyons-Weiler, J. (2021). Plan B Public Health Infrastructure and Operations Oversight Reform for America. Intl J Vaccine Theory, Practice, and Research, 1(2). The particular part of the transcript that may be most relevant is found in the section entitled, Mass Vaccination Programs Are Not Founded on Solid Ethics.

54. Steenhuysen, J. *As pressure for coronavirus vaccine mounts, scientists debate risks of accelerated testing.* (2020). Edited by Michele Gershberg and Bill Rigby. Retrieved from Reuters, https://www.reuters.com/article/us-health-coronavirus-vaccines-insight/as-pressure-for-coronavirus-vaccine-mounts-scientists-debate-risks-of-accelerated-testing-idUSKB-N20Y1GZ.

55. Hotez, P. J., and Caplan, A. L. *Vaccines Did Not Cause Rachel's Autism: My Journey as a Vaccine Scientist, Pediatrician, and Autism Dad.* Baltimore: Johns Hopkins University Press, 2018. https://books.google.ca/books?id=NkpyDwAAQBAJ.

56. Belluz, J. *This autism dad has a warning for anti-vaxxers.* (2019). Retrieved from Vox, https://www.vox.com/science-and-health/2018/10/16/17964992/vaccine-autism-book-peter-hotez.

57. Gutierrez, M. *California vaccine bill clears Assembly panel despite emotional backlash from parents.* (2019). From "Politics". Retrieved from Los Angeles Times, https://www.latimes.com/politics/la-pol-ca-vaccine-exemption-bill-hearing-20190620-story.html.; Gutierrez, M. *Anti-vaccine activist assaults California vaccine law author, police say.* (2019). From "California". Retrieved from the Los Angeles Times, https://www.latimes.com/california/story/2019-08-21/richard-pan-confronted-anti-vaccine-activist.

58. Stack, L. *A Brief History of Deadly Attacks on Abortion Providers.* (2015). Retrieved from the New York Times, https://www.nytimes.com/interactive/2015/11/29/us/30abortion-clinic-violence.html.

59. *Thought leader.* Retrieved from Oxford, Lexico, https://www.lexico.com/en/definition/thought_leader.

60. Rampton, S., and Stauber, J. *Trust Us, We're Experts PA: How Industry Manipulates Science and Gambles with Your Future.* New York: Penguin Publishing Group, 2002. https://books.google.ca/books?id=_OEPBt16JscC.

61. *How to Be a Troll.* (2020). Retrieved from wikiHow, https://www.wikihow-fun.com/Be-a-Troll.

62. *Electronic Support for Public Health - Vaccine Adverse Event Reporting System (ESP:VAERS) (Massachusetts).* (2007). Edited by Harvard Pilgrim Healthcare Inc. From "Digital Healthcare Research". Retrieved from US Department of Health & Human Services, Agency for Healthcare Research and Quality, https://digital.ahrq.gov/ahrq-funded-projects/electronic-support-public-health-vaccine-adverse-event-reporting-system.

63. *Dorit Reiss.* From "People". Retrieved from UC Hastings Law, https://www.uchastings.edu/people/dorit-reiss/.

64. The paper is now in press as Tarsell et al., 2021, in print, *Science, Public Health Policy & the Law,* Accepted 1/13.

65. Moyer, M. W. *Anti-Vaccine Activists Have Taken Vaccine Science Hostage.* (2018). From "Opinion". Retrieved from the New York Times, https://www.nytimes.com/2018/08/04/opinion/sunday/anti-vaccine-activists-have-taken-vaccine-science-hostage.html.

66. Szumilas, M. "Explaining odds ratios." *J Can Acad Child Adolesc Psychiatry,* vol. 19, no. 3 (Aug 2010): 227–9. https://www.ncbi.nlm.nih.gov/pubmed/20842279 PMID:20842279.

67. Skowronski, D. M., De Serres, G., Crowcroft, N. S., Janjua, N. Z., Boulianne, N., Hottes, T. S., Rosella, L. C., Dickinson, J. A., Gilca, R., Sethi, P., Ouhoummane, N., Willison, D. J., Rouleau, I., Petric, M., Fonseca, K., Drews, S. J., Rebbapragada, A., Charest, H., Hamelin, M. E., Boivin, G., Gardy, J. L., Li, Y., Kwindt, T. L., Patrick, D. M., Brunham, R. C., and Canadian, S. T. "Association between the 2008–09 seasonal influenza vaccine and pandemic H1N1 illness during Spring-Summer 2009: four observational studies from Canada." *PLoS Med,* vol. 7, no. 4 (Apr 6 2010): e1000258. doi:10.1371/journal.pmed.1000258 PMID: 20386731.

68. Zhang, A., Stacey, H. D., Mullarkey, C. E., and Miller, M. S. "Original Antigenic Sin: How First Exposure Shapes Lifelong Anti-Influenza Virus Immune Responses." *J Immunol,* vol. 202, no. 2 (Jan 15 2019): 335–40. doi:10.4049/jimmunol.1801149 PMID: 30617114.

69. Skowronski, D. M., Hamelin, M. E., De Serres, G., Janjua, N. Z., Li, G., Sabaiduc, S., Bouhy, X., Couture, C., Leung, A., Kobasa, D., Embury-Hyatt, C., De Bruin, E., Balshaw, R., Lavigne, S., Petric, M., Koopmans, M., and Boivin, G. "Randomized controlled ferret study to assess the direct impact of 2008–09 trivalent inactivated influenza vaccine on A(H1N1)pdm09 disease risk." *PLoS One,* vol. 9, no. 1 (2014): e86555. doi:10.1371/journal.pone.0086555 PMID: 24475142.

70. And indeed, at the University of British Columbia, scholarly ethics guidelines state that you cannot suppress research findings, your own or those of others.

71. I wrote to Dr. Skowronski during the writing of this chapter. I politely explained to her what we do in my laboratory and how this led us to be considered as anti-vaxxers. I reflected on how her situation as described in the Moyer's article had helped me understand some of the peer pressures faced by vaccine researchers. She never wrote back, more or less as expected.

72. Cernic, M. In *Ideological Constructs of Vaccination*. pg. 20–21: Wirral, UK: Vega Press Limited, 2018. https://books.google.ca/books?id=LHz-swEACAAJ.

73. Ibid, 20.

74. Wilson, D. *The Term "Conspiracy Theory"—an Invention of the CIA*. Retrieved from Project Unspeakable, https://projectunspeakable.com/conspiracy-theory-invention-of-cia/.

Chapter 7

1. Shakur, A., Davis, A. Y., and Hinds, L. S. *Assata: An Autobiography*. L. Hill, 2001. https://books.google.ca/books?id=kVVp9RLqlYwC, https://www.goodreads.com/quotes /568550-people-get-used-to-anything-the-less-you-think-about.

2. A more conventional view is that exosomes can be either harmful or beneficial. Here is how exosomes are conventionally defined: "Exosomes are defined as nanometre-sized vesicles, being packages of biomolecules ranging from 40–150 nanometres in size that are released by virtually every cell type in the body. Once thought to be a kind of refuse disposal system for cells, exosomes are now known to be far more important than that. Exosomes have been shown to be key mediators of cell to cell communication, delivering a distinct cargo of lipids, proteins and nucleic acids that reflects their cell of origin." – Edgar, J. R. (2016). Q&A: What are exosomes, exactly? BMC Biology, 14, 46. Retrieved from https://bmcbiol.biomedcentral.com/articles/10.1186/s12915-016-0268-z#citeas.

3. Some of the odder and perhaps more entertaining "conspiracy" theories; Hamill, J. (2017). It's a UF-SNOW: Antarctic UFO hunters spot alien ship hidden in a cave near the South Pole. Retrieved from The Sun: https://www.thesun.co.uk/tech/2701963 /antarctic-ufo-hunters-spot-alien-ship-hidden-a-cave-near-the-south-pole/.

4. Minsky, M. *The Emotion Machine: Commonsense Thinking, Artificial Intelligence, and the Future of the Human Mind*. New York: Simon & Schuster, 2007. https://books.google .ca/books?id=OqbMnWDKIJ4C.

5. As this is my ethic identity, the use of terms like "kike" or "Jewboy" are very offensive. Anyone using them on me or my family had better either be another member of the tribe, or know me super well and say it with a smile.

6. Sadly, some vaccine-hesitant individuals and groups have hewed pretty close to the same habit in their description of what they term "extreme leftist" individuals or policies. By this, they typically mean someone like Senator Bernie Sanders or Representative Alexandra Ocasio-Cortez, both of whom self-describe as "democratic socialists." Calling them extreme leftists only reinforces for me the obvious fact that such people have never lived outside of the United States, a country with a distinctly atrophied political spectrum. In Canada, for example, both Sanders and Ocasio-Cortez would fit comfortably in our own "soft" socialist political party, the New Democratic Party or NDP, which only the most ardent right-wing polemicist could describe as extreme left.

7. Sears's father, William, wrote one of the most cited books on pediatric medicine: Sears, W., et al. *The Baby Book, Revised Edition: Everything You Need to Know About Your Baby from Birth to Age Two*. Boston: Little, Brown and Company, 2013.

8. Sears, R. W. *The Vaccine Book: Making the Right Decision for Your Child*. 2nd ed.: Boston: Little, Brown, 2011. https://books.google.ca/books?id=wL4D0Xmnz3wC.

9. McFarland, G., La Joie, E., Thomas, P., and Lyons-Weiler, J. "Acute exposure and chronic retention of aluminum in three vaccine schedules and effects of genetic and environmental variation." *J Trace Elem Med Biol,* vol. 58 (Mar 2020): 126444. doi:10.1016/j. jtemb.2019.126444 PMID: 31846784.

10. Kupferschmidt, K. *Top Israeli immunologist accused of promoting antivaccine views.* (2019). From "News". Retrieved from Science Magazine, https://www.sciencemag.org/news/2019/11/top-israeli-immunologist-accused-promoting-antivaccine-views.

11. Caplan, A. L., Hoke, D., Diamond, N. J., and Karshenboyem, V. "Free to choose but liable for the consequences: should non-vaccinators be penalized for the harm they do?". *J Law Med Ethics,* vol. 40, no. 3 (Fall 2012): 606–11. doi:10.1111/j.1748-720X.2012.00693.x PMID: 23061588.

12. Graham, M., and Rodriguez, S. *Facebook says it will finally ban anti-vaccination ads.* (2020). From "News". Retrieved from CNBC, https://www.cnbc.com/2020/10/13/facebook-bans-anti-vax-ads.html.

13. Stone, J. *Coronavirus vaccine: Labour calls for emergency censorship laws for anti-vax content.* (2020). From "Politics". Retrieved from Independent, https://www.independent.co.uk/news/uk/politics/coronavirus-vaccine-covid-anti-vax-labour-censor-b1723009.html.

14. Poitras, J. *Cardy hopes to pass contentious vaccination bill by summer.* (2020). From "News". Retrieved from CBC News, https://www.cbc.ca/news/canada/new-brunswick/nb-dominic-cardy-vaccination-bill-1.5567980.

15. Pedersen, K., Szeto, E., and Tomlinson, A. *Nearly half of Canadians are concerned about vaccine safety. Here's why. Social Sharing.* (2020). From "Marketplace". Retrieved from CBC News, https://www.cbc.ca/news/health/anti-vaccine-myths-biases-1.5429845.; Reuters, T. *Vaccine mistrust leaves populations vulnerable, global study shows.* (2019). From "Health". Retrieved from CBC News, https://www.cbc.ca/news/health/vaccines-trust-1.5181208.

16. *Wellcome Global Monitor 2018.* (2019). From "Report Summary". Retrieved from The Wellcome Trust, https://wellcome.org/reports/wellcome-global-monitor/2018.

17. Schiff, A. B. *Schiff Sends Letter to Amazon CEO Regarding Anti-Vaccine Misinformation.* (2019). From "Press Releases". Retrieved from Congressman Adam Schiff: Representing California's 28th District, https://schiff.house.gov/news/press-releases/schiff-sends-letter-to-amazon-ceo-regarding-anti-vaccine-misinformation.; Fingas, J. *Congressman asks Amazon to stop suggesting anti-vaccination content.* (2019). Retrieved from Engadget, https://www.engadget.com/2019-03-01-schiff-asks-amazon-to-stop-anti-vaccination-book-recommendations.html.

18. *MeWe: The Social Network Built on Trust, Control and Love.* (2021). Retrieved from MeWe, https://mewe.com/.; Binder, M. *What is MeWe? Everything you need to know about the social network competing with Parler.* (2021). From "Tech". Retrieved from Mashable, https://mashable.com/article/what-is-mewe-network-explainer/.

19. *Dark web.* (2019). Retrieved from Wikipedia, https://en.wikipedia.org/wiki/Dark_web.

20. Schiff, A. B. *Schiff Sends Letter to Google, Facebook Regarding Anti-Vaccine Misinformation.* (2019). From "Press Releases". Retrieved from Congressman Adam Schiff: Representing California's 28th District, https://schiff.house.gov/news/press-releases/schiff-sends-letter-to-google-facebook-regarding-anti-vaccine-misinformation.

21. Boycott, Divestment, and Sanctions (BDS) is a movement seeking to redress concerns for Israeli governmental policies targeting Palestinians in the West Bank and Gaza. It is a nonviolent group advocating for using economic means to encourage Israel to change policy.; What is BDS? (undated). Retrieved from BDS: https://bdsmovement.net/what-is-bds

22. Bailey, M., dir. *The Pathological Optimist.* (2017; GathrFilms, 20174), Documentary. IMDB: https://www.imdb.com/title/tt6391732/.

23. Gregory, L. M., dir. *1986: The Act*. (2020; 7th Chakra Films, July 10, 2020), Documentary. IMDB: https://www.imdb.com/title/tt12708236/.

24. Ratcliffe, S. "Upton Sinclair 1878–1968." In *Oxford Essential Quotations*. 4th ed.: Oxford, UK: Oxford University Press, 2016. https://www.oxfordreference.com/view/10.1093 /acref/9780191826719.001.0001/q-oro-ed4-00010168#:~:text=American%20 novelist%20and%20social%20reformer,on%20his%20not%20understanding%20it.

25. Britany Valas, one of the organizers of *One Conversation*, suggested that the official reason Roe gave for backing out had more to do with scheduling than being terrified of anti-vaxxers. However, it seems from the email exchange between Gorki and Roe that he certainly frightened her about her fate if she did attend.

26. *National Vaccine Information Center*. Retrieved from National Vaccine Information Center, https://www.nvic.org/.

27. The origin of Barbara Loe Fisher's concerns were due to what she believed was damage to her son from a DPT vaccine. She later helped found the National Vaccine Information Center; Barbara Loe Fisher. (2021). Retrieved from National Vaccine Information Center: https://www.nvic.org/about/barbarafisherbio.aspx.

28. Robert F. Kennedy Jr. (2021). Retrieved from Wikipedia: https://en.wikipedia.org /wiki/Robert_F._Kennedy_Jr.; Wikipedia shows a long list of Kennedy's environmental work. Wikipedia's tendency is to keep harping on Kennedy being an anti-vaxxer, but does note his efforts and victories in environmental law.

29. https://childrenshealthdefense.org/news/robert-f-kennedy-jr-speaks-at-berlin-rally-for -freedom-and-peace/.

30. Retrieved from Children's Health Defense, https://childrenshealthdefense.org/.

31. Ibid.

32. Children's Health Defense Team. "Planned Surveillance and Control by Global Technocrats: A Big-Picture Look at the Current Pandemic Beneficiaries." *Int. Journal of Vaccine Theory, Practice, and Research,* vol. 1, no. 2 (2021): 143–71. https://ijvtpr.com /index.php/IJVTPR/article/view/7/16.

33. Hillel, A. *Spasmodic Dysphonia*. Retrieved from Johns Hopkins Medicine, https://www .hopkinsmedicine.org/health/conditions-and-diseases/spasmodic-dysphonia#:~:text =Spasmodic%20dysphonia%20is%20a%20voice,able%20to%20talk%20at%20all.

34. *About Dr. Phil*. Retrieved from CBS Television Distribution and CBS Interactive Inc., https://www.drphil.com/about-dr-phil/.

35. *SB-277 Public health: vaccinations*. (2015). Retrieved from California Legislative Information, https://leginfo.legislature.ca.gov/faces/billNavClient.xhtml?bill_id=201520160SB277.; *Legislation will protect every student's right to be safe at school by closing California vaccine loophole*. (2015). Retrieved from Dr. Richard Pan: California State senator, https://sd06.senate.ca.gov/news/2015-04-22-senators-richard-pan-and-ben-allen% -E2%80%99s-sb-277-passes-senate-education-committee.

36. Interview with author, Nov. 19, 2020.

37. Many of the international fighters who came to join the Kurdish militias, the YPG and YPJ, were a mixed bag of political beliefs: Some came because they believed in the goals of the revolution; others wanted only to fight the Islamic State; still others were simply adventurers, or FaceBook warriors. In the fighting in Rojava in 2019, members of the International Freedom Battalion (IFB) were on the front lines in Tel Tamer facing the Turkish army and their jihadi allies. IFB is unabashedly anarchist or even Marxist. The most effective ambulance-medic service was provided by the Free Burma Rangers, a

conservative, Christian group. The two groups respected and liked each other, providing an interesting perspective on alliances that might form in the future in other struggles.

38. *Shop*. Retrieved from VASHIVA, https://vashiva.com/shop/.

39. Kennedy, R. F. *Critical Questions for Dr. Shiva About His Attempts to Splinter the Health Freedom Movement.* (2020). From "News". Retrieved from Children's Health Defense, https://childrenshealthdefense.org/news/critical-questions-for-dr-shiva-about-his -attempts-to-splinter-the-health-freedom-movement/.

40. Note regarding dissentions in vaccine movement: Divisions in revolutions/movements are not unusual. For example, see the Conway Cabal during the U.S. revolution. The Conway Cabal. (2021). Retrieved from UShistory: https://www.ushistory.org/march /other/cabal.htm.

41. Kuntz, T. *Dare to Question: One Parent to Another.* CreateSpace Independent Publishing Platform, 2018. https://books.google.ca/books?id=7i2ctAEACAAJ.

Chapter 8

1. Paine, T. *Being an Answer to Mr. Burke's Attack on the French Revolution - part 7 of 16.* (1791). From "The Rights of Man". Retrieved from Independence Hall Association, https://www.ushistory.org/paine/rights/c1-016.htm.

2. Cernic, M. In *Ideological Constructs of Vaccination.* pg. 147: Wirral, UK Vega Press Limited, 2018. https://books.google.ca/books?id=LHz-swEACAAJ.

3. I am fully aware that none of these is a monolithic entity and that significant points of dispute exist within Christianity and Islam, and to a somewhat lesser extent in Judaism. Further, in using these religions as the examples, I do not intend any disrespect to the other large religions such as Hinduism, Buddhism, or Sikhism, nor to the myriad smaller religious faiths, including those of the Native peoples of the Americas. At least to me, they are all valuable aspects of human culture and history.

4. *Religion.* (2001). Retrieved from Wikipedia, https://en.wikipedia.org/wiki/Religion.

5. Lewis, D. *Science, Religion and the Golden Rule: In God We Trust.* pg. 1: New York: Skyhorse, 2021.

6. Ibid, 8.; As this book goes to press, I have not yet received a comment on the subject of religion and vaccinology by anyone of the Muslim faith.

7. Bloom, P. *Scientific Faith Is Different From Religious Faith.* (2015). From "Science". Retrieved from The Atlantic, https://www.theatlantic.com/science/archive/2015/11 /why-scientific-faith-isnt-the-same-as-religious-faith/417357/.

8. Ibid.

9. Wilson, E. O. *Consilience: The Unity of Knowledge.* New York: Vintage Books, 1999. https://books.google.ca/books?id=-YsWNfTXU7oC.

10. Cernic, M. In *Ideological Constructs of Vaccination.* pg. 27: Wirral, UK: Vega Press Limited, 2018. https://books.google.ca/books?id=LHz-swEACAAJ.

11. Ibid, 29.

12. Rogers, T. M. "The Political Economy of Autism." Postgraduate theis for PhD Doctorate, University of Sydney, 2018. Retrieved from https://ses.library.usyd.edu.au /handle/2123/20198.

13. *Vaccinology.* Retrieved from Oxford, Lexico, https://www.lexico.com/definition/vaccinology.

14. The Master of Public Health (MPH) degree (in the School of Population and Public Health) at the University of British Columbia, for example, teaches and "integrates

learning in epidemiology; biostatistics; the social, biological and environmental determinants of health; population health; global health; disease prevention and health systems management with skill-based learning in a practicum setting." – Master of Public Health (MPH). (undated). Retrieved from The University of British Columbia: https://www.spph.ubc.ca/programs/mph/.

15. Petrik, M. S., Wong, M. C., Tabata, R. C., Garry, R. F., and Shaw, C. A. "Aluminum adjuvant linked to Gulf War illness induces motor neuron death in mice." *Journal of Neuromolecular Medicine,* vol. 9, no. 1 (2007): 83–100. doi:10.1385/nmm:9:1:83 PMID: 17114826.

16. Drs. Dan Perl and Michael Strong were the scientists in question. Perl is with the Uniformed Services University; Strong is currently the head of the Canadian Institutes of Health Research.

17. Jews tend to be a bit more tolerant in the sense that you can reject pretty much the whole Torah and still be Jewish. This is, in large part, because Jewishness is tied up in historical, cultural, and ethnic identity that is in many ways quite different from the other religions considered. Notably, there are a few things you can't do and remain Jewish. One of these is to convert to another religion.

18. Cult: ". . . a cult is a social group that is defined by its unusual religious, spiritual, or philosophical beliefs, or by its common interest in a particular personality, object or goal. This sense of the term is controversial, having divergent definitions both in popular culture and academia, and has also been an ongoing source of contention among scholars across several fields of study. It is usually considered a pejorative."—Cult. (2021). Retrieved from Wikipedia: https://en.wikipedia.org/wiki/Cult.

19. Ardelean, C. F., Becerra-Valdivia, L., Pedersen, M. W., Schwenninger, J. L., Oviatt, C. G., Macias-Quintero, J. I., Arroyo-Cabrales, J., Sikora, M., Ocampo-Diaz, Y. Z. E., Rubio, C., Ii, Watling, J. G., De Medeiros, V. B., De Oliveira, P. E., Barba-Pingaron, L., Ortiz-Butron, A., Blancas-Vazquez, J., Rivera-Gonzalez, I., Solis-Rosales, C., Rodriguez-Ceja, M., Gandy, D. A., Navarro-Gutierrez, Z., De La Rosa-Diaz, J. J., Huerta-Arellano, V., Marroquin-Fernandez, M. B., Martinez-Riojas, L. M., Lopez-Jimenez, A., Higham, T., and Willerslev, E. "Evidence of human occupation in Mexico around the Last Glacial Maximum." *Nature,* vol. 584, no. 7819 (Aug 2020): 87–92. doi:10.1038/s41586-020-2509-0 PMID: 32699412.; Montaigne, F. *The Fertile Shore.* (2020). Retrieved from *Smithsonian Magazine,* https://www.smithsonianmag.com/science-nature/how-humans-came-to-americas-180973739/.

20. McLaren, D., Fedje, D., Dyck, A., Mackie, Q., Gauvreau, A., and Cohen, J. "Terminal Pleistocene epoch human footprints from the Pacific coast of Canada." *PLoS One,* vol. 13, no. 3 (2018): e0193522. doi:10.1371/journal.pone.0193522 PMID: 29590165.

21. Tarlach, G. *Did the First Americans Arrive Via A Kelp Highway?* (2017). From "Dead Things". Retrieved from Discover Magazine, https://www.discovermagazine.com/planet-earth/did-the-first-americans-arrive-via-a-kelp-highway.

22. Holen, S. R., Demere, T. A., Fisher, D. C., Fullagar, R., Paces, J. B., Jefferson, G. T., Beeton, J. M., Cerutti, R. A., Rountrey, A. N., Vescera, L., and Holen, K. A. "A 130,000-year-old archaeological site in southern California, USA." *Nature,* vol. 544, no. 7651 (Apr 26 2017): 479–83. doi:10.1038/nature22065 PMID: 28447646.

23. Tuttle, R. H. *The Fossil Evidence.* (2021). From "Human evolution". Retrieved from Britannica, https://www.britannica.com/science/human-evolution/The-fossil-evidence.

24. Ferraro, J. V., Binetti, K. M., Wiest, L. A., Esker, D., Baker, L. E., and Forman, S. L. "Contesting early archaeology in California." *Nature,* vol. 554, no. 7691 (Feb 7

2018): E1-E2. doi:10.1038/nature25165 PMID: 29420468.; Hovers, E. "Archaeology: Unexpectedly early signs of Americans." *Nature,* vol. 544, no. 7651 (Apr 26 2017): 420–21. doi:10.1038/544420a PMID: 28447633.

25. Cernic, M. *Ideological Constructs of Vaccination.* Wirral, UK: Vega Press Limited, 2018. https://books.google.ca/books?id=LHz-swEACAAJ.

26. https://www.abc10.com/article/news/verify/verify-changes-who-definition-herd-immunity -not-secret/507-f90c0199-c88e-4c66-8313-b4ae6e2a72ad.

27. Scientists usually use statistical methods to calculate probability or "p" values. In most cases, at least in the biological sciences, a p value gives the likelihood that some observation is correct. Basically, what one is doing is comparing a "null" hypothesis that something does not change in response to some treatment to the alternative hypothesis that it does. A p-value less than or equal to 0.05 ($\leq$ 0.05), or less than five times in a hundred, is statistically significant and indicates evidence for the alternative hypothesis. Anything larger than (>) 0.05 is evidence for the null hypothesis, 0.06, for example. Values like 0.01 mean that only one time in a hundred might the null hypothesis be correct; 0.001 means only one time in a thousand, etc. But many factors can determine statistical significance, such as the number of samples in the study. One misuse of statistical probability is called "p hacking," in which a scientist will keep moving data around, taking things in and out, seeking to make something seem significant even when it is really not. Those on the pro-vaccine side often accuse the vaccine-skeptical of doing this all the time. Indeed, this does happen, but those on the pro side are not always immune from the temptation to do the same.

28. Cernic, personal communication to the author.

29. Grant tenure has become common, in which a faculty member can only retain their academic position if they continue to have grant support.

30. The Vaccine Confidence Project (VCP): https://www.vaccineconfidence.org/. This organization is the brainchild of Dr. Heidi Larson, an anthropologist. Larson is a professor of anthropology in Risk and Decision Science, at the London School of Hygiene & Tropical Medicine [https://www.lshtm.ac.uk/aboutus/people/larson.heidi]. The VCP website states that "Dr. Larson previously headed Global Immunization Communication at UNICEF, chaired GAVI's Advocacy Task Force, and served on the WHO SAGE Working Group on vaccine hesitancy. The Vaccine Confidence Project is a WHO Centre of Excellence on addressing Vaccine Hesitancy." The website goes on to highlight Larson's "Research focus: Larson's research focuses on the analysis of social and political factors that can affect uptake of health interventions and influence policies. Her particular interest is on risk and rumour management from clinical trials to delivery—and building public trust. She served on the FDA Medical Countermeasure (MCM) Emergency Communication Expert Working Group, and is currently Principal Investigator for a global study on acceptance of vaccination during pregnancy; an EU-funded (EBODAC) project on the deployment, acceptance and compliance of an Ebola vaccine." The four webinars put on by VCP over the course of 2020 were meant to address any future hesitancy to accept COVID-19 vaccines when these vaccines became available. In this series of presentations, the emphasis was on compliance and reassurance, with no consideration whatsoever about possible adverse effects, except insofar as such might impact compliance. The four webinars combined are perhaps the most comprehensive example of what true vaccine faith and cult-like behavior actually looks like.

31. The period after a king named Alexander Yanai was the last period. BC. He conquered non-Jews, and that led to proselytizing. It was not pushed after this but was not

discouraged, e.g., the Khasars. In Europe, it was stopped in the Middle Ages in Christian Europe, as converting a Christian to Judaism could get the whole community killed. To this day, if a non-Jew, goes to a rabbi and asks to convert, the automatic answer is no. Basically, one has to ask numerous times.

32. Weindling, P., Von Villiez, A., Loewenau, A., and Farron, N. "The victims of unethical human experiments and coerced research under National Socialism." *Endeavour,* vol. 40, no. 1 (Mar 2016): 1–6. doi:10.1016/j.endeavour.2015.10.005 PMID: 26749461.; *Nazi Medical Experiments.* (2006). Edited by Washington United States Holocaust Memorial Museum, DC. Retrieved from Holocaust Encyclopedia, https://encyclopedia.ushmm. org/content/en/article/nazi-medical-experiments.; *Unit 1855.* (2017). Retrieved from Wikipedia, https://en.wikipedia.org/wiki/Unit_1855.; *Unit 731.* (2003). Retrieved from Wikipedia, https://en.wikipedia.org/wiki/Unit_731.

33. Kennedy, S. *The Tuskegee Syphilis Study (1939–1972).* (2014). Retrieved from IMARC, https://www.imarcresearch.com/blog/bid/351998/The-Tuskegee-Syphilis-Study -1932-1972.

34. "The Proof as to War Crimes and Crimes Against Humanity." In *Trials of War Criminals Before the Nuernberg Military Tribunals Under Control Council Law No. 10, Nuremberg, October 1946-April, 1949.* pg. 181–82: U.S. Government Printing Office, 1949. https: //www.loc.gov/rr/frd/Military_Law/pdf/NT_war-criminals_Vol-II.pdf.

35. *Universal Declaration of Human Rights.* (1948). Retrieved from United Nations, https: //www.un.org/en/universal-declaration-human-rights/.

36. *The Belmont Report.* (1979). Edited by Office for Human Research Protections. From "Regulations & Policy". Retrieved from US Department of Health & Human Services, https://www.hhs.gov/ohrp/regulations-and-policy/belmont-report/index.html.

37. Ibid, 5.

38. Ibid, 7.

39. Ibid.

40. Irving, D. N. "What Is Bioethics? (Quid Est 'Bioethics'?)." In *Life and Learning Conference.* edited by Joseph W. Koterski.pg. 1–84. Georgetown University: University Faculty for Life, 2000. https://www.catholicculture.org/culture/library/view.cfm? recnum=3320, https://books.google.ca/books/about/Life_and_Learning_Ten.html?id =55biRQAACAAJ&hl=en&output=html_text&redir_esc=y.

41. Ibid, 25.

42. *Federal Policy for the Protection of Human Subjects.* (1991). Retrieved from US Food & Drug Administration, https://www.fda.gov/science-research/clinical-trials-and -human-subject-protection/federal-policy-protection-human-subjects.

43. *WMA Declaration of Helsinki – Ethical Principles for Medical Research Involving Human Subjects.* (1964). Retrieved from World Medical Association, https://www.wma.net/policies -post/wma-declaration-of-helsinki-ethical-principles-for-medical-research-involving -human-subjects/.

44. "World Medical Association Declaration of Helsinki: ethical principles for medical research involving human subjects." *Bulletin of the World Health Organization: the International Journal of Public Health 2001,* vol. 79, no. 4 (2001): 373–74. https://www .who.int/bulletin/archives/79(4)373.pdf.

45. *About.* (2021). Retrieved from The Council For International Organizations Of Medical Sciences, https://cioms.ch/about/.; *International Ethical Guidelines for Biomedical Research Involving Human Subjects.* The Council For International Organizations Of Medical Sciences, 1993. https://books.google.ca/books?id=RvVpAAAAMAAJ.

46. *What Are "Biologics" Questions and Answers.* (2018). Retrieved from US Food & Drug Administration, Center for Biologics Evaluation and Research (CBER), https://www.fda.gov/about-fda/center-biologics-evaluation-and-research-cber/what-are-biologics-questions-and-answers.

47. Hickey, K. J. *The PREP Act and COVID-19: Limiting Liability for Medical Countermeasures.* (2020). Retrieved from Congressional Research Service, https://crsreports.congress.gov/product/pdf/LSB/LSB10443.; Sigalos, M. *You can't sue Pfizer or Moderna if you have severe Covid vaccine side effects. The government likely won't compensate you for damages either.* (2020). From "Health and Science". Retrieved from CNBC, https://www.cnbc.com/2020/12/16/covid-vaccine-side-effects-compensation-lawsuit.html.

48. *Government of Canada Announces pan-Canadian Vaccine Injury Support Program.* (2020). Retrieved from Public Health Agency of Canada, https://www.canada.ca/en/public-health/news/2020/12/government-of-canada-announces-pan-canadian-vaccine-injury-support-program.html.

49. *Benito Mussolini.* From "Quotes". Retrieved from Goodreads Quotes, https://www.goodreads.com/author/quotes/221166.Benito_Mussolini.

50. Soderbergh, S., dir. *Contagion.* (2011; Warner Bros., September 9, 2011). IMDB: https://www.imdb.com/title/tt1598778/.

Chapter 9

1. *Time* Magazine, October 31, 1977; John Osborne, December 12, 1929—December 24, 1994. John James Osborne was an English playwright, screenwriter, actor, and critic of the establishment. The stunning success of his 1956 play *Look Back in Anger* transformed English theater.

2. Retrieved from CBC News, https://www.cbc.ca/.

3. *One Conversation. PART 1 "One Conversation", LIVE from Atlanta!* (2018). Retrieved from YouTube, https://www.youtube.com/watch?v=hc5yHU61jVk.

4. The story is as follows: In late 2011, I proposed to Green College, a resident college at the University of British Columbia, to hold what was termed a "thematic lecture series" on vaccine safety. I proposed finding ten speakers, one per week, who would come in, give a standard academic lecture, and then answer questions. After the talk, the speaker, organizers, and attendees would all go out for dinner. The idea was duly written up, presented to the appropriate vetting committee at Green College, and in due course was approved. It even had a reasonable budget to bring in speakers from outside UBC. All good. The director at Green was pleased, as was I, and I began sending out invitations. Shortly after being approved, Green pulled out. The reason was that the director had received a letter from the then-head of the School of Population and Public Health claiming that having such a series would harm public health by calling into question the benefits of vaccines. This head proposed instead that Green hold a series on the benefits of vaccines and noted that he would be only too happy to provide "trusted" speakers. Green College, always worried about their reputation and hence funding, had to withdraw the series. It was at this point that I realized that vaccine safety was not a topic that could easily be discussed rationally.

5. Orac. (2021). Retrieved from RESPECTFUL INSOLENCE, https://respectfulinsolence.com/.

6. *Wayne State University.* (2021). Retrieved from The World University Rankings, https://www.timeshighereducation.com/world-university-rankings/wayne-state-university.

7. *David Gorski.* (2021). Retrieved from Wayne State University, School of Medicine—Cancer Biology Program, https://cancerbiologyprogram.med.wayne.edu/profile/dz8037.; *David Gorski, MD, PhD, FACS.* (2021). Retrieved from Wayne State University, The Michael and Marian Ilitch Department of Surgery, https://wsusurgery.com/?faculty=david-gorski.; *Meet Dr. David Gorski.* (2021). Retrieved from Wayne State University, Barbara Ann Karmanos Cancer Institute, https://www.karmanos.org/karmanos/video-library/meet-dr-david-gorski-795.; *David H. Gorski, MD, PhD—Managing Editor.* (2021). Retrieved from Science-Based Medicine, https://sciencebasedmedicine.org/editorial-staff/david-h-gorski-md-phd-managing-editor/.

8. *About SBM.* (2021). Retrieved from Science-Based Medicine, https://sciencebasedmedicine.org/about-science-based-medicine/.

9. *Provider Profile: David Gorski, M.D., Ph.D.* (2021). Retrieved from Wayne State University, Barbara Ann Karmanos Cancer Institute, https://www.karmanos.org/karmanos/karmanos-physician-directory/gorski-david-7481.

10. *David Gorski.* (2021). Retrieved from Wayne State University, School of Medicine—Cancer Biology Program, https://cancerbiologyprogram.med.wayne.edu/profile/dz8037.

11. This is a funny case of synchronicity, as riluzole is the sole effective treatment for ALS, albeit with quite marginal effects on symptoms and longevity.

12. *David H Gorski, MD, PhD, FACS.* Retrieved from Google Scholar, https://scholar.google.ca/citations?hl=en&user=7ASTqGoAAAAJ&view_op=list_works&sortby=pubdate.

13. *David H. Gorski, MD, PhD – Managing Editor.* (2021). Retrieved from Science-Based Medicine, https://sciencebasedmedicine.org/editorial-staff/david-h-gorski-md-phd-managing-editor/.; Project: *Inhibition of breast cancer-induced angiogenesis by a diverged homeobox gene.* Gorski, D. H. (2002). Agency: Breast Cancer Research Program. Award Number: DAMD17-03-1-0292. Proposal Number: BC021524. Funding Type: Idea Award. Fund Status: Funded; Retrieved from US Department Of Defense-Congressionally Directed Medical Research Programs, https://cdmrp.army.mil/search.aspx.; Project: *Repurposing a Drug for Amyotrophic Lateral Sclerosis to Treat Triple-Negative Breast Cancer.* Gorski, D. H. (2014). Agency: Breast Cancer Research Program. Award Number: W81XWH-15-1-0468. Proposal Number: BC142052. Funding Type: Breakthrough Award-Funding Level 1. Fund Status: Funded; Retrieved from US Department Of Defense-Congressionally Directed Medical Research Programs, https://cdmrp.army.mil/search.aspx.; Wallner, P. E. *New Jersey Commission on Cancer Research: 2020 Annual Report.* (2000). Retrieved from New Jersey Department of Health & Senior Services, https://www.njleg.state.nj.us/OPI/Reports_to_the_Legislature/cancer_research_2000.pdf.; *Advanced Clinical Research Award.* (2007). Retrieved from American Society of Clinical Oncology, https://www.asco.org/node/145280.; Project: *Mechanism of angiogenesis inhibition by a homeobox gene.* Gorski, D. H. (2008). Agency: National Institute of Health. Award Number: 7385003. Proposal Number: 7R01CA111344-04. Funding Type: Research Project (R01). Fund Status: Funded; Retrieved https://grantome.com/grant/NIH/R01-CA111344-04.; *Scientists receive Karmanos Strategic Research Initiative Grants for promising research.* (2012). From "School of Medicine News". Retrieved from Wayne State University, https://today.wayne.edu/medicine/news/2012/08/22/scientists-receive-karmanos-strategic-research-initiative-grants-for-promising-research-28449.; *David Gorski MD, PhD, FACS.* Retrieved from Doximity, https://www.doximity.com/pub/david-gorski-md.

14. *Dr. David Gorski, MD.* From "Doctors". Retrieved from US News, https://health
.usnews.com/doctors/david-gorski-22437.; *Dr. David Gorski, General Surgeon.* Retrieved
from RateMDs, https://www.ratemds.com/doctor-ratings/329899/Dr-David-Gorski
-Detroit-MI.html.

15. https://en.wikipedia.org/wiki/David_Gorski#/media/File:Gorski1.jpeg.

16. *Orac.* (2020). From "Characters of Blake's 7". Retrieved from Wikipedia, https:
//en.wikipedia.org/wiki/Characters_of_Blake%27s_7#Orac.

17. Kruger, J., and Dunning, D. "Unskilled and unaware of it: how difficulties in recognizing one's own incompetence lead to inflated self-assessments." *J Pers Soc Psychol,* vol. 77,
no. 6 (Dec 1999): 1121–34. doi:10.1037//0022–3514.77.6.1121 PMID: 10626367.

18. Heckenlively, K. *PLAGUE - An Alliance of the Free Peoples of Middle-Earth.* (2014).
Retrieved from Age of Autism, https://www.ageofautism.com/2014/07/plague-an-alliance
-of-the-free-peoples-of-middle-earth.html.

19. *Dorit Reiss.* From "People". Retrieved from UC Hastings Law, https://www.uchastings
.edu/people/dorit-reiss/.

20. Orac. *Christopher Shaw uses the results of an abusive FOIA request to intimidate a scientist.* (2020). Retrieved from RESPECTFUL INSOLENCE, https://respectfulinsolence
.com/2020/06/26/christopher-shaw-uses-abusive-foia/.

21. Ibid.

22. Jenner, E. "Dr. Jenner, in Reply to Mr. Fermor." *Med Phys J,* vol. 6, no. 32 (Oct 1801):
325–26. https://www.ncbi.nlm.nih.gov/pubmed/30491060 PMID: 30491060.

23. *ABOUT SKEPTICAL RAPTOR.* Retrieved from SKEPTICAL RAPTOR, https://www
.skepticalraptor.com/skepticalraptorblog.php/about/.

24. *Dorit Reiss.* From "People". Retrieved from UC Hastings Law, https://www.uchastings
.edu/people/dorit-reiss/.

25. *About us.* (2021). Retrieved from VCS Foundation, https://www.vcs.org.au/about-us
/our-vision/.

26. Hawkes, D., Benhamu, J., Sidwell, T., Miles, R., and Dunlop, R. A. "Revisiting
adverse reactions to vaccines: A critical appraisal of Autoimmune Syndrome Induced
by Adjuvants (ASIA)." *J Autoimmun,* vol. 59 (May 2015): 77–84. doi:10.1016/j.
jaut.2015.02.005 PMID: 25794485.

27. Caulfield, T. *The Vaccination Picture.* Penguin Canada, 2017. https://books.google.ca
/books?id=bJYlDwAAQBAJ.

28. This section was written first, but the article came out before, so I am citing it here to
avoid any concerns about what is termed "self-plagiarism" or using one's own work again
for another purpose.; Shaw, C. A. (2020). Weaponizing the Peer Review System. *Intl J
Vaccine Theory, Practice, and Research,* 1(1), 11–26.

29. Dutta Majumder, P. "Henry Oldenburg: The first journal editor." *Indian J Ophthalmol,*
vol. 68, no. 7 (Jul 2020): 1253–54. doi:10.4103/ijo.IJO_269_20 PMID: 32587145.

30. Enserink, M. *How to avoid the stigma of a retracted paper? Don't call it a retraction.*
(2017). From "News". Retrieved from Science Magazine, https://www.sciencemag.org
/news/2017/06/how-avoid-stigma-retracted-paper-dont-call-it-retraction.

31. *Research.* Retrieved from Meta-Research Innovation Center at Standfard, https://metrics
.stanford.edu/research.

32. Ioannidis, J. P. "The Mass Production of Redundant, Misleading, and Conflicted
Systematic Reviews and Meta-analyses." *Milbank Q,* vol. 94, no. 3 (Sep 2016): 485–
514. doi:10.1111/1468–0009.12210 PMID: 27620683.; Ioannidis, J. P. "Why Science
Is Not Necessarily Self-Correcting." *Perspect Psychol Sci,* vol. 7, no. 6 (Nov 2012):

645–54. doi:10.1177/1745691612464056 PMID: 26168125.; Ioannidis, J. P. "Why most published research findings are false." *PLoS Med,* vol. 2, no. 8 (Aug 2005): e124. doi:10.1371/journal.pmed.0020124 PMID: 16060722.

33. Seralini, G. E., Clair, E., Mesnage, R., Gress, S., Defarge, N., Malatesta, M., Hennequin, D., and De Vendomois, J. S. "Long term toxicity of a Roundup herbicide and a Roundup-tolerant genetically modified maize." *Food Chem Toxicol,* vol. 50, no. 11 (Nov 2012): 4221–31. doi:10.1016/j.fct.2012.08.005 PMID: 22999595.

34. *Seralini's team wins defamation and forgery court cases on GMO and pesticide research.* (2015). Retrieved from GMOSeralini, https://www.gmoseralini.org/seralinis-team -wins-defamation-and-forgery-court-cases-on-gmo-and-pesticide-research/.

35. Inbar, R., Weiss, R., Tomljenovic, L., Arango, M. T., Deri, Y., Shaw, C. A., Chapman, J., Blank, M., and Shoenfeld, Y. "WITHDRAWN: Behavioral abnormalities in young female mice following administration of aluminum adjuvants and the human papillomavirus (HPV) vaccine Gardasil." *Vaccine,* (Jan 9 2016). doi:10.1016/j.vaccine.2015.12.067 PMID: 26778424.

36. Inbar, R., Weiss, R., Tomljenovic, L., Arango, M. T., Deri, Y., Shaw, C. A., Chapman, J., Blank, M., and Shoenfeld, Y. "Behavioral abnormalities in female mice following administration of aluminum adjuvants and the human papillomavirus (HPV) vaccine Gardasil." *Immunol Res,* vol. 65, no. 1 (Feb 2017): 136–49. doi:10.1007/s12026-016-8826-6 PMID: 27421722.

37. Rampton, S., and Stauber, J. *Trust Us, We're Experts PA: How Industry Manipulates Science and Gambles with Your Future.* New York: Penguin Publishing Group, 2002. https://books.google.ca/books?id=_OEPBt16JscC.

38. Sagan, C. *The Demon-Haunted World: Science as a Candle in the Dark.* New York: Random House Publishing Group, 2011. https://books.google.ca/books?id=Yz8Y6KfXf9UC.

39. Popova, M. *The Baloney Detection Kit: Carl Sagan's Rules for Bullshit-Busting and Critical Thinking.* (2014). Retrieved from Brain Pickings, https://www.brainpickings .org/2014/01/03/baloney-detection-kit-carl-sagan/.

Chapter 10

1. Plato. *"We can easily forgive a child who is afraid of the dark; the real tragedy of life is when men are afraid of the light."* From "Quotes". Retrieved from Goodreads Quotes, https: //www.goodreads.com/quotes/19198-we-can-easily-forgive-a-child-who-is-afraid-of.

2. Burke, E. *"No power so effectually robs the mind of all its powers of acting and reasoning as fear.".* From "Quotes". Retrieved from Goodreads Quotes, https://www.goodreads.com /quotes/167843-no-power-so-effectually-robs-the-mind-of-all-its.

3. https://commons.wikimedia.org/wiki/File:Women_wearing_face_masks_in_ Copenhagen_(51087395881).jpg.

4. From Wikipedia: "I. F. Stone (Isidor Feinstein Stone, December 24, 1907–June 18, 1989) was a politically progressive American investigative journalist, writer, and author.

    He is best remembered for I. F. Stone's Weekly (1953–71), a newsletter ranked 16th among the top hundred works of journalism in the U.S., in the twentieth century, by the New York University journalism department, in 1999; and second place among print journalism publications." It was from reading the Weekly that I learned about the US war in Vietnam.

Seymour Hirsh (1937–) is an American journalist perhaps best known for exposing the American massacre of Vietnamese civilians at My Lai in 1968.

Robert Woodward (1943–) and Carl Bernstein (1944–) were the two *Washington Post* reporters whose stories on the break-in at the Watergate Hotel in 1972 in Washington, DC, eventually led to the resignation of President Richard Nixon in 1974.

5. Witness for Peace was a mostly American, mostly Christian, antiwar group that opposed America's attempts to destabilize the Sandinista revolution in Nicaragua. The group would deliberately put their members into border villages hoping that the presence of foreign activists would prevent the Contras from attacking. This tactic seems to have mostly worked.

6. The Contras were a hodge-podge collection of former soldiers from the army of the dictator Anastasio Somoza, along with various mercenaries. They launched endless attacks from neighboring Honduras with the goal of bringing down the Sandinista government of Daniel Ortega through terror. The level of brutality of the Contras was well known, even though (then-) President Ronald Reagan had termed them the "equivalent to our [US] Founding Fathers."

7. I'm not sure how the story was broadcast, or if it was at all. But the events described here are as they actually occurred.

8. Börjesson, K., and Vidal, G. *Into the Buzzsaw: Leading Journalists Expose the Myth of a Free Press*. Amherst, NY: Prometheus Books, 2004. https://books.google.ca/books?id=N-MaAQAAIAAJ.

9. Operation Tailwind was a combined US Special Forces, Army of the Republic of Vietnam (ARVN) assault into Laos that began on September 11, 1970. The official reason was to disrupt North Vietnamese forces controlling a key road. CNN journalists/producers April Oliver and Mike Smith claimed that there were additional angles to the story. One was the use of sarin gas, a nerve agent. The second was that a goal of the operation was to kill US soldiers who had defected. Both producers were fired by CNN.

10. Recombinant bovine growth hormone in milk as reported by Jane Akre and Steve Wilson for a Fox news station, WTVT, of Tampa, Florida. Their story is documented in *Into the Buzzsaw*, ibid. Both were ultimately fired by the station when they would not change the story.

11. Benjamin (Ben) Swann (1978–) is an American television journalist whose broadcast *Reality Check* explores various issues, most recently aspects of vaccines as well as the COVID-19 pandemic. He works for affiliates Fox News and RT America of the Russian state-owned TV network RT. These associations are enough for much of the left to discredit his reports without any critical evaluation of what he is actually saying. Sharyl Attkisson (1961–) is an American journalist and television correspondent, hosting the Sinclair Broadcast Group TV show *Full Measure*. Both have won numerous awards for their reporting.

12. *Fox News*. Retrieved from https://www.foxnews.com/.; *Breitbart*. Retrieved from https://www.breitbart.com/.

13. Guidance for Crown Corporations. *Crown Corporations and Boards of Directors*. (2019). From "Directors of Crown corporations: an introductory guide to their roles and responsibilities". Retrieved from Government of Canada, https://www.canada.ca/en/treasury-board-secretariat/services/guidance-crown-corporations/directors-crown-corporations-introductory-guide-roles-responsibilities.html#cro.

14. Rody, B. *CBC formalizes new branded content unit.* (2020). From "Digital, Social, Television". Retrieved from Media In Canada, https://mediaincanada.com/2020/09/17

/cbc-formalizes-new-branded-content-unit/.; Nardi, C. *CBC plan for branded content has staff feeling betrayed and warning of 'fake news'.* (2020). From "Canada Politics". Retrieved from National Post, https://nationalpost.com/news/politics/cbc-staff-say-they-feel-betrayal-and-question-their-trust-in-management-due-to-new-branded-content-plans.

15. Shaw, C. A. *Five Ring Circus: Myths and Realities of the Olympic Games.* New Society Publishers, 2008. https://books.google.ca/books?id=STMm0jVpoHwC.

16. I can't count the number of times a CBC reporter would question me about some aspect of the Vancouver Olympic Games, get some good quotes, and then use the softer ones to show they had checked off the box marked "balance." Then the reporter would turn off the microphone and say, "I know you are right, but I don't think my editor will let me put that in."

17. *Marketplace: Inside the Anti-Vaccination Movement: Why more are falling for their dangerous message.* (2020). From "CBC Marketplace". Retrieved from CBC News, https://www.cbc.ca/player/play/1676913219653.

18. *What is Media Framing?* (2015). Retrieved from Critical Media Review, https://critical-mediareview.wordpress.com/2015/10/19/what-is-media-framing/.

19. Mnookin, S. *The Panic Virus: The True Story Behind the Vaccine-Autism Controversy.* New York: Simon & Schuster, 2012. https://books.google.ca/books?id=l_KwCOs3QhsC.

20. *Timothy Caulfield.* Retrieved from University of Alberta - Faculty of Law, https://www.ualberta.ca/law/faculty-and-research/health-law-institute/people/timothycaulfield.html.

21. *Germany coronavirus: Hundreds arrested in German 'anti-corona' protests.* (2020). Retrieved from BBC News, https://www.bbc.com/news/world-europe-53959552.; Depuydt, S. *What Really Happened in Berlin? CHD's Senta Depuydt Was There.* (2020). From "Advocacy Policy". Retrieved from Children's Health Defense, https://childrenshealthdefense.org/advocacy-policy/what-really-happened-in-berlin-chds-senta-depuydt-was-there/.

22. Grabish, A. *Manitoba's education minister under fire for comments on right to refuse vaccines in midst of pandemic.* (2020). Retrieved from CBC News, https://www.cbc.ca/news/canada/manitoba/kelvin-goetzen-vaccine-post-1.5712564.

23. In much of what follows, I will lean on my experience in the military to provide some needed context. Some may see in this a reliance on a metaphor that is not warranted in terms of public health. My answer to such critiques is that the media, as well as the medical community, now routinely use such expressions as humanity fighting "a war" or "a battle" against COVID-19. With such examples, I believe it only fair to continue using comparisons to war in much of the following.

24. Cernic, M. In *Ideological Constructs of Vaccination.* pg. 35: Wirral, UK: Vega Press Limited, 2018. https://books.google.ca/books?id=LHz-swEACAAJ.

25. Physicians who dissent about the mainstream narrative on vaccines are subject to being investigated and even delicensed. Recent examples include Dr. Bob Sears of San Diego and Dr. Paul Thomas of Portland.; Associated Press. *Pediatrician's license suspended in Oregon over vaccines.* (2020). Retrieved from Modern Healthcare, https://www.modern-healthcare.com/physicians/pediatricians-license-suspended-oregon-over-vaccines.; Hamilton, M. *Dr. Bob Sears, critic of vaccine laws, could lose license after exempting toddler.* (2016). From "LA Now". Retrieved from the Los Angeles Times, https://www.latimes.com/local/lanow/la-me-ln-oc-vaccine-doctor-20160908-snap-story.html.

26. The New Democratic Party (NDP) is the ruling party in British Columbia that pretends to be democratic and socialist. Called by some the No Difference Party, since many of

the policies of the former governing Liberal Party (not very liberal at all) were then pursued by the NDP when they came into power in 2018.

27. Cable Public Affairs Channel. *British Columbia update on COVID-19 – June 2, 2020.* (2020). Retrieved from Youtube, https://www.youtube.com/watch?v=hfkex_EepTQ.

28. Krugel, L. *'Dealing with a lot:' Suicide crisis calls mount during COVID-19 pandemic.* (2020). From "Coronavirus". Retrieved from CTV News, https://www.ctvnews.ca /health/coronavirus/dealing-with-a-lot-suicide-crisis-calls-mount-during-covid-19 -pandemic-1.5215056.; Moore, O. *Suicides up sharply on Toronto subway during pandemic.* (2020). Retrieved from The Global and Mail, https://www.theglobeandmail .com/canada/toronto/article-suicides-on-the-ttc-have-risen-sharply-over-the-last-eight -months/.; Mcintyre, R. S., and Lee, Y. "Projected increases in suicide in Canada as a consequence of COVID-19." *Psychiatry Res,* vol. 290 (Aug 2020): 113104. doi:10.1016/j .psychres.2020.113104 PMID: 32460184.; McKeen, A. *More young men in Western Canada died than expected last year — and not just because of COVID-19.* (2021). From "News". Retrieved from Toronto Star, https://www.thestar.com/news/canada/2021/01/04/more -young-men-in-western-canada-died-than-expected-last-year-and-not-just-because-of -covid-19.html.; Sajan, B. *'Worst case scenario': Crisis workers seeing spike in domestic violence concerns during pandemic.* (2020). From "Vancouver". Retrieved from CTV News, https://bc.ctvnews.ca/worst-case-scenario-crisis-workers-seeing-spike-in-domestic -violence-concerns-during-pandemic-1.4875911.; Wells, N. *Advocates share fear of worsening overdose crisis in 2021, want national safe supply.* (2020). From "Health". Retrieved from CTV News, https://www.ctvnews.ca/health/advocates-share-fear-of -worsening-overdose-crisis-in-2021-want-national-safe-supply-1.5235074.; D'amore, R. *Domestic disturbance calls jump amid coronavirus, as many advocates feared.* (2020). From "News". Retrieved from Global News, https://globalnews.ca/news/7309496 /domestic-crime-canada-coronavirus/.

29. Seneca, L. A. *Letters from a Stoic: Epistulae Morales Ad Lucilium.* Edited by G.L. Iunior and R. Campbell. New York: Penguin Books Limited, 1969. https://books.google.ca /books?id=33u8PGadq1QC.

30. Pearce, K. *Pandemic simulation exercise spotlights massive preparedness gap.* (2019). From "Health Security". Retrieved from The Hub - Johns Hopkins University, https://hub .jhu.edu/2019/11/06/event-201-health-security/.; *Event 201. Public-private cooperation for pandemic preparedness and response.* (2019). From "Event 201 Recommendations". Retrieved from The Johns Hopkins Center for Health Security, https://www.centerfor -healthsecurity.org/event201/recommendations.html.

31. *What we do.* From "About WHO". Retrieved from World Health Organization, https://www.genevaenvironmentnetwork.org/environment-geneva/organizations/world -health-organization/.

32. *World Health Organization.* From "Intergovernmental Organizations". Retrieved from Geneva Environment Network, https://www.genevaenvironmentnetwork.org /environment-geneva/organizations/world-health-organization/.

33. York, G. *Questions surfacing about history of WHO's director Tedros Adhanom Ghebreyesus.* (2020). From "World". Retrieved from The Global and Mail, https://www.theglobeandmail.com/world/article-questions-surfacing-about-history-of-whos -director-tedros-adhanom/.

34. https://upload.wikimedia.org/wikipedia/commons/0/0b/Mukhisa_Kituyi%2C _Houlin_Zhao%2C_Tedros_Adhanom_Ghebreyesus_with_Sophia_-_AI_for_Good _Global_Summit_2018_%2841223188035%29_%28cropped%29.jpg.

35. Rincón, E. *Bill Gates' Strange Relationship with China and the Coronavirus.* (2020). Retrieved from Panam Post, https://en.panampost.com/emmanuel-rincon/2020/05/06/bill-gates-strange-relationship-with-china-and-the-coronavirus/.

36. Huet, N., and Paun, C. *Meet the world's most powerful doctor: Bill Gates.* (2017). Retrieved from Politico, https://www.politico.eu/article/bill-gates-who-most-powerful-doctor/.

37. *Voluntary contributions by fund and by contributor, 2018.* (2019). From "Seventy-Second World Health Assembly". Retrieved from World Health Organization, https://www.who.int/about/finances-accountability/reports/A72_INF5-en.pdf?ua=1.; *Voluntary contributions by fund and by contributor, 2019.* (2020). From "Seventy-Third World Health Assembly". Retrieved from World Health Organization, https://www.who.int/about/finances-accountability/reports/A73-INF3-en.pdf.; *Bill & Melinda Gates Foundation.* (2020). From "Contributors—Funding by contributor—Updated Until Q3-2020". Retrieved from World Health Organization, https://open.who.int/2020–21/contributors/contributor?name=Bill%20%26%20Melinda%20Gates%20Foundation.

38. Mehra, M. R., Desai, S. S., Ruschitzka, F., and Patel, A. N. "RETRACTED: Hydroxychloroquine or chloroquine with or without a macrolide for treatment of COVID-19: a multinational registry analysis." *Lancet,* (May 22 2020). doi:10.1016/S0140-6736(20)31180–6 PMID: 32450107.; Mehra, M. R., Desai, S. S., Kuy, S., Henry, T. D., and Patel, A. N. "Cardiovascular Disease, Drug Therapy, and Mortality in Covid-19." *N Engl J Med,* vol. 382, no. 25 (Jun 18 2020): e102. doi:10.1056/NEJMoa2007621 PMID: 32356626.

39. Neergaard, L. *Q&A: What the WHO pandemic declaration means.* (2020). From "News". Retrieved from Kimberley Bulletin, https://www.kimberleybulletin.com/news/qa-what-the-who-pandemic-declaration-means/.

40. Oller, J. W., Shaw, C. A., Tomljenovic, L., Karanja, S. K., Ngare, W., Clement, F. M., and Pillette, J. R. "HCG Found in WHO Tetanus Vaccine in Kenya Raises Concern in the Developing World." *J Open Access Library Journal,* vol. 4, no. 10 (2017): 1–32. doi:10.4236/oalib.1103937.

41. Oller, J. W., Shaw, C. A., Tomljenovic, L., Karanja, S. K., Ngare, W., Clement, F. M., and Pillette, J. R. "Addendum to "HCG Found in Tetanus Vaccine": Examination of Alleged "Ethical Concerns" Based on False Claims by Certain of Our Critics." *International Journal of Vaccine Theory, Practice, and Research,* vol. 1, no. 1 (July 15 2020): 27–50. https://ijvtpr.com/index.php/IJVTPR/article/view/3/11.

42. *Former Directors-General.* (2021). From "WHO headquarters leadership team". Retrieved from World Health Organization, https://www.who.int/director-general/who-headquarters-leadership-team/former-directors-general/.; Brundtland is a medical doctor who served for a time as Norway's health minister, and later for a considerable time as the prime minister.

43. Raghavan, C. (2001). *WHO unduly influenced by large pharma companies, complains Nader.* Retrieved from Business & Human Rights Resource Centre: https://www.business-humanrights.org/de/neuste-meldungen/who-unduly-influenced-by-large-pharma-companies-complains-nader/.

44. Patel, M. K., Dumolard, L., Nedelec, Y., Sodha, S. V., Steulet, C., Gacicdobo, M., Kretsinger, K., McFarland, J., Rota, P. A., and Goodson, J. L. *Progress towards regional measles elimination—worldwide, 2000–2018.* (2019). From "Weekly epidemiological record". Retrieved from World Health Organization, https://www.who.int/immunization/monitoring_surveillance/burden/estimates/measles/WER9449-eng-fre.pdf.

45. Ibid, 590.

46. *Developing together the vision and strategy for immunization 2021–2030.* (2019). From "Draft Zero in developing together the vision and strategy for immunization (2021–2030)". Retrieved from World Health Organization, https://www.who.int/immunization/ia2030_Draft_Zero.pdf?ua=1.

47. *World Health Assembly.* Retrieved from World Health Organization, https://www.who.int/about/governance/world-health-assembly.

48. https://www.who.int/teams/immunization-vaccines-and-biologicals/strategies/ia2030.

49. Ibid.

50. *Developing together the vision and strategy for immunization 2021–2030.* (2019). From "Draft Zero in developing together the vision and strategy for immunization (2021–2030)". 15, Retrieved from World Health Organization, https://www.who.int/immunization/ia2030_Draft_Zero.pdf?ua=1.

51. Ibid, 16.

52. *Immunization Agenda 2030 - A global strategy to leave no one behind.* (2020). Retrieved from World Health Organization, https://www.who.int/immunization/IA2030_draft_4_WHA.pdf.

53. Ibid, 5.

54. Ibid, 25.

55. Ibid, 4.

56. Ibid, 5.

57. Ibid, 8.

58. Paul, Y., and Dawson, A. "Some ethical issues arising from polio eradication programmes in India." *Bioethics,* vol. 19, no. 4 (Aug 2005): 393–406. doi:10.1111/j.1467-8519.2005.00451.x PMID: 16222855.

59. *Immunization Agenda 2030 - A global strategy to leave no one behind.* (2020), 11.

60. Ibid, 15.

61. Ibid.

62. Ibid, 7.

63. Ibid, 8.

64. Huxley, A. *Brave New World: By a Novel a Novel.* London: Chatto and Windus, 1932. https://books.google.ca/books?id=i_T3vwEACAAJ.

65. *Wellcome.* Retrieved from The Wellcome Trust, https://wellcome.org/.

66. Retrieved from Vaccine Confidence Project, https://www.vaccineconfidence.org/.

67. *Capo.* Retrieved from Merriam-Webster, https://www.merriam-webster.com/dictionary/capo.; *Consigliere.* Retrieved from Merriam-Webster, https://www.merriam-webster.com/dictionary/consigliere.

Chapter 11

1. Frederick Douglass (1818–1895) was a remarkable man by all accounts. Born into slavery in Maryland, he escaped to become one of the major social activists of his day, fighting for the abolition of slavery, for women's suffrage, and for human freedom in general as an eloquent and powerful writer and orator. A good summary of his life can be found here: https://en.wikipedia.org/wiki/Frederick_Douglass.

2. Paine, T. *Of the Origin and Design of Government in General, with Concise Remarks on the English Constitution.* (1776). From *Common Sense.* Retrieved from Independence Hall Association, https://www.ushistory.org/paine/commonsense/sense2.htm.; Thomas

Paine (1737–1809) was born in England but immigrated to the American colonies in 1774 and quickly found work as a journalist with the *Pennsylvania Magazine,* which he helped run. The pamphlet *Common Sense* (1776), which made the persuasive case for American independence, was his response to the battles at Lexington and Concord in 1775. It could be rightly said that Paine, probably more than any other writer of his time, became the most prominent advocate for American independence with *Common Sense* and his various essays during the Revolutionary War (*American Crisis*). After the war, Paine expanded his views on human freedom generally with his seminal book, *The Rights of Man,* written in support of the French Revolution. A short biography of Paine's life and works can be found here: https://www.britannica.com/biography/Thomas-Paine.

3. Carlin, G. ". . . and rights aren't rights if someone can take them away. They're privileges. That's all we've ever had in this country is a bill of TEMPORARY privileges; and if you read the news, even badly, you know the list get's shorter, and shorter, and shorter.". From "Quotes". Retrieved from Goodreads Quotes, https://www.goodreads.com/author/quotes/22782.George_Carlin?page=9.; *George Carlin.* (2006). Retrieved from Wikipedia, https://en.wikipedia.org/wiki/George_Carlin.

4. *The Declaration of Independence and Natural Rights.* (2001). From "Natural Rights". Retrieved from Constitutional Rights Foundation, https://www.crf-usa.org/foundations-of-our-constitution/natural-rights.html.

5. *Civil Rights Act of 1964.* (2010). From "Black History". Retrieved from History, https://www.history.com/topics/black-history/civil-rights-act#:~:text=The%20Civil%20-Rights%20Act%20of%201964%2C%20which%20ended%20segregation%20-in,proposed%20by%20President%20John%20F.

6. Mello, M. M., Silverman, R. D., and Omer, S. B. "Ensuring Uptake of Vaccines against SARS-CoV-2." *N Engl J Med,* vol. 383, no. 14 (Oct 1 2020): 1296–99. doi:10.1056/NEJMp2020926 PMID: 32589371.

7. Goodwin, L. *Eventually, getting the COVID-19 vaccine could be required for many.* (2020). Retrieved from Boston Globe, https://www.bostonglobe.com/2020/12/15/nation/will-covid-19-vaccine-be-mandatory-future/.

8. *Universal Declaration of Human Rights.* Retrieved from United Nations, https://www.un.org/en/universal-declaration-human-rights/.

9. *VCC responses to Bill 87 Committee: Where there is Risk there Must be Choice.* (2017). From "Summary of our Concerns Bill 87". Retrieved from Vaccine Choice Canada, https://vaccinechoicecanada.com/exemptions/heather-frasers-statement-ontario-bill-87-education-sessions-committee/.

10. It's a bit more nuanced than this, but the basic idea is still worth remembering: Ben Franklin's Famous 'Liberty, Safety,' Quote Lost its Context in 21st Century. (2015). Retrieved from NPR: https://www.npr.org/2015/03/02/390245038/ben-franklins-famous-liberty-safety-quote-lost-its-context-in-21st-century.

11. Agamben, G. *State of Exception.* Translated by Kevin Attell. Chicago: University of Chicago Press, 2005. https://books.google.ca/books?id=7t9vmAEACAAJ.

12. *Privilege (law).* (2006). Retrieved from Wikipedia, https://en.wikipedia.org/wiki/Privilege_(law)#:~:text=A%20privilege%20is%20a%20certain,or%20on%20a%20-conditional%20basis.&text=By%20contrast%2C%20a%20right%20is,from%20-the%20moment%20of%20birth.

13. *Giorgio Agamben.* (2019). From "Division of Philosophy, Art, & Critical Thinking". Retrieved from The European Graduate School, https://egs.edu/biography/giorgio-agamben/.

14. A "false flag" incident is one in which an entity, such as a government, that wants a particular outcome will use an event (or even stage one) that will be blamed on someone else. A now-classical example is the sinking of the US Navy vessel *Maine* in Havana Harbor, which led the United States to accuse Spain of the explosion as the reason for going to war. The consequence of the war was the acquisition by the United States of the islands of Puerto Rico and islands in the Marianas, and the occupation by US forces of the Philippines.

15. The Reichstag fire in 1933 largely destroyed the building housing the German parliament. The fire was promptly blamed on communists and led to the Enabling Act, which gave Adolf Hitler the powers to rule by decree.

16. The attacks of September 11, 2001, in New York City and Washington led the government of President George W. Bush to declare that the culprit was Osama bin Laden's Al Qaeda network.

17. Dark sites: During the "war against terror" starting in 2001, the U.S maintained a series of overseas sites in various countries to which they would send suspected terrorists. At these sites, the prisoners could be tortured without any oversight by civilian authorities.; Priest, D. *CIA Holds Terror Suspects in Secret Prisons.* (2005). From "Politics". Retrieved from the Washington Post, https://www.washingtonpost.com/archive/politics/2005/11/02/cia-holds-terror-suspects-in-secret-prisons/767f0160-cde4-41f2-a691-ba989990039c/.

18. Snowden, E. J. *Permanent Record.* New York: Metropolitan Books/Henry Holt, 2019. https://books.google.ca/books?id=DV5PyAEACAAJ.; Greenwald, G., Macaskill, E., and Poitras, L. *Edward Snowden: the whistleblower behind the NSA surveillance revelations.* (2013). From "World". Retrieved from The Guardian, https://www.theguardian.com/world/2013/jun/09/edward-snowden-nsa-whistleblower-surveillance.; Pue, W. "The war on terror: Constitutional governance in a state of permanent warfare." *Osgoode Hall Law Journal,* vol. 41 (2003): 267–92.

19. *Uniting And Strengthening America By Providing Appropriate Tools Required To Intercept And Obstruct Terrorism (USA Patriot Act) Act Of 2001.* (2001). From "World". Retrieved from congress.gov, https://www.congress.gov/107/plaws/publ56/PLAW-107publ56.pdf.

20. Agamben, G. "State of Exception." Translated by Kevin Attell. pg. 18: Chicago: University of Chicago Press, 2005. https://books.google.ca/books?id=7t9vmAEACAAJ.

21. Ibid, 14.

22. After losing its parliamentary majority in the 2019 federal election, the Liberal Party under Prime Minister Justin Trudeau assumed a minority government status in that they were forced to seek votes from other parties for legislation they wished to pass. In a parliamentary system, as in Canada, if a party in power loses a crucial vote in Parliament, it is considered to have "lost the confidence of the House" and is dissolved by the governor general, triggering a new election. Trudeau and the Liberals sidestepped this in the spring of 2020 during the early days of the Covid-19 pandemic by simply not allowing Parliament to sit. Their new gun control measures were enacted by what is known as an "Order in Council," or in other words, simply the equivalent of an American president's "executive order."

23. Cecco, L. *Twelve hours of terror: how the Nova Scotia shooting rampage unfolded.* (2020). From "World". Retrieved from *The Guardian,* https://www.theguardian.com/world/2020/apr/23/nova-scotia-shooting-canada-new-details.

24. Palango, P., Maher, S., and Gormley, S. *The Nova Scotia shooter case has hallmarks of an undercover operation.* (2020). From "News". Retrieved from Maclean's, https://www.macleans.ca/news/canada/the-nova-scotia-shooter-case-has-hallmarks-of-an-undercover-operation/.

25. The fact that Canada is more a legalistic and paper democracy than an actual one is the subject of another book that discusses the civil rights abuses during the lead-up to the 2010 Olympic Games in Vancouver: Shaw, C. A. (2008). *Five Ring Circus: Myths and Realities of the Olympic Games.* Gabriola, BC: New Society Publishers.

26. Weber, B. *COVID-19 crowd limits make it a 'great time to be building a pipeline': Alberta minister.* (2020). Retrieved from Calgary Herald, https://calgaryherald.com/business/energy /covid-19-crowd-limits-make-it-a-great-time-to-be-building-a-pipeline-alberta-minister.

27. Benjamin, G., and Brown, S. *New Brunswick's mandatory vaccination bill voted down.* (2020). Retrieved from Global News, https://globalnews.ca/news/7080555/bill-11/.

28. Facebook and other social media sites have moved to censoring comments considered to be "misinformation" both about vaccines and COVID-19, by having so-called "independent fact checkers" verify any information imparted.

29. Schiff, A. B. *Adam B. Schiff (Member of Congress) to Mark Zuckerberg.* (2019). Retrieved from Congressman Adam Schiff: Representing California's 28th District, https://schiff .house.gov/imo/media/doc/Vaccine%20Letter_Zuckerberg.pdf.

30. Agamben, G. *Clarifications.* (2020). Retrieved from European Journal of Psychoanalysis, https://www.journal-psychoanalysis.eu/coronavirus-and-philosophers/.

31. Richardson, V. *WHO retreats on 'no evidence' claim, says antibodies will protect most from reinfection.* (2020). Retrieved from the Washington Times, https://www.washington -times.com/news/2020/apr/27/who-walks-back-no-evidence-claim-coronavirus -immun/?utm_campaign=shareaholic&utm_medium=facebook&utm_source=social -network&fbclid=IwAR30cwryALWm0V8yalgYftiqAbCAeUVcKK65LrQ -JygvHhqp4tMON_niG0jU.

32. This is the reason that command-and-control centers are usually the first targets in modern warfare, forcing subunits to fight on without knowing what other units are doing. A good example of this was in the 2003 American-led invasion of Iraq, which, as in the 1991 Gulf War, attempted to rapidly achieve "full spectrum dominance" by Coalition forces.

33. Browne, R. *Facebook to remove misinformation about Covid vaccines.* (2020). Retrieved from CNBC, https://www.cnbc.com/2020/12/03/facebook-to-remove-misinformation-about -coronavirus-vaccines.html.; Kelly, M. *Twitter says it will start removing COVID-19 vaccine misinformation.* (2020). Retrieved from The Verge, https://www.theverge .com/2020/12/16/22179074/twitter-coronavirus-misinformation-covid19-vaccine -vaccination-label.

34. Jarrett, M., and Sublett, C. "The Anti-Vaxxers Movement and National Security." *The Infragard Journal,* vol. 2, no. 1 (2019): 24–29. https://www.infragardnational.org /wp-content/uploads/2019/07/InfraGard_June_2019_Article3.pdf.

35. Ibid, 26.

36. Ibid, 28.

37. The "Dark Web" as defined by Wikipedia with their references left in (https://en.wiki-pedia.org/wiki/Dark_web#Definition): "The dark web is the World Wide Web content that exists on darknets: overlay networks that use the Internet but require specific software, configurations, or authorization to access. Through the dark web, private strap-hanger networks can communicate and conduct business anonymously without divulging identifying information, such as a user's location. The dark web forms a small part of the deep web, the part of the Web not indexed by web search engines, although sometimes the term deep web is mistakenly used to refer specifically to the dark web." D'Amore, R. *Coronavirus misinformation is spreading — what is Canada doing about it?* (2020). From "News". Retrieved from Global News, https://globalnews.ca

/news/7249102/coronavirus-canada-misinformation-strategy/.; Thompson, E. *Federal government open to new law to fight pandemic misinformation*. (2020). From "Politics". Retrieved from CBC News, https://www.cbc.ca/news/politics/covid-misinformation -disinformation-law-1.5532325.; as a side bar, it is worth noting that in any parliamentary "democracy," a majority government that holds the most legislative seats functions basically as a "constitutional dictatorship," almost the purest modern example of how a state of exception can come to pass. Briefly, in such a circumstance a majority government can do whatever they want for the term of the government, that is, until the next election. Unlike in the United States, there are no midterm elections to give voters an opportunity to correct governmental malfeasance. In principle, one can appeal to the highest court, the Supreme Court of Canada, but an appeal typically requires enormous amounts of money and often time to move forward.

38. *Vaccine Choice Canada (VCC) v. Denis Rancourt*, CV-20-00643451-0000 (Ontario Superior Court of Justice, 2020). Retrieved from https://vaccinechoicecanada.com /wp-content/uploads/vcc-statement-of-claim-2020-redacted.pdf.

39. Ibid, 4.

40. Ibid, 5.

41. Ibid, 7.

42. Ibid.

43. Ibid, 17.

44. Gandhi's march to the sea in 1930 took him and his followers from Ahmedabad to Dandi on the Arabian Sea as a protest against the British rule in India and the government's 1882 Salt Act. A good summary of the march and its antecedents and outcomes can be found here: https://www.history.com/topics/india/salt-march#:~:text=The%20 -Salt%20March%2C%20which%20took,distance%20of%20some%20240%20miles.

45. Paine, T. *Of the Origin and Design of Government in General, with Concise Remarks on the English Constitution*. (1776). From *Common Sense*.; *John Laurens*. (2020). Retrieved from Britannica, https://www.britannica.com/biography/John-Laurens.

46. Agamben, G. "State of Exception." Translated by Kevin Attell. pg. 50: Chicago: University of Chicago Press, 2005. https://books.google.ca/books?id=7t9vmAEACAAJ.

47. Arendt, H., and Elon, A. *Eichmann in Jerusalem: A Report on the Banality of Evil*. New York: Penguin Publishing Group, 2006. https://books.google.ca/books?id=yGoxZE -dw36oC.

48. Paine, T. *The Crisis*. (1776). From "The American Crisis". Retrieved from Independence Hall Association, https://www.ushistory.org/paine/crisis/c-01.htm.

49. Hoylman, B. *Senate Bill S2276*. (2019). Retrieved from The New York State Senate, https://www.nysenate.gov/legislation/bills/2019/s2276.

50. Hoylman, B. *Senate Bill S298B*. (2019). Retrieved from The New York State Senate, https://www.nysenate.gov/legislation/bills/2019/s298.

51. Paulin, A. *Bill No. A00973*. (2019). Retrieved from New York State Assembly, https: //nyassembly.gov/leg/?default_fld=&leg_video=&bn=A00973&term=2019&Summary =Y&Actions=Y#jump_to_Actions.

52. Fahy, P. *Bill No. A06564C*. (2019). Retrieved from New York State Assembly, https: //nyassembly.gov/leg/?default_fld=&leg_video=&bn=A06564&term=2019&Summary=Y.

53. Personal communication with John Gilmore, Autism Action Network.

54. Ambra Fedrico, personal communication.

55. Mateja Cernic, personal communication.

Chapter 12

1. Gates, B. *The Vaccine Race, Explained: What you need to know about the COVID-19 vaccine.* (2020). Retrieved from Gates Notes, https://www.gatesnotes.com/health/what-you-need-to-know-about-the-covid-19-vaccine.

2. For clarity, an epidemic is a disease outbreak that affects a large number of people within a community or a region or even a country. For example, a part of Los Angeles, all of greater LA, the county, or even as much as California or even the entire United States; in contrast, a pandemic is an epidemic that's spread over multiple countries or continents.

3. *The Echo Machine: Vaccines and War are Necessities.* (2020). Retrieved from Brasscheck TV, https://www.brasscheck.com/video/the-echo-machine/.

4. Transcript. *Bill Gates: U.S. Senate Committee Hearing on Strengthening American Competitiveness.* (2007). Retrieved from Microsoft News, https://news.microsoft.com/speeches/bill-gates-u-s-senate-committee-hearing-on-strengthening-american-competitiveness/.

5. A summary of the outcome of the legal cases is cited here: https://corporatefinanceinstitute.com/resources/knowledge/strategy/microsoft-antitrust-case/.

6. Klein, C. *Andrew Carnegie Claimed to Support Unions, But Then Destroyed Them in His Steel Empire.* (2019). Retrieved from History, https://www.history.com/news/andrew-carnegie-unions-homestead-strike.

7. Harper, like Gates, had been notorious as a micromanager with a firm grip on all of his subordinates, a martinet in fact. Harper's public relations firm advised him to try to seem more personable to ordinary working class Canadians. Part of the advice was to wear sweaters rather than suits and demonstrate his affection for his family. This last was crucial since prior to the makeover, Harper had been mocked for giving his kids only handshakes. Harper was also advised to talk about the national sport, hockey, usually a plus with Canadians. In other words, the PR firm was trying to make him out to be the kind of guy people could relate to personally, the sort of guy you'd want to have a beer with to chew the fat. The PR strategy didn't work, as Harper and his Conservative Party got clobbered by Justin Trudeau and the Liberals. Trudeau, as we have seen since, didn't really care about people all that much either, but he seemed to be all about "sunny ways" and being people-focused and sincere about environmental and women's issues. Trudeau, a former high school drama teacher, kept the spin going for quite a time.

8. *Microsoft Timeline.* Retrieved from the Washington Post, https://www.washingtonpost.com/wp-srv/business/longterm/microsoft/timeline.htm.

9. *Microsoft announces change to its board of directors.* (2020). Retrieved from Microsoft News, https://news.microsoft.com/2020/03/13/microsoft-announces-change-to-its-board-of-directors/.

10. *About our Alliance.* (2020). Retrieved from GAVI: The Vaccine Alliance, https://www.gavi.org/our-alliance/about.; *GAVI.* (2013). Retrieved from Wikipedia, https://en.wikipedia.org/wiki/GAVI.

11. *Creating a world in which epidemics are no longer a threat to humanity.* (2021). Retrieved from CEPI, https://cepi.net/about/whyweexist/.; *Coalition for Epidemic Preparedness Innovations: Mission.* (2020). Retrieved from Wikipedia, https://en.wikipedia.org/wiki/Coalition_for_Epidemic_Preparedness_Innovations#Mission.

12. *Trustee of the Wellcome Trust.* (2016). Retrieved from Bill & Melinda Gates Foundation, https://www.gatesfoundation.org/How-We-Work/Quick-Links/Grants-Database/Grants/2016/07/OPP1151904.

13. Larson is also the author of a new book on the benefits of vaccines: Larson, H. J. (2020). *Stuck: How Vaccine Rumors Start – and Why They Don't Go Away.* New York: Oxford University Press.

14. *Innovative Medicines Initiative.* (2016). Retrieved from Bill & Melinda Gates Foundation, https://www.gatesfoundation.org/How-We-Work/Quick-Links/Grants-Database /Grants/2016/08/OPP1149597.

15. One example is this: https://www.bloomberg.com/news/articles/2021-05-21/bill-gates -s-carefully-curated-dad-geek-image-unravels-in-two-weeks; there are many more to choose from, and there certainly will be many others before this book is released.

16. National Geographic Staff. *Watch: Bill Gates on how to end this pandemic—and prepare for the next.* (2020). From "Science | Coronavirus Coverage". Retrieved from *National Geographic*, https://www.nationalgeographic.com/science/2020/09/bill-gates-how-to-end -this-pandemic-and-prepare-for-the-next/.

17. Schwab, T. *Bill Gates's Charity Paradox: A Nation investigation illustrates the moral hazards surrounding the Gates Foundation's $50 billion charitable enterprise.* (2020). Retrieved from The Nation, https://www.thenation.com/article/society/bill-gates -foundation-philanthropy/.

18. Levine, M. *Gates and the COVID Vaccine: A Case of Philanthropic Overreach?* (2020). Retrieved from *Nonprofit Quarterly*, https://nonprofitquarterly.org/gates-and -the-covid-vaccine-a-case-of-philanthropic-overreach/.

19. https://www.thenation.com/article/society/bill-gates-foundation-philanthropy/.

20. Schwab, T. *Journalism's Gates keepers.* (2020). From "Criticism". Retrieved from Columbia Journalism Review, https://www.cjr.org/criticism/gates-foundation-journalism-funding .php.

21. *Our Mission.* (2021). Retrieved from World Economic Forum, https://www.weforum .org/about/world-economic-forum.

22. Shiva, V. *Bill Gates' Global Agenda and How We Can Resist His War on Life.* (2020). Retrieved from Independent Science News, https://www.independentsciencenews.org /biotechnology/bill-gates-global-agenda-and-how-we-can-resist-his-war-on-life/.

23. Gates, B. *The Next Outbreak? We're Not Ready.* (2015). From "TED2015". Retrieved from TED, https://www.ted.com/talks/bill_gates_the_next_outbreak_we_re_not_ready.

24. In an article in *Vice* in March 2020 at the beginning of the pandemic, Marie Solis interviewed Naomi Klein, author of the 2007 book *Shock Doctrine*, demonstrating how the pandemic neatly fit the pattern; Solis, M. (2020). Coronavirus is the Perfect Disaster for 'Disaster Capitalism'. Retrieved from *Vice*: https://www.vice.com/en/article/5dmqyk /naomi-klein-interview-on-coronavirus-and-disaster-capitalism-shock-doctrine; Klein, N. (2007). *The Shock Doctrine: The Rise of Disaster Capitalism.* Toronto: Random House of Canada Ltd.

25. The trajectory of capitalism as an economic system has often been debated from Karl Marx to the present. Whether or not its endless desire for consumption and growth sows the seeds of its own eventual destruction is also debated.

26. Urhahn, J. *Bill Gates's Foundation Is Leading a Green Counterrevolution in Africa.* (2020). Retrieved from Jacobin, https://jacobinmag.com/2020/12/agribusiness-gates -foundation-green-revolution-africa-agra.

27. From Wikipedia, https://en.wikipedia.org/wiki/Potemkin_village: In politics and economics, a Potemkin village is any construction (literal or figurative) whose sole purpose is to provide an external facade to a country that is faring poorly, making people believe that the country is faring better.

28. Angell, M. *The Truth About the Drug Companies: How They Deceive Us and What to Do About It*. New York: Random House Publishing Group, 2004. https://books.google.ca /books?id=5DKwxAnhTygC.

29. Robert F. Kennedy Jr., personal communication.

30. Angell, M. "The Truth About the Drug Companies: How They Deceive Us and What to Do About It." pg. 20: New York: Random House Publishing Group, 2004. https: //books.google.ca/books?id=5DKwxAnhTygC.

31. Ibid, 216.

32. Moynihan, R., and Cassels, A. *Selling Sickness: How the World's Biggest Pharmaceutical Companies are Turning Us All Into Patients*. Vancouver: Greystone Books, 2005. https: //books.google.ca/books?id=rfAKkgEACAAJ.

33. Ireland, N. *Getting a flu shot during the COVID-19 era: Here's what you need to know.* (2020). Retrieved from CBC News, https://www.cbc.ca/news/health/flu-vaccine-covid-19 -twindemic-what-you-need-to-know-1.5709559.

34. Alan Cassels, personal communication.

35. Ibid.

36. Ibid.

37. Mikovits, J., and Heckenlively, K. *Plague of Corruption: Restoring Faith in the Promise of Science*. New York: Skyhorse, 2020. https://books.google.ca/books?id=P6GCxQEACAAJ.

38. Heckenlively, K., and Mikovits, J. *Plague: One Scientist's Intrepid Search for the Truth about Human Retroviruses and Chronic Fatigue Syndrome (ME/CFS), Autism, and Other Diseases*. New York: Skyhorse, 2017. https://books.google.ca/books?id=xbBgjwEACAAJ.

39. As defined by the Mayo Clinic, CFS is defined as "… a complicated disorder character-ized by extreme fatigue that lasts for at least six months and that can't be fully explained by an underlying medical condition. The fatigue worsens with physical or mental activ-ity, but doesn't improve with rest." See: https://www.mayoclinic.org/diseases-conditions /chronic-fatigue-syndrome/symptoms-causes/syc-20360490

40. Bookchin, D., and Schumacher, J. *The Virus and the Vaccine: Contaminated Vaccine, Deadly Cancers, and Government Neglect*. New York: St. Martin's Press, 2005. https: //books.google.ca/books?id=QaubtBGQzjsC.; Debbie Bookchin's father, Murray Bookchin, was the American communalist/anarchist philosopher whose quote opens this book. This may not be a simple coincidence: Murray Bookchin was ever dedicated to human freedom and how best to achieve it. Debbie Bookchin seems to be carrying on this tradition by showing how capitalism and government connivance take it away.

41. Alcorn, K. *Was AIDS epidemic caused by 1950s polio vaccine trials?* (2001). Retrieved from NAM Aidsmap, https://www.aidsmap.com/news/mar-2001/was-aids-epidemic -caused-1950s-polio-vaccine-trials.; Secko, D. *Polio vaccine-AIDS theory dead: Study finds that Kisangani chimpanzees contain a virus unrelated to HIV-1.* (2004). Retrieved from The Scientist, https://www.the-scientist.com/research-round-up /polio-vaccine-aids-theory-dead-50186.

42. Gatti, A. M., and Montanari, S. "New quality-control investigations on vaccines: micro-and nanocontamination." *International Journal of Vaccines and Vaccination,* vol. 4, no. 1 (2016): 00072. https://siksik.org/wp-content/uploads/vaccins/02-2017-Medcrave -Nanocontamination.pdf.

43. The Corvelva results are summarized here: Kaur, B. (2019). The "vaccinegate" of Italy. Retrieved from Down to Earth: https://www.downtoearth.org.in/news/health/the -vaccinegate-of-italy-63235.

44. Oller, J. W., Shaw, C. A., Tomljenovic, L., Karanja, S. K., Ngare, W., Clement, F. M., and Pillette, J. R. "HCG Found in WHO Tetanus Vaccine in Kenya Raises Concern in the Developing World." *J Open Access Library Journal,* vol. 4, no. 10 (2017): 1–32. doi:10.4236/oalib.1103937.; Oller, J. W., Shaw, C. A., Tomljenovic, L., Karanja, S. K., Ngare, W., Clement, F. M., and Pillette, J. R. "Addendum to 'HCG Found in Tetanus Vaccine': Examination of Alleged 'Ethical Concerns' Based on False Claims by Certain of Our Critics." *International Journal of Vaccine Theory, Practice, and Research,* vol. 1, no. 1 (July 15 2020): 27–50. https://ijvtpr.com/index.php/IJVTPR/article/view/3/11.

45. Rogers, T. M. "The Political Economy of Autism." Postgraduate thesis for PhD Doctorate, University of Sydney, 2018. Retrieved from https://ses.library.usyd.edu.au /handle/2123/20198.

46. Ibid, 61.

47. Ibid, 66.

48. Ibid.

49. Ibid, 103.

50. Ibid.

51. Ibid, 167.

52. See also James Corbett's report on the WHO and their failed past pandemics—Corbett, J. (undated). Who the Heck is the WHO? James Corbett Explains. Retrieved from Brasscheck TV: https://www.brasscheck.com/video/who-the-heck-is-the-who/.

53. Farber, C. *AIDS and the AZT Scandal: SPIN's 1989 Feature, "Sins of Omission": The story of AZT, one of the most toxic, expensive, and controversial drugs in the history of medicine.* (2015). Retrieved from SPIN, https://www.spin.com/featured/aids-and -the-azt-scandal-spin-1989-feature-sins-of-omission/.

54. *Number of Patients Affected by Vioxx.* From "Vioxx". Retrieved from Drugwatch, https: //www.drugwatch.com/vioxx/#:~:text=Number%20of%20Patients%20Affected%20 -by,60%2C000%20deaths%2C%20FDA%20investigator%20Graham.

55. Husten, L. *Merck Pleads Guilty and Pays $950 Million for Illegal Promotion of Vioxx.* (2011). Retrieved from Forbes, https://www.forbes.com/sites/larryhusten/2011/11/22/merck -pleads-guilty-and-pays-950-million-for-illegal-promotion-of-vioxx/?sh=3317c00f20f4.

56. Mikulic, M. *Global vaccine market revenues from 2014 to 2020 (in billion U.S. dollars)*.* (2019). From "Pharmaceutical Products & Market". Retrieved from Statista, https://www .statista.com/statistics/265102/revenues-in-the-global-vaccine-market/.

57. *Section 7 – Life, liberty and security of the person.* (2021). From "The Canadian Charter of Rights and Freedoms". Retrieved from Government of Canada, https://www.justice .gc.ca/eng/csj-sjc/rfc-dlc/ccrf-ccdl/check/art7.html.

58. In military terms, "vital ground" is that piece of terrain that cannot be lost lest the battle and then the war be lost.

Chapter 13

1. Moore, R. *We live in a world of stories.* Retrieved from rkm's cyberjournal, https://cyber -journal.org/.

2. Shaw, C. A. "The Age of COVID-19: Fear, Loathing, and the 'new normal'." *International Journal of Vaccine Theory, Practice, and Research,* vol. 1, no. 2 (January 5 2021): 98–142. https://ijvtpr.com/index.php/IJVTPR/article/view/11/15.

3. *Naming the coronavirus disease (COVID-19) and the virus that causes it.* (2020). From "Coronavirus disease (COVID-19)—Technical guidance". Retrieved from World Health Organization, https://www.who.int/emergencies/diseases/novel-coronavirus-2019 /technical-guidance/naming-the-coronavirus-disease-(covid-2019)-and-the-virus-that -causes-it.

4. King, A. *Common Cold Coronaviruses Tied to Less Severe COVID-19 Cases.* (2020). From "News & Opinion". Retrieved from The Scientist, https://www.the-scientist.com /news-opinion/common-cold-coronaviruses-tied-to-less-severe-covid-19-cases-68146.

5. Hewings-Martin, Y. *How do SARS and MERS compare with COVID-19?* (2020). Retrieved from Medical News Today, https://www.medicalnewstoday.com/articles /how-do-sars-and-mers-compare-with-covid-19.

6. For clarity, an epidemic is a spread of disease within a region or country; a pandemic is one that involves various countries.

7. The Joint Mission. *Report of the WHO-China Joint Mission on Coronavirus Disease 2019 (COVID-19).* (2020). Retrieved from World Health Organization, https://www.who .int/docs/default-source/coronaviruse/who-china-joint-mission-on-covid-19-final -report.pdf.; Feldwisch-Drentrup, H. *How WHO Became China's Coronavirus Accomplice.* (2020). Retrieved from Foreign Policy, https://foreignpolicy.com/2020/04/02/china -coronavirus-who-health-soft-power/.

8. Oller, J. W., Jr. "Weaponized Pathogens and the SARS-CoV-2 Pandemic." *International Journal of Vaccine Theory, Practice, and Research,* vol. 1, no. 2 (January 5 2021): 172–208. https://ijvtpr.com/index.php/IJVTPR/article/view/16/17 .

9. Shaw, C. A. "The Age of COVID-19: Fear, Loathing, and the 'new normal'." *International Journal of Vaccine Theory, Practice, and Research.*

10. Hou, C.-Y. *Virologists Escorted Out of Lab in Canada.* (2019). From "News & Opinion". Retrieved from The Scientist, https://www.the-scientist.com/news-opinion/virologists -escorted-out-of-lab-in-canada-66164.; Pauls, K. *Chinese researcher escorted from infectious disease lab amid RCMP investigation.* (2019). From "Manitoba". Retrieved from CBC News, https://www.cbc.ca/news/canada/manitoba/chinese-researcher-escorted -from-infectious-disease-lab-amid-rcmp-investigation-1.5211567.

11. Shaw, C. A. "The Age of COVID-19: Fear, Loathing, and the 'new normal'." *International Journal of Vaccine Theory, Practice, and Research.*

12. *Players.* (2019). From "Event 201 Players". Retrieved from The Johns Hopkins Center for Health Security, https://www.centerforhealthsecurity.org/event201/players.

13. *Covid 19 deaths in China.* (2021). Retrieved from Google, https://www.google .com/search?q=covid+19+deaths+in+china&rlz=1C1CHBF_enCA816CA816& -oq=covid+19+deaths+in+china&aqs=chrome..69i57.9533j1j7&sourceid= -chrome&ie=UTF-8.

14. *Wuhan / Population.* (2018). Retrieved from Google, https://www.google.com /search?q=population+wuhan&rlz=1C1CHBF_enCA816CA816&oq=population+ wuhan&aqs=chrome..69i57.18868j1j4&sourceid=chrome&ie=UTF-811.

15. *Hubei / Population.* (2015). Retrieved from Google, https://www.google.com /search?q=population+hubei&rlz=1C1CHBF_enCA816CA816&oq=Population+ -Hubei&aqs=chrome.0.0j0i22i30l3j0i390.21424j1j4&sourceid=chrome&ie=UTF-8.

16. *China / Population.* (2019). Retrieved from Google, https://www.google.com /search?q=population+china&rlz=1C1CHBF_enCA816CA816&oq=population+ -china&aqs=chrome..69i57j0l3j0i395l4.4314j1j7&sourceid=chrome&ie=UTF-8.

17. Talmazan, Y., and Baculinao, E. *As Covid-19 runs riot across the world, China controls the pandemic.* (2020). Retrieved from NBC News, https://www.nbcnews.com/news/world/covid-19-runs-riot-across-world-china-controls-pandemic-n1246587.

18. Oller, J. W., Jr. "Weaponized Pathogens and the SARS-CoV-2 Pandemic." *International Journal of Vaccine Theory, Practice, and Research.*

19. Maron, D. F. *'Wet markets' likely launched the coronavirus. Here's what you need to know.* (2020). From "Coronavirus Coverage". Retrieved from National Geographic, https://www.nationalgeographic.com/animals/article/coronavirus-linked-to-chinese-wet-markets.

20. Jacobsen, R. *Could COVID-19 Have Escaped from a Lab?* (2020). Retrieved from Boston Magazine, https://www.bostonmagazine.com/news/2020/09/09/alina-chan-broad-institute-coronavirus/.

21. Zhan, S. H., Deverman, B. E., and Chan, Y. A. "SARS-CoV-2 is well adapted for humans. What does this mean for re-emergence?". *bioRxiv*, (2020): 2020.05.01.073262. doi:10.1101/2020.05.01.073262 %J bioRxiv.

22. Y. A. Chan, personal communication.

23. Quay, S. C. *A Bayesian analysis concludes beyond a reasonable doubt that SARS-CoV-2 is not a natural zoonosis but instead is laboratory derived.* (2021). Retrieved from Zenodo, https://zenodo.org/record/4477081.

24. Oller, J. W., Jr. "Weaponized Pathogens and the SARS-CoV-2 Pandemic." *International Journal of Vaccine Theory, Practice, and Research.*

25. Ibid, 77.

26. Blaylock, R. L. "Excitotoxicity (Immunoexcitotoxicity) as a Critical Component of the Cytokine Storm Reaction in Pulmonary Viral Infections, Including SARS-Cov-2." *International Journal of Vaccine Theory, Practice, and Research,* vol. 1, no. 2 (January 5 2021): 223–42. https://ijvtpr.com/index.php/IJVTPR/article/view/14/19.

27. Vanderwerf, J. D., and Kumar, M. A. "Management of neurologic complications of coagulopathies." Chap. 40 In *Handbook of Clinical Neurology.* edited by Eelco F. M. Wijdicks and Andreas H. Kramer.pg. 743–64: Elsevier, 2017. https://www.sciencedirect.com/science/article/pii/B9780444635990000405.

28. Mokhtari, T., Hassani, F., Ghaffari, N., Ebrahimi, B., Yarahmadi, A., and Hassanzadeh, G. "COVID-19 and multiorgan failure: A narrative review on potential mechanisms." *J Mol Histol,* vol. 51, no. 6 (Dec 2020): 613–28. doi:10.1007/s10735-020-09915-3 PMID: 33011887.

29. Montalvan, V., Lee, J., Bueso, T., De Toledo, J., and Rivas, K. "Neurological manifestations of COVID-19 and other coronavirus infections: A systematic review." *Clin Neurol Neurosurg,* vol. 194 (Jul 2020): 105921. doi:10.1016/j.clineuro.2020.105921 PMID: 32422545.; Wang, H. Y., Li, X. L., Yan, Z. R., Sun, X. P., Han, J., and Zhang, B. W. "Potential neurological symptoms of COVID-19." *Ther Adv Neurol Disord,* vol. 13 (2020): 1756286420917830. doi:10.1177/1756286420917830 PMID: 32284735.

30. Sheraton, M., Deo, N., Kashyap, R., and Surani, S. "A Review of Neurological Complications of COVID-19." *Cureus,* vol. 12, no. 5 (May 18 2020): e8192. doi:10.7759/cureus.8192 PMID: 32455089.

31. Baig, A. M. "Neurological manifestations in COVID-19 caused by SARS-CoV-2." *CNS Neurosci Ther,* vol. 26, no. 5 (May 2020): 499–501. doi:10.1111/cns.13372 PMID: 32266761.

32. Pereira, A. "Long-Term Neurological Threats of COVID-19: A Call to Update the Thinking About the Outcomes of the Coronavirus Pandemic." *Front Neurol,* vol. 11 (2020): 308. doi:10.3389/fneur.2020.00308 PMID: 32362868.; Wu, Y., Xu, X., Chen,

Z., Duan, J., Hashimoto, K., Yang, L., Liu, C., and Yang, C. "Nervous system involvement after infection with COVID-19 and other coronaviruses." *Brain Behav Immun,* vol. 87 (Jul 2020): 18–22. doi:10.1016/j.bbi.2020.03.031 PMID: 32240762.

33. Chao, Y. X., Gulam, M. Y., Chia, N. S. J., Feng, L., Rotzschke, O., and Tan, E. K. "Gut-Brain Axis: Potential Factors Involved in the Pathogenesis of Parkinson's Disease." *Front Neurol,* vol. 11 (2020): 849. doi:10.3389/fneur.2020.00849 PMID: 32982910.

34. Sultana, S., and Ananthapur, V. "COVID-19 and its impact on neurological manifestations and mental health: the present scenario." *Neurol Sci,* vol. 41, no. 11 (Nov 2020): 3015–20. doi:10.1007/s10072-020-04695-w PMID: 32865638.

35. Swann, B. *TRUTH: Seasonal FLU TWICE as Deadly as Coronavirus?* (2020). Retrieved from Youtube, https://www.youtube.com/watch?v=ohO8eAwi_po.

36. Howdon, D., Oke, J., and Heneghan, C. *Death certificate data: COVID-19 as the underlying cause of death.* (2020). Retrieved from The Centre for Evidence-Based Medicine, https://www.cebm.net/covid-19/death-certificate-data-covid-19-as-the-underlying-cause-of-death/.

37. Bendavid, E., Mulaney, B., Sood, N., Shah, S., Ling, E., Bromley-Dulfano, R., Lai, C., Weissberg, Z., Saavedra-Walker, R., Tedrow, J., Tversky, D., Bogan, A., Kupiec, T., Eichner, D., Gupta, R., Ioannidis, J. P. A., and Bhattacharya, J. "COVID-19 Antibody Seroprevalence in Santa Clara County, California." (2020): 2020.04.14.20062463. doi:10.1101/2020.04.14.20062463 %J medRxiv.

38. Schwalbe, N. *We Could Be Vastly Overestimating the Death Rate for COVID-19—Here's Why.* (2020). Retrieved from Our World, https://ourworld.unu.edu/en/we-could-be-vastly-overestimating-the-death-rate-for-covid-19-heres-why.

39. Adam, D. *Special report: The simulations driving the world's response to COVID-19.* (2020). From "News Article". Retrieved from Nature, https://www.nature.com/articles/d41586-020-01003-6.; Onge, P. S., and Campan, G. *The Flawed COVID-19 Model That Locked Down Canada.* (2020). Retrieved from The MEI, https://www.iedm.org/the-flawed-covid-19-model-that-locked-down-canada/.; Ferguson, N. M., Ghani, A. C., Donnelly, C. A., Hagenaars, T. J., and Anderson, R. M. "Estimating the human health risk from possible BSE infection of the British sheep flock." *Nature,* vol. 415, no. 6870 (Jan 24 2002): 420–4. doi:10.1038/nature709 PMID: 11786878.

40. https://youtu.be/hfkex_EepTQ (June 2, 2020). A question was posed by a reporter about one Covid-19 case who was in an intensive care unit. The question was posed at timestamp 33:45; Henry's reply at timestamp 34:28 has her stating that, in relation to Covid-19 cases, "…very severe as influenza can be." Henry also refers earlier in the clip to have antibody work done on blood samples to find out how many people had had the disease. What this further suggests is that B.C. has been using, and merging, serology tests with PCR data to get Covid-19 case numbers.

41. Ghoussoub, M. *B.C. records 31 deaths and 1,330 new cases of COVID-19 over 3 days.* (2021). Retrieved from CBC News, https://www.cbc.ca/news/canada/british-columbia/covid-19-update-jan-18-1.5878116.

42. It is certainly possible that I am missing something here, as epidemiology is not my field. However, having been through the myriad studies cited in Chapter 3 and taken into consideration the comments of Dr. Lyons-Weiler and others, the problem may lie more with the set of working assumptions that some epidemiologists bring to their studies. Given the "religious" nature that some of them clearly hold in regard to vaccine safety studies, I am strongly beginning to suspect that an agenda, rather than an unbiased search for truth, is the main driving force.

43. Demographics in overall infections U.S.: https://covid.cdc.gov/covid-data-tracker/#
    -demographics; Canada: https://health-infobase.canada.ca/covid-19/epidemiological
    -summary-covid-19-cases.html.

44. *United States / Population.* (2019). Retrieved from Google, https://www.google.com
    /search?q=us+population&rlz=1C1CHBF_enCA816CA816&oq=US+population&aqs
    =chrome.0.0i433i457j0i433j0l2j0i433l2j0l2.2958j0j7&sourceid=chrome&ie=UTF-8.

45. In both countries, the alleged COVID-19 deaths have roughly doubled since December
    2020.

46. *Prevalence.* (2021). From "Alzheimer's and Dementia - Facts and Figures". Retrieved
    from Alzheimer's Association, https://www.alz.org/alzheimers-dementia/facts-figures#:~:
    -text=HAVE%20INCREASED%20146%25.-,Prevalence,with%20Alzheimer's%20
    -dementia%20in%202020.

47. Gu, Y. *A closer look at U.S. deaths due to COVID-19.* (2020). Retrieved from The
    Johns Hopkins News-Letter, https://www.jhunewsletter.com/article/2020/11/a-closer
    -look-at-u-s-deaths-due-to-covid-19.

48. Burrell, C. J., Howard, C. R., and Murphy, F. A. "Pathogenesis of Virus Infections."
    [In eng]. *Fenner and White's Medical Virology,* (2017): 77–104. doi:10.1016/B978-0-12-
    375156-0.00007–2 PMID: PMC7150039.; According to this source, "Viral virulence
    is influenced by viral genes in four categories: (1) those that affect the ability of the virus
    to replicate, (2) those that affect host defense mechanisms, (3) those that affect tropism,
    spread throughout the body and transmissibility, and (4) those that encode or produce
    products that are directly toxic to the host."

49. Nolan, T., Hands, R. E., and Bustin, S. A. "Quantification of mRNA using real-time
    RT-PCR." *Nat Protoc,* vol. 1, no. 3 (2006): 1559–82. doi:10.1038/nprot.2006.236
    PMID: 17406449.

50. Cassels, A. *With the COVID-19 test, positivity doesn't mean infectious.* (2020). From "Issue
    analysis". Retrieved from FOCUS, https://www.focusonvictoria.ca/issue-analysis/43/.

51. Jaafar, R., Aherfi, S., Wurtz, N., Grimaldier, C., Hoang, V. T., Colson, P., Raoult, D.,
    and La Scola, B. "Correlation between 3790 qPCR positives samples and positive
    cell cultures including 1941 SARS-CoV-2 isolates." *Clin Infect Dis,* (Sep 28 2020).
    doi:10.1093/cid/ciaa1491 PMID: 32986798.

52. *Understanding cycle threshold (Ct) in SARS-CoV-2 RT-PCR: A guide for health protec-
    tion teams.* (2020). Retrieved from Public Health England, https://assets.publishing
    .service.gov.uk/government/uploads/system/uploads/attachment_data/file/926410
    /Understanding_Cycle_Threshold__Ct__in_SARS-CoV-2_RT-PCR_.pdf.

53. Rebuttal to the December 11, 2020, Food and Drug Administration's Response to
    Citizen Petition and Petition for Administrative Stay of Action (Docket Number: FDA-
    2020-P-2225) (February 10, 2021).

54. Kelly, R. P., Shelton, A. O., and Gallego, R. "Understanding PCR Processes to Draw
    Meaningful Conclusions from Environmental DNA Studies." *Sci Rep,* vol. 9, no. 1 (Aug
    20 2019): 12133. doi:10.1038/s41598-019-48546-x PMID: 31431641.

55. *American Type Culture Collection.* Retrieved from https://www.atcc.org/.

56. Menage, J. "Rapid Response to: Rapid roll out of SARS-CoV-2 antibody testing-a
    concern." *BMJ,* vol. 369 (Jun 24 2020): m2420. https://www.bmj.com/content/369
    /bmj.m2420/rr-5.; Zhuang, G. H., Shen, M. W., Zeng, L. X., Mi, B. B., Chen, F. Y.,
    Liu, W. J., Pei, L. L., Qi, X., and Li, C. "[WITHDRAWN: Potential false-positive
    rate among the 'asymptomatic infected individuals' in close contacts of COVID-19
    patients]." *Zhonghua Liu Xing Bing Xue Za Zhi,* vol. 41, no. 4 (Mar 5 2020): 485–88.

doi:10.3760/cma.j.cn112338-20200221-00144 PMID: 32133832.; Note that this article was withdrawn, not retracted, by the authors. The editor provided the following statement: "Editor office's response for Ahead of Print Article withdrawn. The article 'Potential false-positive rate among the 'asymptomatic infected individuals' in close contacts of COVID-19 patients' was under strong discussion after pre-published Questions from the readers mainly focused on the article's results and conclusions were dependant on theoretical deduction, but not the field epidemiology data and further researches were needed to prove the current theory. Based on previous discussions, the article was decided to be offline by the editorial board from the pre-publish lists."

57. Dr. Henry went from a little-known figure within the BC Ministry of Health, as most provincial health officers are, into a media star with the advent of the COVID-19 pandemic. Her almost daily media briefings became the go-to source for information about the pandemic and made her one of the most respected personalities in the province. Her star has faded a bit since March 2020 due to the length of the pandemic and her role in the then-endless control measures. At present, the state of emergency extensions was in the twenty-first such order and has since gone to an indefinite status as allowed by provincial legislation in Bill 19. Currently, Henry's popularity is pretty much a "U"-shaped curve: you either love her or hate her, and there is not much in between. Some of the growing antipathy can be traced to her comments near Christmas 2020 when, perhaps trying to be funny, she told the parents of young children that Santa would be coming down the chimney wearing a mask and that they should not leave cookies and milk out for him lest they expose him to COVID-19. A lot of people of my acquaintance found her advice/humor creepy and potentially scary to young children. Back in April 2020, Henry had also proposed that people not have sex with their partners, long-term or otherwise, without a mask on if facing each other face to face. It seemed hard to believe it at the time, but she counseled the use of "glory holes" for sex, that is, putting a penis through a hole in some object to a partner, so as not to have full bodily contact: https://globalnews.ca/news/7204384/coronavirus-glory-holes-sex/. (How this was supposed to work for a lesbian couple was not made clear.) It was comments such as these that convinced a notable part of the population that Henry was not only "winging it" about the pandemic, but also showing the signs of a bureaucrat given far too much power.

58. "Individuals commit the sunk cost fallacy when they continue a behavior or endeavor as a result of previously invested resources (time, money or effort)... This fallacy, which is related to loss aversion and status quo bias, can also be viewed as bias resulting from an ongoing commitment." See: Sunk cost fallacy. (2021). Retrieved from Behavioral Economics: https://www.behavioraleconomics.com/resources/mini-encyclopedia-of-be/sunk-cost-fallacy/.

59. https://www.cbc.ca/news/health/pcr-tests-false-positives-myth-reality-1.6034273.

60. *Understanding "False Positives": Serology Testing.* (2020). Retrieved from COVID Explained, https://explaincovid.org/other/understanding-false-positives/.

61. Ludvigsson, J. F. "The first eight months of Sweden's COVID-19 strategy and the key actions and actors that were involved." *Acta Paediatr,* vol. 109, no. 12 (Dec 2020): 2459–71. doi:10.1111/apa.15582 PMID: 32951258.

62. Claeson, M., and Hanson, S. "COVID-19 and the Swedish enigma." *Lancet,* vol. 397, no. 10271 (Jan 23 2021): 259–61. doi:10.1016/S0140-6736(20)32750-1 PMID: 33357494.

63. *Why Is COVID-19 More Contagious Than The Flu?* (2020). Retrieved from Henry Ford Health System, https://www.henryford.com/blog/2020/11/why-is-covid-more-contagious-than-flu.

64. Eustaeter, B. *'Another layer of protection': Feds now recommend three-layer masks with filters.* (2020). From "Coronavirus". Retrieved from CTV News, https://www.ctvnews.ca/health/coronavirus/another-layer-of-protection-feds-now-recommend-three-layer-masks-with-filters-1.5172995.

65. These may not be harmful to humans, but they are not precisely environmentally friendly: Roberts, K. P., et al. (2020). Coronavirus face masks: an environmental disaster that might last generations. Retrieved from The Conversation: https://theconversation.com/coronavirus-face-masks-an-environmental-disaster-that-might-last-generations-144328.

66. Ferguson, D. *Rape survivors say they are being stigmatised for not wearing masks.* (2020). From "Society". Retrieved from The Guardian, https://www.theguardian.com/society/2020/aug/10/survivors-say-they-are-being-stigmatised-for-not-wearing-masks.

67. MacIntyre, C. R., Seale, H., Dung, T. C., Hien, N. T., Nga, P. T., Chughtai, A. A., Rahman, B., Dwyer, D. E., and Wang, Q. "A cluster randomised trial of cloth masks compared with medical masks in healthcare workers." *BMJ Open,* vol. 5, no. 4 (Apr 22 2015): e006577. doi:10.1136/bmjopen-2014-006577 PMID: 25903751.

68. MacIntyre, C., Tham, D., Seale, H., and Chughtai, A. *Shortages of Masks and the Use of Cloth Masks as a Last Resort.* 2020. https://bmjopen.bmj.com/content/5/4/e006577.responses#covid-19-shortages-of-masks-and-the-use-of-cloth-masks-as-a-last-resort.

69. Sparks, D. *How do the different types of masks work?* (2021). From "COVID-19: How much protection do face masks offer?". Retrieved from Mayo Clinic News Network, https://newsnetwork.mayoclinic.org/discussion/covid-19-how-much-protection-do-face-masks-offer-2/#:~:text=As%20the%20name%20indicates%2C%20the,are%20-intended%20to%20be%20disposable.

70. Enserink, M. *Evidence-based medicine group in turmoil after expulsion of co-founder.* (2018). Retrieved from Science Magazine, https://www.sciencemag.org/news/2018/09/evidence-based-medicine-group-turmoil-after-expulsion-co-founder.

71. Jefferson, T., and Heneghan, C. *Masking lack of evidence with politics.* (2020). Retrieved from The Centre for Evidence-Based Medicine, https://www.cebm.net/covid-19/masking-lack-of-evidence-with-politics/.

72. Heneghan, C., and Jefferson, T. *Landmark Danish study finds no significant effect for facemask wearers.* (2020). Retrieved from The Spectator, https://www.spectator.co.uk/article/do-masks-stop-the-spread-of-covid-19-.

73. Bundgaard, H., Bundgaard, J. S., Raaschou-Pedersen, D. E. T., Von Buchwald, C., Todsen, T., Norsk, J. B., Pries-Heje, M. M., Vissing, C. R., Nielsen, P. B., Winslow, U. C., Fogh, K., Hasselbalch, R., Kristensen, J. H., Ringgaard, A., Porsborg Andersen, M., Goecke, N. B., Trebbien, R., Skovgaard, K., Benfield, T., Ullum, H., Torp-Pedersen, C., and Iversen, K. "Effectiveness of Adding a Mask Recommendation to Other Public Health Measures to Prevent SARS-CoV-2 Infection in Danish Mask Wearers: A Randomized Controlled Trial." *Ann Intern Med,* (Nov 18 2020). doi:10.7326/M20-6817 PMID: 33205991.

74. Jones, N. R., Qureshi, Z. U., Temple, R. J., Larwood, J. P. J., Greenhalgh, T., and Bourouiba, L. "Two metres or one: what is the evidence for physical distancing in covid-19?". vol. 370 (2020): m3223. doi:10.1136/bmj.m3223 %J BMJ.

75. Bourouiba, L., Dehandschoewercker, E., and Bush, John W. M. "Violent expiratory events: on coughing and sneezing." *Journal of Fluid Mechanics,* vol. 745 (2014): 537–63. doi:10.1017/jfm.2014.88.

76. Claeson, M., and Hanson, S. "COVID-19 and the Swedish enigma." *Lancet,* vol. 397, no. 10271 (Jan 23 2021): 259–61. doi:10.1016/S0140-6736(20)32750-1 PMID: 33357494.

77. Leonhardt, M. *63% of Americans have been living paycheck to paycheck since Covid hit.* (2020). Retrieved from CNBC, https://www.cnbc.com/2020/12/11/majority-of-americans -are-living-paycheck-to-paycheck-since-covid-hit.html.; Tranjan, R. *The Rent Is Due Soon - Financial Insecurity and COVID-19.* (2020). Retrieved from Canadian Centre for Policy Alternatives, https://www.policyalternatives.ca/sites/default/files/uploads /publications/2020/03/Rent%20is%20due%20soon%20FINAL.pdf.

78. See article in *Irish Times* reporting on a company called ROQG. ROQU group stands for Robert O. Quirke Group and their Covid-19 status app: https://www.prnewswire .com/news-releases/irish-based-roqu-group-launches-world-first-health-passport-digital -platform-to-support-increased-global-covid-19-testing-301120037.html. Quirke is described as a financier. A Google search of companies (https://www.dnb.com /business-directory/company-profiles.roqu_group_limited.7d261896a9388d -b7a102d0d421a87336.html) reveals that ROQU has two employees and €80,000 in the bank. This is not much for a company developing a major app to track COVID-19 status. The *Irish Times* didn't do more than dutifully report the story on ROQU, a story that has the smell of a shell company about it.

79. *Overdose Deaths Accelerating During COVID-19.* (2020). From "CDC Newsroom Releases". Retrieved from US Centers for Disease Control and Prevention, https://www .cdc.gov/media/releases/2020/p1218-overdose-deaths-covid-19.html.; Soke, S. *Parallel Pandemics: Opioid overdoses surge during COVID-19.* (2020). Retrieved from University of Alberta, https://www.ualberta.ca/public-health/news/2020/december/parallel-pandemics -opioid-overdoses-surge-during-covid19.html.

80. Abramson, A. *How COVID-19 may increase domestic violence and child abuse.* (2020). Retrieved from American Psychological Association, https://www.apa.org/topics/covid-19 /domestic-violence-child-abuse.

81. Rosenbaum, L. "The Untold Toll - The Pandemic's Effects on Patients without Covid-19." *N Engl J Med,* vol. 382, no. 24 (Jun 11 2020): 2368–71. doi:10.1056 /NEJMms2009984 PMID: 32302076.

82. Lyons-Weiler, J. "Balance of Risk in COVID-19 Reveals the Extreme Cost of False Positives." *International Journal of Vaccine Theory, Practice, and Research,* vol. 1, no. 2 (January 05 2021): 209–22. https://ijvtpr.com/index.php/IJVTPR/article/view/15/24.

83. Loftus, P., and Pulliam, S. *People Harmed by Coronavirus Vaccines Will Have Little Recourse.* (2020). Retrieved from The Wall Street Journal, https://www.wsj.com/articles /people-harmed-by-coronavirus-vaccines-will-have-little-recourse-11602432000.; Halabi, S., Heinrich, A., and Omer, S. B. "No-Fault Compensation for Vaccine Injury - The Other Side of Equitable Access to Covid-19 Vaccines." *N Engl J Med,* vol. 383, no. 23 (Dec 3 2020): e125. doi:10.1056/NEJMp2030600 PMID: 33113309.; Sigalos, M. *You can't sue Pfizer or Moderna if you have severe Covid vaccine side effects. The govern-ment likely won't compensate you for damages either.* (2020). Retrieved from CNBC, https: //www.cnbc.com/2020/12/16/covid-vaccine-side-effects-compensation-lawsuit.html.

84. Alberer, M., Gnad-Vogt, U., Hong, H. S., Mehr, K. T., Backert, L., Finak, G., Gottardo, R., Bica, M. A., Garofano, A., Koch, S. D., Fotin-Mleczek, M., Hoerr, I., Clemens,

R., and Von Sonnenburg, F. "Safety and immunogenicity of a mRNA rabies vaccine in healthy adults: an open-label, non-randomised, prospective, first-in-human phase 1 clinical trial." *Lancet,* vol. 390, no. 10101 (Sep 23 2017): 1511–20. doi:10.1016/S0140-6736(17)31665–3 PMID: 28754494.

85. Bahl, K., Senn, J. J., Yuzhakov, O., Bulychev, A., Brito, L. A., Hassett, K. J., Laska, M. E., Smith, M., Almarsson, O., Thompson, J., Ribeiro, A. M., Watson, M., Zaks, T., and Ciaramella, G. "Preclinical and Clinical Demonstration of Immunogenicity by mRNA Vaccines against H10N8 and H7N9 Influenza Viruses." *Mol Ther,* vol. 25, no. 6 (Jun 7 2017): 1316–27. doi:10.1016/j.ymthe.2017.03.035 PMID: 28457665.

86. https://www.nejm.org/doi/full/10.1056/nejmoa2024671.

87. Corbett, K. S., Flynn, B., Foulds, K. E., Francica, J. R., Boyoglu-Barnum, S., Werner, A. P., Flach, B., O'connell, S., Bock, K. W., Minai, M., Nagata, B. M., Andersen, H., Martinez, D. R., Noe, A. T., Douek, N., Donaldson, M. M., Nji, N. N., Alvarado, G. S., Edwards, D. K., Flebbe, D. R., Lamb, E., Doria-Rose, N. A., Lin, B. C., Louder, M. K., O'dell, S., Schmidt, S. D., Phung, E., Chang, L. A., Yap, C., Todd, J. M., Pessaint, L., Van Ry, A., Browne, S., Greenhouse, J., Putman-Taylor, T., Strasbaugh, A., Campbell, T. A., Cook, A., Dodson, A., Steingrebe, K., Shi, W., Zhang, Y., Abiona, O. M., Wang, L., Pegu, A., Yang, E. S., Leung, K., Zhou, T., Teng, I. T., Widge, A., Gordon, I., Novik, L., Gillespie, R. A., Loomis, R. J., Moliva, J. I., Stewart-Jones, G., Himansu, S., Kong, W. P., Nason, M. C., Morabito, K. M., Ruckwardt, T. J., Ledgerwood, J. E., Gaudinski, M. R., Kwong, P. D., Mascola, J. R., Carfi, A., Lewis, M. G., Baric, R. S., Mcdermott, A., Moore, I. N., Sullivan, N. J., Roederer, M., Seder, R. A., and Graham, B. S. "Evaluation of the mRNA-1273 Vaccine against SARS-CoV-2 in Nonhuman Primates." *N Engl J Med,* vol. 383, no. 16 (Oct 15 2020): 1544–55. doi:10.1056/NEJMoa2024671 PMID: 32722908.

88. Corbett, K. S., Edwards, D., Leist, S. R., Abiona, O. M., Boyoglu-Barnum, S., Gillespie, R. A., Himansu, S., Schäfer, A., Ziwawo, C. T., Dipiazza, A. T., Dinnon, K. H., Elbashir, S. M., Shaw, C. A., Woods, A., Fritch, E. J., Martinez, D. R., Bock, K. W., Minai, M., Nagata, B. M., Hutchinson, G. B., Bahl, K., Garcia-Dominguez, D., Ma, L., Renzi, I., Kong, W.-P., Schmidt, S. D., Wang, L., Zhang, Y., Stevens, L. J., Phung, E., Chang, L. A., Loomis, R. J., Altaras, N. E., Narayanan, E., Metkar, M., Presnyak, V., Liu, C., Louder, M. K., Shi, W., Leung, K., Yang, E. S., West, A., Gully, K. L., Wang, N., Wrapp, D., Doria-Rose, N. A., Stewart-Jones, G., Bennett, H., Nason, M. C., Ruckwardt, T. J., Mclellan, J. S., Denison, M. R., Chappell, J. D., Moore, I. N., Morabito, K. M., Mascola, J. R., Baric, R. S., Carfi, A., and Graham, B. S. "SARS-CoV-2 mRNA Vaccine Development Enabled by Prototype Pathogen Preparedness." *bioRxiv,* (2020): 2020.06.11.145920. doi:10.1101/2020.06.11.145920 %J bioRxiv.

89. *How to calculate risk.* Retrieved from BMJ Best Practice, https://bestpractice.bmj.com /info/toolkit/learn-ebm/how-to-calculate-risk.

90. *Moderna's COVID-19 Vaccine Candidate Meets its Primary Efficacy Endpoint in the First Interim Analysis of the Phase 3 COVE Study.* (2020). From "Press Release". Retrieved from Moderna, https://investors.modernatx.com/news-releases/news-release-details /modernas-covid-19-vaccine-candidate-meets-its-primary-efficacy.

91. *Conclusion.* (2020). From "Estimated Influenza Illnesses, Medical visits, Hospitalizations, and Deaths in the United States—2018–2019 influenza season". Retrieved from US Centers for Disease Control and Prevention, https://www.cdc.gov/flu/about/burden/2018–2019 .html#:~:text=vaccination%20uptake11.-,Conclusion,2012%E2%80%932013%20 -influenza%20season1.

92. Doshi, P. *Peter Doshi: Pfizer and Moderna's "95% effective" vaccines—let's be cautious and first see the full data.* (2020). Retrieved from BMJ Opinion, https://blogs.bmj.com/bmj/2020/11/26/peter-doshi-pfizer-and-modernas-95-effective-vaccines-lets-be-cautious-and-first-see-the-full-data/.

93. Mulligan, M. J., Lyke, K. E., Kitchin, N., Absalon, J., Gurtman, A., Lockhart, S., Neuzil, K., Raabe, V., Bailey, R., Swanson, K. A., Li, P., Koury, K., Kalina, W., Cooper, D., Fontes-Garfias, C., Shi, P. Y., Tureci, O., Tompkins, K. R., Walsh, E. E., Frenck, R., Falsey, A. R., Dormitzer, P. R., Gruber, W. C., Sahin, U., and Jansen, K. U. "Phase I/II study of COVID-19 RNA vaccine BNT162b1 in adults." *Nature,* vol. 586, no. 7830 (Oct 2020): 589–93. doi:10.1038/s41586-020-2639-4 PMID: 32785213.

94. Ibid.

95. *Johnson & Johnson Posts Interim Results from Phase 1/2a Clinical Trial of its Janssen COVID-19 Vaccine Candidate.* (2020). Retrieved from Johnson & Johnson, https://www.jnj.com/johnson-johnson-posts-interim-results-from-phase-1-2a-clinical-trial-of-its-janssen-covid-19-vaccine-candidate.

96. Kemp, A. *AZD1222 vaccine met primary efficacy endpoint in preventing COVID-19.* (2020). Retrieved from AstraZeneca, https://www.astrazeneca.com/media-centre/press-releases/2020/azd1222hlr.html.

97. *Safety and Immunogenicity Study of 2019-nCoV Vaccine (mRNA-1273) for Prophylaxis of SARS-CoV-2 Infection (COVID-19).* (2020). Retrieved from US National Library of Congress, ClinicalTrial.gov, https://clinicaltrials.gov/ct2/show/NCT04283461.

98. *Vaccines and Related Biological Products Advisory Committee Meeting, December 17, 2020* (2020). By US Food & Drug Administration. Retrieved from ModernaTX, Inc., https://www.fda.gov/media/144434/download.

99. *Vaccines and Related Biological Products Advisory Committee Meeting, December 17, 2020* (2020). Chapter "5.2.6 Safety". pg. 31. By US Food & Drug Administration. Retrieved from ModernaTX, Inc., https://www.fda.gov/media/144434/download.

100. *Vaccines and Related Biological Products Advisory Committee Meeting, December 17, 2020* (2020). Chapter "5.2.6 Safety". pg. 32. By US Food & Drug Administration. Retrieved from ModernaTX, Inc., https://www.fda.gov/media/144434/download.

101. EUA guidelines (https://www.fda.gov/media/142749/download): "The chemical, biological, radiological, or nuclear (CBRN) agent referred to in the March 27, 2020 EUA declaration by the Secretary of HHS (SARS-CoV-2) can cause a serious or life threatening disease or condition"; "Based on the totality of scientific evidence available, including data from adequate and well controlled trials, if available, it is reasonable to believe that the product may be effective to prevent, diagnose, or treat such serious or life-threatening disease or condition that can be caused by SARS-CoV-2"; "The known and potential benefits of the product, when used to diagnose, prevent, or treat the identified serious or life-threatening disease or condition, outweigh the known and potential risks of the product"; "There is no adequate, approved, and available alternative to the product for diagnosing, preventing, or treating the disease or condition."

102. Mulligan, M. J., Lyke, K. E., Kitchin, N., Absalon, J., Gurtman, A., Lockhart, S., Neuzil, K., Raabe, V., Bailey, R., Swanson, K. A., Li, P., Koury, K., Kalina, W., Cooper, D., Fontes-Garfias, C., Shi, P. Y., Tureci, O., Tompkins, K. R., Walsh, E. E., Frenck, R., Falsey, A. R., Dormitzer, P. R., Gruber, W. C., Sahin, U., and Jansen, K. U. "Phase I/II study of COVID-19 RNA vaccine BNT162b1 in adults." *Nature,* vol. 586, no. 7830 (Oct 2020): 589–93. doi:10.1038/s41586-020-2639-4 PMID: 32785213.

103. *Vaccines and Related Biological Products Advisory Committee Meeting, December 10, 2020* (2020). pg. 32. By US Food & Drug Administration. Retrieved from Pfizer and BioNTech, https://www.fda.gov/media/144245/download.

104. Polack, F. P., Thomas, S. J., Kitchin, N., Absalon, J., Gurtman, A., Lockhart, S., Perez, J. L., Perez Marc, G., Moreira, E. D., Zerbini, C., Bailey, R., Swanson, K. A., Roychoudhury, S., Koury, K., Li, P., Kalina, W. V., Cooper, D., Frenck, R. W., Jr., Hammitt, L. L., Tureci, O., Nell, H., Schaefer, A., Unal, S., Tresnan, D. B., Mather, S., Dormitzer, P. R., Sahin, U., Jansen, K. U., Gruber, W. C., and Group, C. C. T. "Safety and Efficacy of the BNT162b2 mRNA Covid-19 Vaccine." *N Engl J Med,* vol. 383, no. 27 (Dec 31 2020): 2603–15. doi:10.1056/NEJMoa2034577 PMID: 33301246.

105. Favaro, A., Philip, E. S., and Jones, A. M. *COVID-19 vaccines have expected side-effects, but experts say they're no cause for concern.* (2020). From "Coronavirus". Retrieved from CTV News, https://www.ctvnews.ca/health/coronavirus/covid-19-vaccines -have-expected-side-effects-but-experts-say-they-re-no-cause-for-concern-1.5222927.; Woodruff, M. *Vaccines against the coronavirus will have side effects – and that's a good thing.* (2020). From "Health". Retrieved from PBS News Hour, https://www.pbs.org /newshour/health/vaccines-against-sars-cov-2-will-have-side-effects-thats-a-good-thing.

106. Polack, F. P., Thomas, S. J., Kitchin, N., Absalon, J., Gurtman, A., Lockhart, S., Perez, J. L., Perez Marc, G., Moreira, E. D., Zerbini, C., Bailey, R., Swanson, K. A., Roychoudhury, S., Koury, K., Li, P., Kalina, W. V., Cooper, D., Frenck, R. W., Jr., Hammitt, L. L., Tureci, O., Nell, H., Schaefer, A., Unal, S., Tresnan, D. B., Mather, S., Dormitzer, P. R., Sahin, U., Jansen, K. U., Gruber, W. C., and Group, C. C. T. "Safety and Efficacy of the BNT162b2 mRNA Covid-19 Vaccine." *N Engl J Med,* vol. 383, no. 27 (Dec 31 2020): 2603–15. doi:10.1056/NEJMoa2034577 PMID: 33301246.

107. Smith, Y. *An Internal Medicine Doctor and His Peers Read the Pfizer Vaccine Study and See Red Flags [Updated].* (2020). Retrieved from Naked Capitalism, https://www .nakedcapitalism.com/2020/12/an-internal-medicine-doctor-and-his-peers-read-the -pfizer-vaccine-study-and-see-red-flags.html.

108. Cobb, M. "60 years ago, Francis Crick changed the logic of biology." *PLoS Biol,* vol. 15, no. 9 (Sep 2017): e2003243. doi:10.1371/journal.pbio.2003243 PMID: 28922352.

109. Pellionisz, A. J. "The principle of recursive genome function." *Cerebellum,* vol. 7, no. 3 (2008): 348–59. doi:10.1007/s12311-008-0035-y PMID: 18566877.

110. Nirenberg, E. *No, Really, mRNA Vaccines Are Not Going To Affect Your DNA.* (2020). Retrieved from Deplatform Disease, https://www.deplatformdisease.com/blog/no-really -mrna-vaccines-are-not-going-to-affect-your-dna.

111. Bahl, K., Senn, J. J., Yuzhakov, O., Bulychev, A., Brito, L. A., Hassett, K. J., Laska, M. E., Smith, M., Almarsson, O., Thompson, J., Ribeiro, A. M., Watson, M., Zaks, T., and Ciaramella, G. "Preclinical and Clinical Demonstration of Immunogenicity by mRNA Vaccines against H10N8 and H7N9 Influenza Viruses." *Mol Ther,* vol. 25, no. 6 (Jun 7 2017): 1316–27. doi:10.1016/j.ymthe.2017.03.035 PMID: 28457665.

112. ALS, Alzheimer's disease, Parkinson's disease, and Huntington's disease neurons contain specific misfolded proteins in what is called a β-sheet confirmation. Considering this, a neuron with the mRNA for the spike protein, just like a muscle cell, may now make abundant spike protein. Will all of it leave the cell? Might some of it form into β-sheet forms? Moreover, what will the resident immune cells, the microglia, likely do to the cell making this extruded protein?

113. Lee, W. S., Wheatley, A. K., Kent, S. J., and Dekosky, B. J. "Antibody-dependent enhancement and SARS-CoV-2 vaccines and therapies." *Nat Microbiol,* vol. 5, no. 10 (Oct 2020): 1185–91. doi:10.1038/s41564-020-00789-5 PMID: 32908214.

114. Tseng, C. T., Sbrana, E., Iwata-Yoshikawa, N., Newman, P. C., Garron, T., Atmar, R. L., Peters, C. J., and Couch, R. B. "Immunization with SARS coronavirus vaccines leads to pulmonary immunopathology on challenge with the SARS virus." *PLoS One,* vol. 7, no. 4 (2012): e35421. doi:10.1371/journal.pone.0035421 PMID: 22536382.

115. *Precautionary principle.* (2004). Retrieved from Wikipedia, https://en.wikipedia.org/wiki/Precautionary_principle.

116. *Johnson & Johnson Posts Interim Results from Phase 1/2a Clinical Trial of its Janssen COVID-19 Vaccine Candidate.* (2020). Retrieved from Johnson & Johnson, https://www.jnj.com/johnson-johnson-posts-interim-results-from-phase-1-2a-clinical-trial-of-its-janssen-covid-19-vaccine-candidate.

117. Kemp, A. *AZD1222 vaccine met primary efficacy endpoint in preventing COVID-19.* (2020). Retrieved from AstraZeneca, https://www.astrazeneca.com/media-centre/press-releases/2020/azd1222hlr.html.

118. Pitt, S. *A coronavirus vaccine may require boosters – here's what that means.* (2020). Retrieved from The Conversation, https://theconversation.com/a-coronavirus-vaccine-may-require-boosters-heres-what-that-means-143370.

119. Braithwaite, T. *Why the Pfizer CEO selling 62% of his stock the same day as the vaccine announcement looks bad.* (2020). From "Financial Times". Retrieved from Financial Post, https://financialpost.com/financial-times/why-the-pfizer-ceo-selling-62-of-his-stock-the-same-day-as-the-vaccine-announcement-looks-bad.; Nagarajan, S. *Moderna's CEO sold nearly $2 million of his stock ahead of the company's emergency use vaccine filing. He's now worth $3 billion.* (2020). From "Financial Times". Retrieved from Market Insider, https://markets.businessinsider.com/news/stocks/moderna-ceo-sold-million-stock-company-emergency-use-vaccine-filing-2020-11-1029830266.

120. Children's Health Defense Team. "Planned Surveillance and Control by Global Technocrats: A Big-Picture Look at the Current Pandemic Beneficiaries." *International Journal of Vaccine Theory, Practice, and Research,* vol. 1, no. 2 (2021): 143–71. https://ijvtpr.com/index.php/IJVTPR/article/view/7/16.

121. Kulldorff, M., Gupta, S., and Bhattacharya, J. *Great Barrington Declaration.* (2020). Retrieved from https://gbdeclaration.org/.

122. See comments of Dr. K. Wittkowski, former long-term head of The Rockefeller University's Department of Biostatistics, Epidemiology, and Research Design: https://www.reddit.com/r/conspiracy/comments/g2yl1c/professor_knut_wittkowski_for_twenty_years_head/.

123. Asymptomatic means just that: no symptoms of the disease. This can arise where the person so considered has not had the disease or where someone who had the disease is no longer infectious.

124. Morens, D. M., Taubenberger, J. K., and Fauci, A. S. "Predominant role of bacterial pneumonia as a cause of death in pandemic influenza: implications for pandemic influenza preparedness." *J Infect Dis,* vol. 198, no. 7 (Oct 1 2008): 962–70. doi:10.1086/591708 PMID: 18710327.

125. Barry, J. M. "The site of origin of the 1918 influenza pandemic and its public health implications." *J Transl Med,* vol. 2, no. 1 (Jan 20 2004): 3. doi:10.1186/1479-5876-2-3 PMID: 14733617.; Gates, F. L. "A Report on Antimeningitis Vaccination and

Observations on Agglutinins in the Blood of Chronic Meningococcus Carriers." *J Exp Med,* vol. 28, no. 4 (Oct 1 1918): 449–74. doi:10.1084/jem.28.4.449 PMID: 19868270.

126. Shaw, C. A. *Neural Dynamics of Neurological Disease.* Boston: John Wiley & Sons, 2017. doi:https://doi.org/10.1002/9781118634523.ch1.; Shaw, C. A., and Marler, T. E. "The lessons of ALS-PDC: Environmental factors in ALS etiology." In *Spectrums of Amyotrophic Lateral Sclerosis: Heterogeneity, Pathogenesis and Therapeutic Directions.* edited by C. A. Shaw and Jessica R. Morrice. New Jersey: John Wiley & Sons, 2021.

127. Zimmerman, H. M. "Progress report of work in the laboratory of pathology during May, 1945. Guam." *US Naval Medical Research Unit Number 2, 1 June. Washington, DC,* (1945).

128. Hammon, W. M., Tigertt, W. D., Sather, G. E., Berge, T. O., and Meiklejohn, G. "Epidemiologic studies of concurrent virgin epidemics of Japanese B encephalitis and of mumps on Guam, 1947–1948, with subsequent observations including dengue, through 1957." *Am J Trop Med Hyg,* vol. 7, no. 4 (Jul 1958): 441–67. doi:10.4269/ajtmh .1958.7.441 PMID: 13559599.

129. Sacks, O. W. *Awakenings.* Vintage Books, 1999. https://books.google.ca/books? -id=8Uw9PIuM8q8C.

130. Sato, F., Tanaka, H., Hasanovic, F., and Tsunoda, I. "Theiler's virus infection: Pathophysiology of demyelination and neurodegeneration." *Pathophysiology,* vol. 18, no. 1 (Feb 2011): 31–41. doi:10.1016/j.pathophys.2010.04.011 PMID: 20537875.; Jubelt, B., and Lipton, H. L. "Enterovirus/Picornavirus infections." Chap. 18 In *Handbook of Clinical Neurology.* edited by Alex C. Tselis and John Booss.pg. 379–416: Elsevier, 2014. https://www.sciencedirect.com/science/article/pii/B9780444534880000183.

131. Bookchin, D., and Schumacher, J. *The Virus and the Vaccine: The True Story of a Cancer-Causing Monkey Virus, Contaminated Polio Vaccine, Deadly Cancers, and Government Neglect.* New York: St. Martin's Press, 2005. https://books.google.ca/books?id=QaubtBGQzjsC.

132. Danish, P. *The Cutter incident: A cautionary tale for Operation Warp Speed.* (2020). From "Opinion". Retrieved from Boulder Weekly, https://www.boulderweekly.com/opinion /the-cutter-incident-a-cautionary-tale-for-operation-warp-speed/.

133. Kurland, L. T., and Mulder, D. W. "Epidemiologic investigations of amyotrophic lateral sclerosis. I. Preliminary report on geographic distribution, with special reference to the Mariana Islands, including clinical and pathologic observations." *Neurology,* vol. 4, no. 5 (May 1954): 355–78. doi:10.1212/wnl.4.5.355 PMID: 13185376.

134. J. Steele, personal communication.

135. *Hantavirus Pulmonary Syndrome (HPS).* (2013). From "Hantavirus". Retrieved from US Centers for Disease Control and Prevention, U.S. Department of Health & Human Services, https://www.cdc.gov/hantavirus/hps/index.html#:~:text=Hanta -virus%20Pulmonary%20Syndrome%20(HPS)%20is,primary%20risk%20for%20 hantavirus%20exposure.

136. Jacobsen, R. *Could COVID-19 Have Escaped from a Lab?* (2020). From "Research". Retrieved from Boston Magazine, https://www.bostonmagazine.com/news/2020/09/09 /alina-chan-broad-institute-coronavirus/.

137. Oller, J. W., Jr. "Weaponized Pathogens and the SARS-CoV-2 Pandemic." *International Journal of Vaccine Theory, Practice, and Research.*

138. Ibid.

139. Ibid.

140. Patel, R. *Garneau won't rule out invoking Emergencies Act to limit pandemic travel.* (2021). From "Politics". Retrieved from CBC News, https://www.cbc.ca/news/politics/garneau-emergencies-act-pandemic-travel-1.5885770.

141. Children's Health Defense Team. "Planned Surveillance and Control by Global Technocrats: A Big-Picture Look at the Current Pandemic Beneficiaries." *Int. Journal of Vaccine Theory, Practice, and Research,* vol. 1, no. 2 (2021): 143–71. https://ijvtpr.com/index.php/IJVTPR/article/view/7/16.

142. Schwab (1938) is a German with advanced degrees in both engineering and economics. How one goes from that to functionally directing the world during the pandemic may be the key unanswered question of this time in history. And the usual question to ask is cui bono, who benefits from Schwab's ascendency?

143. *Shaping the Future of the Internet of Bodies: New challenges of technology governance.* (2020). Retrieved from World Economic Forum, http://www3.weforum.org/docs/WEF_IoB_briefing_paper_2020.pdf.

144. Schwab, K. and Malleret, T. *Covid-19: The Great Reset.* Amazon Digital Services LLC - KDP Print US, 2020. https://books.google.ca/books?id=kruwzQEACAAJ.; Incidentally, the book describes Malleret as "…the Managing Partner of the Monthly Barometer, a succinct predictive analysis provided to private investors, global CEOs, and opinion- and decision-makers" (p. 6). Yes, just another person who cares for the health of the Earth and all of her people.

145. Schwab, K. *Now is the time for a 'great reset'* (2020). Retrieved from World Economic Forum, https://www.weforum.org/agenda/2020/06/now-is-the-time-for-a-great-reset.

146. Schwab, K. and Malleret, T. *Covid-19: The Great Reset.* Amazon Digital Services LLC - KDP Print US, 2020. https://books.google.ca/books?id=kruwzQEACAAJ.

147. Ibid.

148. *Your Manuscript Is Good and Original, But What is Original Is Not Good; What Is Good Is Not Original.* (2013). Retrieved from Quote Investigator, https://quoteinvestigator.com/2013/06/17/good-original/.

149. *Responding to the COVID-19 Crisis (Option 2).* (2021). From "Davos 2021". Retrieved from World Economic Forum, https://www.weforum.org/events/the-davos-agenda-2021/sessions/responding-to-the-covid-19-crisis-western-hemisphere.

150. Some of those on the political right tend to see in such measures a form of socialism or communism, but those who think so need to go back to Political Science 101. What we have here with the Great Reset is nothing other than true fascism in its purest twenty-first-century form, a form not all that much different from the Nazi version in the twentieth century, where major corporations and governments collude to take capital and decision making from below and concentrate it upward into the hands of those already rich and powerful beyond measure. The new WEF brand of fascism now has a happy, green, and caring face.

151. Children's Health Defense Team. "Planned Surveillance and Control by Global Technocrats: A Big-Picture Look at the Current Pandemic Beneficiaries." *International Journal of Vaccine Theory, Practice, and Research,* vol. 1, no. 2 (2021): 143–71. https://ijvtpr.com/index.php/IJVTPR/article/view/7/16.

152. Ibid, 13.

153. Ibid, 1.

154. Fitts, C. A. *The State of Our Currencies — Just a Taste.* (2020). From "The Solari Report". Retrieved from Truth Comes to Light, https://truthcomestolight.com/catherine-austin-fitts-the-state-of-our-currencies/.

155. Children's Health Defense Team. "Planned Surveillance and Control by Global Technocrats: A Big-Picture Look at the Current Pandemic Beneficiaries." *International Journal of Vaccine Theory, Practice, and Research,* vol. 1, no. 2 (2021): 14. https://ijvtpr .com/index.php/IJVTPR/article/view/7/16.

156. Abraham Maslow (1908–1970), an American psychologist, characterized human "needs" at a range of levels he depicted as a pyramid with the bottom level being physiological basics such as food, shelter, sex, etc. Moving up in the pyramid are family, then wealth, and finally up to "self-realization," the last presumably a higher state of awareness.

157. *Psychic driving.* (2012). Retrieved from Wikipedia, https://en.wikipedia.org/wiki /Psychic_driving.

158. Schwab, K. and Malleret, T. *Covid-19: The Great Reset.* Amazon Digital Services LLC-KDP Print US, 2020. https://books.google.ca/books?id=kruwzQEACAAJ.

159. https://www.businessinsider.com/bill-gates-wanted-jeffrey-epstein-help-him-nobel -peace-prize-2021-5.

160. This is hardly the first time in history that we have seen such a collaboration for world domination. One thinks of Genghis Khan, the Mongol visionary, who together with his master tactician, Subedei, pretty much conquered most of Asia, the Middle East, and vast swaths of Europe. Such investigation would also, if it could dig deep enough, lead us to those even higher up than Gates and Schwab. How do we know that such people exist? Simple: to have orchestrated the massive takedown of Bill Gates that we've recently seen, someone or a collection of someones with more wealth and power must be giving the orders.

161. Nelson, K. *COVID-19: Pursuing Truth to Protect Our Liberties.* (2020). From "Vaccinations". Retrieved from The Weston A. Price Foundation, https://www.westonaprice .org/health-topics/covid-19-pursuing-truth-to-protect-our-liberties/.

162. Lena, T. *The Great Reset for Dummies.* (2020). Retrieved from Tessa Fights Robots, https://tessa.substack.com/p/great-reset-dummies.

163. *The Webinar You've Been Waiting For: Alison McDowell, Joseph Gonzalez, and Incite Seminars Present "Level-Up".* (2020). Retrieved from Wrench in the Gears, https: //wrenchinthegears.com/2020/08/25/the-webinar-youve-been-waiting-for-alison -mcdowell-joseph-gonzalez-and-incite-seminars-present-level-up/.

164. Brand, R. *The Great Reset - Conspiracy or Fact?* (2021). Retrieved from Youtube, https: //www.youtube.com/watch?v=d4AduA8mgro.

165. Woodruff, J. *FULL INTERVIEW: Dr. Fauci on rising COVID-19 cases, a future vaccine and what the U.S. needs to do.* (2020). From "PBS Newshour". Retrieved from Youtube, https://www.youtube.com/watch?v=8Su5C_YefBU.

166. Kulldorff, M., Gupta, S., and Bhattacharya, J. *Great Barrington Declaration.* (2020). Retrieved from https://gbdeclaration.org/.

167. https://www.librti.com/victoria-declaration.

168. Shaw, C. A. *Neural Dynamics of Neurological Disease.* Boston: John Wiley & Sons, 2017. doi:https://doi.org/10.1002/9781118634523.ch1.

169. Shapiro, A. *America's Biggest Owner Of Farmland Is Now Bill Gates.* (2021). From "Editor's Pick". Retrieved from Forbes, https://www.forbes.com/sites/arielshapiro/2021/01/14 /americas-biggest-owner-of-farmland-is-now-bill-gates-bezos-turner/?sh =46d1fd236096.

170. Kennedy, R. F. *Bill Gates and Neo-Feudalism: A Closer Look at Farmer Bill.* (2021). Retrieved from Children's Health Defense, https://childrenshealthdefense.org/defender /bill-gates-neo-feudalism-farmer-bill.

Chapter 14

1. Masters, M. P. In *Identified Flying Objects: A Multidisciplinary Scientific Approach to the UFO Phenomenon.* pg. 17: Masters Creative LLC, 2019. https://books.google.ca /books?id=ixOXDwAAQBAJ.

2. Arendt, H. *The origins of totalitarianism.* New York: Harcourt, Brace and Co., 1951.

3. Stockton, F. R. *The Lady, or The Tiger?* (1882). Retrieved from East of the Web, http: //www.eastoftheweb.com/short-stories/UBooks/LadyTige.shtml.

4. Keane, T. *When a COVID-19 vaccine finally arrives, the government must make people take it.* (2020). From "Magazine". Retrieved from Boston Global, https://www.bostonglobe .com/2020/10/15/magazine/when-covid-19-vaccine-finally-arrives-government-must -make-people-take-it/?fbclid=IwAR2yZv0kmm83vYkeUbndXS0PG6EgQ4EIiXQMq _GwR_gXVPJ5I6c6Muyfr7M.

5. Kennedy, R. F., Hyman, M., and Herbert, M. R. *Thimerosal: Let the Science Speak: The Evidence Supporting the Immediate Removal of Mercury-a Known Neurotoxin-from Vaccines.* New York: Skyhorse, 2014. https://www.skyhorsepublishing.com/9781634504423 /thimerosal-let-the-science-speak/.

6. Shaw, C. A. *Neural Dynamics of Neurological Disease.* Boston: John Wiley & Sons, 2017. doi:https://doi.org/10.1002/9781118634523.ch1.

7. Morrice J. R., Smith M., Shan X., Libbrecht M. W., Hancock R. E. W., Gregory-Evans C. Y., and Shaw C. A. "Enhancer regulatory elements are novel risk factors for sporadic ALS." (2021, in preparation).

8. Shaw, C. A. *Neural Dynamics of Neurological Disease.*

9. Rogers, T. M. "The Political Economy of Autism." pg. 60–61. Postgraduate theis for PhD Doctorate, University of Sydney, 2018. Retrieved from https://ses.library.usyd.edu.au /handle/2123/20198.

10. Ibid, 50.

11. Ibid, 84.

12. Institute of Medicine, Board on Health Promotion and Disease Prevention, and Immunization Safety Review Committee. *Immunization Safety Review: Thimerosal-Containing Vaccines and Neurodevelopmental Disorders.* Edited by M.C. McCormick, A. Gable and K. Stratton. Washington, DC: National Academies Press, 2001. https: //books.google.ca/books?id=IHSwG1mw6ukC.

13. Angell, M. *The Truth About the Drug Companies: How They Deceive Us and What to Do About It.* New York: Random House Publishing Group, 2004. https://books.google.ca /books?id=5DKwxAnhTygC.; Husten, L. *Merck Pleads Guilty and Pays $950 Million for Illegal Promotion of Vioxx.* (2011). Retrieved from Forbes, https://www.forbes.com/sites /larryhusten/2011/11/22/merck-pleads-guilty-and-pays-950-million-for-illegal-promotion -of-vioxx/?sh=8b15d5c20f4e.

14. Gandhi's march to the sea is the perfect example of presenting any ruling authority with choices that are equally fatal for that authority. Mahatma Gandhi created the original version in trying to overturn the British Salt Tax Act of 1887 as part of a goal to overcome British rule in India. Leaving from Ahmedabad in March 1930, Gandhi and his followers finally arrived at Dandi on the Arabian Sea on April 5, having walked 241 miles in twenty-four days. As journalists and supporters watched, Gandhi gathered salt. According to History Stories, Gandhi proclaimed, "With this, I am shaking the foundations of the British Empire." The British now faced the choice of what to do: arrest Gandhi and expose the absurdity and injustice of the law, or let him get away

with his defiance to the Raj. They chose to arrest him, which in turn led to a growing resistance and eventual independence.; Andrews, E. *When Gandhi's Salt March Rattled British Colonial Rule.* (2019). Retrieved from History, https://www.history.com/news /gandhi-salt-march-india-british-colonial-rule.

15. Kruger, J., and Dunning, D. "Unskilled and unaware of it: how difficulties in recognizing one's own incompetence lead to inflated self-assessments." *J Pers Soc Psychol,* vol. 77, no. 6 (Dec 1999): 1121–34. doi:10.1037//0022–3514.77.6.1121 PMID: 10626367.

16. There are various definitions of fascism, but strictly speaking, it means a form of governance in which the government has become indistinguishable from the corporations such that the State has been given over to the latter. Fascism, historically, is marked by official bigotry against some part of the population, an expansion of governmental powers over the civil and human liberties of citizens with attendant suppression of these rights, and wars of conquest, or at least domination.

17. A still better depiction might be to add a moral dimension.

18. Eisenstein, C. *The Conspiracy Myth.* (2020). Retrieved from Charles Eisenstein, https: //charleseisenstein.org/essays/the-conspiracy-myth/.

19. Allen, D. *The Reaction of the Left to Lockdown.* (2020). Retrieved from Expressive Egg, https://expressiveegg.org/2020/08/09/the-reaction-of-the-left-to-lockdown/.

20. In regard to the position of much of the left in North America on human freedom positions on medical/vaccine rights, it has become very clear that the left of today bears little or no resemblance to the left of the past. Our grandparents and great-grandparents fought, and often bled, to create labor unions to give workers some voice and power. Our parents fought for civil rights and against illegal wars. Our own generations took on the giant multinational corporations and the World Trade Organization in the fight against economic globalization. Today's left, particularly in Vancouver, is a pathetic shell of their predecessors, almost solely engaged in identity politics, the attempt to stifle those they disagree with (either externally or internally), and to substitute name-calling and fury for analysis and debate. The consequence is obvious: such solipsistic pursuits become ever more self-centered to the point that the left in Vancouver and elsewhere is rendered ineffective. If one were to try to devise a left incapable of working for social change, the recipe would deliver an outcome much like the one now extant.

21. Bourdry, D., Hoop, D.K., 2021, "Messianic mad men, medicine and the media war on empirical reality: Discourse analysis of mainstream media Covid-19 propaganda." *Intl J Vaccine Theory, Practice and Research,* 2021; an anarchist publication, https://winteroak .org.uk/the-great-reset/; and even including a confusing article by the normally solid Naomi Klein, trying in a rambling way to make a case for a left-wing/right-wing conspiracy about the WEF while acknowledging that the WEF's vision is a huge problem for the world: https://theintercept.com/2020/12/08/great-reset-conspiracy/. In this piece, Klein wanders away from her usually reliable economic critiques of capitalism to rather gratuitously slam those whom she considers anti-vaxxers. One is left to suspect some jealousy that anti-vaxxers figured out Schwab and the Great Reset before she did.

22. *SWOT Analysis.* (2020). Retrieved from Business Balls, https://www.businessballs.com /strategy-innovation/swot-analysis/.

23. https://www.weforum.org/projects/cyber-polygon.

24. Bozni100. *George Carlin - You have no rights.* (2011). Retrieved from Youtube, https: //www.youtube.com/watch?v=m9-R8T1SuG4.

25. Eisenstein, C. *The Conspiracy Myth.* (2020).

26. Schwab, K. and Malleret, T. *Covid-19: The Great Reset*. Amazon Digital Services LLC - KDP Print US, 2020. https://books.google.ca/books?id=kruwzQEACAAJ.

27. Paine, T. *Appendix to the Third Edition*. (1776). From "Common Sense". Retrieved from Independence Hall Association, https://www.ushistory.org/paine/commonsense/sense6.htm.

28. *Butcher's bill*. (2007). Retrieved from Urban Dictionary, https://www.urbandictionary.com/define.php?term=Butcher%27s%20Bill.; Hedges, C. *Wages of Rebellion*. New York: Nation Books, 2015. https://books.google.ca/books?id=cy6NswEACAAJ.

29. Cernic, M. In *Ideological Constructs of Vaccination*. pg. 9: Wirral, UK: Vega Press Limited, 2018. https://books.google.ca/books?id=LHz-swEACAAJ.

30. A. Siri, personal communication.

31. *Section 7—Life, liberty and security of the person*. (2021). From "The Canadian Charter of Rights and Freedoms". Retrieved from Government of Canada, https://www.justice.gc.ca/eng/csj-sjc/rfc-dlc/ccrf-ccdl/check/art7.html.

32. QAnon. (2021). Retrieved from Wikipedia: https://en.wikipedia.org/wiki/QAnon. In its most basic form, adherents of QAnon have linked the COVID-19 events to attempts by a global elite to seize control (this part correct, in my view) and to a gigantic and Satanic pedophile ring praying on children and containing the likes of the late Jeffrey Epstein, Bill Clinton, Prince Andrew, and a host of others (some of the characters named have indeed been accused of pedophilia). Finally, they believe that Donald Trump is here to save us all from these evil forces. My opinion is that QAnon is either a CONTELPRO-type invention or simply a search by people for explanations for what seems like the collapse of civilization and a way of life. Similar belief systems have arisen for other people facing the destruction of their way of life.

33. As a general rule, dramatic events draw in a host of groups and personalities, many with widely disparate reasons for being involved. For example, in the case of the struggle for the autonomous Kurdish region of Rojava in their fight against the Islamic State, volunteers came from all over the world. Some of these were revolutionaries who identified with the Kurdish cause and/or the anarchist social experiment aspects of the struggle. Others could have cared less about the politics but were focused on fighting the IS and often came with a frank animosity toward Islam. The latter was strange in that most Kurds are themselves Sunni Muslims. Some of the volunteers came just for the adventure, and some came just to post on Facebook. The latter grouping showed a basic fact of social media life, that those who posted the most about their exploits generally did the least actual work or fighting. And, as always, in every struggle, some were just grifters.

34. Mikulic, M. *Global vaccine market revenues from 2014 to 2020 (in billion U.S. dollars)*. (2019). From "Pharmaceutical Products & Market". Retrieved from Statista, https://www.statista.com/statistics/265102/revenues-in-the-global-vaccine-market/.; In 2014 were $32.2 billion according to Statista, and an anticipated $54.2 billion in 2019 and $59.2 billion in 2020 before COVID-19. Claims that vaccines are not enormously profitable don't seem to hold much water given these numbers. These profits are in addition to the vast amount of money the same top companies make from all of their various products.

35. Davis, N. *What is the fourth industrial revolution?* (2016). From "Global Agenda". Retrieved from World Economic Forum, https://www.weforum.org/agenda/2016/01/what-is-the-fourth-industrial-revolution.; The first three industrial revolutions: the First Industrial Revolution featured the shift to water and steam power for transportation and industrial production; the Second featured electric power for mass production; the Third was built on electronic and information technology (IT).

36. *Frederick Douglass.* Retrieved from Quotes, https://www.quotes.net/quote/39692.

37. Gall, G. L., and McLpellan, A. A. *Peace, Order and Good Government.* (2006). Retrieved from The Canadian Encyclopedia, https://www.thecanadianencyclopedia.ca/en/article /peace-order-and-good-government.

38. Children's Health Defense. *Robert F. Kennedy, Jr.: Int'l. Message for Freedom and Hope.* (2020). Retrieved from Youtube, https://www.youtube.com/watch?v=NpMWDCX1yMI.

39. *Duluth Reader.* Retrieved from Duluth Reader, https://duluthreader.com/home.

40. Warchus, M., dir. *Pride.* (2014; CBS Films Distribution, September 12, 2014). IMDB: https://www.imdb.com/title/tt3169706/.; Kellaway, K. *When miners and gay activists united: the real story of the film Pride.* (2014). Retrieved from The Guardian, https://www .theguardian.com/film/2014/aug/31/pride-film-gay-activists-miners-strike-interview.

41. Dirik, D. *Solidarity is not a one-way charity undertaking by privileged activists, but a multidimensional process that contributes to the emancipation of everyone involved.* (2016). Retrieved from ROAR Magazine, https://roarmag.org/essays/privilege-revolution -rojava-solidarity/.

42. Children's Health Defense. *Robert F. Kennedy, Jr.: Int'l. Message for Freedom and Hope.* (2020).

43. Paine, T. *Appendix to the Third Edition.* (1776). From *Common Sense.*

44. Paine, T. *The Crisis.* (1776). From "The American Crisis". Retrieved from Independence Hall Association, https://www.ushistory.org/paine/crisis/c-01.htm.

45. Brady, K. *Thousands turn out in Berlin to protest coronavirus measures.* (2020). From "News". Retrieved from Deutsche Welle, https://www.dw.com/en/thousands-turn-out -in-berlin-to-protest-coronavirus-measures/a-54756290.; *What Really Happened in Berlin? CHD's Senta Depuydt Was There.* (2020). From "Advocacy Policy". Retrieved from Children's Health Defense, https://childrenshealthdefense.org/advocacy-policy /what-really-happened-in-berlin-chds-senta-depuydt-was-there/.

46. Eidi, H., Yoo, J., Bairwa, S. C., Kuo, M., Sayre, E. C., Tomljenovic, L., and Shaw, C. A. "Early postnatal injections of whole vaccines compared to placebo controls: Differential behavioural outcomes in mice." *J Inorg Biochem,* vol. 212 (Nov 2020): 111200. doi:10.1016/j.jinorgbio.2020.111200 PMID: 33039918.

47. Bairwa, S. C., Shaw, C. A., Kuo, M., Yoo, J., Tomljenovic, L., & Eidi, H. (2021). "Cytokines profile in neonatal and adult wild-type mice post-injection of US pediatric vaccination schedule." *Brain, Behavior, & Immunity-Health,* 15, 100267.

48. *About the Journal.* Retrieved from *Int. Journal of Vaccine Theory, Practice, and Research,* https://ijvtpr.com/index.php/IJVTPR.

49. Oller, J. W., Jr., and Shaw, C. A. "From Superficial Damage to Invasion of the Nucleosome: Ranking of Morbidities by the Biosemiotic Depth Hypothesis." *International Journal of Sciences,* vol. 8, no. 06 (2019): 51–73. doi:10.18483/ijSci.2069.

Epilogue

1. https://en.wikipedia.org/wiki/Thomas_Paine#/media/File:Portrait_of_Thomas_Paine .jpg.

Sources for Figures and Tables

Figures

1.1.

- https://commons.wikimedia.org/wiki/File:William_of_Ockham.png
- https://en.wikipedia.org/wiki/Francis_Bacon#/media/File:Somer_Francis_Bacon.jpg
- https://commons.wikimedia.org/wiki/File:Edward_Jenner._Oil_painting._Wellcome_V0023503.jpg
- https://commons.wikimedia.org/wiki/File:Karl_Popper.jpg
- https://commons.wikimedia.org/wiki/File:Thomas-kuhn-portrait.png
- https://commons.wikimedia.org/wiki/File:Andrew_Wakefield_with_Justyna_Socha_Warsaw_2019.jpg
- https://commons.wikimedia.org/wiki/File:Carl_Sagan,_1994.jpg

2.1.

- https://wellcomecollection.org/works/krgb7nyy/items?canvas=11
- https://www.ncbi.nlm.nih.gov/pmc/articles/PMC5650807/?page=1

3.3.

- Committee to Review Adverse Effects of Vaccines; Institute of Medicine. Adverse Effects of Vaccines: Evidence and Causality. Stratton K, Ford A, Rusch E, Clayton EW, editors. Washington (DC): National Academies Press (US); 2011 Aug 25. PMID: 24624471. https://pubmed.ncbi.nlm.nih.gov/24624471/

3.4.

- Two sources of these data are Project Tycho (https://www.tycho.pitt.edu/) or the Historical Record of the United States

- (https://www.census.gov/library/publications/1975/compendia /hist_stats_colonial-1970.html). These data have not been published previously (Morrice, J. and Shaw, C.A).

4.1.

- Adapted from Lord, C. and Jones, R. M., "Annual research review: re-thinking the classification of autism spectrum disorders." *J Child Psychol Psychiatry,* vol. 53, no. 5 (2012): 490-509. doi:10.1111/j.1469-7610.2012.02547.x. PMID: 22486486)

4.2.

- https://childrenshealthdefense.org/news/the-2020-addm-report-on -u-s-autism-prevalence-three-reasons-why-the-popular-narrative-was -misleading/

5.1.

- https://en.wikipedia.org/wiki/Ethylmercury
- https://commons.wikimedia.org/wiki/File:Methylmercury.png

5.2.

- Petrik, M. S., Wong, M. C., Tabata, R. C., Garry, R. F., and Shaw, C. A., "Aluminum adjuvant linked to Gulf War illness induces motor neuron death in mice." *Journal of Neuromolecular Medicine,* vol. 9, no. 1 (2007): 83-100. PMID: 17114826

5.3.

- Petrik, M. S., Wong, M. C., Tabata, R. C., Garry, R. F., and Shaw, C. A., "Aluminum adjuvant linked to Gulf War illness induces motor neuron death in mice." *Journal of Neuromolecular Medicine,* vol. 9, no. 1 (2007): 83-100. PMID: 17114826

5.5.

- Adapted from McFarland, G., La Joie, E., Thomas, P., and Lyons-Weiler, J., "Acute exposure and chronic retention of aluminum in three vaccine schedules and effects of genetic and environmental variation." *J Trace Elem Med Biol,* vol. 58, no. (2020): 126444. doi:10.1016/j. jtemb.2019.126444. PMID: 31846784. https://www.sciencedirect.com/ science/article/pii/S0946672X19305784

5.6.

- https://www.healthlinkbc.ca/sites/default/files/pdf/immunization
  -infants-children.pdf

6.1.

- https://en.wikipedia.org/wiki/Paul_Offit#/media/File:Paul_Offit.jpg
- https://en.wikipedia.org/wiki/Stanley_Plotkin#/media/File:Photo
  _Plotkin1.jpg
- https://en.wikipedia.org/wiki/Peter_Hotez#/media/File:Peter
  _Hotez_2019_Texas_Book_Festival.jpg (Larry D. Moore, CC BY-SA
  4.0, Wikimedia Commons.)

7.1.

- https://childrenshealthdefense.org/news/
  robert-f-kennedy-jr-speaks-at-berlin-rally-for-freedom-and-peace/

9.1.

- https://en.wikipedia.org/wiki/David_Gorski#/media/File:Gorski1.jpeg.

10.1.

- https://commons.wikimedia.org/wiki/File:Women_wearing_face_
  masks_in_Copenhagen_(51087395881).jpg

10.2.

- https://upload.wikimedia.org/wikipedia/commons/0/0b/Mukhisa
  _Kituyi%2C_Houlin_Zhao%2C_Tedros_Adhanom
  _Ghebreyesus_with_Sophia_-_AI_for_Good_Global_
  Summit_2018_%2841223188035%29_%28cropped%29.jpg.

10.3.

- https://www.who.int/teams/immunization-vaccines-and-biologicals
  /strategies/ia2030

10.4.

- https://www.who.int/teams/immunization-vaccines-and-biologicals/
  strategies/ia2030

12.1. and 12.2.

- https://www.thenation.com/article/society/bill-gates-foundation
  -philanthropy/

12.4.

- https://health-infobase.canada.ca/src/data/covidLive/covid19 -download.csv.

13.1.

- https://health-infobase.canada.ca/src/data/covidLive/covid19 -download.csv

14.1.

- https://en.wikipedia.org/wiki/Thomas_Paine#/media/File:Portrait_of _Thomas_Paine.jpg

Tables

2.1.

- https://www.vaccines.gov/basics/types
- http://vk.ovg.ox.ac.uk/vk/types-of-vaccine
- National Research Council (US) Division of Health Promotion and Disease Prevention. Vaccine Supply and Innovation. Washington (DC): National Academies Press (US); 1985. 2, Vaccines: Past, Present, and Future. Available from: https://www.ncbi.nlm.nih.gov/books /NBK216821/
- Clem, A. S., "Fundamentals of vaccine immunology." *J Glob Infect Dis*, vol. 3, no. 1 (2011): 73-8. doi:10.4103/0974-777X.77299. PMID: 21572612. https://www.ncbi.nlm.nih.gov/pmc/articles/PMC3068582/

4.1.

- https://www.psychiatry.org/
- https://tpb.psy.ohio-state.edu/5681/readings/CH%202%20 Beauchaine%20&%20Klein%20DSM%20final.pdf
- https://myaspieworld.home.blog/whats-in-a-name
- https://www.autismag.org/dsm-5

5.1.

- Adapted from Tomljenovic, L., "Aluminum and Alzheimer's disease: after a century of controversy, is there a plausible link?" *J Alzheimers Dis*, vol. 23, no. 4 (2011): 567-98. doi:10.3233/JAD-2010-101494. PMID: 21157018)